FIREARMS INVESTIGATION,
IDENTIFICATION AND
EVIDENCE

A BOX MAGAZINE SIGNATURE

These two cases are from .22 long rifle cartridges fired through a Colt 'Woodsman' automatic pistol. Observe that both show, by the long, flattened-S scratch running lengthwise on the side of each, that they were fired in a repeating or automatic firearm functioning from a box magazine.

These scratches are scored into the side(s) ot the case by the magazine lip(s) as the cartridge is shoved ahead by the breechblock on its forward stroke. They are caused by the sharp edge(s) of the magazine lips, which are under varying tensions in different magazines; also by the pressure from the magazine follower, which is greatest when the magazine is full of cartridges—the scratches may not show at all on the last two or three cartridges in the magazine. They can appear on both or on only one side of the cartridge case.

Similar scratches may appear on center fire military cartridges fired in automatic or semi-automatic machine rifles and machine guns loading from a box magazine. They are often found on automatic pistol cases; some cases being covered with such scratches where the cartridges had been shoved into and out of the magazine several times before being fired.

Box magazines are found on many models of .22 caliber, trombone action rifles, also on some center fire automatic and bolt action rifles; located ahead of the trigger guard and feeding directly into the breech of the gun. On most automatic pistols the magazine is fitted into the butt stock—but it is a box magazine just the same. All box magazines are liable to mark their cartridges due to one cause or another.

In repeating arms having a tubular magazine underneath the barrel the cartridge is carried up to the breech by a 'carriage' or 'lifter'; in those arms feeding from the butt stock of the rifle the cartridge is fed directly into the chamber by spring pressure.

Neither of these two last systems is likely to mark up its cartridges.

FIREARMS
INVESTIGATION
IDENTIFICATION

and

EVIDENCE

By

MAJOR GENERAL JULIAN S. HATCHER

Ordnance Department, U. S. Army, Retired
Director of Technical Staff, National Rifle Association

LIEUTENANT COLONEL FRANK J. JURY, U. S. A. R.

Officer in Charge, Firearms Identification Laboratory, New Jersey State Police
Distinguished Marksman (Rifle) U. S. Army; Distinguished Pistol Shot,
New Jersey State Police
Member, Captain, and/or Coach of Teams Competing at National Match 1927,
1928, 1929, 1938, 1939, 1940, 1955 & 1956

JAC WELLER

Graduate and Licensed Engineer
Special Police Investigator, Firearms Consultant
Ordnance and Historical Research, Honorary Curator West Point Museum

Edited by
THOMAS G. SAMWORTH

PHILADELPHIA, PENNSYLVANIA
RAY RILING ARMS BOOKS COMPANY
2006

ISBN 0-9603094-97

RAY RILING ARMS BOOKS CO.
Publisher
6844 Gorsten Street
Philadelphia, Pennsylvania 19119, U.S.A.
www.rayrilingarmsbooks.com

PREFACE

A book listing three authors needs a word of explanation. In the late winter of 1953, after the acceptance by *the American Rifleman* of two articles on Modern Firearms Identification by Jury and Weller, General Hatcher approached them in behalf of his original Publisher. General Hatcher's TEXTBOOK OF FIREARMS INVESTIGATION, IDENTIFICATION AND EVIDENCE, which originally appeared in 1935, was out of print. For almost twenty years it had been practically the only reference book in its field. However, it needed revising and bringing up to date. General Hatcher was eminently capable of doing this; however, he had neither the time nor the inclination. New research was necessary as well as several thousand miles of leg work. Jury and Weller were chosen to do these; they enormously appreciate the honor of coauthorship with one of the truly great Firearms men of all time.

The book as it is now appearing is more than 90% new material. This preface is being written by Jury and Weller. We had the original TEXTBOOK to use as we saw fit. Occasionally, we found it impossible to do half so well as General Hatcher had done long ago. Some of his original sketches, numerous incidents and anecdotes, and a few explanatory passages appear exactly, or almost exactly, as they did in the first book. However, it's mainly a new text. The shortcomings and errors, which we hope are few, are completely ours. General Hatcher did far more than his share not only with the old manuscript but also in providing information and contacts we would not have been able to get unassisted. His advice and counsel were invaluable.

As we look back over the period of preparation of this volume, we remember so many pleasant hours spent with literally dozens of Firearms men. A mere mention of their names, rank, and organization just isn't enough. There is something about those men that is rather wonderful. Almost all who love guns and are constantly working with them are grand fellows. We don't know why; they just are. We are indebted to them in widely varying amounts for material and technical assistance, but we want them all to know how much we enjoyed knowing them personally.

Our greatest technical aid came from Director J. Edgar Hoover and his Federal Bureau of Investigation. Before we undertook the book at all, General Hatcher obtained for us interviews with Special Agents Donald J. Parsons and Marion E. Williams, who are in charge respectively of the Physical Laboratory and the Firearms Identification Section. These men not only read critically the manuscript in its entirety, but aided us in many other things.

The late Colonel Calvin Goddard generously gave us the benefit of his unique experience, checked Chapters 1 and 16, and helped with constructive criticism. We could not have written the chapter on Explosives without the aid of Colonel Henry Marsh of the Hercules Powder Company. He changed what would have been just another run-of-the-mill, layman's account of a complex subject to about the best thing in our book.

In the summer of 1953 we visited briefly various people in England. We are indebted to Sir Gerald Burrard, the foremost British Expert, not only for a full and complete picture of the science in his country, but also his own personal comparison microscope. Colonel Oswald A. Watts, Chief Inspector of the Royal Smallarms Manufactory at Enfield, Colonel E. Labbett of the School of Infantry, and Keith Neal, Esq. gave their aid freely.

The American arms and ammunition manufacturing companies, without exception, were most cooperative. They did more for us than we thought possible. We are particularly indebted to Mr. Roy C. Swan, and Charles C. Peterson of Remington, Harold O. Austin, Harry Steins, and Frederick Miller of Smith and Wesson, Arnold Goodwin, Harold Descault, Arthur Donovan, and William R. Henry of Colt, Thomas R. Robinson and Henry McCluskey of Marlin, Harry Sefried and Robert Hillberg of High Standard, W. H. Coxe of DuPont, Ross Blanchard of Harrington & Richardson and, in addition to Colonel Marsh already mentioned, Dr. Leonard Farmer and A. R. Ely of Hercules. George W. Spence, a professional hand-loader and extremely clever Firearms man in general, gave us freely of his specialized knowledge. B. D. Munhall of the H. P. White Laboratories kindly extended his hospitality and felicitations.

We owe so much to so many police officers throughout the country; we sincerely appreciate their aid. Colonel Russell A. Snook, Superintendent of the New Jersey State Police, graciously gave his approval to our entire scheme and to the inclusion of data from the Department files. Without his backing the book could not have been completed. We want to thank particularly our close friends and associates, Sergeant Frank J. Pasch and Detective 1/c William Lewis of the New Jersey State Police, Chief James Campbell, Jr. of the Princeton Township Police, Chief John H. Smith and Lieutenant Frank T. Bird of the Princeton Borough Police, and Assistant Prosecutor (now Judge) Arthur S. Lane, Jr., and Detective Lieutenant Dennis R. Dore of the Mercer County Prosecutor's office. These men couldn't have been more considerate.

Director William E. Kirwin and Andrew Hart of the New York State Scientific Laboratory and the late Lieutenant George Roche and Sergeant John F. Collins of the Massachusetts State Police Ballistics Bureau extended to us the hospitality of their laboratories and gave unstintingly of their time in providing material, taking pictures, and the like. We appreciate the aid of Lieutenants Robert Pardua and Leslie Smith of the New York City Ballistics Squad. Lieutenant Pardua read a part of this manuscript and gave valuable suggestions. Lieutenant William Del Torre of the Philadelphia Ballistics Bureau was particularly kind and generous. We spent two pleasant and informative days with Detectives Edward Culkin, Francis E. Bailey, and Earl Laird, of the Boston Municipal Police Firearms Investigation Bureau. But these men are not all. Dr. J. H. Mathews, Director of the University of Wisconsin, furnished full information regarding their comparison camera. J. J. Rehling, State Toxicologist of Alabama, contributed pertinent photographs and data. Ralph W. Prouty of the Oregon Crime Detection Laboratory and Fred R. Rymer of the Texas Department of Public Safety sent in several outstanding case histories. Firearms Examiner John Davis of the Oakland California Police Department, Criminologist David Q. Burd of the California Department of Justice, and others helped from afar. Technical assistance of unique value came from Master Sergeant George A. (Sandy) Gordon, far ranging Expert of the U. S. Military Police. He knows more than any other man in the world about the identification of that chameleon of criminal firearms, the Cal. 45 U. S. Model 1911A1 Automatic Pistol.

We thank these men one and all for their technical aid and also for so many pleasant hours spent in their company. Much of the credit should be theirs, but the mistakes are ours.

Frank J. Jury

Mercer County, New Jersey
2 December 1956

Jac Weller

United States Department of Justice
Federal Bureau of Investigation
Washington 25, D. C.

September 1, 1954

FOREWORD

This revised edition of "Textbook of Firearms Investigation, Identification and Evidence," a technical treatise by Major General Julian S. Hatcher, is a source of instructive and up-to-date information on subjects of great interest to the profession of law enforcement.

This book encompasses a tremendously broad scope and contains a store of reference data. It places emphasis on the sound evaluation of firearms evidence in criminal cases. This work also sets forth the related role which can be played only by the impartial expert who devotes full time to mastering the intricacies of the science in which he specializes.

Both as text for the student and as reference for the expert, this revised volume will be welcome in a field in which up-to-the-minute data has, in the past, been inadequate in scope.

John Edgar Hoover
Director
Federal Bureau of Investigation

CONTENTS

CHAPTER 1

THE HISTORY OF FIREARMS IDENTIFICATION

THE place of the Firearms Identification Expert is now firmly established in this country and abroad. His work has become routine in police investigation. His evidence is accepted without question in courts. Science has come to the aid of Justice to an astonishing degree. Dramatic cases of the apprehension and conviction of criminals by firearms evidence are more frequent than the public realizes. However, those who have given their minds and so much of their lives to the development of this science take even more satisfaction from the cases where they have prevented the trial and conviction of innocent men for crimes they did not commit.

Here is a New Jersey case of a type well known throughout the country. A well-to-do housewife was murdered with a .32 caliber revolver in her home and money taken. A transient worker, poor, old, ignorant, and probably moronic, was picked up in the vicinity with a gun of the type that was used to commit the crime. Under questioning, he broke down and confessed. A first degree homicide conviction was almost certain. Investigation of the case was practically closed. However, the crime bullet when examined in the Firearms Identification Laboratory of the New Jersey State Police was found not to have come from the suspect's revolver. The investigation was continued and the real murderer finally apprehended. The old man was guilty of no worse crime than carrying a concealed weapon. These circumstantial evidence cases, frequently accompanied by unrepudiated confessions, and even eyewitness identification, involving men of this type plague the execution of justice. The comparison microscope alone sometimes prevents the execution of innocent people.

A gun involved in the accidental shooting of a soldier was recently identified in an unusual manner. A young soldier was killed by a .30 caliber bullet in battalion blank cartridge maneuvers. The shot could have been fired by any one of a large number of fellow soldiers

1

all armed with M1 Garand rifles. In this case there was neither a bullet nor cartridge case. The boy had been shot in the seat of his trousers with a bullet emerging from the top of his head and continuing on its way. A large number of rifles were taken from the soldiers involved in the maneuver and impounded for further study. Investigation of the incident limited the number of possible rifles to eight. These rifles were sent to the FBI laboratory for examination.

There were no established laboratory procedures to determine which one of the eight rifles had fired the fatal shot. It was known that the rifles were cleaned prior to the beginning of the maneuvers and that one specific soldier and only one had fired each weapon. Following control tests to establish a workable procedure a single clean patch was run through each barrel. These patches, each carefully identified, were then reduced to ash and the ash was examined spectrographically. Based upon the presence of an excessive amount of copper present in the ash from the patch passed through the bore of one of the rifles, it was possible to determine that a bullet had been fired from this particular rifle since it was last cleaned. The results of this examination assisted materially in the investigation of the case.

Early Identification

Crimes solved by firearms identification are commonplace today. However, the science is still young. A relatively few years ago it was in a dreadful muddle, even though we can now trace its beginnings back a long way. Projectile identification is older than most people realize. The cast lead bullets thrown by slingers attached to Roman legions had individual legion emblems upon them. English archers of the Fourteenth Century frequently marked their arrows distinctively. A particular fatal pellet or arrow could sometimes be traced to a specific unit.

In the early days of firearms, bullet wounds were easily distinguished from those made by other arms. Sometimes, in retaliation for the firearms death of an important person, all prisoners captured with firearms or showing evidence of having carried them recently were summarily executed. Armor encased knights resented the employment against them of weapons that failed entirely to respect their investment in equipment and training.

Perhaps the most dramatic and tragic identification of a bullet ever made was that in connection with the fatal wounding of Stonewall Jackson on May 2, 1862. Dr. Hunter McGuire removed a spherical ball .675″ in diameter from the great man's right hand and amputated his left arm. The bullet could only have come from a Confederate musket since this smoothbore weapon had been abandoned for use in the Union Army the year before. General Jackson's death by mistake at the hands of his own men was confirmed.

A similar identification of the bullet that killed Union General Sedgewick was made. It came from a hexagonal-bored Whitworth rifle brought into the Confederacy from England. The shot is said to have been fired by Sergeant Grace of the Fourth Georgia Infantry at a range of 800 yards.

A century ago, in the days of muzzleloaders when each civilian hunter usually made his own bullets and perhaps even his shot, crude firearms investigation was of importance to the even cruder police procedures. Projectile evidence was frequently introduced into court. However, the jury was usually able to evaluate that evidence for themselves. The juries of that time knew far more of firearms than they do today. Variations in home-made muzzleloading ammunition were more obvious than in factory mass-produced modern bullets.

The admission of the so-called expert to interpret to the jury the meaning of specific firearms evidence precedes by several decades the existence of any real experts. From the 1870s on, the courts sometimes allowed sheriffs, police officers, and particularly qualified civilians to give their opinions. Some of these men, because of natural smartness and common sense, gave good opinions. A crime bullet showing four rifling grooves cannot have come from a five-groove barrel, a .38-40 bullet (actual diameter about .403") cannot have come from a .38 Special revolver (actual diameter about .357"). So long as these early experts kept within the limit of their knowledge and equipment, they served a very valuable purpose. Unfortunately, these individuals sometimes exceeded in their testimony the range of their knowledge. Perfectly conscientious, they arrived at conclusions astonishing to us today.

The First Experts

Further, what knowledge a few men had before the end of the Nineteenth Century was largely personal; there was not any interchange of ideas through publications or identification societies. Scientific literature on the subject did not exist. In fact, nobody thought of it as a science. The first, and for several years the only worthwhile article to appear in print on the subject in this country, was by Dr. Albert Llewellyn Hall in *The Buffalo Medical Journal* in June of 1900. In the midst of a full medical practice, he had found time to give a

BULLET EVIDENCE

Modern firearms identification is at present based primarily upon the individual markings, or striae, scored into the bearing surfaces of the fired bullet as it passes through the barrel of the firearm.

The two bullets shown above are from .38 S. & W. Special cartridges manufactured by the Remington Arms Company. The one on the right has not been fired, having been taken from the case undamaged or unmarked except for the casual markings acquired as it was processed through the various stages of manufacture.

The bullet on the left was fired into water and recovered. Those lengthwise scratches now appearing on its lower half (the bearing surfaces) are the striations put there by the tool marks left in the bore when this particular revolver barrel was manufactured. These striae are highly individual to this one particular barrel—and they form the means of comparison and identification by which the bullets fired from this revolver can be positively identified.

good deal of his keen mind and leisure to experiments and thinking about wounds and the identification of crime bullets. He explains in some detail many of the basic principles of the science as we know them today. The article, however, was not widely circulated.

The next significant contribution to the Science of Firearms Identification in the country was made by the personnel of Frankford Arsenal in 1907. Some soldiers rioted in Brownsville, Texas. Thirtynine .30 caliber rifle cartridge cases as well as a number of bullets

THE BORE OF A REVOLVER BARREL

Here is a photomicrograph of a .38 caliber revolver bore with the lands designated as 'L' and the grooves as 'G.' These lands and grooves are what put the striae on all bullets fired through the barrel.

The bore of a barrel is reamed to final size before rifling, whether by hook, scrape, broach or button. The rifling head, by the first three methods, removes metal from the surface of the bore, forming the grooves of the barrel. The remaining surfaces form the lands. As the above photomicrograph shows, the marks on the lands are those left by the reaming operation, at right angles to the axis of the bore.

It is the effect of these marks on the lands that leaves the striae in the grooves of the bullet, and not, as is commonly supposed, the tool marks left in the grooves of the barrel. The grooves leave their marks on the lands of the bullet.

The grooves of a barrel will leave adequate marks on a bullet—when the bullet fills or 'bottoms' in the grooves. However, this is not always the case. Especially in metal case bullets, such as the .45 Auto, 9mm Luger, 8mm Mauser and the like, the bullet more often than not does not completely fill the bore. Thus many bullets do not show striae left by the rifling cutter at all.

and a considerable number of suspect rifles were sent to the Arsenal. The staff there studied this crime carefully and worked out for themselves methods of identifying the cartridge cases with individual suspect rifles. Of the 39 cases, 33 were divided into four groups and each group tied to one of the suspect rifles. The other six cartridge cases could not be identified with any of the rifles submitted for the tests. These investigators had learned to identify a weapon by fired cartridge case markings, probably principally those left on the primers. They felt, however, that similar conclusions in connection with bullets were not possible. This report was extremely interesting and rewarding for any who may have read it; however, it also received no great publicity and was buried in the report of the Chief of Ordnance for that year.

On the basis of Dr. Hall's published work and the report of the Frankford Arsenal officers, we undoubtedly led the world in the new science at this time. However, beginning about 1912, Professor Balthazard of the University of Paris learned probably independently most of what we knew of cartridge case identification and went further. He began to identify a weapon by its bullets. The method was slow, crude, and expensive but it worked. The Professor took photographs of a crime bullet and also a test bullet from a suspect weapon. Six or more photographs of each bullet were taken equally spaced radially around its circumference. These were considerably enlarged and then compared.

Since this method of identification required extreme photographic competence particularly when the crime bullet was mutilated, it was not widely used. It had the fundamental handicap of projecting onto a flat film surface a curved surface from the bullet. The spacing of the photographs at 60° or even closer did do a good deal to remove this handicap. If an identity was suspected, a new set of photographs could be made of one or the other of the bullets, revolving it into a corresponding position. In other words, all the 60° pictures could be taken from exactly the same point on the circumference of both the test and crime bullets.

Even before Professor Balthazard used his photographic method of comparison, men widely separated geographically had thought of another procedure which in various forms has been tried many times. Briefly, this consists of causing both a crime and a test bullet to transfer to some other medium a record of their surface markings and then compare these two records. Usually the bullets were rolled on some plastic surface such as lead or carbon paper with white paper below. This method never worked well. It fell down completely in cases of deformed bullets. Even when using special equipment and skill in rolling the bullets, no very definite pattern will be left; comparison between two patterns even from the same bullet are not conclusive. This method, however, keeps on cropping up, even among some detectives today.

The Charlatans

The decade following the invention of photographic and plastic impression comparison was the heyday of charlatans. Very few judges knew anything about firearms; they had heard vague rumors of marvelous developments. The world was willing to accept anything said to be scientific. Almost anyone was admitted to testify in court as an expert. Many had little knowledge but a natural court presence and a great deal of gall. For $50.00 a day—a lot of money in those days—they would go cheerfully into court and swear to most anything. With a pair of outside calipers, an ordinary hand magnifying glass, and a steel scale graduated to 1/32 of an inch, they would cheerfully swear away the life of an innocent man or free for further depredations against society the most atrocious criminal. There were few real experts to check up on them; they made their own rules and calmly put forth opinions based entirely on who was paying their fee. It must seem fantastic to the layman that this could happen in the Twentieth Century in our enlightened courts of justice, yet it did happen and happened frequently. The few men who were capable of giving reasonable opinions were sometimes made to appear unreliable by their more glib opponents.

The Stielow Case

One of the most important single cases in Firearms Identification History in this country was that involving two men who were supposed to have killed their employer and his housekeeper in upper New York State. This is the famous Stielow case. Eventually, C. E. Waite, who claimed to have worked with the Department of Justice during World War I, came into the investigation. It is sufficiently interesting to describe in some detail.

At daybreak on March 22, 1915, Charlie Stielow, an illiterate, good-natured tenant farmer, on going out of his house discovered a woman in her nightdress lying dead on his doorstep, shot through the heart. Her footsteps in the light fall of snow led from the nearby house of Charles B. Phelps, the owner of the farm.

The dead woman was housekeeper for the aged Phelps who was reputed to keep large sums of money in the house. Stielow found the kitchen door of the Phelps house open, a bullet hole in the glass and his employer, a man of 70, unconscious on the kitchen floor with three bullet wounds in his body. A lighted lamp was burning low on a nearby table. Stielow at once ran to a neighbor's house and gave the alarm. It was quite evident that robbery was the motive, as the bureau where Mr. Phelps kept his money was rifled and his wallet was missing.

The murder was committed in a small rural community. The local authorities, unused to coping with such crimes as homicide, had allowed the curious crowd to trample the foot prints and destroy most of the clues before any real constructive work was done toward solving the mystery.

Stielow lived in the tenant house with his wife and two children, his wife's mother and his wife's brother, Nelson Green. The shooting had been done with a .22 caliber firearm. Both Stielow and Green swore at the inquest that neither had a gun of any kind, but it turned out later that Stielow owned a cheap .22 caliber revolver and a .22 caliber rifle. He had given these guns to another brother-in-law to hide.

After 10 days had elapsed without any progress, Nelson Green was arrested. At two o'clock the next morning detectives secured his signature to a confession that Stielow and he had committed the murders. Stielow was thereupon arrested and held in jail separately

FIRED CARTRIDGE CASE EVIDENCE

Supplementing the identification evidence supplied by the bullet's passage through the barrel is also that available through the case from the fired cartridge.

The cases above show the embossings put into their primers when these cartridges were fired. Such markings are highly individual to the firearm used; they are of two types—marks embossed upon the flat surface of the primer cup when it is forced back against the breechblock or recoil shield of the arm, and impressions punched into the imprint of the hammer nose or striker.

Such fired-primer evidence is what is generally known as a "breechblock signature" (also bolt-face, breech-face or breech markings). Some experts consider it secondary to the bullet striae evidence but it can be of equal value and importance and in many cases is the only physical evidence available for use.

from Green. He made a request to be allowed to talk to his wife as he said he had a heavy burden on his soul. He was, therefore, accused of the crime to which he finally confessed.

The two confessions checked generally, though neither told anything that was not common knowledge to everybody who had been in the vicinity. However, each man accused the other of instigating the crime and firing the fatal shots. The confessions stated that they had planned to rob Phelps and had waited outside until he was in bed, then rapped on the kitchen door and shot him down when he answered the knock. As they went toward the bedroom where they expected to find the money, the housekeeper ran out of her room through the kitchen and out of the door, closing it behind her. The two men ran after her, shot her through the glass, ran to the bedroom, made a search until they found the money, and left the house. They heard and saw the housekeeper pounding and clawing at the front

BOLT FACE SIGNATURE OF A MODERN COMMERCIAL CARTRIDGE

Above is shown the bolt face of a Remington Model 30 sporting rifle chambered for the .257 Roberts cartridge, together with a cartridge case fired in it. The action is an Enfield, identical in design, dimensions and fitting to the British action shown on page 116.

The tool markings on this bolt face are extensive and clearly visible to the eye, and they emboss their features strongly upon the face of the primer. However, any bolt face or breech block finished off as this one is puts an additional marking upon the primer in the form of a heavily embossed ring running around the circumference of the firing pin imprint. This is caused by the chamfored edge to the firing pin hole running through the bolt face, as clearly shown above. Not all firing pin holes are so chamfered. Such an embossed ring will show additional matching marks when viewed through the comparison microscope.

A similar embossed ring around the firing pin imprint can also be caused by excessive chamber pressure pushing back the firing pin and forcing the soft primer metal into the firing pin hole to a more or less degree.

door of the tenant house and screaming to be admitted. Ignoring her, they entered the rear door of the house, discussed the screams with their women folks, and went to bed. They said they got about two hundred dollars; each man accused the other of having kept it.

At the trial, Stielow repudiated this confession, and when questioned about the "heavy burden" he had spoken of, said it was the fact that he had lied about not having any guns. An expert for the prosecution testified that under the microscope he had found nine abnormal defects in the flare of the muzzle of Stielow's gun, and found nine corresponding peculiar scratches on the four bullets taken from the bodies. The scratches were not visible to the naked eye, he said, and were first detected under the lens. He gave it as his opinion that all the bullets were fired from Stielow's gun and could have been fired from no other.

On cross-examination it was brought out that the enlarged photograph of the bullets showed to the jury by the "expert" did not reproduce the nine marks; it was, for some unknown reason, a photograph of the opposite side of the bullet.

Asked why the uneven ridges at the very extremity of the barrel should mark the bullet, the "expert" replied: "The cylinder fitted so tightly against the rear of the barrel that there was no leakage of gas at the breech. The full force of the gas following the bullet out of the muzzle expands the lead bullet as it leaves the muzzle, fills in any depressions existing at the outer edge of the bore and receives scratches from the elevations existing between said depressions."

On the basis of the "expert" testimony that the bullets had been fired from Stielow's revolver and could have been fired from no other, the jury found Stielow guilty of murder in the first degree and sentenced him to die in the electric chair some time during the week of September 5, 1915. Green was told that Stielow had been sentenced to death and that if he would confess he might get off with twenty years. Accordingly, he pleaded guilty and was sent to Auburn Penitentiary.

But Stielow did not go to the electric chair. His lawyer was convinced of his innocence and began a series of appeals. A Humanitarian Cult of New York City began taking an interest in the case. Members of this Cult soon found some slight evidence that two tramps and horse traders who had been in the vicinity of the house might know something of the murders. While this investigation was going on, the case had been reviewed by the Court of Appeals; a motion for a new trial was denied in February, 1916. In June of that year a second application for a new trial, and in July a third, were denied; in October a fourth appeal met the same fate. All the judges found that Stielow had been fairly tried and justly convicted. During this time the ignorant farmer had been adjudged guilty by a jury and 10 judges who had subsequently passed on his case in connection with the numerous appeals.

In the meantime, the Humanitarian Cult had been strengthening the evidence against the two horse peddlers. Finally, the Governor had one of the two suspected horse traders brought before him; his answers were evasive and unsatisfactory. On December 4, 1916, the Governor, unsatisfied of Stielow's guilt, commuted his sentence to life imprisonment. He also appointed a prominent Syracuse lawyer, George H. Bond, to conduct a special investigation into the case of the murder of Phelps and his housekeeper, Miss Wolcott. Mr. Bond drafted Mr. Waite from the Attorney General's office to aid him.

Under searching analysis, the confessions of Green and Stielow, once considered solid evidence, appeared in a new light. Such glaring inconsistencies were discovered as to make it almost self-evident that the men who made the confessions could not have had first hand knowledge of the crime. After shooting down Phelps, as told in the confession, Stielow was supposed to have picked up the lamp and advanced from the kitchen through the dining room and a small hall. He was just about to enter the old man's room to search it, when Miss Wolcott ran out of her room, through the dining room, toward the kitchen door. The two men followed her. By the time they reached the kitchen, she was outside and the door was closed behind her. They shot her through the glass door and heard a scream. If the confession is to be credited, the two men fired at the fleeing woman, but did not even bother to see whether or not she was hit. They let her run screaming into the night to arouse the neighborhood, while they returned to ransack the premises and make a careful search for the money.

Miss Wolcott, when she ran towards the kitchen, passed close to the enormous Stielow who was carrying a lighted lamp and wore no disguise. His mere bulk would have identified him; it is incredible that she would have failed to recognize this man who had been working on the farm under her direct observation. Yet to credit the confession, we must believe that the housekeeper fled directly to the home of her slayer for sanctuary, crying out to him by name to open the door, knowing as she must have that he was not there but in the house from which she had just run.

At the least calculation, it would have taken an expert cracksman 10 minutes to make the search of the bedroom as detailed in the confession. Yet when Stielow and Green emerged from the house after finding the money, Stielow is supposed to have heard Miss Wolcott crying, "Charlie, please let me in! I am dying!"

FIRING PIN DIFFERENCES

Occasionally, especially on older type weapons, the firing pins, even on similar type guns, are quite different in shape, as is clearly shown in the above illustration, from a photograph by Capt. E. C. Crossman. This shows the firing pins of two Colt Frontier revolvers of the same caliber.

Differences such as this are clearly visible to the eye, and form negative proof, for obviously a cartridge fired by either of these guns can be seen and shown to have such an appearance that it could not possibly have come from the other gun. Evidence of this kind is often extremely useful in quickly establishing innocence, or in eliminating suspected weapons.

Miss Wolcott had been shot directly through the heart with a jagged and deformed bullet which had been battered out of shape by first passing through a pane of glass. Yet this woman of 50 is supposed to have run 100 yards after she was shot, and at least 10 minutes later to have been screaming loudly enough to be heard in the next house.

Apparently no one had even taken the trouble to measure the distance from the floor to the hole in the glass of the kitchen door, and check it against the statements of the defendants. The confessions stated that Miss Wolcott was outside, with the door closed behind her when she was shot. Outside the door, which opened inward, there were several steps leading down to the ground. The threshold was less than two inches wide. No one could stand on the threshold while the door was closed.

The hole in the glass was exactly 3'8½" from the floor, and showed plainly by its shape that the bullet had gone through on a line, and not on an angle. The course of the bullet through the body also indicated that the shot had been fired on a nearly horizontal line. Miss

Wolcott was 4'11½" tall. Once she was outside and the door closed, a bullet could not hit her below the shoulder. The height of the wound in her body was exactly that of the distance from the kitchen floor to the hole in the glass. The conclusion was irresistible that she was shot either while the door was partly open and she was on the level of the kitchen floor, or at the moment of leaving it and turning to leap from the steps. Whoever fired the shot was probably not pursuing her as Stielow's confession stated, but standing over the body of Phelps behind the partly opened kitchen door.

Firing tests were made with Stielow's revolver in the presence of Captain Jones, the revolver Expert of the New York Detective

These three .22 long rifle cases were fired in a rifle which had a most distinctive nose (surface or profile) on its firing pin. Note the double impression it puts into the rim of every case fired—an imprint readily matched by the eye alone.

Bureau, Dr. Otto Schultz, Police Surgeon, Inspector Faurot, Bond, Waite, and others. Before firing, the revolver was handed to Captain Jones and his opinion asked as to how long a time had elapsed since it had been fired. "Certainly three or four years, apparently a longer time," Jones said after he had examined the heavy incrustations in the barrel, discolored with age. According to Stielow, the gun had not been used for seven years.

The first test was made by placing a sheet of paper over the gun and firing a cartridge of the same type as those found in a box in Stielow's possession. Instead of there being no leakage of gas at the breech end of the barrel, as the "expert" had testified, the paper was set on fire. Afterward, bullets were fired into cotton and recovered. The test bullets were then taken to the Bausch & Lomb plant at Rochester for scientific examination by Dr. Max Poser, an expert in microscopic research.

Neither Dr. Poser nor anyone else could find a trace of the peculiar scratches which had doomed Stielow, either under high or low powered magnification. Even more amazing was the fact that one of the lands on the murder bullet was abnormal, equalling the combined widths of two normal lands and one groove, while there were five lands and five grooves all of normal width on the test bullets from Stielow's gun.

The killing of Phelps and his housekeeper undoubtedly had been done with a defective gun in which the rifling tool had presumably been broken, or the rifler had failed to complete his operation and the error had escaped the inspector's notice. The distinction between the two sets of bullets was glaring, and there was no possible chance that the fatal bullet could have come from Stielow's gun.

The micrometer completed the evidence. Test bullets from Stielow's gun were found to be .0018″ larger in diameter than the crime bullets that were not too deformed for proper measurement. These were not opinions; they were facts.

Bond and Waite had no further doubt that Stielow was innocent; the Governor accordingly pardoned him. The two horse traders were brought to trial and made a full confession, telling many facts which had not been known before but which were easily checked and found to be true.

After the conclusion of the case, "Judge" Waite had the germ of an idea. He traveled around the country from one firearms company to another, accumulating data in connection with the manufacturing standards of weapons currently in production and also obsolete arms. He consulted retired foremen in connection with their notes; he accumulated a mass of data and a large collection of firearms of all descriptions. His ammunition collection was very extensive. He spent some time in Europe getting the details of foreign firearms manufacture at that time. However, Mr. Waite's knowledge of things scientific was by his own admission negligible. He had the idea, but it remained for three associates to really build up within a period of a few months and for the first time a real Science of Firearms Identification.

His first associate and partner was the late Major Calvin Goddard, a doctor of medicine in the Medical Corps in the First World War who had transferred to the Ordnance Corps. Major Goddard had not only a brilliant scientific mind, but also a knowledge of firearms and ordnance that has known few equals. Two other associates—they were never partners in the enterprise—were Philip O. Gravelle, a trained microscopist and gifted photographer and John E. Fisher, a man of independent means with long experience in design and precise machine work. Mr. Fisher had formerly been employed by the Bureau of Standards.

These three men were working in the small laboratory in New York classifying, firing, and testing the weapons, ammunition, and data collected by "Judge" Waite. Actually Major Goddard was living adjacent to the laboratory as well. The three of them worked very hard indeed at pure research since concrete cases did not at first arrive.

The most significant contribution to the Science of Firearms Identification was the introduction into that field of the comparison microscope. One of these was obtained and put into service in April of 1925. The comparison microscope is actually two instruments with an optical bridge between them so that one-half the field seen by the observer comes from one microscope, and the other half from the other microscope. This arrangement had been used in textile research for several years where Mr. Gravelle came in contact with it; however, it was a brand new idea for critical bullet examination.

"Judge" Waite persuaded Remington to make for them special bullet mounts on two mechanical stages for their microscopes that allowed one bullet to be mounted under each instrument and moved around easily in any way that the operator saw fit. The bullets must be pointed in the same direction. One actually sees through the comparison eyepiece a forward portion of one bullet and a rear portion of the other. The fine markings (called striae or striations) left by passage through a rifled bore can be seen very clearly on each bullet. If one remains stationary while the other is slowly revolved, a point of matching patterns will be apparent if the bullets came from the same barrel. For the first time a test bullet from a suspect weapon could be compared directly and relatively simply with a crime bullet. The result was as conclusive as fingerprints.

Waite visited the *Saturday Evening Post* in Philadelphia in the early spring of 1925 and came away from a conference with George H. Lorimer with a check for $2,000 and the assignment of a staff writer to produce two articles which appeared in the *Saturday Evening Post* in June of that year entitled, "Fingerprinting Bullets." These two articles brought the new science into the ken of the general public, although many police departments were hesitant to accept it. Meanwhile, Major Goddard was writing article after article for technical magazines and working out procedures for micro-comparison of bullets and the photographing of them. He was also tabulating the data then available about firearms. They were operating as the Bureau of Forensic Ballistics, which was a private enterprize but would cooperate with any police department.

Waite died in November of 1926; Colonel Goddard continued the Bureau. Cases were slow in coming in at first. Police departments began to find out, however, that for the first time in history an expert had a real scientific basis for his findings. Actually, Colonel Goddard was stating facts rather than giving opinions. His testimony was accepted without question in courts throughout the East.

On February 14, 1929 a crime occurred in Chicago which changed the course of Firearms Identification in this country. Six men were murdered in cold blood in a garage; it was called in the press the St. Valantine's Day Massacre. Five of these individuals were members of one gang; the sixth man was an innocent bystander. A rival gang sent several men, some of them disguised as policemen, into the garage; they fired 70 bullets and several charges of buckshot at the six victims, killing all except one of them almost instantly. The other died later in the hospital.

A "silk-stocking" coroner's jury was appointed in Chicago with Bert A. Massey, vice-president of the Colgate-Palmolive-Peet Company, as foreman. This jury obtained the services of Colonel Goddard who arrived to examine the evidence. There were 70 cartridge cases, all .45 caliber ACPs of commercial manufacture, 70 bullets, whole or fragmented, and some buckshot. Colonel Goddard, after appropriate examination, determined conclusively that two weapons and only two were used to fire the .45 ACP cartridges. They were Thompson sub-machine guns, one with a 50-shot drum magazine and the other using a 20-shot clip. There was also one 12-bore shotgun. The Thompson submachine guns were found in the possession of one of the gangsters later apprehended in Michigan. The case was solved though all participants were not brought to trial.

Colonel Goddard made such an astonishing impression upon Mr. Massey and others of the jury that he was asked to bring his Bureau to Chicago, all expenses to be paid by Massey. Colonel Goddard and this public benefactor discussed the matter at length and decided instead of confining the laboratory to firearms studies only, that it be enlarged to include other branches of scientific crime detection.

Colonel Goddard visited thirteen countries in Western Europe and spent considerable time in each of the important metropolitan Scientific Crime Detection Laboratories and Medico-Legal Institutes there before he returned to this country late in 1929.

The Scientific Criminal Investigation Laboratory opened in April of 1930 with the finest of equipment and the best trained personnel ever assembled in one place until that time. This was associated with Northwestern University and later became a department of Northwestern University. Students were accepted in various courses early in 1931. Colonel Goddard was head of this department with full professorial rank and also editor of *The American Journal of Police Science,* which he launched in 1930. It was later combined with the *Journal of Criminal Law and Criminology* and continued as the *Journal of Criminal Law, Criminology and Police Science,* still published by Northwestern University.

The FBI National Police Academy was established by the Federal Bureau of Investigation in 1935. The curriculum of this school included lectures on Firearms Identification. There was no connection between the FBI school and the courses at Northwestern; however, both courses stimulated interest in the science. Northwestern's school provided the means for the publication of Colonel Goddard's many articles. Unlike investigators in Europe, his discoveries and procedures were freely available to anyone in the field.

Northwestern University has discontinued their school; the physical equipment from it was sold to the Chicago Police Department. However, the school conducted by the Federal Bureau of Investigation is now known as the FBI National Academy. The curriculum has been expanded to include all phases of police operations and is considered the finest of its type in the world. The purpose of the Academy is to train state, county, and municipal law enforcement officers as instructors and executives in the field of law enforcement. The FBI National Academy course is 12 weeks in length and includes a course of 10 weeks with two additional weeks for specialized training. Application for an officer to attend this school must be made by the head of his organization. The courses are presented without charge; however, each officer must maintain himself while in attendance.

At least two other men made significant early contributions to Firearms Identification. Lieutenant Van Amberg of the Massachusetts State Police, working with limited scientific background and less equipment, was of great value to the cause of justice in the famous Sacco-Vanzetti and other pre-comparison microscope cases. The great Captain Edward C. Crossman, of Los Angeles was a real Expert in California from the early 1920s.

Experts and Procedures

By the early 1930s the Science of Firearms Identification was firmly established on a scientific basis. Changes since that time have been in emphasis rather than in basic procedure. At first, a great deal of time was expended in the examination and measurement of individual crime bullets, even though there were no suspect weapons.

CENTER FIRE FIRING PIN IMPRINTS

The firing pins on center fire arms are all pretty much the same in construction and appearance—having a diameter of about 0.075″ with the end rounded hemispherically; this on modern guns, while on the older black powder firearms larger diameter pins were used.

To the eye, the imprints from center fire firing pins are likely to look pretty much alike, but the above illustration will show why such a firing pin can generally be identified from the marks on the primer. This microphotograph, taken by the late Capt. E. C. Crossman, shows the tool marks left on the end of a firing pin—and these same marks will be impressed into the soft metal of the primer at the time of the explosion.

It was thought that identification of the weapon as to make and model would aid the solution of the crime. Further, Experts examined crime weapons minutely. The "Goddard" helixometer, actually developed by Fisher, was used to measure accurately the pitch of the rifling. Various bore scopes were made so that a microscopic

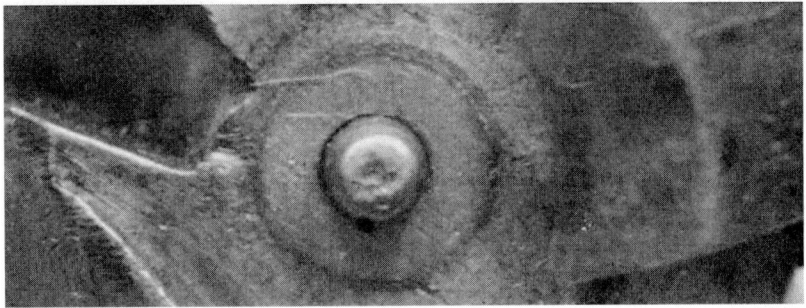

A DISFIGURED FIRING PIN

This is the bolt face of a .22 Hornet rifle, with the striker (firing pin) protruding. Above it is a cartridge case fired in it, showing the distinctive marking on the primer.

This type of mutilation of the firing pin tip results from the escape of gas through a very fine perforation in the primer, or what is usually known as an incipient pierced primer. There are various degrees of primer perforations. In severe cases, a disc may be punched out of the primer, and be blown into the firing pin hole. Such defects are due to an imperfect firing pin, or to very thin, hard, or defective metal in the primer cup.

Individual markings such as this are instantly recognizable, and often enable a positive identification to be made by eye alone.

examination of the inside of a barrel was possible. These things are not now considered very important.

Deputy Sheriff C. A. Peterson of Miami worked with and perfected a geared camera capable of taking strip photographs of bullets in such a way that the photograph would show in a single plane all markings on the circumference of a bullet. This camera was used by several large metropolitan police departments for some time for both crime and test bullets. It had the advantage that many copies of photographs of a crime bullet could be sent out to other departments. However, the photograph once received was virtually useless unless the receiving department also had a geared camera and the special skill and know-how to work with this apparatus. The geared camera cannot take the place of the comparison microscope. Few present day Experts speak favorably of it at this time.

Luke S. May of Seattle perfected his Magnascope about 1936. It combines two lens systems so a photograph can be taken of a crime bullet and a test bullet in absolute coincidence. This apparatus had a great deal to recommend it; however, unfortuntely, it was not simple. It required special skill and photographic ability. It was expensive, slow and inconvenient for the average Expert. It is not practical for most laboratories.

Professor Allen R. Moritz perfected an arrangement of a plastic band which would reproduce all the irregularities on the surface of a bullet. He then mounted this plastic band in a regular microscope slide, and illuminated it from beneath. Bottom illumination adds a great deal to the clarity and definition of anything examined microscopically. A band from a crime bullet and one from a test bullet could be compared under a standard comparison microscope. However, most experts considered this refinement unnecessary. Bullets from the same barrel can be compared directly with positive results. The plastic strips introduce another element of uncertainty for court presentation; satisfactory strips from mutilated crime bullets are difficult to make.

Modern Techniques

Since World War II, the trend in Firearms Identification has been towards simplicity and efficiency. Most laboratories today have several times as many cases as they did in 1940. Minute measurement of crime bullets and arms has been found to be of little practical importance. For instance, few Experts today waste time trying to accurately determine make and model of a crime pistol from the crime bullet or cartridge case. The officer in charge of the investigation is told the caliber and given a list of possible weapons. He is asked to submit any similar arm. You will generally get all that turn up, regardless of caliber, anyhow. The most important job of any laboratory is matching crime bullets with test bullets from suspect weapons, not forecasting what type, make, and model of arm will ultimately be proved to have fired the crime bullets.

The courts today accept the findings of a competent laboratory. The Expert in charge of a Firearms Identification Unit is far too conscientious to risk his reputation on anything save scientifically positive results. If there is any doubt at all, the case will not be brought to trial on the basis of firearms evidence.

The old days of charlatans and other unreliable "experts" testifying glibly to most anything at $50.00 a day are done forever. Scientific facts are not open to question. A bullet did or did not come from a given firearm. If one Expert and his comparison microscope say it did, every Expert and his microscope will also independently reach the same conclusion. Today, independent Firearms Consultants, when they appear for the defense, very rarely contest the scientific findings of the prosecution Expert. They may, however, disagree with the interpretation of those findings, if a crime bullet and suspect weapon match, there is not any question that the bullet came from the gun. A considerable knowledge of firearms may reveal several possibilities in the way the shot was fired, however.

The routine of a Firearms Laboratory is not all dramatic positive identifications. Far more frequently the work done will be negative in character. A crime may be committed in a metropolitan area with a crime bullet obtained from a corpse or found on the premises. The police will almost immediately pull in several likely suspects with somewhat similar modus operandi, frequently carrying firearms. These guns are sent to the laboratory. The usual report paraphrased in non-technical language says: "This gun did not do it." Test bullets from each weapon received in this way will be compared not only with the specific crime bullet but every other one on file of a similar caliber. Occasionally a match will be obtained.

The history of Firearms Identification abroad has been quite similar to that in this country. Britain, France, Germany, and Holland have all had their experts of note. Major Sir Gerald Burrard of England is perhaps the best known over here. The Continental authorities have too often in their published work been more interested in taking credit for themselves than in furthering the science. Duplication of methods and procedures have been arrived at independently, sometimes by several men almost simultaneously. The average police investigator the world over is likely to be clever rather than academically and linguistically trained.

Forensic Ballistics

Firearms Identification is also known as "Forensic Ballistics" or just "Ballistics." In the country at large, and particularly among police officers, these latter words have caught on. The Firearms Investigation Laboratory may be the official name of your unit; however, the chances are that most patrolmen and detectives will refer to it as "Ballistics." The science is popularly known as "Forensic Ballistics."

Ballistics really means a study of the motion of a projectile. Interior Ballistics is the study of the motion of that projectile within the firearm. Exterior Ballistics is similarly the study of the motion after it leaves the firearm. Forensic is an adjective referring to formal eloquence or consequently formal court discussion. More broadly, "Forensic" has become almost synonymous with "legal." For instance, "Forensic Medicine" means "Legal Medicine." "Forensic Ballistics" was coined by Colonel Goddard in an article with that title written for *Army Ordnance* in 1925. He used this name in the title of his bureau in New York before going to Chicago.

Some police departments have bowed to the popular trend and actually call their laboratories "Forensic Ballistics Laboratories." Others officially title them Firearms Identification Laboratories, or maybe Scientific Criminal Detection Laboratories. In the case of the Federal Bureau of Investigation, it is the Firearms Unit of the FBI Laboratory. In New York City the same work is done by the "Ballistics Squad." In court, it is best for the Expert to stick firmly to Firearms Identification. Most of us would have difficulty qualifying as real Experts in the extremely complicated science of projectile motion. However, we are publicly known as Ballistics Experts. There probably is not much that anyone can do about it.

View of a crime bullet as taken 'all around' with the geared camera. The full, round bearing surface is shown on a flat plane.

CHAPTER 2

EARLY HISTORY OF FIREARMS

ONE who desires to become a firearms Expert cannot know too much of general firearms history. The opposing attorneys in trials involving firearms evidence are likely to endeavor to discount the testimony of an Expert in any way possible. They have resorted in some cases to the most irrelevant and even absurd questions. However, the ability to answer all questions of a general firearms nature, even though they have nothing directly to do with regular identification procedures, will be of value in establishing in the minds of the jury the competence and completeness of knowledge of the Expert under examination.

This is a typical "zip" gun made of a piece of tubing in which a .22 long rifle cartridge will fit fairly snugly. It is entirely home-made, using the tubing, two types of tape, a pin, a key, some whittled wood, and the rubber bands. The cartridge is placed in the rear end of the barrel. The firer seizes the key and pulls it back against the pressure of the rubber bands. Upon release of this key, suitably filed, it strikes the unsupported rear of the cartridge case and fires the weapon. The weapon is extremely unsafe; however, it is easily made. This particular specimen actually seriously injured a policeman without doing damage to the shooter. These guns are usually found in the hands of juveniles, although sometimes men well along in life employ them.

Though it is not absolutely necessary, many firearms Experts throughout the country are chosen in part for their jobs because of their general knowledge of firearms. There are usually gaps, however, in their detailed knowledge of firearms history. The independent experts of the past who contributed so much to the Science in its early days were without exception men with many years of experience in many different firearms fields.

Old guns are occasionally used in modern crime and come into the laboratory for examination and testing. Extremely old types of weapons of modern manufacture are also met with. Far too many crimes are committed by juveniles using home-made weapons generally referred to as "zip" guns. These usually fire modern ammunition but from a contraption that is primitive in the extreme. Sometimes one runs across an entire weapon and ammunition of a very early type but made today. The literally matchlock pistol shown was designed and made by a boy in Boston and fired with black powder also produced by him.

One of the authors at the age of twelve had his first brush with police authority in a sleepy Southern town over a hand cannon that he had made himself. Weapons of this type had not been in use in Europe for 450 years. However, it would shoot reasonably accurate and certainly mortally at close range.

Those used to modern firearms sometimes forget the potentialities of an old gun in the hands of an expert. We have a target shot with a rifled wheellock arquebus more than three hundred years old at a range of 100 yards showing all five bullets in a group measuring 3″ vertical and 7/8″ horizontal.

A large muzzleloading shotgun can be even more destructive at close range than a modern 12-bore.

The Earliest Firearms

A composition similar to gunpowder was known and used in war a long time ago. The Gentoo Laws, believed to be coeval with those of Moses, refer in some translations to firearms and cannon. These translations are probably in error; however, weapons of the basic firearms type were used against the Macedonian Army of Alexander the Great in Northern India about 330 B. C. Certainly the Chinese were employing war rockets and firecrackers early in the Christian era. Some cannon found in the East by the Portuguese in 1498 were said to be many centuries old at that time. However, the Eastern

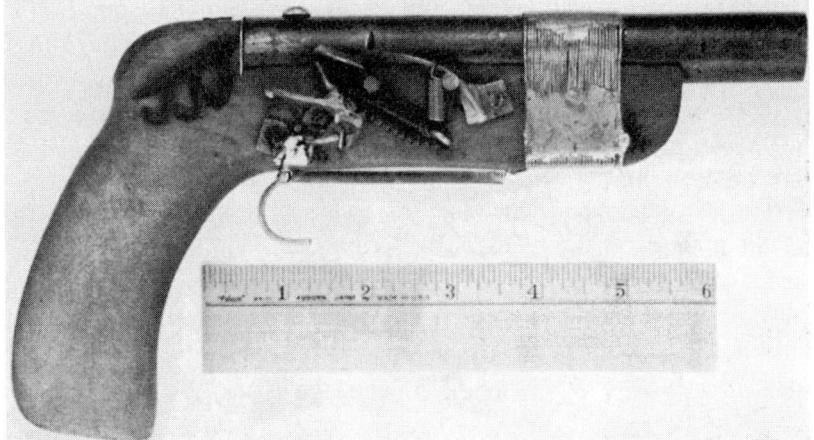

A modern matchlock pistol of more than usual ingenuity. A juvenile in Massachusetts made this weapon, which is self-explanatory, save for the ignition system.
A kitchen match with a short length of wood attached is mounted in the Y-shaped
metal bracket and is then held back by the trigger against spring tension. When the
trigger is pulled, the spring tension causes the head of the match to impinge against
the frizzen, which is actually a small piece of nail file, soldered to a length of wire,
thus striking the match and thereby igniting the powder inside the file cut in the
barrel. The maker of this pistol was also able to make himself gunpowder. This
weapon actually worked five times out of five when tried out experimentally in
the Massachusetts State Police Laboratory. The complicated system of springs, even
though undoubtedly easily put out of adjustment, was in adjustment when the
weapon was confiscated. This weapon would certainly be dangerous if properly loaded.

nations made no great use of this knowledge and did not develop
for any widespread use on land or sea any even reasonably efficient
weapons of this type.

It remained for Europeans to develop military gunpowder and the
weapons in which it was to function. With these weapons, the nations of Western Europe were able to conquer the Asiatics even
when outnumbered several dozen to one. Firearms played a more
important part in the spread of Western civilization over the globe
than many people realize. This was particularly true with the coming
of Europeans to North and South America.

Sometime in the middle of the Thirteenth Century Europeans
began to think seriously of the potentialities of explosives. The Moors
and Arabs brought a knowledge of them from the East and actually
used substances like gunpowder at the siege of Seville in 1247. This
is a full twenty years before Roger Bacon, an Englishman, wrote
his famous "Opus Majus" in which he gives a detailed description of
gunpowder. He was experimenting with the production of chemically
pure niter or potassium nitrate by crystalization from an aqueous
solution. By mixing the niter so produced with sulphur and charcoal,

a true explosive was manufactured, perhaps for the first time.

How soon this explosive was confined in the bore of a gun is not known. It is probable, however, that Friar Bacon, a monk of a philosophical rather than a mechanical turn of mind, did not carry his experimenting so far. Primitive grenades may have been used before firearms in Europe.

The Germans claim, without any valid basis, that a more or less nebulous character by the name of Berthol Schwartz not only invented gunpowder, but also used it in firearms. Unfortunately, they cannot even establish with certainty the century in which Schwartz lived; however, his name is as good as any other in relation to the first firearms. They probably appeared shortly before the end of the Thirteenth Century but well after the writing of Bacon's famous work.

The South German and Flemish towns began to add cannon to their defenses very early in the Fourteenth Century. Amberg had at least one cannon by 1301; Ghent had them by 1313. Their use in siege operations both by the garrison and the attackers was well established almost immediately. Portable cannon, however, were not apparently used in any great numbers for at least three decades. Edward III of England seems to have been a pioneer in their employment; he used them in his campaigns against the Scots in 1327 and employed them actively at Crecy in 1346. The Crecy cannon were breechloaders but of small casualty-producing power, although they did scare the horses of the French knights.

A few firearms capable of being carried and fired by one man began to be used in battle in the 1360s. They were known throughout Europe by the end of the century. However, they were auxiliary weapons only; they were never numerous in any army. Agincourt, fought in 1415, was decided without firearms on either side because of the torrential rains the night before and the miry condition of the field.

A hand cannon was far inferior to the bow. The former weapon is, however, most interesting. A tube of metal closed at one end was loaded with gunpowder and one or more projectiles. The tube had a touchhole, or vent, over the powder and was mounted in a straight spar of wood. The soldier would carry this arrangement in one hand and a smouldering match in the other. He would aim the whole spar at the target and then place the smouldering match in the touchhole, thus igniting the powder charge and throwing the projectile or projectiles with considerable velocity in the general direction of the target. This arm obviously had to be of relatively low power and light weight since it had to be handled with one hand; all recoil had to be absorbed without even the convenience of a pistol-type grip. Often this same sort of weapon was made larger and mounted in a six to ten foot shaft, which was stuck into the ground. This was really more of a small piece of artillery than a one-man mobile weapon.

Even this first hand cannon had, however, an astonishing social and political influence. The common man, even though weak physically and relatively unskilled with arms, could use it. Even these early firearms would pentrate armor at close range. Firearms from the very beginning have done more to make men equal than all the laws ever passed. A wealthy man before the age of gunpowder could use his money to protect himself against personal injury in war by the purchase of fine armor. A strong and skillful man could greatly improve his chances of coming out of a fray alive by developing his physical strength and skill by practice with arms. Undoubtedly, the English archers of the Fourteenth Century were infinitely more to be feared than any soldier armed with a hand cannon; however, years of experience, a great deal of skill, and unusual muscular power were needed to produce a fine archer. Most any yokel could be taught to load and fire a hand cannon in half a day.

The writers made up a weapon similar to the old hand cannon from a piece of scrap barrel, a threaded plug, a few bits of metal and an Enfield stock, in a total elapsed time of about two hours. We

Example of a "late" hand cannon—one made according to old principles but of new materials.

tested it for accuracy and penetration. It can be fired as accurately as one can fire a pistol from the usual quick-draw position. It will penetrate at least 4" of timber and undoubtedly would have toppled any armored knight from his horse dead if hit in a vital spot.

The hand cannon historically, however, was not really satisfactory. It was clumsy, so inaccurate as to be useless save at the very closest range, and uncertain because if one looked at the target one could

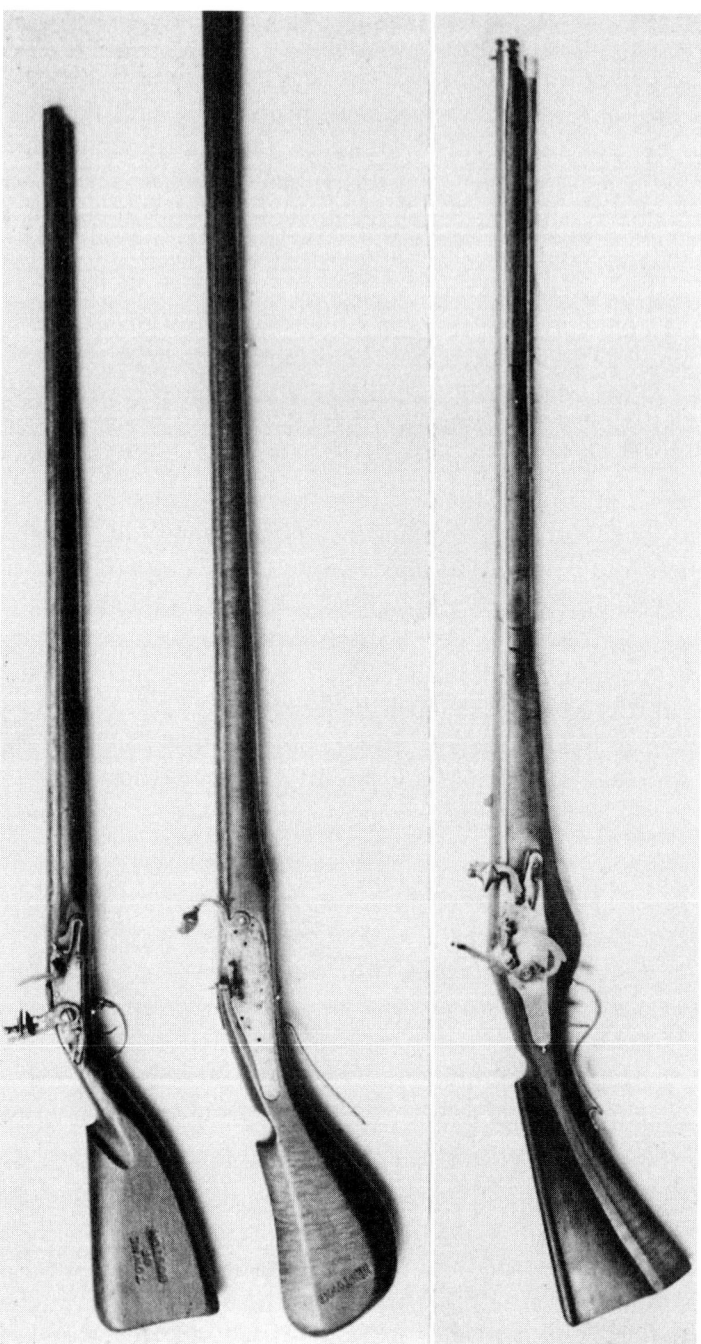

(Descriptions are at bottom of opposite page.)

easily miss the vent with the glowing end of match held separately in the other hand. The basic gunpowder idea would not have achieved much battlefield importance in small arms if a cheap, sure, and handy answer had not been found for the shortcomings of the hand cannon.

The Fifteenth Century Matchlock

The simple expedient of mounting a pivoted lever on the gun stock to convey the glowing match end precisely and at will to the vent solved most of the hand cannon problems. This new weapon, called the matchlock, came into being about the end of the first quarter of the Fifteenth Century. A man now could use both hands in controlling his weapon; it was less clumsy and capable of better accuracy. A two-hand hold allowed the recoil to be absorbed against the shoulder and chest. It could also be taken up by padded sleeves. The matchlock could be made both longer, heavier, and more powerful than any of the hand cannons. Further, it was theoretically possible to make a matchlock pistol light enough to be operated reasonably precise with one hand while on horseback.

The first matchlock was a very simple contrivance; a lever was arranged so that the trigger end needed to be moved only a short distance to cause a large movement of the cock end. In general, matchlocks of this type had the cock holding the match facing forward. Soon, however, a true trigger was worked out. The cock, under the impetus of a spring, moved smartly and surely directly into the pan of powder as soon as the spring release, or trigger, was pulled to the rear. This was probably an adaptation of the regular release mechanism from the cross bow. These cocks generally faced to the rear. An added improvement to the matchlock was a "fence" behind the touchhole so that particles of the priming powder and slow match would not be blown into the shooter's eye. At about the same time, the touchhole, or vent, was moved from the top of the barrel to the side. Further, covers were arranged for the pans con-

---

The upper weapon is a very early dog-catch flintlock musket, re-stocked in Boston but probably made in England originally about 1670. This weapon was also called a caliver in the days when muskets meant a larger form of usually monopod-supported shoulder weapon. The center weapon is a matchlock musket of an early type, in which the serpentine is directly connected with a long trigger. The fence to protect the shooter's face can be seen plainly. This weapon also was probably re-stocked in Massachusetts, apparently in Newtown. Each of these may have been a part of the common arms of these two early Massachusetts villages. Courtesy Smithsonian Institution. Lower is an early wheellock musket, probably of Dutch manufacture, now in the Tower of London and believed to have been used in the English Civil War. The wheel of this weapon is partly concealed in the massive lock mechanism.

taining the priming compound, which remained in place until the trigger was pulled.

The matchlock by the end of the Fifteenth Century was a creditable weapon. It had the one great disadvantage of the glowing match which had to be almost continuously adjusted and kept burning. However, it had as yet achieved no great battlefield recognition. In England the archer was still supreme; his weapon had greater range, accuracy and rapidity of fire. On the Continent the pike was queen of battles. A hedge of pikes could do more than firearms against armored horsemen.

Sixteenth and Seventeenth Century Small Arms

By slow degrees over almost 200 years firearms developed from a mere auxiliary weapon to the principal arm of all the infantry. Although other ignition forms were known and used in small numbers throughout the entire period, the matchlock was standard until very late in the Seventeenth Century. Perhaps the greatest single step in the whole period was the appearance of the musket in the hands of the excellent Spanish infantry. The word musket is of Spanish or French origin; when cannon were named for birds of prey as Falcon, Falconette, and such, the smallest was named for mousquette, a small hawk, i.e., musket.

The Sixteenth Century Spanish musket was not, however, a small weapon. It was powerful enough to smash through any armor and weighed up to 35 pounds. It was fired from a rest. Infantry so armed were not really mobile; each company contained both musketeers and pikemen. The pikes protected all from shock attack by cavalry. The muskets could inflict casualties at a distance. This last was of particular importance in combating a new type of cavalry arms.

In the last half of the Sixteenth Century hand guns had reached a state of development allowing their battlefield use. These were called petronels and were probably first used by German mercenaries known as Landsknechts. These long heavy pistols were fired by each rank of horsemen in turn. After firing, they wheeled to the rear and loaded. Pikemen unsupported by musketeers were overcome with ease and safety by these tactics. Whereas some of the petronels were matchlocks, many were wheellocks.

In addition to the adjusting of the match itself, the matchlock had several other shortcomings. It could not be used at all in foul weather. Its employment required time to strike a light with flint and steel, kindle a small blaze, and then ignite the lengths of slow match to be used in the weapons of the musketeers. The smoke, glow, and sound of a number of these matches sputtering violently precluded surprise. A matchlock musket required rather formal battlefield conditions; however, it was efficient under those conditions, as well as cheap and rugged.

The wheellock on the other hand was none of these. Contrary to the usual belief, it did not supersede the matchlock. The first wheellock was probably produced within a score of years following the first matchlock. The two continued to exist for upwards of 250 years side by side. The wheellock had the priceless advantage of being ready to fire the first shot immediately. It was, however, expensive, delicate, and not capable of firing nearly so many rounds before it had to be cleaned.

The ignition arrangement consisted of mounting a piece of iron pyrites in contact with a serrated steel wheel right above the priming charge. Before pushing the cock containing the iron pyrites against the wheel, a clock-like mechanism was wound up against a spring and the trigger release set. The iron pyrites was then moved into contact with the wheel, and the weapon was ready to fire. Pulling the trigger caused the wheel to revolve rapidly sending a tremendous shower of sparks into the priming charge.

Although somewhat cumbersome, the wheellock could be brought into action almost as quickly as a modern weapon. It was certainly a vast improvement over the matchlock in this respect. Some of these fine wheellock weapons would even stand a certain amount of damp weather before they failed to function on the first discharge. The pan, if really well made, was almost rain-tight; the arm could be carried in any position. The petronel for use in one hand on horseback was, of course, greatly improved by this method of ignition.

Before the introduction of the wheellock pistol, the average citizen was largely at the mercy of bands of robbers. He could now carry in his saddle holster, or attached to his belt, a one-hand weapon capable at a moment's notice of smashing a nice little round hole through the finest armor. A wheellock pocket pistol could kill the most powerful unarmored foot pad. The matchlock had introduced democracy of survival to the battlefield; the wheellock extended the equality of man to everyday life.

Another Sixteenth Century development in firearms was the rifling of the insides of gun barrels. Few of us who work in Firearms Identification stop to realize that rifling, which is the basis of our Science, is centuries old. We do not mean to imply, however, that rifling was used to any considerable extent, or even that the principles of rifling were clearly understood. The spiral groove arrangement was probably copied from the cross bow bolts and arrows which were fin stabilized and sometimes fletched so they revolved in their passage through the air. The early grooves in firearms bores are believed to have been spiral in shape; however, some gun makers believed that the grooves themselves were what was needed and not any particular spinning motion in the projectile. Peculiar as it may seem to us today, they were not entirely wrong. We tested a short time ago for accuracy a breechloading metallic cartridge pistol with straight grooves and found it to shoot quite well with spherical ammunition at close range. The grooving prevents a haphazard and entirely unpredictable rolling motion in the bore and produces quite a characteristic pattern on the projectile itself. The limit of real accuracy is about 15 yards.

The Seventeenth Century in firearms, as in politics, is a confused one. Europe was expanding to America, to Africa, and to Asia. The various maritime nations of Western Europe had their whole economies, ways of life, and standards of living changed by each new naval engagement. The entire century saw very few weeks of peace. Firearms development was continuous, but most sporadic.

By slow stages, the heavy matchlock musket was lightened, shortened, and made more portable. Its armor penetration decreased in consequence. However, the Seventeenth Century light musket, or caliver as it was sometimes called, would blast a hole through the strongest breastplate at close range. It was still cheap to make, required little or no repair, and would function for a long time in battle. These were the first weapons brought to America by the British, French and Dutch.

However, the matchlock form of ignition was found to be totally unsuitable for wilderness conditions. The wheellock, on the other hand, was extremely expensive, required care to operate, and real skill to repair. It did not function so well in continuous firing. Not many wheellock muskets were ever brought to America. The early colonists would have undoubtedly ruined these expensive weapons and lost in the forests the spanners used to wind up the mechanism.

A new form of ignition employing flint and steel was used in small numbers in Europe all through the Sixteenth Century; it was ideal

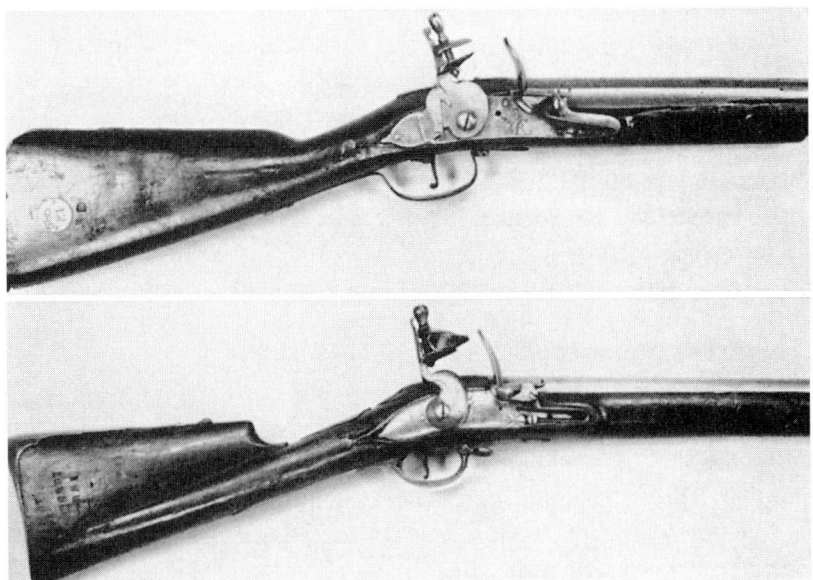

Upper—An early dog-catch weapon of the late Seventeenth Century, bearing the cipher of William of Orange. This lock had an internal sear. Weapons of this type were the first to be used in the British Army where real similarity was aimed at. It was a solid, substantial piece which took a plug bayonet. Lower—The earliest known Brown Bess musket, dated 1717 and brass mounted. This weapon has been the subject of considerable controversy since it is ten years ahead of the next known brass-mounted Brown Bess. It is thought by one of the authors that this weapon may have been assembled in the 1730s, using an early lock intended for iron-mounted Brown Bess or even pre-Brown Bess weapons. However, since it is in the Tower of London, there can be no question as to its complete authenticity.

for conditions over here. Within 25 years of the permanent settlement of Englishmen in Virginia the flintlock was the most numerous weapon in their hands. More than half a century was to pass before this was true in any European army.

Collectors divide flintlocks into at least three classes. The basic idea of striking sparks into the priming pan by a blow of flint on an inclined steel surface was the same in all. We now think of the snaphaunce as having the inclined steel surface, or frizzen, separate from the pan cover. In the English, or Miguelet, lock the two are integral but with a sear protruding through the lock plate and engaging (externally) the cock which held the flint. The true flintlock as we think of it today has an internal sear arrangement and an integral frizzen and pan cover. In the Seventeenth Century these terms probably did not mean what they do today.

Close to the end of the Seventeenth Century, the bayonet was evolved for military use in several different countries, probably independently. Until this little item of equipment was thought of and placed in the hands of the infantry, muskets were not really practical for more than a part of the soldiers in any company. The bayonet, rather than the musket, meant the end of the pike.

Armor, which survived in the form of helmets, breast and back plates, and certain other pieces for pikemen until almost 1700 was finally discarded for regular infantry use. The musket could beat it. It was worn to some extent in attenuated form, however, by cavalry as long as sabres were a principal weapon of the opposition.

The Age of the Flintlock

The flintlock musket was by the beginning of the Eighteenth Century the principal and almost the only weapon of the infantry. By 1704, the date of the Battle of Blenheim where the Duke of Marlborough with a coalition army defeated the generals of Louis XIV, the entire infantry of both sides was armed with flintlock muskets. There were probably still many thousands of the old matchlocks in arsenals; however, they were not considered fit for first line infantry use in the field. There were probably also wheellocks both in government armories and in the homes and castles of the wealthy. These continued to be used for many decades, but only for specialized work. Actually, the wheellock rifle continued as an instrument of precision for hunting long after its delicacy and cost had removed it from the field of battle.

As we have seen, the pike had finally been abandoned for the bayonet, although the latter weapon was not yet of a type that allowed the musket to be fired with the blade in place. The sleeve bayonet, not in common use until about 1720, was superior to the older plug bayonets, although the older type had several advantages. If it fitted tightly, the whole musket and bayonet more nearly resembled the pike in rigidity and strength. The plug bayonet made a better sidearm. A soldier probably had to remove either type before loading comfortably, although he could, of course, fire one shot with the sleeve bayonet fixed.

The American Colonies entered, as we have seen, the flintlock period before Europe. The whole of our history belongs to the age of firearms. Our ancestors brought these with them. They used firearms from the very beginning as their basic weapon. Antiquarians have found some half-armor, swords, and the like here in small numbers; however, these were more or less for show, or at best for garrison duty. The conditions in the new world were such that armor and edged weapons just were not effective. The early settlers were in more or less continuous bicker with the original Indians. Most of the actual fighting, however, was done between individuals or small groups. Shock tactics were unknown. The firearm was the obvious answer for these conditions. Even during the Seventeenth Century the rifle was extremely popular over here both as a weapon of defense against the Indians and as a provider of fresh meat for the family. Yet smoothbore guns and muskets were always far more numerous.

The British Brown Bess musket is perhaps the most famous of all flintlock weapons. It was used for almost a century and a half. Several million were made. It was basically reliable, reasonably powerful and, in the hands of trained men, a quick firer. Its accuracy was poor; however, it would continue to function throughout a battle. The piece of flint had to be changed fairly frequently, but this required only a few seconds. It would not function in a rainstorm, but Western European battles were rarely fought in rainstorms, save in exceptional instances such as when the Scottish Highlanders with their predominantly edged weapons fought the English and lowland infantry at Falkirk in 1746. The rainstorm which started at the beginning of the battle was extremely hard on the efficiency of the Brown Bess muskets, but did not hamper the Highland claymores at all. The Scots won a victory that would have been utterly impossible on a dry day.

The Brown Bess began to arrive in this country early in the Eighteenth Century. We used some of them in battle 150 years later in 1861. These muskets and the early United States flintlocks copied from similar French models show clearly the advance in manufacturing techniques and skills during the flintlock period. Arms made before 1700 are sometimes wonderously wrought and beautifully ornamented; however, they were all non-standard to an astonishing degree. The Eighteenth Century flintlock muskets, whereas not interchangeable, were remarkably similar in each model.

The later flintlock sporting weapons are believed by some people to be the most beautiful firearms ever made. Disassembly of a fine English double flintlock gun is a joy to anyone of a mechanical frame of mind. When one realizes that all this was produced entirely with hand labor and relatively primitive tools, one's respect for the skill of Joe Manton and his fellow gunmakers is extreme. The restrained ornamentation of this period has never been surpassed. These weapons for the wealthy, however, were always rare and played no significant part in the history of nations.

The flintlock musket changed the world. It was able to continue to function in desert, in jungle, in the arctic, and in the tropics. Lots of them continued in use long after they were abandoned in favor of more modern weapons in the civilized countries. Ammunition for them could be produced in savage lands. Flint is found the world over. Black powder is a mixture of almost universally available ingredients. Lead balls could be dispensed with and stones fired instead, if necessary. There are instances of African natives killing lions with iron bolts from a flintlock muzzleloader in recent times. A modern magazine rifle would be useless to those people as soon as the initial supply of ammunition was exhausted.

During the Eighteenth Century, the pistol and carbine in the hands of the cavalry began to be as important as the saber and lance. The rise of firearms slowly cut down, however, the battlefield importance of horsemen. Unbroken infantry armed with flintlock muskets could withstand a charge of an equal number of cavalry. Firearms played a more important part in civilian life than ever before. The broadsword and other edged weapons disappeared from the general population, although the gentleman class in Europe continued to carry rapiers for some time. They did this, however, more in anticipation of duels with social equals than for protection against criminals. The fine flintlock pistols of both large and small sizes were extremely efficient. The big ones for belt or holster use could be relied upon to give almost a certain shot, even in a rainstorm. The smaller weapons could be carried in the pocket and would kill the strongest brigand. Robbers throughout all nations and all ages are out for plunder. They do not want to run the risks involved in encountering an armed man on an equal footing.

The Percussion Lock Firearms

A Scotch clergyman by the name of James Forsythe discovered in 1805 that a chemical compound, fulminate of mercury, upon being placed next to the vent and struck a blow would fire a modified flintlock weapon. At first, the employment of this compound for ignition was accomplished in several different forms of locks. In the pill lock weapons a small pill of this compound is placed under an auxiliary hammer which is struck a blow by a regular hammer. The tubelock used copper tubes containing the same compound which lay in a groove at right angles to the bore and horizontal when the gun was pointed in the normal way. A hammer blow either direct or through an auxiliary plunger would set off the weapon. These tubelocks had real efficiency for some forms of large shotguns.

The percussion cap, well known to all of us even today, was developed perhaps before 1815, yet it did not come into general use until after 1821. Many people have claimed the invention of the percussion cap. Undoubtedly, it was perfected separately by several different individuals in different places but using the same basic approach. Percussion ignition was accepted by sportsmen reasonably soon. However, military flintlocks continued to be made here and in Europe until after 1840. Our regular infantry in the Mexican War in 1847 were still armed with flintlock muskets.

The percussion lock was in general military use for about 25 years. Its use in sporting arms was not more than twice that long. Yet over 90 per cent of all surviving muzzleloading weapons are of this type. They were efficient in their day. Some of them will still shoot quite well and are still used both in backward and civilized areas for reasons of necessity or personal choice.

Several features usually associated with modern firearms were known and widely used in the percussion era. Many of the modern basic designs were more or less standardized. For instance, the double-barreled shotgun has undergone little change in external form since the percussion days. Percussion revolvers reached a relatively high state of perfection.

The breechloading idea is as old as firearms. For special purposes wheellock and flintlock breechloaders were quite satisfactory. However, with the coming of the percussion cap and semi-modern industrial procedures there were literally dozens of practical breechloaders using self-consuming cartridges. In this country the Sharps was the best known; in Britain the Wesley Richards was considered the most satisfactory.

Rifling as already pointed out is quite old; it is probably Fifteenth Century. Until the percussion era there were two methods of obtaining satisfactory results. The European rifles generally used an oversize ball forced down the bore with a heavy rod and sometimes a mallet. In this country we perfected the well known patched-ball principle. Neither was really satisfactory for a standard military arm, though the American type rifles were wonderful for use under our wilderness conditions.

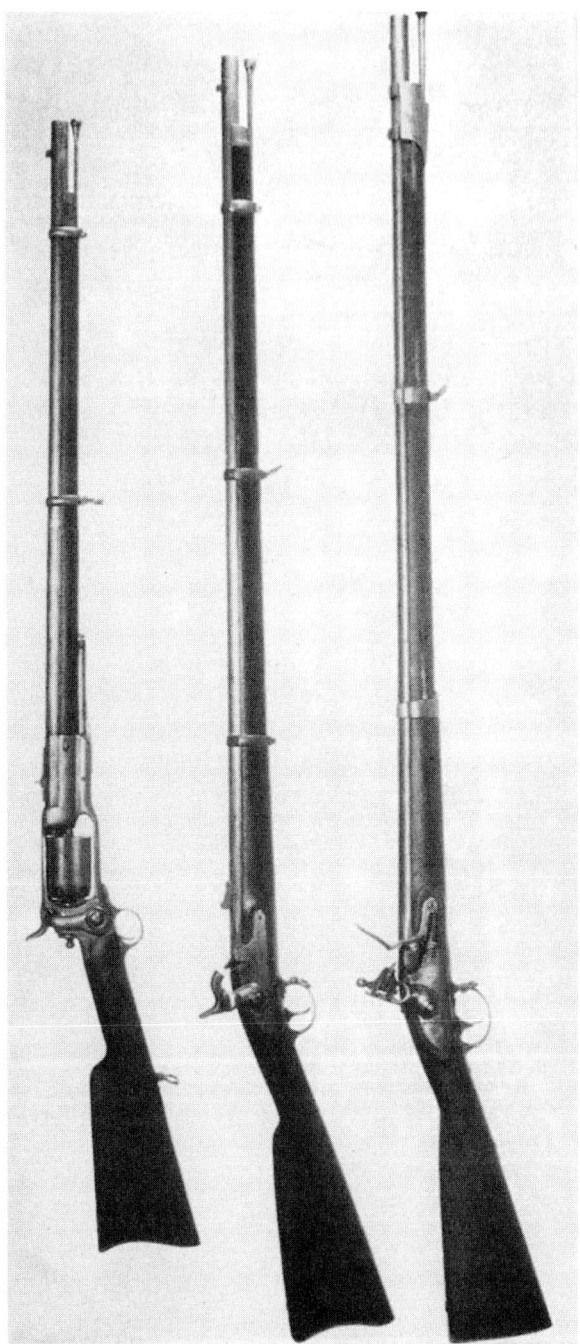

Three characteristic weapons of the American Civil War. The top is a Colt revolving rifle issued at the start of the war in considerable numbers and of full musket bore, but entirely unsatisfactory for use in the field. In the middle, is a standard Model 1863 Springfield muzzleloading Cal .58 Minie rifle, the principal rifle used. At the bottom, a standard flintlock Cal .69 smoothbore musket, U. S. Model 1821. A few of these weapons saw service in the early days of the Civil War. They were the standard weapon of our regular infantry in the Mexican War.

Towards the middle of the Nineteenth Century, however, several men evolved, perhaps independently, the idea of using an elongated or cylindro-conoidal bullet slightly below bore size with a hollow base. The explosion of the powder charge expanded the base of the bullet into the grooves of the bore. The range and accuracy of these muzzleloading, expanding-bullet rifles of our Civil War period is truly amazing to one not familiar with these weapons. Quite good groups can be made at a thousand yards on reasonably small targets by shooters familiar with precise muzzleloading rifles. Even the common soldier armed with the regulation Enfield or Springfield rifle in the Civil War could lay down accurate fire on a large target such as a battery of field guns at ranges beyond half a mile. If he hit a man in a vital spot at this range, he would kill him.

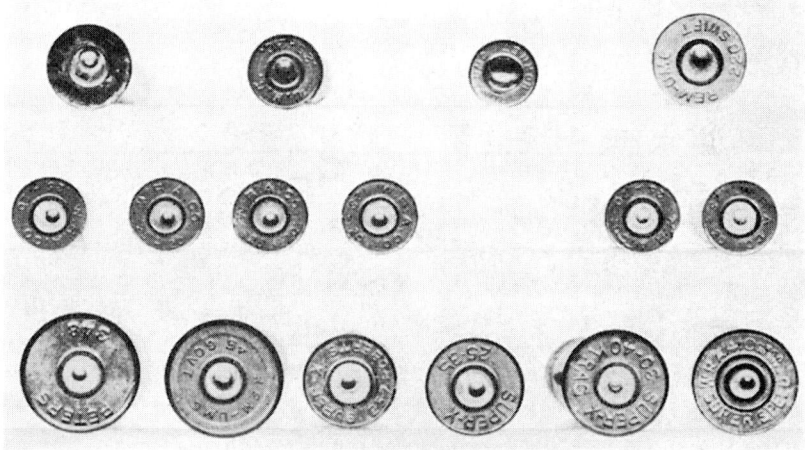

PRIMER IMPRINT EVIDENCE

Often, in court trials in past years, some not-too-well-informed 'expert' has taken the witness stand and identified or attempted to connect a suspected firearm by the fact that "it was defective in that the firing pin struck off-center, like defendant's gun does." Such evidence and information should be looked into very carefully before it can be accepted, as above illustration will show.

Observe the fired cases in the bottom row; all six of these show the firing pin imprint more or less off center, yet all of these cartridges were fired in high-grade firearms in perfect condition and the cartridge next to be fired in any one of them might show a perfectly centered imprint. This strike of the firing pin on exact center of the primer is governed by the manner in which the cartridge lies in the chamber. Not all loaded cartridges are exactly the same in diameter (or in any dimension, for that matter) and a 'minimum' cartridge in a 'maximum' chamber will generally show an off-centered imprint upon being fired.

The center row shows six cartridges fired consecutively in the same rifle. Note the variations in the positions of the firing pin imprint, caused by the manner in which these cartridges positioned in the chamber of the rifle.

However, the top row shows four fired cases from cartridges that were fired in rifles which were not properly adjusted, or which were worn in parts and breeched-up in a loose manner. This caused the firing pin to consistently strike well off center. Such firearms are uncommon, but they do have a 'defective firing pin' and they distinctively mark their fired cartridges in this off-center manner.

CHAPTER 3

BREECHLOADING METALLIC CARTRIDGE FIREARMS

E VEN though occasionally muzzleloaders and various percussion lock breechloaders are met with in the field of Firearms Identification, more than 99 per cent of investigations involve metallic cartridge breechloaders. These are not always of modern types. As a matter of fact, a rather large percentage of the weapons coming into the New Jersey Firearms Identification Laboratory were made in the Nineteenth Century.

We will endeavor in this discussion to cover briefly all the various major types, even though many of them are not often actually encountered. A knowledge of the various possibilities, particularly when only an unusual bullet remains, is essential. Opposing attorneys may easily ask questions about unusual arms even though the actual information is not pertinent to the case on trial. For instance, an attorney once asked a so-called independent expert questions in connection with heavy caliber revolvers; the expert stated that revolvers could have bores of from .22 caliber to .45 caliber only. This, of course, was in error. The prosecution had available, but was not permitted to introduce as evidence, a Webley-Enfield revolver with a .476 bore, and a W. W. Greener revolver of .577 bore. The jury was impressed, however, with the shallowness of the so-called expert's knowledge even though the judge upheld the objection to producing the irrelevant weapons. The crime was committed by a .455 Smith & Wesson revolver which has the same bore diameter as their .45 caliber weapons.

Pistols

All firearms intended to be used in one hand are by definition pistols. (See Webster's New International dictionary.) Revolvers are more technically revolving pistols. Automatics are really semi-automatic pistols or autoloading pistols, but let us just call them automatics, like almost everyone else does. Revolvers and automatics constitute more

40

than 90 per cent of the pistols in the world. However, single-shot, multi-barreled, and manually operated repeating pistols do exist.

Since hand guns are far more frequently met with in crime than all other firearms combined, we will discuss these first. In general size, shape, and even appearance, pistols were pretty well set before metallic cartridge operation was really practical. The only significant external change was brought about by automatic operation; automatics did not become common until the Twentieth Century.

The revolving pistol was used in enormous quantities in our Civil War; these were usually, although not always, percussion cap weapons. Smith & Wesson brought out a bottom-break rimfire metallic cartridge revolver in 1858. Following the Civil War, the conditions in the West and South led to a great emphasis on revolvers and shooting with revolvers. The Western movies and more recently television have made revolvers known to even four-year-olds. They are typically American and are more popular in this country than elsewhere.

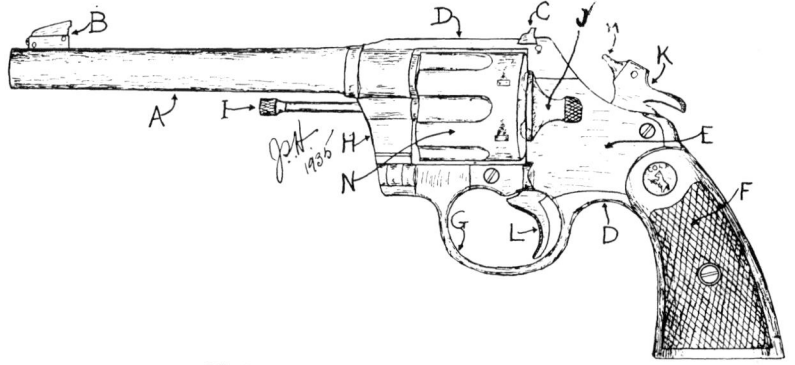

THE MODERN TYPE OF REVOLVER

Showing an example of a Solid Frame, or Swing-out Cylinder, or Side Ejection revolver, such as used today by the police, military and target shooters. This is the only type of revolver that is safe with modern, smokeless ammunition of any appreciable power.

The principle parts are: A—barrel, B—adjustable front sight, C—adjustable rear sight, D—frame, E—side plate, F—grip, butt or handle, G—trigger guard, H—crane, I—ejector rod, J—cylinder latch, K—hammer, L—trigger, M—hammer nose, or firing pin, and N—cylinder (of six chambers which are not visible).

Illustration shows a target model of revolver fitted with movable sights; for police or military use the gun would have fixed sights.

The basic revolver design is well known and needs little description. There are a few variations that should be mentioned. Frames can be solid, as in the famous Colt Frontier Model which uses a side ejector, or the modern Smith & Wesson and Colt revolvers in which the cylinders swing out for unloading and loading. Revolver frames can also be top-break like the many relatively low-powered weapons

which flooded the country around the turn of the century. A powerful
revolver of this type like the Smith & Wesson .44 Russian or the
heavy caliber Webleys is also met with occasionally. Less frequently
one will see a bottom-break Smith & Wesson or even something
that comes apart entirely like the Merwin and Hulbert.

Revolvers are known as hammerless if the hammer is completely
enclosed. The method of cocking and releasing the hammer deter-

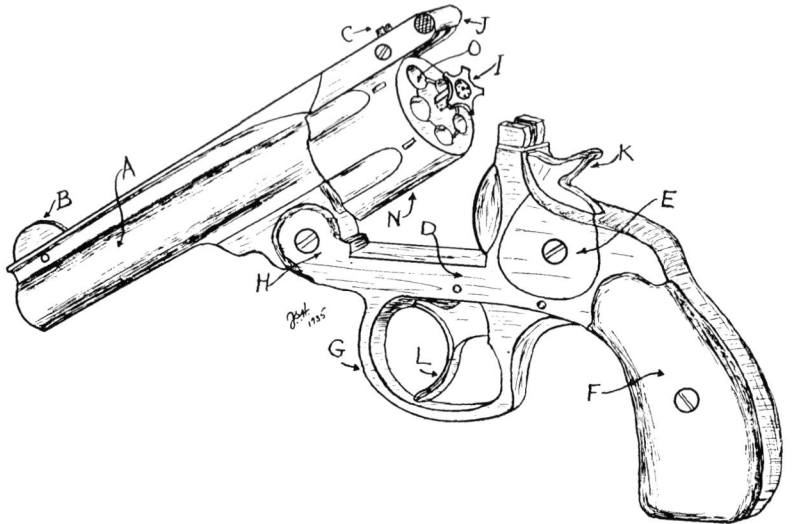

A HINGED FRAME REVOLVER

The above "Break Frame," "Hinged Frame," "Top Break" or "Tip Up" model
of revolver is an earlier type than the Solid Frame models and is not suited to fire
the more modern and powerful revolver cartridges. Although falling steadily in
popular favor they are still made and sold by the thousands in the .22 rim fire
calibers and in the less powerful center fire calibers. This type of handgun is used
mainly as a pocket or self-defense gun—and is often referred to as the "bureau
drawer" gun.

A is the barrel, B—front sight, C—rear sight, D—frame, E—side plate, F—grip,
butt or handle, G—trigger guard, H—hinge, I—ejector (which is really an extractor
but which withdraws the fired cases sufficiently that they fall clear of the action),
J—latch or barrel catch, K—hammer, L—trigger, M—(hammer nose or firing pin
is not visible), N—cylinder, and O—chambers.

mines the action. The old Frontier Colt and many of its contemporaries were single action, which meant that the hammer had to be
cocked by hand before a shot could be fired. The cylinder was revolved by the cocking of the hammer. In this type of weapon, the
trigger can be removed and the gun fired by manually releasing the
hammer at full cock.

A double action revolver is one that can be fired by pulling the
trigger only. The trigger normally moves through a greater distance

and under heavier pressure than in single action revolvers. Most modern as well as many obsolete revolvers are of this type. They can usually also be fired single action by cocking the hammer manually between each shot. Generally, modern target shooters follow this procedure even in rapid fire; some do not. A few models can be fired double action only; the British Enfield No. 2 Mark 1 Star .380 is of this type; there is no full cock notch on the hammer at all.

In general today revolvers fire centerfire cartridges, save for the many .22 cal. rimfire models; however, many crimes are still committed with .32 rimfire weapons as well as some with other rimfire cartridges.

The so-called "automatic revolver" does actually exist, although mystery story writers probably do not know about it. An English Army officer and expert pistol shot by the name of Fosberry collaborated with Webley to produce a weapon in which recoil served to push the entire barrel and action back over the butt stock in such a manner as to cock the hammer and revolve the cylinder preparatory for a second shot. We have done a good deal of shooting with these weapons and do not believe that any criminal who has any experience with them is very likely to so arm himself. The trouble is that even if the action works when the weapon is extended at arm's length with a stiff elbow, as in target shooting, they are almost certain not to work from a usual quick-draw position or when the gun is held at all loosely. A similar weapon was produced in this country by the Union Arms Company. It is even rarer than the original Webley-Fosberry.

The Iver Johnson Company, and others, made and advertised so-called "Automatic" revolvers. However, these were simply double action topbreak revolvers; the "Automatic" referred to the ejection of the shells, which were automatically ejected when the revolver was opened.

Revolvers may be found in almost any caliber. We hesitate to say the .577 W. W. Greener already mentioned is the largest; it is the largest we have seen; however, the Montenegrin Gasser has a larger cartridge. The most frequently met with arm of any single type is the cheap top-break revolver firing the .32 Smith & Wesson cartridge and made by at least two dozen firms over a period of more than half a century, concluding about the time of World War II. These weapons can be obtained second-hand almost anywhere in the country for, in some instances, as little as the average family now pays for a toy pistol for junior to imitate Hopalong Cassidy. These

A typical thug's gun; a double action, top-break revolver of the cheapest type, .22 caliber, rim fire and actually used in a holdup and murder of a police officer.

.32 revolvers usually cannot be traced, are not now manufactured, are frequently in poor condition and unsafe; however, they kill and kill and kill. More crimes are committed with them than with all other types of revolvers in the country by at least threefold. They are the so-called "Saturday night" guns that seem to be ever-present in certain sections of the population throughout the whole United States.

The very high proportion of these low-priced .32 revolvers to better guns reaching the firearms laboratories can be better appreciated when it is realized that one company alone made over three million of them before 1939. Few of these were exported.

Most of the .32 revolvers coming into the New Jersey Laboratory are in unbelievably poor condition due to misuse and neglect. After all, gun lovers, who take care of their guns, rarely buy guns of this type and seldom use their pistols in crimes.

In digression, it might be interesting to note that modern high-grade weapons are seldom actually used in crime. It would seem that those persons who love firearms and provide themselves with the best very seldom commit any crimes at all. Sometimes a weapon of this type will be stolen and used by a criminal; however, not often. Even a large laboratory will not have occasion to fire a .357 Magnum revolver or a new .44 Special revolver more than once every few years. Neither Boston nor Philadelphia can remember ever having such a crime weapon.

Unlike revolvers, the automatic is of quite recent origin. No successful weapon of this type was made until the 1890s when suddenly Borchardt, Luger, Mauser, Browning within 10 years produced designs that are still with us today. The Borchardt-Lugers are the most common foreign high-power crime automatic. The big Mauser is still a superb weapon. However, the designs evolved many years ago by a Mormon inventor have swept the world; John Moses Browning was the greatest firearms engineer of all time, in pistols as well as larger weapons.

Browning began to work on automatic pistols in 1896; his patents were granted beginning in 1897. He employed two basic types of operation. There was the straight blow-back in which the barrel is stationary and the cartridge case acts directly on the unlocked breech to eject the fired case and cock the firing pin or hammer. The breechblock then returns under the action of the recoil spring to load another round into the chamber. This action was used for relatively low power cartridges like the .25 ACP, 32 ACP and .380 ACP which in Europe are called 6.35mm, 7.65mm and 9mm Brownings.

Browning's other type of operation was for more powerful cartridge such as the .38 ACP, the .45 ACP and later the 9mm Luger; for these a locked breech and recoiling barrel were used. After a travel of about a quarter of an inch, the barrel has dropped down sufficiently to unlock the breech block which then continues to the rear as in the blow-back.

Browning's blow-back models of 1900, 1903, 1906, 1911, and 1922 as made by Fabrique Nationale d'Arms Belgique and Pocket Model 1903 and 1907 as made by Colt embody almost every worthwhile feature of weapons of their type. Double action firing and the chambering of more powerful cartridges in these blow-back actions occurred after Browning's death in 1927; however, essentially all blow-back automatics are Browning's or adaptations of Browning's different designs with occasionally a minor new feature.

The actual makers of blow-back automatics are too numerous to list. The only large producer here was Colt; however, Smith & Wesson, and Harrington & Richardson as well as others turned out weapons of this basic type. The Remington and Savage automatics, never very popular, incorporated certain delaying features but were essentially blow-backs. Abroad there were hundreds of firms, particularly in Belgium, Germany and Spain, producing these. Photographs of actual crime automatics with captions as to operation will give some idea of their diversity.

In the field of recoil-operated automatics Browning had several rivals. The most numerous pistol ever made is Browning's U. S. Service Model Automatic 1911 and 1911A1. The French and Belgian Model 1935 and 1935A as well as the Polish Radom and Russian Tokarev and several others embody many features of this weapon. However, in addition to the Lugers and big Mausers already men-

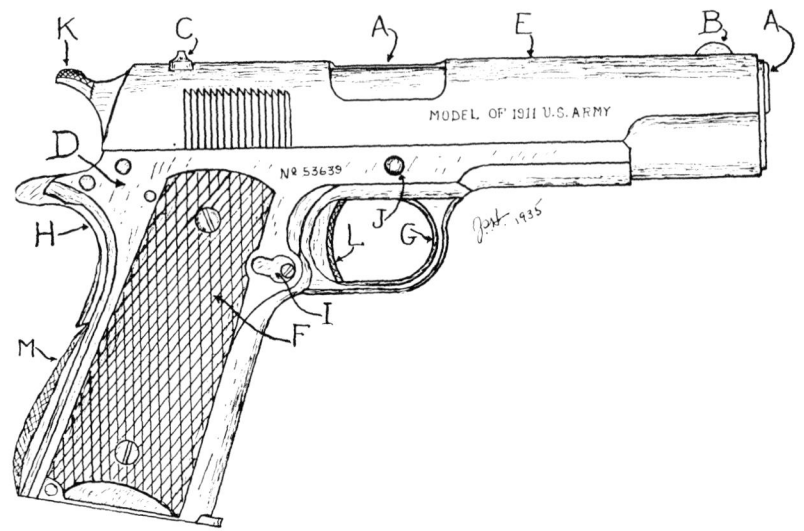

EXAMPLE OF AN AUTOMATIC PISTOL

Here is shown an Automatic Pistol, or "Automatic"; the .45 Colt, one of the official side arms of all U. S. Forces. This handgun carries its supply of loaded cartridges inside the butt and these cartridges are fed, one by one, into the chamber of the pistol by the action and force of the preceding cartridge being fired.

The principle parts are: A—barrel (which is mostly concealed inside the slide), B—front sight, C—rear sight, D—frame or receiver, E—slide, F—grip, butt or handle, G—trigger guard, H—grip safety (there is also a thumb safety on opposite side, which is not shown), I—end of the magazine release (which is operated from opposite side of gun), J—end of slide lock, or action catch (which holds the action open and is operated from opposite side), K—hammer, L—trigger, and M—mainspring housing.

tioned there are several other non-Browning high power automatics that may occasionally be used today by criminals. These include, in approximate order of their prevalence in this country: The German P-38; the Japanese automatics, usually called indiscriminately Nambus; and the Austro-Hungarian Frommer, Steyr and Roth-Steyr. There are others, although their use in crime is rare. In some localities such as New York City war souvenirs such as Lugers, Radoms, P38s and military blow-backs are used almost as often as the top-break .32 Smith & Wesson revolvers in crime.

In addition to the centerfire automatics, there are now .22 caliber rimfire automatics, all of fairly recent origin. They were originally designed for target or sporting work and were not ordinarily easily concealed on the person, however, barrels now have been shortened. Their lack of noise suits them to certain kinds of crime. The most common .22 automatics today are the various models of the Colt Woodsman, the High Standard automatic in several models, and finally, the new Ruger. There are also two types of the big Colt Service Ace and a few foreign .22 rimfire pocket models.

Pistols proper, as distinguished from revolvers and automatics, are generally although not always of a top-break construction. The rolling block action made by Remington was adapted to pistols for military and sporting use. The Martini action was used in the Bavarian Army for a pistol that was said to be quite accurate and efficient. Several other single-shot rifle actions have been adapted to pistols. We illustrate some of these that we have tested. Although these weapons shoot well, they are about as handy to carry on one's person as a standard-size frying pan. One of the best known and most numerous single-shot pistols is the Stevens Off-hand. Another was the Smith & Wesson Perfection. Both of these are .22 rimfire and primarily target weapons but are not so heavy and clumsy as the European .22 free pistols. Other weapons of similar types came into this country from abroad, particularly from Germany in the late 1930s.

At one time, top-break action pistols were very popular in England for producing the so-called "howdah" pistol originally intended to be carried while tiger hunting in a quick-draw holster in the castle arrangement perched up on top of an elephant. The idea was that these pistols would serve as a secondary battery to be quickly brought into action if the hunter missed the tiger with his double rifle. These weapons are beautiful but extremely expensive if in good condition; further, they recoil viciously since they fired .577 spherical balls with a charge equivalent to a 20-gauge shotgun shell. They are frequently groove-rifled but are also met with both smoothbore and rifled according to the Lancaster oval bored system which resembles a smoothbore on visual inspection.

These same weapons were sometimes carried by British officers as side arms, particularly against the Somali Fuzzy-Wuzzys who got themselves into such a fervor before battle that they were almost impossible to stop; it is said that a Fuzzy-Wuzzy at Omdurman ran for a distance of more than 100 yards after his entire head had been taken off by a field-gun shell. Such a guy would be as hard to stop as the

Philippine Moros which led to the abandoning of the .38 caliber revolver for side arms in the United States Army.

Shotgun pistols of this same general type were formerly quite frequently met with in this country and were made and offered for sale in the 1920s. They were advertised variously as home defense weapons and for hunting from automobiles. One of the writers recalls several pleasant hunts with a 20-bore weapon of this type for crows some years ago in a sparsely populated area. Unfortunately, such a pastime

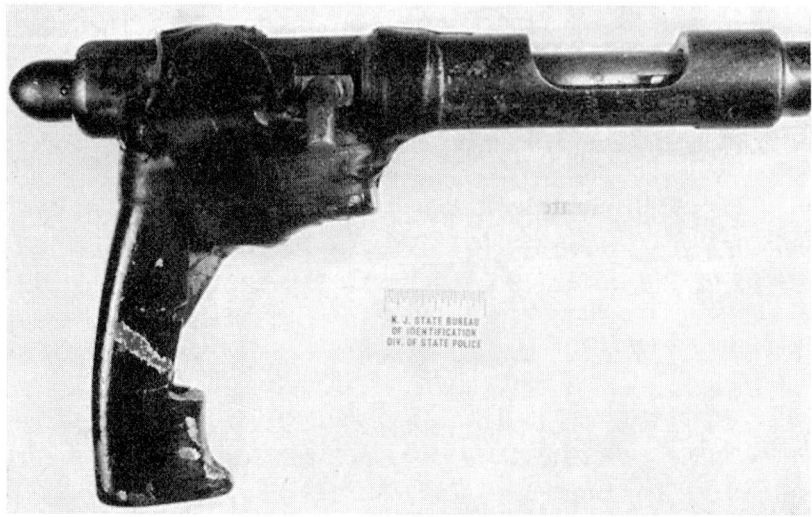

A home-made firearm taken from a prisoner who used it in a holdup and robbery in Trenton, New Jersey. Originally this was a Marlin Model 81 Repeating bolt action rifle. The handle (butt) is from a cheap electric drill. As a result of sawing off the barrel, distinctive marks were imparted to bullets fired through it. Any bullets fired before sawing off could not be identified as coming through this barrel after being sawed off, due to the many burrs remaining on end of barrel.

is now no longer legal under two counts. First, the weapon itself has to be registered and a transfer tax paid. Second, few states now allow shooting from an automobile. Unfortunately, these shotgun pistols came into the hands of criminals and were devastating, not only against human beings but also against so-called bullet-proof glass and light armor at close range.

No discussion of pistols would be complete without the mention of Deringers. At least one of our three assassinated presidents was killed with them. However, none are made today. Their characteristic shape needs no description; they were most commonly .41 rimfire, although Lincoln was killed with a muzzleloader.

By far the most common single-shot pistols met with in crime, how-

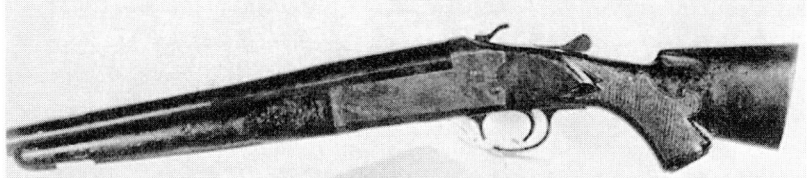

A TYPICAL MURDER WEAPON

This is a Stevens Model 107B 16-gauge single barrel shotgun, with the barrel sawed off to about 12 inches. The Tenite plastic stock was also sawed off, as shown.

In this particular case, the assailant threatened the victim with this gun, which he had forgotten to load. Then he finished the job with a 14 inch carving knife.

Nothing but a hacksaw and a few minutes' time was needed to make this conversion from a legal firearm to an illegal weapon.

ever, are home-made. These are of two varieties. Sometimes someone, not necessarily a criminal or an adolescent, will butcher a shotgun or rifle. The shotgun with cut-off barrel and stock is quite common in crime. More frequently a weapon is entirely hand-made. A tragedy still in the minds of the writers occurred some time ago. A 14 year old boy wanted very badly a real gun, which for some reason he could not obtain. He converted a toy pistol to handle the 12-gauge shotgun shell. He used very considerable ingenuity and some mechanical skill in the making of this contraption; then fired it with disastrous results; the breech, of course, let go, tearing the gun to pieces and amputating the boy's right arm well up towards the elbow. He did survive. Purely from the evidence of his work on this pistol, we believe that he has the type of mind that might easily produce some-

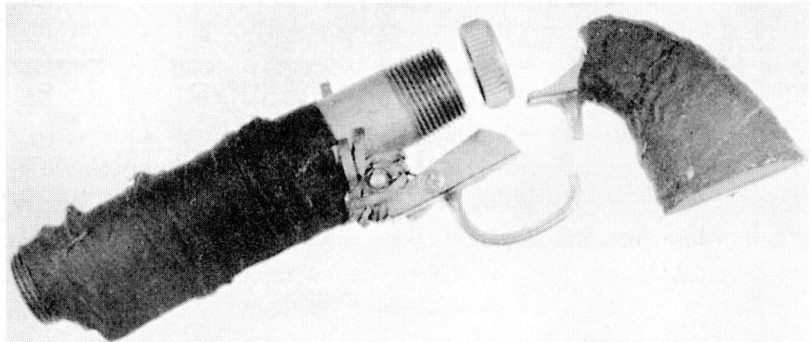

This home-made weapon is in a class above a "zip" gun. Although crude, it shows considerable evidence of ingenuity. The 14-year-old maker reworked the hammer of a toy pistol with considerable cleverness and had a good idea in connection with barrel and breeching, although he left far too much space unsupported around the primer of the 12-gauge shotgun shell which he fired in this weapon. With the explosion of the propellent charge, the entire base of the cartridge case gave way and came to the rear, while the shot charge was moving forward, inflicting a very serious wound, shattering the forearm and necessitating amputation.

thing fine in ordnance in spite of this terribly misguided initial attempt.

Usually, however, these "zip" guns handle .22 caliber rimfire cartridges which are not so powerful. Unfortunately, these home-made jobs are very common in certain sections of large cities. They are used by juvenile gangs and on occasion by adults. The firing mechanism is generally a key or some other similar object pulled sharply against the base of the rimfire cartridge by rubber bands. One weapon illustrated herewith dangerously and critically wounded a police officer. Strange as it may seem, the users of these weapons are seldom hurt, although no Expert likes to fire them.

Shotguns

The shotgun is undoubtedly the most common of all firearms in the United States; even today in some areas, it is pretty much a tool for ordinary everyday use rather than a sporting weapon. Throughout the length and breadth of the nation, some form of shotgun is likely to stand in a corner of almost every farmer's home. The mere presence of this weapon may be one of the factors leading to rather honest behavior of our rural people in general. Shotguns are sometimes involved in criminal investigation; however, most of the human shootings that they do are accidental. Occasionally, a criminal will use a shotgun because he can find nothing else, or because he appreciates its devastating qualities. Any form of shotgun at close range is a deadly and terribly powerful weapon.

Fundamentally, shotguns, like pistols, divide into three classes. First, there are the single and double-barreled top-break weapons. Second, there are the mechanical repeaters which occupy very closely the same position as revolvers do among hand guns. Finally, there are the automatics which, of course, are actually semi-automatics.

Many books have been written about single and double-barreled fowling-pieces. They can still be purchased for as little as $25.00 for a new single-barreled gun and about $60.00 for a double; these are perfectly satisfactory modern weapons made of strong materials under precision interchangeable manufacture. However, one can also pay $2,000 for a beautiful hand-made English weapon of extreme excellence. The mechanism of the latter is superb, the workmanship really astonishing, and the materials are the very best grade obtainable. However, the cheap weapon would probably in most instances perform almost as well. Double guns in general can be of the older side-by-side or the more recently popular over-and-under type. They subdivide into boxlocks and sidelocks. Almost all new doubles are hammerless; however, very many old hammer guns are still being used. Further, some of the new single-barreled top-break shotguns still retain the hammer as a doubtful safety feature.

Repeating shotguns are found most commonly in the so-called pump action which has been popular in this country for many years; It is still popular and a first-rate weapon. It will fire almost as fast as an automatic and has the great advantage that many sportsmen are used to handling them. Grade for grade, they are much cheaper than a double and a bit cheaper than an automatic. They can be either

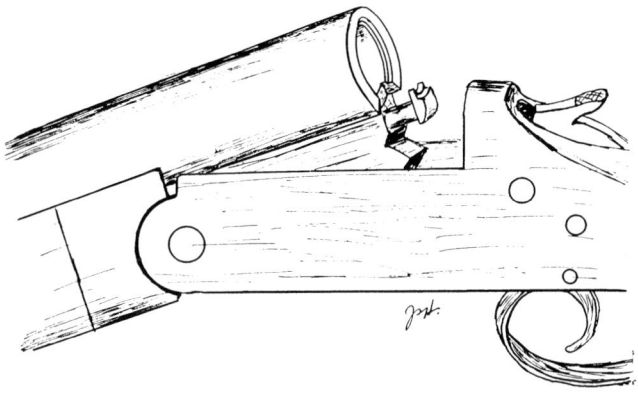

THE CONVENTIONAL SINGLE BARREL SHOTGUN

Sketch of the breech section of a typical single barrel, hammer shotgun. Note the extractor on this shotgun. When the gun is opened, a cam or leverage connection in the forearm pushes the extractor out to the position shown. When the gun is closed, this extractor recedes until it forms a portion of the bottom wall of the chamber. In the present-day method of single barrel construction this extractor is actuated by a stiff, always-loaded, coil spring which pushes out the shell about a quarter-inch every time the breech is opened.

Some of these single barrel guns have what is called an "automatic ejector," consisting of this same extractor actuated by a spring action which throws the shell clear of the gun whenever it is opened—and such action throws both loaded or empty shells out of the chamber. Under these circumstances, this spring-loaded extractor is called an ejector.

When the gun is closed, either extractor or ejector helps to support a small portion of the cartridge wall at the base of the case; the fit is seldom perfect and an impression of such place faced by the moving segment will generally be seen on the shell fired.

hammer guns or hammerless. The hammer feature was retained because of supposed safety, but is no longer popular. A crime was committed a few months ago with an old lever action Winchester shotgun. Many people do not realize that Winchester ever made a lever action shotgun; however, this one turned up in the automobile of a migrant farm worker. The firearms used in this type of crime are rusty, broken, unsafe, and antiquated but they continue to be used

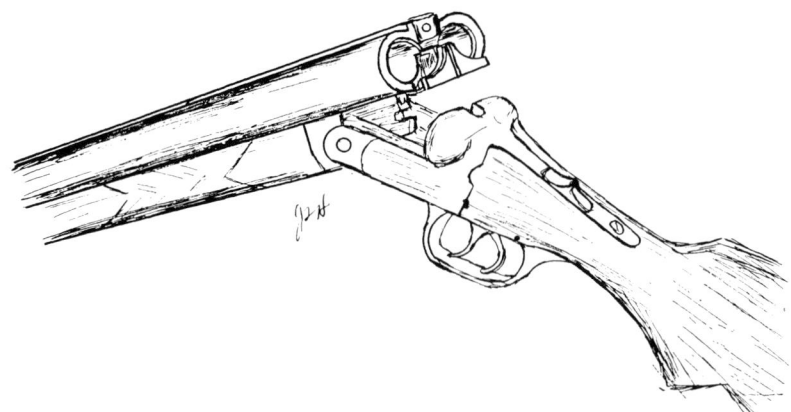

THE STANDARD MODEL OF DOUBLE BARREL SHOTGUN

The hammer action, in double barrel form, is today unknown. ALL double shotguns are now hammerless models, with inside hammers. This sketch shows a modern double gun and this particular one is fitted with "ejectors" as shown above.

These ejectors are divided, one for each barrel, and they work independently of each other. They are operated by a pair of lock mechanisms, located in the forearm, each of which is tripped and operates only when that particular barrel is fired. When one or both barrels be fired and the gun opened, those ejectors function and the empty cases are kicked out clear of the gun. If the gun is not fired but is opened, the loaded cases are pushed back sufficiently to permit being picked out by the fingers and changed to different loads or the gun left empty.

A visual indication as to whether or not a double shotgun is fitted with ejectors can be told by the appearance of the movable barrel segment; if it works in one solid part for both barrels the gun is an "extractor model" and merely extracts the shells part way; but if there be a division in the center of the segment (as shown above) permitting each half to work independently of the other, then the gun has "ejectors."

Whether divided or not, this mechanism forms a portion of the chamber when the gun is closed, and it may or may not show its markings on the fired cases—the reason being that any double gun fitted with ejectors is likely to be rather well made and fitted and that fit is probably a good one which may give no indication until after a long period of usage.

while the many superb weapons of sportsmen are seldom involved in crime.

A more recent innovation in the repeating shotgun field is the bolt action. There seems to be little reason for this action in a shotgun other than economy. We believe that an automatic-ejector single can be fired almost as fast and certainly more pleasantly than the bolt action, since the bolt itself means an unreasonable lengthening of the weapon in order to get decent length barrels.

Finally, there are the autoloading shotguns generally known as automatics. Almost all of these used in the country are recoil operated, with an abnormally long length of recoil. The barrel and block go back together about three inches. The barrel then returns and the bolt pushes in the next shell. These weapons make a ghastly racket in the action; however, men who are used to them do not ever hear the noises, nor are they conscious of anything unusual. A man who is fond of an automatic will defend it most vigorously. The weapon in general is reasonably expensive. They are not by any means trouble-free under all field conditions.

Some of the disagreeable features of the long recoil shotguns may be eliminated soon. One of the authors has fired recently an automatic with fixed barrel using an adaptation by Carbine Williams of his old floating chamber idea; like most everything else he has done, it seemed to work perfectly.

It is interesting to note that a certain field officer in action early in the Korean War carried an automatic shotgun with a 15-inch barrel suspended around his shoulder so that it hung where ordinarily the .45 automatic pistol would have been. He found this far more effective at close range than any pistol. The tremendous smashing effect of a 12-gauge shotgun can scarcely be over-estimated. It is probable that at really close range, say under 10 feet, this weapon would kill any animal existing on earth. The effect at close range on human tissue is far greater than any other small arm including the .50 caliber machine gun.

Rifles

When men fight men in large groups and armies, they have used rifles almost exclusively in one form or another since the American Civil War. Fortunately, these weapons are not used nearly so much in individual crime, largely because of the inconvenience of carrying them and the general impossibility of concealing them on the person. However, from time to time, criminals will deliberately choose rifles, particularly for the assassination type of crime. There are always hunting accidents and illegalities leading out of hunting in various forms in which rifles figure, even in New Jersey, which bans all hunting with rifles save with special permits.

Single and double rifles of the top-break type exist. Actually, however, these weapons are very rare in crime in this country. They are extremely expensive, if in at all good condition. They are frequently chambered for the most powerful cartridges which are not really suitable for most American game. Needless to say, all of them are far more powerful than would be needed for shooting human beings.

Occasionally, however, one will come upon a combination rifle and shotgun, either such as was made in this country before World War II, allowing a rifle barrel to be substituted for the shotgun barrel in a standard single-shot hammer weapon, or in the even more frequently found and still manufactured over-and-under rifle-shotgun combination with .22 rimfire rifle on top and .410 shotgun underneath. An early variety of the latter arm had a folding stock.

Multi-barreled sporting arms were common in Europe for a period of several decades previous to World War II. These weapons consisted of from two to as many as five different barrels, all brazed or sweated together to form a most outlandish looking arrangement. In general, the workmanship on these combination guns was beautiful but gaudy. The three-barrel weapon was the most common; two shotgun barrels were side by side and a rifle barrel underneath. This is an entirely practical gun, but one that the average American has a prejudice against. At a distance of 30 feet, few people will recognize the difference between this gun and a standard double-barreled shotgun. It would, of course, be possible to use one of these in some prepared crime and have witnesses swear that a certain man could not have killed an individual with a rifle since they saw him carrying a shotgun and only a shotgun. In actuality, however, these weapons have not been involved in many deliberate crimes. Maybe criminals feel the same way that sportsmen do. If you want a rifle, you want a rifle. On the other hand, if you want a shotgun, you want a shotgun. One of the writers personally has done a good deal of hunting where both weapons were required for different purposes; although possessing an excellent German three-barreled gun, he has much preferred to take a standard rifle slung on his back and a shotgun in his hands.

Repeating single-barreled rifles are of two types: First, those operated manually; second, the automatics and semi-automatics. In manually operated rifles, the whole world has now swung very strongly towards the bolt action; however, in this country at least, the lever action sporting weapons, such as those made by Winchester, Savage and Marlin, are extremely popular. Slide or pump-action rifles are also numerous over here although those made by Remington are the only common ones above .22 caliber rimfire.

The bolt actions predominate both in crime and in the hands of sportsmen. Previous to the adoption of the M1 Garand in our service all armies were using manually operated bolt action repeaters, predominately of the Mauser type. These military rifles with more or less remodeling—sometimes the original rifle was produced new as a sporter but of the military type—form the most numerous class of high-power rifle in this country. Those of domestic manufacture are our 1903 Springfield, the tremendously popular U. S. Enfield Model 1917 and Remington and Winchester commercial arms following closely the Mauser-Springfield design. The foreign bolt actions are mostly Mausers and copies of them ranging from military models of the late 1880s to new commercial FN rifles are for sale today.

There are two other bolt actions that are still in the hands of hunters to a considerable extent and, therefore, might be met with in crime. First, there are many Krag rifles originally introduced into the U. S. Army in 1892 and still used for training purposes in the First World War. These weapons were sold to members of the National Rifle Association for as little as $1.50. It might be well to mention that the Russian Three Line (Moisin) rifle made in this country during World War I was similarly sold. The Russian cartridge is hard to obtain but the rifle can be rechambered easily (if unwisely) for the .30-06. They are still accurate, first-rate, reliable weapons. The .30-40 Krag cartridge is entirely respectable, although perhaps not quite so modern as it might be.

Second, there are bolt actions of the basic Mannlicher design in which the bolt handle passes through the bridge of the action. Several military weapons brought to this country in fair numbers are of this type. There are also the high-priced and extremely beautiful commercial Mannlicher-Schoenauer rifles. Occasionally, one will run across still other bolt actions such as the British Lee-Metfords and Lee-Enfields and the straight-pull Canadian Rosses and Austrian Steyrs.

Even rarer in intentional crime are the military single-shot arm and commercial adaptations of them. There were hundreds of different patents taken out in the 1860s, 1870s and even early 1880s. No real description is possible because of their multiplicity. The Allin-Springfield trap-door rifle used as late as the Spanish War but made originally in part from components of our Civil War muskets is the best known. The rolling block Remington was manufactured as late as World War I for our allies. We understand that the Jewish Army is still armed in part with British Martini-Henrys. Both the Remington Rolling Block and the Martini-Henry are being sold in considerable quantities in this country today and at prices within reach of all.

Some of these early single-shot actions were altered to repeaters by the installation of box or tubular magazines. The Mauser 1871, the French Gras and the Dutch Beaumont were so changed. One of the writers operated a repeating Martini-Henry for several rounds without trouble, but it was a Goldberg rig.

The term automatic rifle cannot be used in the way that we used "automatic" to refer to a special type of pistol. There are too many real full-automatic rifles. Even the term autoloader to refer to a rifle from which one shot and only one can be fired with one pull of the trigger may be misleading. Many autoloading rifles can be converted to fire full-automatic by a relatively simple mechanical operation on their mechanism; some have a control switch to fire either way at the option of the shooter.

Centerfire autoloading rifles have been manufactured in the past commercially in this country by both Remington and Winchester. The Standard Arms Company of Delaware made a few of these. The Remington model 8 was a locked breech long recoil weapon chambered for moderately power cartridges. The Standard rifle was gas operated and handled .30-30 class ammunition. The Winchester rifles were weight-delayed blow-backs and, even though chambered for less powerful cartridges, were much more popular; they were highly thought of for police duty and are still sometimes so used, particularly in penitentiaries. The .22 rimfire autoloaders, made by many firms, are always with us.

High power autoloaders are usually, although not always, military weapons. The three methods of possible operation are delayed blow-back, recoil operation, and gas operation. The first has not been used extensively, although Pedersen did produce experimentally a beautiful weapon of this type. Recoil unlocking of the breech by movement of the barrel has not achieved the success in shoulder weapons it did in pistols. The Johnson autoloading and full automatic rifles operate in this fashion. They are short-recoil weapons.

Gas pressure from the cartridge is used to operate the M1 Garand and most other successful autoloading rifles. This gas is taken from the barrel near the muzzle in the Garand. The M1 carbine uses a short-stroke piston and takes gas from about $3\frac{1}{2}''$ in front of the breech. The latter arm is more important in recent crime than all other autoloaders combined; its handiness and 15-shot capacity as well as its fairly powerful .30 caliber cartridge seem to appeal to some younger criminals. It will fire as fast as the trigger can be pressed and released. A separate trigger pull is necessary, in the M1 model for each shot.

Machine Guns

Very few heavy and medium machine guns have ever been used in crime. However, many full automatic weapons—sometimes referred

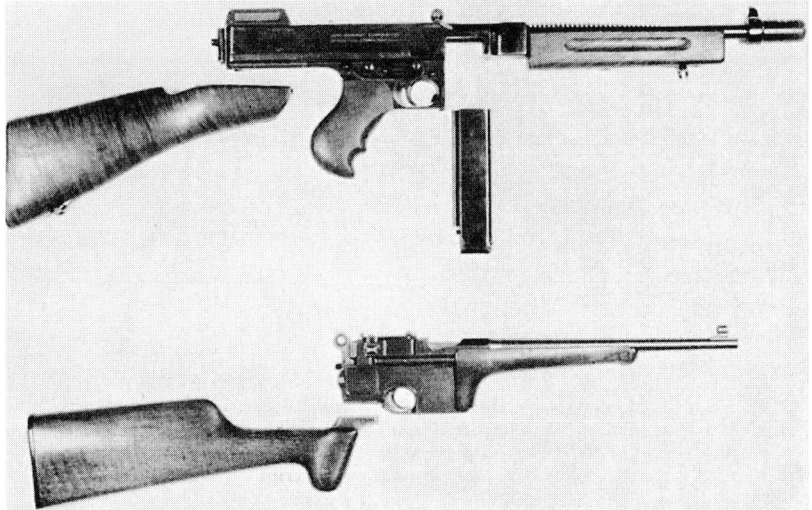

A commercial Thompson sub-machine gun made by Auto Ordnance Corporation. Below, a rare form of Mauser Model 1902 pistol which, in spite of the additional length over the standard model, was no more difficult to carry around, for instance in a brief case, since the weapon disassembled as shown. It was, however, more of a rifle than a pistol. Courtesy Smithsonian Institution.

to as machine rifles or machine pistols—have been so used. Further, almost all autoloading rifles and automatic pistols have at one time or another been made full automatic—so that by squeezing the trigger, the weapon would fire continuously until the trigger is released or the supply of cartridges exhausted. Various criminals have gone to work for the arms manufacturing companies in order to learn how to make these adjustments, which are never particularly simple and are without exception entirely undesirable. A full automatic weapon not designed for full automatic fire will not be nearly so dangerous in the hands of an inexperienced person as the same weapon unaltered. The M1 carbine is sometimes found in civilian hands produced or altered by the Government to the M2 model which is a full automatic or semi-automatic at the pleasure of the shooter. This weapon is rare; our armed forces found it impractical and its manufacture has been abandoned.

There are, however, a number of full automatics sometimes met with in the hands of criminals that were designed to be so used. Before World War II, the most frequently met with was the Thompson .45 ACP sub-machine gun. This weapon was greatly feared by the population in general during the Prohibition era. They were called "typewriters" and were for several years highly prized equipment for

A Japanese machine gun of the best type. This is popularly called the "Nambu Light" and is reasonably common in this country at the present time. One of these was taken in by the New Jersey State Police during 1953 in perfect working condition, although the owner probably had no criminal intent. Closely allied models are chambered for the 6.5 Jap and the 7.7 Jap.

underworld executioners. They were less than perfect, however, save for extremely close range work.

Another weapon sometimes used by hardened criminals before World War II was the Browning Automatic Rifle, or BAR, which is, in the opinion of these writers, an exceedingly good weapon of its type. Its power, however, was not often necessary for crime; it was like going fishing for trout with 64-pound-test line. It seems to have disappeared from the arsenals of criminals after the breaking up of the Barker-Karpis gang. To a lesser extent than the BAR, the Lewis gun figured in investigations. Several were confiscated in raids, chambered for both the .30-06 and the .303 British cartridges. These full-power light machine guns had pretty well disappeared from organized crime by 1940. They have not figured prominently since.

Conceivably, economic conditions or social unrest could change the machine gun situation. Hundreds of thousands of citizens know in a general way how to handle these weapons. It is a simple matter to steal them from armories or museums. Returning G.I.s brought back the full automatic weapons of our enemies and allies in fairly large numbers. Police Firearms Laboratories often receive in firing condition the German Mg34 and the Mg42 machine guns; the Japanese Nambu Light and the British Bren are less common.

The full powered machine rifle is heavy, bulky, noisy and unnecessary. The machine pistol or sub-machine gun, however, is even more practical for crime than the old Thompson. In general, sub-machine guns are straight blow-back operated, whereas the true machine guns firing rifle-type ammunition are either gas or recoil operated. The German Schmeisser (popularly, the burp gun) is the best known and most common today. Quite a number of others including the U. S. M3 (called the grease gun) several models of the British Sten, and even Russian weapons of this type have been picked up from criminals.

The term Machine Pistol includes not only the standard sub-machine guns but also a very few real pistols—weapons for one hand use— capable of full automatic fire such as the big Mauser Model 1932 and copies of it made in Spain. The French Union automatic is a copy of our Model 1911A1 save for having a tangent rear sight, a detachable shoulder stock, and provision for full automatic fire. At the other extreme are the so-called German machine pistols MP43, MP44, and MP45 firing the 8mm short cartridge which is more powerful than our M1 carbine cartridge; they are good weapons but resemble auto-matic rifles rather than sub-machine guns, even though called machine pistols in the German Ordnance Department terminology.

Miscellaneous Firearms

A good many weapons used in crime do not fit exactly into any of the above categories. Something very close to the juvenile "zip" gun was made by the United States Government for use in occupied coun-tries of Europe during the recent war. A picture of this gun, called the "Liberator," is shown. This weapon was made at a total cost of around $1.75 and fired the .45 ACP cartridge, single-shot, down a smooth bore. The whole weapon occasionally blew up. It was, how-ever, even more deadly than the .45 automatic at close range since the bullet invariably tumbled.

Along the same idea, certain of the flare guns used in the two World Wars have been fired occasionally with shotgun shells of the proper size rather than the flare ammunition. Certainly such an arrangement is not desirable; however, it seems to work out without destroying the guns in most cases. Perhaps the most fantastic capture from a criminal in recent times was a bazooka and two rounds of live am-munition. Whether he planned to use it in breaking into a building is a matter for conjecture. It would have made a mess out of a standard commercial armored car, but a BAR would do the same. Occasionally such weird things as harpoon guns and old stanchion weapons with tremendous bores, relics of the days when water-fowl were hunted from a large rowboat, are encountered.

Air weapons also are sometimes met with, although they are not

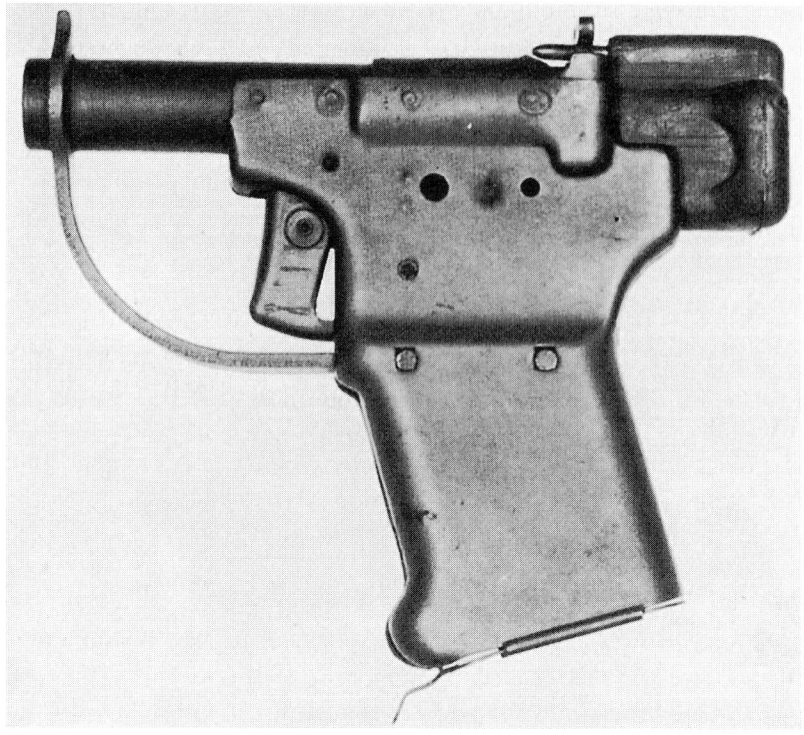

The .45 Liberator

true firearms. These are of two types. First, there is a weapon functioning completely by compressed air contained in a chamber. Second, and more frequently encountered, is a weapon with compressed springs which, at the release of the sear, all of a sudden drive air into the bore to the rear of the projectile which is discharged. Actually, the pneumatic chamber air gun is at least two centuries old. It was widely used for decades for illegal hunting in Europe. Some specimens, at present met with occasionally, have fake flintlocks attached to resemble conventional guns, because air guns were illegal. The air gun was, of course, intended to kill animals without the noise of an explosion.

The largest air gun ever made was installed on the USS Vesuvius at the time of the Spanish War and took part in the bombardment of Morro Castle in the Havana Harbor. This weapon fired a dynamite projectile to a distance of about two miles. Dynamite was too sensitive to shock to be projected from a normal cannon. Its effectiveness was from the tremendous blast of the dynamite going off rather than from projectile penetration. This idea has not been used to any considerable extent for large weapons since that time. It was not a success. However, torpedoes are discharged from submarines by use of compressed air; actually, they are very large, low velocity air guns. The torpedo, however, achieves its range and its final velocity from its own internal engines.

The common air rifle is frequently met with in small boys' pranks and in vandalism claims. Occasionally some child will lose an eye because of these weapons firing the well-known BB shot of .177″ in diameter. These shot, even at a range of a few feet, will seldom penetrate human beings, although very small animals can be killed with them.

Several years ago, air rifles and air pistols were developed with rifled bores to fire elongated projectiles made of light metal with considerable velocity and a certain amount of penetration. These weapons are accurate and can be reasonably dangerous. They are made both in the .177 caliber and .22 caliber. Some experiments were recently conducted by the writers in which penetration was measured at a range of 10 feet in a quarter of beef. These pellets would actually penetrate, but to a distance of less than half an inch in the dead tissue. Perhaps live tissue is more easily penetrated; any volunteers? An air pistol would be a very poor weapon with which to commit a crime, although it is conceivable that a shot in the neck could prove fatal.

Similar rifled pistols and rifles firing skirted pellets are in current production using CO_2 gas in small cylinders as the propellant.

At this writing there is considerable consternation and excitement over the nationally advertised Kruger pistol, a small plastic replica of the Luger. This gun fires No. 6 shot using the common paper cap as the propellant. It is a single shot and not at all easy or quick to reload. The power of this pellet is considerably less than that of the pellet shot from a common air gun, and very much less than the power of a pellet from a sling shot.

Silencers

Silencers on weapons firing conventional ammunition have been used in one form or another for several decades. The original Maxim silencer was an over-sized tube screwed onto the end of a barrel, with baffles allowing the bullet to pass through appropriate holes but retaining the blast gases from the muzzle and very considerably cutting down the sound of the weapon where the projectile had a velocity below the speed of sound. The impossibility of completely silencing a weapon with a greater velocity than the speed of sound will be discussed in detail later.

During World War II, a much more effective silencer was made for the OSS and other special troops by taking a regular weapon and drilling holes in the barrel at 90° from each other radially. These four rows of holes usually extend approximately half-way from the muzzle to the breech. The gases that exhaust through these holes impinge on a stainless steel wire screen wrapped tightly about the barrel. As an added silencing feature, these same weapons have a chamber containing stainless steel mesh washers to the extent of several score mounted in front of the muzzle proper but in a separate chamber so that the bullet passes through the holes in the middle of the washers. Powder gases escaping from the muzzle behind the bullet are to a large extent caught in this wire mesh. The Thompson sub-machine gun was manufactured in this manner. There are also silenced barrels that can be installed on any standard M3 sub-machine gun. The High Standard Company put out its regular .22 automatic silenced in this manner. Some Colt .45s Model 1911A1 were equipped with new long barrels of this type. These silencers are illegal and were made for Government orders only. They are extremely clumsy and ruin the appearance of the weapons, but when properly adjusted do not hurt its accuracy. The M1 carbine was made into a silenced weapon by this system; this was only partially successful because of the 1900 feet per second muzzle velocity. The British Sten silencer used rubber rings in front of the barrel. Any one of these weapons might be met with in the hands of a criminal; some of them already have been used in crime. We will devote considerable space to a discussion of identifying bullets from these weapons, even in the absence of the gun itself.

Perhaps the most successful of all silenced weapons is a special single-shot .32 auto. caliber pistol. In all the automatics, the silencing works only in connection with the noise of explosion of the propellant charge inside the cartridge. Frequently, as for instance with the M3, the hammering of the breech is even more noticeable in the firing of the gun than the sound of the explosion. Even the High Standard .22 automatic makes a good deal of noise when the mechanism operates; one never notices this noise when firing in the normal way since the explosion conceals it. However, it does become very noticeable indeed when the explosion is silenced.

CHAPTER 4

HISTORY OF AMMUNITION

THE first form of ammunition for firearms was loose powder and various projectiles. Actually some forms of near gunpowder probably ante-date by several centuries the first firearm. These were used in rockets, flaming mixtures and perhaps in bombs.

The first projectiles were probably stone balls of appropriate size and perhaps irregularly shaped. However, some of the early manuscript illustrations show a javelin-like spear protruding from the end of the firearm. By the Fifteenth Century, ammunition was reasonably stable. Loose powder was usually loaded in at the muzzle with or without a wad separating it from the projectile or projectiles which could again be secured in place by more wadding. Essentially this procedure is still used in muzzleloaders today.

Far more projectiles have been made of lead than all other substances combined. The spherical shape was early arrived at and was the only one of any real significance until the middle of the last century. However, in the patents of James Puckle of London issued in 1718, we have a rather strange wording stating that round bullets were to be used against Christians and square bullets against Turks. The idea was to inflict horribly wounds on the Turks. However, the accuracy of the square piece of lead from a square barrel would have been extremely poor.

Although this square bullet idea was fantastic, there were other variations both in bullet material and shape that worked. In early times, bullets were made of copper, tin, zinc, iron, and even silver. The lighter materials were known to give higher velocity. For artillery cast iron shot soon became standard, though lead balls were known to give greater long-range penetration. Throughout the history of firearms, various elongated projectiles have been tried. It was known that if their bases were hollow and their center of gravity in the first third of their length, these would shoot quite well from smoothbore

63

weapons. They tended to fly through the air point foremost—like Asiatic arrows which have no feathers on the shaft but rely on a heavy head and a light bamboo shaft to give stability and accuracy.

The cylindro-conoidal hollow base bullets for muzzleloading rifles which came into prominence about a century ago were the first really practical single projectiles for small arms other than the spherical lead bullets. Both these types are occasionally used in crime even today. Since both are usually cast, it is possible to identify a crime bullet with a particular mold because of the irregularities of almost every mold when examined microscopically. A case of this type actually came into the New Jersey State Police Firearms Laboratory a couple of years ago where a deer had been killed out of season with a muzzleloading Mississippi rifle firing a patched bullet. The striae on the bullet from the bore were insignificant; however, the suspect mold was also available and the bullet found to have been cast by that mold. A test bullet was cast with the mold and then compared to the crime bullet with positive results.

Some form of consumable cartridges have been in use since the early days of firearms. The idea is so simple that many different people must have evolved it separately. These early cartridges consisted of a paper capsule or bundle containing the required amount of powder and a bullet. Several of these could be placed together in a cartridge box. Such an arrangement greatly facilitated loading since the whole could be pushed down the bore of a musket and smashed with a blow of the ramrod so that powder came out through the touchhole and into the pan of the piece. With some it was necessary to bite or tear the rear end of the cartridge and to pinch out a small amount into the pan before inserting it in the muzzle. Fine powder was preferable for the priming charge in sporting pieces.

Metallic Cartridges

The earliest form of metallic cartridge was undoubtedly the spare breeches for various forms of Fourteenth and Fifteenth Century breech-loading artillery and small arms. These breeches could be loaded and the piece fired as rapidly as the breech could be changed and the weapon primed. None of these was particularly satisfactory; after the advent of grained powder they were usually not sufficiently strong to stand the increased strain. This idea persisted, however, throughout the entire flintlock era; museums contain specimens of various adaptations of this principle. A man could fire faster because he or someone else had done most of the work of loading beforehand.

With the advent of the percussion cap, the metallic cartridge really became practical. Ignition was now possible merely by a blow delivered directly or through an intermediate lock part to the metallic cartridge. Some of the early percussion cartridges were made of brass, iron, or steel and were really little more than spare breeches with nipples for percussion caps attached. Throughout the percussion era, breechloaders were firing both metallic and combustible cartridges which required separate outside priming. However, a Swiss by the name of Pauli working in France produced the first self-primed metallic cartridge. Colonel Goddard had in his collection a breech-loading hammerless shotgun made by this inventor in Paris a century and a quarter ago. Both the gun and the cartridges for it are astonishingly like those in use today. The idea, however, was not widely adopted for 40 years.

Another idea of considerable importance for a time was a self-primed combustible cartridge. In this country the Volcanic Arms Company brought out a cartridge of low power which contained both the primer and the propellant charge inside a hollow bullet. The mechanism they employed is similar to the old Henry which was developed from it; it is the antecedent of the present lever action Winchesters.

Abroad, this same principle was employed by Johann Von Dreyse in his famous needle gun. The percussion cap was located in the base of a bullet contained in a consumable paper cartridge. The firing pin had to penetrate completely through the entire powder charge in order to set off the cap. This is a horrible action, but must have worked. The early Prussian victories over Austria, Denmark, and France are attributed to this weapon.

The first successful use of metallic cartridges in war was by the Union forces in our Civil War. The Spencer and Henry breechloading repeating carbines firing rimfire copper cartridges with groove diameter bullets of approximately .52 caliber and .44 caliber respectively, were particularly effective. At this time, various inventors were perfecting self-primed cartridges of other types. The pinfire, titfire, centralfire and finally centerfire types were manufactured and used. In this country, we used at first rimfire designs but adopted the present centerfire arrangement quite early. Pinfire and titfire cartridges were never popular here, although the former are still in use in Europe.

Centerfire Cartridge Case Manufacture

At first cartridge cases were formed in several different ways. Sometimes an iron base would be soldered into a thin brass tube. One of

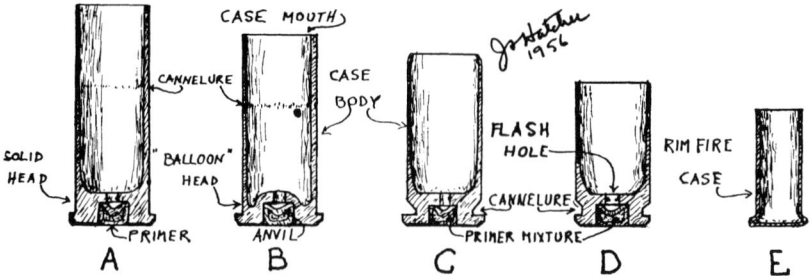

TYPICAL HANDGUN CARTRIDGE CASES

A is the case for the .38 S & W Special cartridge. It is of the rim type, as distinguished from the rimless case shown in D and the semi-rim (or semi-rimless) case shown in C.

B is a case for the .38 long Colt cartridge. It is of the rim type, and has the obsolescent "balloon head" construction, which does not stand as high a pressure with safety as does the more modern "solid head" construction shown in A, C, and D.

C is for the .38 A C P cartridge. It looks like a rimless case, with a cannelure for the extractor, but the rim is actually a bit larger in diameter than the body of the case, so that the rim can stop the forward motion of the cartridge when it is inserted in the chamber.

D is for the 9 mm Luger. It is a true rimless case, with the rim just the same diameter as the body of the case. The square front edge of the case is what stops the case when it is chambered.

E is a rim-fire case, the .22 long rifle. It is made of thin brass, or sometimes of copper, and the priming mixture is contained in the fold of the rim.

Major Boxer's designs called for a .577 Snider cartridge made of strips of brass or tinfoil coiled together with paper and secured to a more or less substantial head. These cartridges were not reloadable. For that reason, they continued to be issued in the British Army for a long time for service in certain possessions; England was not anxious for the native population to possess reloadable breechloading cartridge cases.

At the other extreme, there were the "Everlasting" lathe-turned brass cartridge cases made in this country for target shooters and good for many thousand reloadings. Rifle and pistol cartridge cases have been produced in all manner of ways with foldings, solderings, and crimpings securing different pieces of metal together; however, today all cases are produced from one brass blank by drawing on modern punch presses. Most of these are the so-called solid head type with little metal drawn from the part of the blank that becomes the head. The cartridge companies still make some cases by the older folded head method. This is a one-piece construction but leaves less metal around the primer. This construction is much stronger than it appears since the grain structure of the brass is right. However, all special loadings of the .44 Special have been slightly handicapped by the folded head or semi-balloon cases which are not quite so strong as the solid variety.

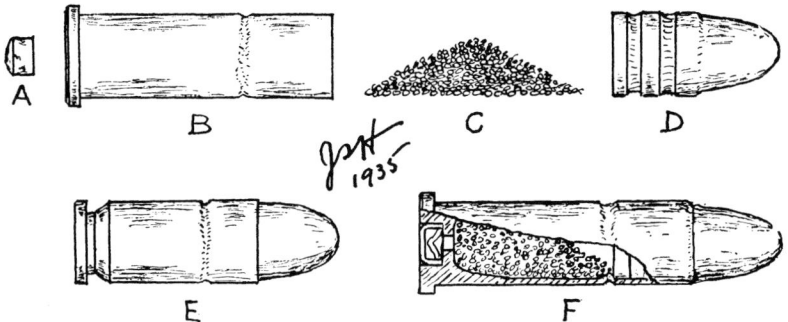

COMPONENT PARTS OF A CARTRIDGE

The assembly which creates the power in any firearm is known as the "cartridge" and the above sketch shows the various components which comprise a loaded cartridge.

A—Primer or "Cap." B—Case or "Shell" or "Empty" or "Brass." C—Powder or "Charge" or "Load." D—Bullet, Projectile or "Ball."

F is an assembled Cartridge, a Round of Fixed Ammunition also often referred to as a "Load"; this particular cartridge is the .38 Smith & Wesson Special, a rimmed type of cartridge and the most popular revolver cartridge in existence amongst target shooters and handloaders. In this particular cartridge, the lead bullet is held in place by crimping over the mouth of the case, a method known as a "crimp."

E shows a rimless cartridge for use in automatic pistols—the .38 Colt Automatic; loaded with a metal cased bullet held in place by a uniform restriction of the front end of the case, a method known as "friction tight" and necessitated by some pistols in which the cartridge is positioned by the front edge of the case. This friction tight method of holding the bullet is not considered as good or as reliable, or as waterproof, as a good tight crimp.

Both of the above cartridges are loaded with smokeless powder and the cases are "cannelured" to prevent the bullets from slipping backwards into the cases.

In future, solid head construction will probably become standard even in these calibers.

The actual processes of manufacture depend upon the cartridge case being made and the particular factory procedure. Brass cups (about 70% copper 30% zinc) of proper diameter and thickness are stamped out. Each of these after appropriate drawings and annealings becomes a cartridge case. Generally in modern high power rifle ammunition the body of the case is left quite hard with only the neck annealed after the last drawing operation.

Bullet Manufacture

There are two basic types of commercial bullets in common use in America today. Lead bullets are used in almost all revolver ammunition and in some low and medium power rifle cartridges. Jacketed bullets are used for automatic pistol ammunition and medium and high power rifle ammunition.

Lead bullets are produced in automatic swedging machines from extruded wire containing the proper percentages of tin and anti-

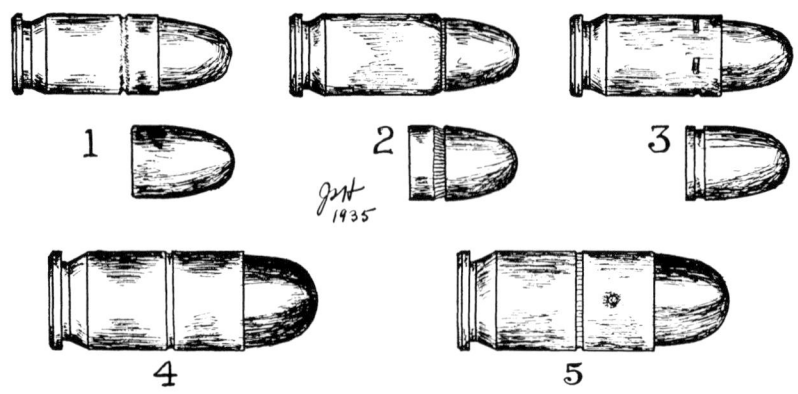

CRIMPS ON AUTOMATIC PISTOL CARTRIDGES

Cartridges designed for use in revolvers are universally made with rimmed cases and their bullets are firmly held in place by means of more or less heavily turning over the mouth of the case into the forward bearing diameter of the bullet—"crimping" it is called. Often, a groove is placed there in the bullet, for such purpose. Generally, the bullet is further retained in place by a cannelure in the case, against which the rear edge of the bullet abuts; no travel or displacement back or forth is possible.

With the automatic pistols, a different condition exists. All modern automatic pistol cartridges, (other than the .22 long rifle) are loaded into rimless-type cases, for certainty of feeding from the magazine and functioning through the mechanism of the arm. They do not have the substantial protruding rim on the base of the cartridge case, hence a different method of positioning the cartridge is necessary. Most automatic pistol cartridges are uncrimped, the front edge of the case being left square and the cartridge is positioned in the chamber by this square edge. The exception to this method is in the calibers using "semi-rimmed" cartridges, where the case looks as if it is rimless but which does have a slight extension to its base which actually is, and acts like, a rim.

Sketch 1, above, is a Remington .380 pistol cartridge, in which the bullet is pushed in against a cannelure in the case and is held firmly by the friction of the neck of the case, a method known as "friction tight." This cartridge positions on the front edge of the case.

2. A later Remington loading of this same .380 cartridge, in which the neck of the case is heavily crimped into a deep cannelure of the bullet. Not the best practice, for this is strictly a rimless case, but it works; the deep rim at the base affords a firm hold by the extractor and this, reinforced by the bullet being seated up against the rifling, positions the cartridge for firing.

The "semi-rimless" automatic pistol cartridges, such as the .25 Colt, .32 Colt, and .38 Colt Automatic Pistol calibers (the 6.35mm, 7.65mm and 9mm calibers by European standards) may properly be crimped in this manner, as they are caught and held in position by the case rim.

3. A .380 pistol cartridge made in Germany. This bullet carries a deep, smooth cannelure, and the case is indented into this cannelure at several places. An excellent method of holding the bullet in place, but not any too good for accuracy. Shooters who load their cartridges will not use such ammunition as the cases are hard to reload properly.

4. A .45 Automatic Pistol Cartridge, in which the bullet is seated against a cannelure in the cartridge case and held friction tight in the neck of the case. This is the usual construction for this cartridge.

5. A .45 Automatic Pistol Cartridge in which the bullet is held in place by the "stab crimp" method, where the case is stabbed into the bullet in three places evenly spaced on its circumference. This method of holding the bullet in place is fairly common, especially in war-time ammunition, but is not rated very highly by shooters. The stab marks show plainly on the bullets after firing.

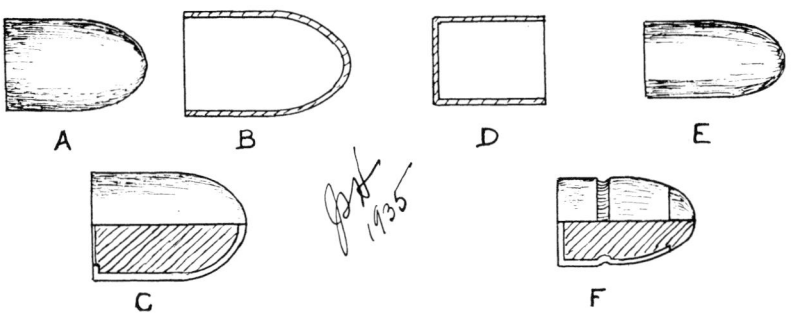

METAL CASE BULLET CONSTRUCTION

The modern, metal cased or metal jacketed bullet is made by enclosing a core or slug of lead alloy in an envelope of harder, tougher and stronger metal. Such bullets are occasionally referred to as compound bullets. The standard practice in America is to make the jacket of cupro-nickel, gilding metal, or Lubaloy. In Europe the jacket is often made of steel, plated with gilding metal or with cupro-nickel, or even coated with a heavy layer of varnish.

The sketches above show method of construction. A is the core of a .45 caliber automatic pistol bullet, B is the jacket, and C is the assembled bullet, made by inserting the core in the jacket and then swaging the two together under heavy pressure, at the same time crimping the edges of the jacket over the base of the core as shown.

The making of a soft point, metal cased bullet is shown at the right. In this case, the jacket is made closed at the base and open at the point and the core is inserted from the front, it then is crimped on to the core as shown. The cannelure, or groove, around the bullet serves not only as a place into which the front of the cartridge case may be crimped, but also as a method of giving the jacket an additional hold on the bullet.

mony for hardening. These bullets are extremely uniform in size and weight. These are in general the bullets most commonly used in crime. Several of the usual types are illustrated here both before and after firing.

Jacketed bullets have been produced in dozens of different types. However, most of these variations have been in connection with hunting bullets for rifles and ordinarily play only an insignificant part in crime. The general idea in hunting is to weaken the front part of the bullet sufficiently to have it expand upon hitting a rabbit at 500 yards, but to build the base sufficiently strong to have it reach the brain of a rhinoceros. This is of course an extreme case and is not possible; yet for half a century sporting ammunition makers have been making almost similar claims. They try to achieve this with different exposed, hollow, and protected points and varying thicknesses of jacket.

The jacketed bullets most frequently met with in crime are simple cupro-nickel or gilding metal envelopes covering lead cores, fired from automatic pistols. The most common are those from the blow-backs—.25 ACP, .32 ACP, and .380 ACP—but the more power-

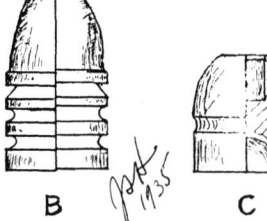

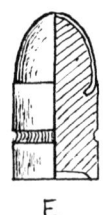

A   B   C   D   E

TYPICAL REVOLVER BULLETS

A represents a factory-made, swaged, lead bullet for the .38 S. & W. Special cartridge. This bullet has a round point, a cupped base and two lubricating grooves: note that these grooves have been rolled into the bullet by knurled rollers. This knurling of the grooves is a sure sign of the factory-made product as such knurls can not be made in a bullet cast in a mould by a handloader.

B is a handloader's product, a hand-cast or "home made" bullet for the .44 S. & W. Special cartridge; the two bottom grooves are for lubricant while the top one is a crimping groove into which the mouth of the case folds. Smooth grooves are generally a sign of a cast bullet. Another feature of these handloader, cast bullets is that they invariably have a fresh, shiny, silvery look about them; whereas the factory-made lead bullet has a duller, leaden look, due to having been rumbled in a barrel after having been swaged.

This sketch of the hand-made bullet shows the line running longitudinally around the center of the bullet where the two halves of the mould come together, a feature that will be found on all moulded bullets. Furthermore, the artist has accented the fact that the two halves of the mould did not exactly meet, a fault found occasionally in old moulds but never with ones that are adjusted properly.

C. This shows a metal cased .44/40 Hi-speed rifle and revolver bullet; its jacket is of copper colored gilding metal and it would be considered to be a copper jacketed bullet by the uninitiated. Metal cased bullets are seldom lubricated and the one groove shown in this bullet is for crimping but its knurling also helps to fasten the jacket to the lead core and prevent "stripping."

This bullet has a flat, hollow point and also a hollow base.

D. An old fashioned outside-lubricated, or "heel" bullet for the .38 rim fire cartridge, the reduced diameter at the base fits tightly into the mouth of the case and is then crimped. The lubricant on such cartridges was always on the outside, the bullet end of the cartridge being dipped into melted lubricant which covered all exposed lead and also filled the lubricating grooves. These were very messy cartridges to handle and are now practically obsolete in all but .22 caliber rim fire ammunition.

E. A modern type of revolver bullet, the "metal-point" .38/44 developed by Smith and Wesson. The entire bearing surface of the rear of the bullet is of lead with one lubricating groove, the point is encased in a cupro nickel jacket. Base is slightly cupped.

ful types are also used in crime such as the 9mm Luger, the .45 ACP and several types of high velocity .30s. These are all made automatically by swedging a cup of metal around a lead core.

Sometimes full metal-jacketed rifle bullets of government design will be met with. The .30-06 was loaded during World War I with a spitzer pointed 150-grain flat-base bullet with a relatively thick jacket. Between the World Wars a great deal of 172-grain boat-tail ammunition was produced; however, most of that made and used in World War II was of flat-base of the lighter weight type. The World War II bullets usually weighed 152 or 153 grains. All these as well as similar military and target bullets produced here and

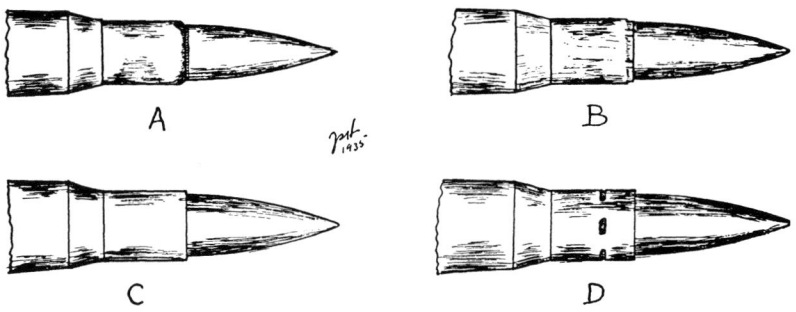

CRIMPS ON MILITARY RIFLE CARTRIDGES

A. The .30 Caliber MI cartridge, with the end of the case neck crimped firmly into a deep cannelure in the bullet.

B. German 7.9mm Mauser cartridge, with the bullet held in place by the "segmental crimp" in which the end of the case neck is pressed into the bullet by the action of several segmental jaws, which close in on the neck from all directions in making the crimp.

C. A .30 Caliber National Match cartridge in which there is no cannelure in the bullet and the latter is held friction tight only. This, in rifle ammunition, puts a heavy strain on the metal in the neck of the case, which will usually crack after a few years of storage, a condition known as "season cracking." This illustration should not lead the reader to believe that all National Match ammunition is made without cannelure, or crimped in this manner, for the design changes from one match to the next in order to try out various methods. The only way to know about National Match ammunition for any particular year is to consult the specifications for the year in question.

D. .303 British Mark VII cartridge, with a groove around the bullet, into which certain segments of the neck of the cartridge case are pressed, as indicated by the sketch.

abroad were made by forming both the jacket and lead filling separately and then swedging them together under heavy pressure. Because of their solid points, strong construction and high velocity they are seldom found in a human target. Sometimes an almost-spent bullet will be recovered in the victim of an accidental shooting. However, a full metal-jacketed rifle bullet will pass lengthwise through a human body and still have enough velocity at exit to lose itself, if is is not stopped by some other substance. These bullets, if recovered at all in a human target, are easily identified. However, if they do not stop in the victim but smash against a wall or the like, they are likely to tear themselves to bits, so that identification requires a good deal of reconstruction beforehand and possible loss in evidence value.

Even rarer in crime than the full metal-jacketed bullets are those of sporting types already referred to. These can be .38-40 and .44-40 bullets with large areas of lead exposed at the points and fired at relatively low velocity. On the other hand remains of a .220 Swift

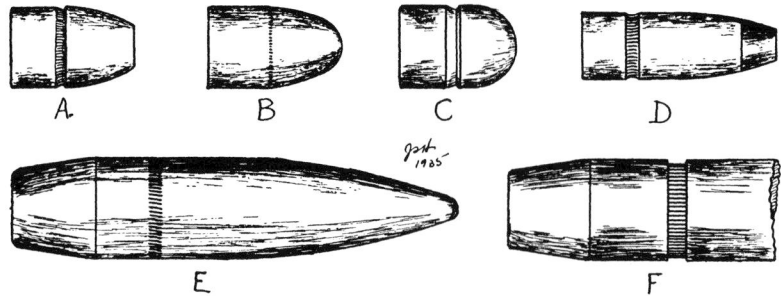

CANNELURES ON METAL CASED BULLETS

A. Remington .44/40 bullet, showing a knurled cannelure, deeper at the front edge so as to conform to the bend formed on the mouth of the case making the crimping operation.

B. A .45 Automatic bullet of Remington make, showing a slight knurled line around the bullet just forward of where the mouth of the cartridge case comes when assembled. This knurling is not, strictly speaking, a cannelure but is really only a line of knurling, put on by the factory for one reason or another—probably to designate this particular batch of bullets from others, or ornamental. This knurling is placed forward of where the mouth of the case comes in assembly and it does not in any way assist in holding the bullet in place. The .45 Automatic cartridge, it will be remembered, is not crimped but the bullet held in place friction tight.

C. A bullet from the .45 Webley Self-loading Pistol cartridge, British ammunition. Note that the cannelure in this bullet, of the same general form as that shown by A is not knurled, but instead is smooth on its surface.

A, B and C are all what are termed "full patch" or "full metal cased" bullets, being made for military or target shooting where an expansion of the bullet is not desired.

D. .33 Winchester soft point bullet, for use in hunting rifle where expansion of the bullet is wanted. This has a wide, knurled cannelure of curved design into which the case mouth is crimped.

E. An early design of .50 caliber machine gun bullet, with a very shallow cannelure lightly knurled with diagonal markings. This proved too shallow to afford a firm seat for crimping the neck of the cartridge case, and it was later made wider and deeper, of rectangular section with the bottom deeply knurled, as shown by F.

Both E and F bullets have the "boat tail" bases to add to their ranging powers and to better sustain velocity. No pistol bullets have been boat tailed, as there is no value to this feature in a bullet of low velocity for short range use.

with a spire point were recovered in a hunting accident recently. Hollow pointed and flat pointed jacketed bullets have been fired from both rifles and automatic pistols.

Regardless of bullet shape and construction, they are produced in approximately the same manner. The solid point bullets are in general swedged about the core with the jacket cup closed in about the base. Those with special points on the other hand have a solid base with the cup closed in at the point.

Velocity rather than construction generally determines the difficulty of working upon crime bullets. At high velocity, most sporting bullets will tear themselves into fragments, even upon striking a human being which, all things considered, is not very resistant compared even to a small bear. Reconstruction of such a bullet is necessary before identification, although a single small bit of jacket may definitely tie the crime to a given weapon.

Centerfire Primers

Priming is accomplished in centerfire cartridges in this country along a basic design originated by an Englishman by the name of Boxer in the period immediately following the Civil War. An illustration shows this in detail. The standard American primer consists of a light metal cup containing both the priming compound and its own anvil. All American centerfire cartridges have a central hole for the communication of the jet of flame from the primer to the propellant charge.

MODERN SMALL ARMS PRIMERS

Here are shown the three world-wide systems of modern, metallic cartridge ignition. All three cases were fired with standard loadings. The principle of ignition is the same with all three types of primers—the sensitive priming compound is struck a sudden, sharp blow and mashed by the firing pin between the base of its container and an 'anvil,' this anvil being arranged for in one of three different ways. In our illustration above, the anvil has been designated as 'A.'

To the left is a .22 Winchester Rim Fire case, the right edge of which shows where the firing pin struck and smashed the case rim together and against the anvil which, in the case of all rim fire cartridges, is that portion of the breech of the rifle (in the rim counterbore) where the firing pin strikes. This .22 W R F case is of brass; notice how thin it is compared with the primer cups of the other two high-power cartridges—which is why all .22 caliber rim fire cases so readily take the chamber and breech face markings when fired.

In the middle is a .30'06 Springfield rimless case which, like all American center fire ammunition, is primed with the Boxer primer. Observe that the anvil in this type of primer is a separate bit of brass, contained in the primer cup and positioned on top of the primer pellet, which latter is smashed between the cup and this anvil when fired. In the Boxer type the anvil is a part of the primer, with a large flash hole positioned centrally above it in the web of the case.

To the right is a .303 British case, primed with the Berdan primer. This Berdan primer is a shallow cup containing the priming compound (pellet) and may or may not have a thin disc of paper covering the pellet. In the Berdan type primer the anvil is an integral part of the brass cartridge case, it has two and sometimes three small flash holes positioned equidistantly around it (which do not show in the example above). The pellet is smashed between the cup and the base of the cartridge case. Practically all European center fire ammunition uses this system of ignition.

Centerfire cartridges produced elsewhere than in the United States are usually primed by a system invented by the American General Hiram Berdan. In this system the anvil is part of the cartridge case; the primer consists only of the cup and priming composition suitably protected. There are usually two non-concentric flash holes, but some cases have three flash holes evenly spaced around the anvil.

Save for the special primer used in .45 caliber A.C.P. ammunition loaded at Frankford arsenal, American primers are of two diameters and of two approximate thicknesses of metal. The small diameter primers are used both for small pistols and small rifles. The cups for the rifle primers are thicker to withstand the heavier firing pin blows and pressures. The priming composition in the rifle primers is generally stronger than that for pistols. The same situation is met with in the larger diameter primers which are similarly for large pistols and large rifles. There are certain irregularities in this arrangement, however. For instance, sometimes pistols such as the .44 Special work better with rifle primers, particularly when loaded with rifle powder. The famous .38 Special revolver ammunition has been loaded both with small and large pistol primers, although at the present time the small size only is used.

The priming composition used almost exclusively throughout the world until around 1900 was fulminate of mercury. The residue from this hurts both the gun barrels and brass cartridge cases. It was, however, cheap and surefire. The next composition to be used widely was non-mercuric but corrosive; it was usually known as chlorate priming, since potassium chlorate was an important ingredient. The new smokeless powders required a hotter, stronger jet of flame to ignite them properly. This chlorate priming had the unfortunate faculty of leaving a residue in the bores that would attract moisture and corrode barrels within a short time. More recently, non-mercuric, non-corrosive priming has been used almost exclusively in civilian ammunition and in some of the Government ammunition. The .30 caliber carbine cartridge was loaded exclusively with non-corrosive priming. During World War II, the Dominion Cartridge Company of Canada made many millions of .30-06 cartridges with non-corrosive priming. However, in general, most .30-06 and .45 ACP Government ammunition was loaded with corrosive priming until about 1948.

There are literally dozens of different types of primers, each a little different from the others. They would be a study in themselves. Before an ammunition company releases a single cartridge of a new type exhaustive tests are made to see that the cartridge case, the bullet, the powder and the primer all match or "balance."

Assembly of Centerfire Ammunition

Modern ammunition manufacture is one of the true marvels of the Machine Age. We make billions of rounds on automatic machinery that is precise, exact, and perfect. Perhaps the outstanding achievements of industry in World War II was the American ammunition. Many officers in action for months and even years never came across a single round of American ammunition that had failed to function. Such reliability was simply unheard of among even our German enemies. Sometimes the Japanese would find more than half the cartridges issued to them defective.

Assembly of automatically inspected and tested components is done on groups of machines all connected together with an ingenuity bordering on the miraculous. Daily production of many millions of perfect rounds is achieved with little handwork, including even inspection and packaging. However, the assembly operations, when broken down, are similar to commercial manufacture and relatively simple.

In all centerfire ammunition the primer is assembled into the cartridge case by means of a press fit of from .0005" to .0015" interference. In military ammunition there is also an upsetting of the brass around the primer pocket to clamp the primer in place. This is usually ring-crimping in this country but can be done also at three or more separate points called staking. Most military and many commercial primers are coated so that the base of the cartridge is water and oil proof.

Powder charges are almost universally determined by volume and dropped into each cartridge case from measuring hoppers. The bullet is then inserted in the case neck. In military rifle ammunition there is usually a crimping groove or cannelure on the bullet into which the neck of the case is forced. This joint is every bit as solid as the primer and primer pocket. The military crimp is quite tight—a .30-06 caliber bullet will require about 70 pounds pull to free it from the case.

Lead bullets are usually crimped into place in commercial revolver ammunition. On the other hand jacketed bullets are generally press-fitted into automatic pistol cartridge cases and held in position by the tension in the case neck, with or without a slight additional crimp at the mouth on the flat surface of the bullet. Commercial rifle ammunition usually follows the military style of crimp, less vigorously

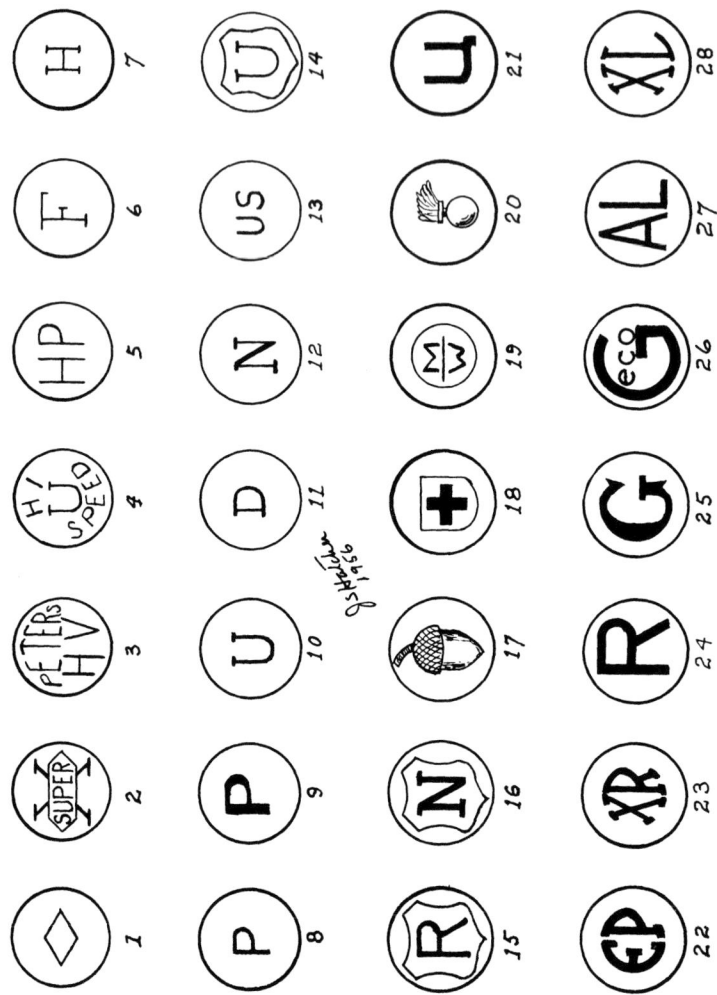

(Descriptive caption at bottom of opposite page.)

applied. Some match cartridges use a tension fit on a smooth bullet.

At one time, various other methods of securing bullets in place have been used, including staking the neck of the cartridge case radially into the bullet; a firearms laboratory will run into World War I .45 ACPs made like this. Many commercial cartridges have a groove pressed into them circumferentially either to form a seat for the bullet on the inside of the case or, as in modern metal-piercing .38 Special ammunition, to provide an extra tightness of crimp.

Rimfire Ammunition

The .22 rimfire cartridge, even though the only important surviving member of the old rimfire family, is without question the most popular ammunition in the world today for civilian purposes. The .22 short and .22 long are still for some reason fairly numerous; however, the .22 long rifle is by far the most common.

There are probably several million rifles chambered for this cartridge, ranging from light boy's bolt actions to fine target arms. A large number of handguns are chambered for it also, including several models being made at the present time. The ammunition is easy to obtain; the weapons themselves run in price from a few dollars for a simple light revolver to almost $100 for a specially equipped target automatic. The .22 is being used fairly extensively in crime today, although the experienced criminal tends to believe that the .22 long rifle is a poor cartridge for his purposes. He frequently chooses, however, a .32 Smith & Wesson or a .25 or .32 ACP, when in actuality, these three cartridges are not greatly more powerful than the .22, particularly when the latter is used in high speed hollow-pointed loading.

Perhaps no cartridge in the world save the 8mm Mauser and the .30-06 have had as much time and money spent in experiments with

1 Western Cartridge Co., (Division of Olin Mathieson Chemical Corporation.)
2 Same, Super-X brand
3 Peters High Velocity, Peters Cartridge Co., (Division of Remington Arms Co.)
4 Remington Hi-Speed, Remington Arms Co., Bridgeport, Conn.
5 Federal Cartridge Co., High Power brand.
6 Federal Cartridge Co., Foshay Tower, Minneapolis, Minn.
7 Winchester Repeating Arms Co. (Division of Olin Mathieson Chemical Corporation.)
8 Peters Cartridge Co., King's Mills, Ohio. (Bought out by Remington 1934)
9 Peters Cartridge Co., Division of Remington Arms Co., current marking, 1956.
10 Remington Arms Co., Bridgeport, Conn.
11 Dominion Ammunition Division, Canadian Industries, Ltd., Montreal.
12 Nobel Explosives Co. (Now a part of Imperial Chemical Industries.)
13 U. S. Cartridge Co., later taken over by Winchester, now discontinued.
14 Rheinisch-Westfalische Sprengstoff-A-G, Nurnberg, Germany.
15 Same
16 Same, Nurnberg brand
17 Same, Eichel (Acorn) brand
18 Swiss Federal Cartridge Works, Thun.
19 Made for Montgomery Ward & Co., by Federal Cartridge Co.
20 Giulio Fiocchi, Lecco, Italy.
21 Russian Ammunition used at International Matches, Caracas, 1954.
22 Made for Montgomery Ward & Co., by Federal Ctg. Co.
23 Sears Roebuck Extra Range brand made by Federal Ctg. Co.
24 Western Auto Stores Revelation Brand made by Federal Ctg. Co.
25 From box marked "Imported from France by J. L. Galeff & Son" (1953).
26 Gustav Genschow, Berlin
27 Federal Ctg. Co., Airline Brand
28 Federal Ctg. Co., XL brand

them as the .22 long rifle. Considering the cost of the ammunition and the general materials of which it is made, the precision of the .22 long rifle is truly astonishing. Groups fired at 50 yards and even 100 yards indoors are very near the ultimate. Various design changes in the products of all major ammunition companies have been made from time to time; these do not ordinarily show externally in the finished product.

The case is of thin copper or brass, formed to extremely close tolerances by automatic machinery; the priming compound also being automatically inserted into the case with unbelievable accuracy as to amount and evenness of application around the inner circumference of the rim. The bullets are cut from lead wire of a special size and composition and then swaged accurately to shape by automatic machinery.

As mentioned above, all .22 caliber components are made by automatic machinery. However, the actual assembly of these components is done by hand, by what is known as the "block-and-gang" method, in which the primed cases are automatically shaken into a perforated steel plate containing some 200-300 chambers made to properly accommodate them. A similar but thinner plate is imposed above this base plate, into which the powder charges are thrown and measured; the two plates are then moved into register and these charges drop into the cases below. A third plate containing the bullets is then substituted for the charge-plate and those bullets are forced down into the case mouths by a gang press. The assembled cartridges are dumped and then go back to the automatic machinery where they are crimped and lubricated. A more complete description of this block-and-gang method of cartridge manufacture will be found on page 98.

A word in connection with the bullets. In general, these weigh about 39 grains for the long rifle and about 29 grains for the shorts and longs. These are almost pure lead and are swaged from extruded lead wire of the proper size. Over the years, a good deal of experimenting has been done with lubrication in one form or another and bullet coatings. In general, the coatings have been an electro-plating of some other metal as, for instance, cadmium on the outside of the lead bullet itself. Various waxes and greases have also been used over the entire surface of the bullet, just in the grooves of the bullet, and finally only in the groove not showing above the crimped-in case.

The so-called lubaloy coating, which is a thin film of copper, has a tendency to destroy rather rapidly the individual identification characteristics found on fired bullets. Since this form of coating is applied to several different caliber lead bullets this situation occurs with all of them; however, it is particularly apparent with the .22s. In about six months, an outer layer of coating and lead may start to flake off. If it is necessary to keep a crime bullet of this type, an effort should be made to preserve it in some manner from all contact with air. However, superficial experiments along this line indicate that this flaking probably cannot be avoided entirely.

Miscellaneous Metallic Cartridges

Although the foregoing discussion includes all the usual ammunition met with in crime, there are very many uncommon varieties left out. Advanced cartridge collectors will show many specimens that even they cannot identify positively. Pistols and rifles for firing most of this are either unknown or not available to the average criminal. A great deal of semi-obsolete ammunition that is no longer being made is used in crime. Most of it conforms, however, to the types described above. The unusual in ammunition is only an advantage to the criminal so long as it is identified incorrectly. As soon as it is properly classified, the very unusualness of it greatly reduces the number of possible suspects.

Cartridges for Shotguns

Shotgun ammunition is fundamentally different from that employed in rifles and pistols. It is in general larger; it is to function from smoothbores capable of withstanding far less pressure. The basic design of shotgun shells must take into account two ever-present facts. First, a shotgun in order to have proper balance must have very light and therefore weak barrels especially towards the muzzle. Second, a charge of shot seems to hang together better and pattern better when fired with a relatively low pressure behind it for the last third of the barrel length. It is, therefore, necessary in shotgun shells to have a fast burning powder that will not give too high an initial pressure. In general, the allowable pressure in a shotgun is not more than about 10,000 to 12,000 pounds per square inch, whereas some sporting rifles regularly have pressures of at least 50,000 pounds per square inch. The S & W .357 Magnum revolver develops as much as 44,000 pounds per square inch with some commercial cartridges.

Shotgun shells have come in for a great deal of experimental work by arms companies and others in recent years. The basic American

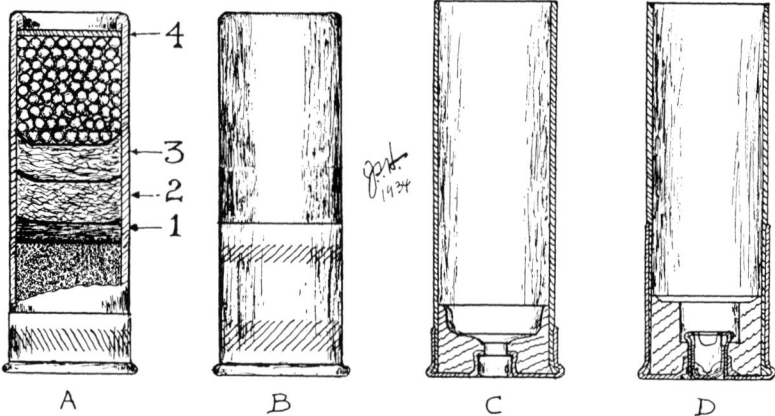

SHOTGUN SHELL CONSTRUCTION

    A   Typical arrangement of a loaded shotgun shell. First comes the powder charge, on top of which there is a wad, No. 1, usually of cardboard about ⅛ inch thick, though at times a dry and greaseless felt wad of the same thickness is used instead of cardboard. Sometimes the cardboard wad is what is known as a "greaseproof" wad, which is covered on both sides with black waterproof paper, or is made throughout of this black paper. The purpose of this wad is to prevent the blast of certain types of powder from blowing the felt wads to pieces, as well as to prevent the seepage of oil from the felt into the powder charge. Next come wads 2 and 3, which are usually felt wads, either 3/16, ¼, or ⅜ of an inch thick, but may be of compressed ground cork mixture or of some other special composition. Various different kinds and trade names of felt wads may be used, such as "Black Edge," "White Felt," "Nitro Felt," etc., and various thicknesses and combinations are used, according to the type of load and the manufacturer's ideas. It is sometimes considered a ballistic advantage for the felt wads to be cupped as shown. This may be done by the rammer used in loading the shell, or in the case of composition wads, they may be moulded in a concave shape. Next comes the load of shot, which is held in place by a thin cardboard wad, No. 4, which usually has the size of the shot, and various other information, according to the fancy of the maker, stamped or printed on it. Sketch "A" also shows what is known as a "Low Brass" shell, in which the brass head extends only a short way up the side.

    B   "High Brass" shell, in which the brass extends further up the side.

    C   "Low Base" shell, in which the paper reinforcement inside the head of the shell is relatively small in amount. This is the usual construction when black or "bulk" smokeless powders are used. The primer shown is the old black powder type, resembling the ordinary rifle or pistol primer in size and construction.

    D   "High Base" shell, in which the paper reinforcement inside the head of the shell extends higher into the powder space, which is thus reduced in volume so as to make it suitable for use with concentrated or "Dense" smokeless powders, which require less space than do the ordinary black powder or the "Bulk" smokeless. This also shows the usual "Battery Cup" type of shotgun primer, commonly used in all but black powder shells.

brass-cardboard design, however, seems to be withstanding fairly well the competition of some European reloadable copper, brass, and aluminum cartridges. Recently, plastic cartridge cases have been used successfully for shotguns; the entire case including the head is plastic.

    In general, shotgun shells have a slightly different primer arrangment because of their less rigid base. The primer itself is usually now mounted into a battery cup which is then pushed into the

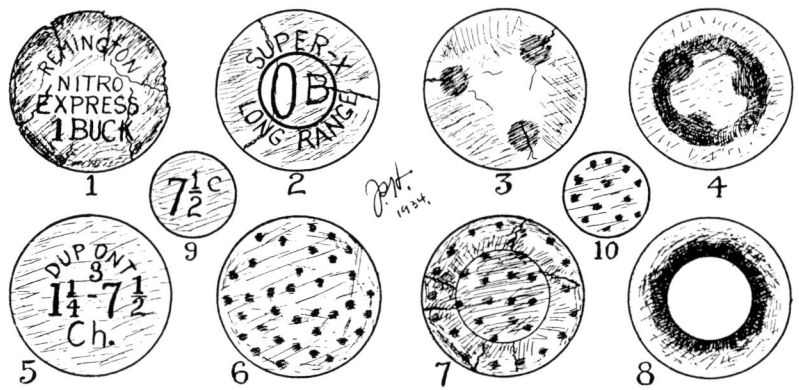

SHOTGUN WADS

1. Top shot wad picked up at the scene of a shooting. It happens that in this case a repeating shotgun was used, and the empty cases were also recovered, but it will be seen how important the information given on this wad might be in any case. This particular wad was badly smoked up, and the bottom layer of paper was blown entirely off.

2. Top shot wad removed from an unfired buckshot load of another make. It will be observed that this wad gives the maker's trade mark, and the size of the shot; namely, No. 0 Buck. This wad is of thin cardboard.

3. Bottom surface of the wad shown as No. 2. Note the marks made by the top layer of three buck shot. There were twelve pellets in this load, in four layers of three pellets each.

4. Under shot wad from the same shell. This wad is made of compressed cork composition, and is deeply cupped on both sides. The impressions of the three pellets forming the bottom layer of buckshot are plainly seen and the wad is about a quarter of an inch thick.

5. Top shot wad from an ordinary trap load. Note the complete details given as to the load; 3 drams of DuPont powder and 1¼ ounces of No. 7½ chilled shot. The maker's name does not appear on this wad; it is from a Peters "Target" shell.

6. Bottom side of No. 5, showing the impressions of the top layer of shot. This is an accurate sketch of the exact manner in which the spots appeared, the somewhat open places indicating where the shot was bridged over so that no pellet touched right there.

7. Top surface of the under shot wad from this same No. 5 and No. 6 shell. This is a felt wad with a paper surface on eachside. The top surface is deeply cupped, or pressed in, by the rammer used to seat the wad firmly in loading the shell. The marks of the individual pellets comprising the bottom layer of shot can be plainly seen, and if the wad were recovered after being fired these marks might show as deep indentations covering the entire surface. Wad is about a quarter of an inch in thickness.

8. Thick cardboard ring, or doughnut shaped wad used over the top of the one round, lead ball in a 12 gauge, single ball load. The end of the paper case is crimped over this ring (making it a top shot wad), thus centering and holding it in place. In the loaded shell, the surface of the ball can plainly be seen through the opening in the wad, thus clearly showing that it is a single ball cartridge.

9. Top shot wad from a .410 shell. Note that no information is printed on this wad except the size and kind of shot; namely, No. 7½ chilled. This is from a 3-inch Western Super-X shell.

10. Bottom surface of No. 9, showing the impressions of the pellets forming the top layer. Some makers print the load specifications on both sides of the top shot wad.

center of the base of the cartridge. In some older types of ammunition still to be met with, the primer is inserted directly into the brass base of the case itself, as in rifle and pistol ammunition.

Little effort is made by shotgun shell manufacturers to keep their powder uniform from lot to lot. They do, however, carefully test each lot when it arrives from the powder manufacturers and determine the exact amount to be loaded into each shell for the specific ballistics required for the particular brand being loaded. The charge, therefore, varies considerably in volume. Where there is less powder, there must be more wadding.

FOLDED CRIMP AS USED IN SHOTGUN AMMUNITION

The most popular end closure for shotgun ammunition at present is the folded crimp shown in this picture, as contrasted with the older system of using an over-shot cardboard wad and crimping it in place by rolling over the edge of the shell. The over-shot wad has the disadvantage of leaving the muzzle just ahead of the shot charge, and more or less disrupting or spreading the pattern due to the impact of the shot pellets against it. The folded crimp shown thus gives more even patterns.

The folded crimp was first introduced in France about 1924, and was hailed with delight, but it was not until about 1939 that it was first introduced into the United States, and still later than that before the ammunition companies' advertising departments began to shout loudly about its virtues.

Shown in the picture are the latest Remington and Peters shells of this type. In making them, the paper case has been left slightly longer that for the old type rolled crimp, and this extra length has been so creased and folded over that it firmly and adequately encloses the shot charge, after which a paper seal is glued over the crimp as shown.

When the shell is fired, these folds open out and the charge of shot leaves the muzzle with no cardboard wad in front of it to disrupt the pattern. Generally the paper seal splits open and remains fastened to segments of the crimp, as shown. Occasionally it is blown to bits or shredded into small fragments in front of the gun.

Lead shot originally was known as "drop" shot, the term coming from the method employed in making it. Such shot was of pure lead and quite soft. Later on, a small proportion of antimony was added to harden the pellets, the resulting shot was known as "chilled" shot; in time this chilled shot became the more popular

THE WINCHESTER-WESTERN FOLDED CRIMP ARRANGEMENT

Winchester and Western shotshells have a different no-wad-over-shot crimp, as pic-
tured above. Both concerns crease and fold-over the front end of the case until it completely encloses the shot charge, but they then add a firm, conventional type of crimp
as in older ammunition, resulting in a very good enclosure and end fastening.

The load and shot-size data is printed on the side of the case. Upon being fired,
some of the case mouths open up in the usual circular manner; others remain in the
scalloped shape shown above. Nothing leaves the muzzle of the gun but the shot
charge and powder wads.

and with the coming of higher velocities in shotgun ammunition
its use became imperative. Today, drop shot is no longer made, the
chilled being all that American manufacturers turn out; they now
refer to it as "high antimony" shot.

Therefore, we no longer have drop shot available, but modern shot
in all the small game or "dropshot" sizes (No. 1 to No. 12) is still
manufactured by molten lead being dropped through a colander
from a high tower and falling into water about 200 feet below. It
is then sorted as to size, selected for pellet perfection and concen-
tricity and inspected by automatic devices; by the time it is loaded
into the case it is about as uniform and perfect a product as can
be imagined.

Buckshot, however, is still manufactured and loaded by individual
methods. The melted lead is cast by hand into gang or multiple
moulds and the pure lead balls then rumbled in a barrel until all
sprue and fin remaining on the cast pellets are rolled away, some

graphite being added in the process.

The entire process of loading shotgun ammunition, in the "drop-shot" sizes, is by automatic machinery. The amount of inspecting which such ammunition undergoes in the process of manufacture is unbelievable, there being four or five inspectors to every machine operator.

However, buckshot cartridges are loaded partly by hand. The cases—primed, powder-loaded, and with powder wads seated by machinery—are brought to a work table where the buckshot pellets are counted into layers and the layers each carefully positioned into the case by hand. A top wad is then put in place and the cartridges taken back to the machines for a final crimping. This machine-hand method of loading results in a certain uniform number of pellets being placed into a buckshot cartridge, a fact which should be kept in mind by the firearms investigator.

The whole subject of wads is practically unknown in rifles and pistols; however, it is very improtant in shotguns. In order to keep the shot charge completely free of the propellant gases in the bore, a minimum thickness of heavy, strong wadding is necessary. Many different substances have been used over the years; experimenting is still going on with new plastics, and such. The thickness of this column of wad varies with the powder, as mentioned above. Wadding because of its variation between different makes and types of ammunition, can be of extreme importance in crime.

The shot charge is located in the cartridge above the wads. It is accurately measured and usually capped at the present time with a special folding of the shell casing itself to get away from any over-shot wad, which was supposed to have created a disturbance in the shot pattern. However, cartridges with the old type cardboard over shot wads are still being made. We are all familiar with the well known sizes of shotgun ammunition. Usually crime weapons will be of 12, 16, or 20 gauge or bore. The .410 shotgun is now quite common; less often 10, 24, 28, and 14 gauge weapons are met with. All these can be loaded with several different weights and sizes of shot. Stoeger's catalogue or a similar book will list the possibilities.

In theory quite a number of loadings can be purchased. Not nearly so many are actually available. In 12-bore about four sizes of shot account for far more than 90% of the rounds produced. The range is even more limited in the other gauges. Shot is standard-

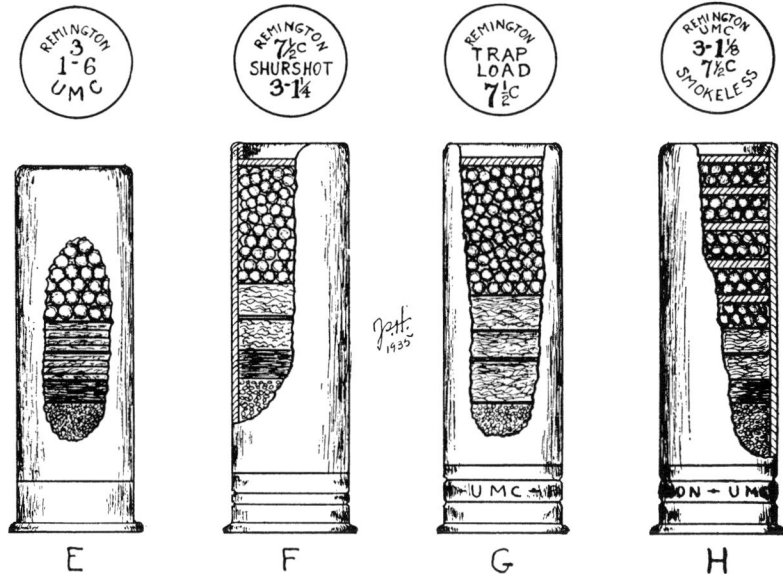

VARIATIONS IN ARRANGEMENT OF WADS IN SHOT SHELLS

Diagrammatic sketch of four shells of the same make. These sketches were made from actual samples, and a careful attempt was made to have the thickness and location of all wads, etc., in exact proportion. Above each shell is a sketch of the top-shot wad, showing the lettering. It will be seen that there is a wide variation in both the kind of wads and their location, as well as in the kind and type of information given on the printed top shot wad, and as these samples are all of the same make, it will be seen that with all the different makes in existence, the number of different combinations is enormous.

E A black powder load, the "New Club," which has one cardboard wad over the powder, and two laminated felt wads on top of that.

F Remington "Shurshot."

G Nitro Club "Trap Load." Note that there is no card wad over the powder.

H Nitro Club "Scatter Load,' with card wads between the layers of shot to cause the charge to scatter quickly for brush shooting, where the range is likely to be short.

ized as to size in this country, although different lots of lead can sometimes be identified spectrographically. So-called chilled shot contains a fairly high percentage of antimony for hardening and may contain small amounts of arsenic. The older pure lead or soft shot is not much used today; it had a tendency to ball; that is, several pellets adhere together in one mass. It gave, therefore, less even patterns and perhaps irregular penetration.

In many shotgun crimes, wads are recovered at the scene of the crime, sometimes right in the body of the victim. Unburned powder also can sometimes be recovered. All these should, of course, be preserved since wads and powder, more than shot, serve to identify at least the maker of the ammunition and sometimes specific lots of ammunition.

One soon learns to determine the gauge of a shotgun, its ammunition, wadding, etc., by eye alone; but in case of doubt, a comparison with unfired components will soon settle the point. However, for the benefit of those who may desire to know the accurate dimensions, we give below the diameters of the standard shotguns bores and also the diameter of the wadding. It will be noted that the diameter of the wadding is greater than that of the gun bore. This is proper and customary, but in actual practice the wad diameters will vary considerably and may be much greater than the figures given, particularly in the case of the soft felt wadding which must act as a gas check, and seal the bore while the shot are traveling out ahead of the powder gases.

Table of Shotgun Gauges

Gauge	Diameter of Bore	Diameter of Wadding
8	0.835 inch	0.845 inch
10	0.775 "	0.784 "
12	0.729 "	0.738 "
16	0.662 "	0.671 "
20	0.615 "	0.623 "
28	0.550 "	0.557 "
.410	0.410 "	0.415 "

Until a few years back, there were two general types of shot in common use; soft, or drop shot, and chilled shot. Soft shot was, as the name implied, ordinary soft lead formed into round pellets. Soft shot is no longer made to any extent, but much of it is still around.

Chilled shot is made from lead specially hardened by the addition of a certain amount of antimony. There is a vast difference between the hard kinds, as anyone can readily tell if he will just bite a few pellets between the teeth. A late development is the plating of shot with copper to harden the surface.

The investigator should familiarize himself with the appearance of the various sizes of shot, for the size can be told very closely by visual inspection after a little experience. With a half ounce or so of the pellets in the palm of the hand, it is possible to state the size of the shot with confidence after a little experience. A more precise way would be to measure the pellets; and for all sizes under No. 1, the diameter in hundredths of an inch subtracted from 17 will give the exact shot number. For example, if the diameter is 10 hundredths of an inch, it is No. 7 shot; if the diameter is 9½, it is No. 7½ shot, and so on. This method applies to unfired shot, but when recovered ones are measured, different diameters must be averaged, because the pellets most likely will be deformed or flattened, and a degree of judgment must thus be used. In the case of such recovered shot, it would be well to weigh a few of the recovered pellets and base the determination of the size on the results.

The following table gives the diameters and weight in grains of the different sizes of shot, including buckshot and round ball.

TABLES OF SHOT SIZES AND WEIGHTS

LEAD BALLS

Name	Diameter in Inches	Diameter in Millimeters	Approx. No. of Balls per Lb.	Weight in Grains
10-Gauge	.710	18.03	13½	520.0
12-Gauge	.645	16.38	17	412.0
16-Gauge	.610	15.49	20½	342.0
20-Gauge	.545	13.84	28½	246.0
24-Gauge	.542	13.76	29½	238.0
28-Gauge	.510	12.95	35	199.0
½-Inch	.500	12.70	38	189.0
.44 S. & W. Russian Gallery	.428	10.87	58	118.0
.44 Game Getter	.425	10.79	60	116.0

SOFT AND CHILLED SHOT

No.	Diameter in Inches	Diameter in Millimeters	Soft Shot No. to the Ounce	Soft Shot Weight in Grains	Chilled Shot No. to the Ounce	Chilled Shot Weight in Grains
1	.16	4.06	71	6.16	73	5.99
2	.15	3.78	86	5.08	88	4.97
3	.14	3.53	106	4.12	109	4.01
4	.13	3.30	132	3.31	136	3.21
5	.12	3.02	168	2.60	172	2.54
6	.11	2.79	218	2.00	223	1.96
7	.10	2.54	291	1.50	299	1.46
7½	.09½	2.41	338	1.29	345	1.26
8	.09	2.28	399	1.10	409	1.06
9	.08	2.03	568	.77	585	.75
10	.07	1.78	848	.52	868	.50
11	.06	1.52	1346	.33	1380	.32
12	.05	1.27	2326	.19	2385	.18
Dust	.04	1.02	4565	.10	....	..

COMPARISON OF AMERICAN AND FOREIGN SHOT SIZES

English	American Eastern Buck-shot	American Western Buck-shot	Belgian	Canadian	Dutch	French	German	Italian	Italian	Spanish Linares English type	Spanish Barcelona Figueroa type	Spanish Sevilla Mata type	Swedish	Turkish	Turkish
L G	000	2	—	—	—	—	II	—	—	—	—	—	—	—	—
M G (mould)	00	3	—	—	—	C.1	—	—	—	—	—	—	—	—	—
S G	0	4	—	—	—	—	III	—	—	—	—	—	—	—	—
Special S G	1	5 or 6	—	S G	—	C.2	IV	—	—	—	—	—	—	18/A	—
S S G	2	7	—	—	—	C.3	V	—	—	—	—	—	—	17/A	—
S S S G	3 or 4	8 or 9	—	—	—	C.4	—	—	—	—	—	—	—	—	—
S S S S G	FF		—	AAAA or 12 Seal	—	C.4 bis	VI	—	—	—	—	—	—	16/A	—
S S S S S G or A A A A A	TT		—	AA	—	—	000000	7/o	7/o	7/o	—	—	—	15/A	—
A A A	T		—	A	—	5/0	00000	6/o	6/o	6/o	—	—	11	14/A or 13/A	4/o

COMPARISON OF AMERICAN AND FOREIGN SHOT SIZES (Continued)

English	American	Belgian	Canadian	Dutch	French	German	Italian	Italian	Linares English type	Barcelona Figueroa type	Sevilla Mata type	Swedish	Turkish	Turkish
AA	BBB	—	BBB	—	4/0	0000	—	5/0	—	—	—	10	3/0	12/A
A or BBBB	BB	5/0	BB	—	2/0	00	—	3/0	—	—	—	9	0	11/A
BBB	B	4/0	B	00	0	0	—	2/0 or 1/0	4/0	1	—	8	1	10/A
BB	1	3/0	1	0	1	1	—	1	3/0	—	—	7	2	9/A
B	2	2/0	2	1	—	1	00	2	2/0	—	—	—	—	8/A
1	3	0	3	3	2	2	0 or 1	3	1	—	—	6	3	7/A
2	—	2	—	4 or 5	3	3	2	—	2	—	—	5	4	6/A
3	4	3	4	G.6	4	4	3	4	4	4	—	4	5	5/A
4	5	4	5	—	5	5	4	5	5	5	—	3	6	4/A
4½	—	—	—	—	—	—	—	—	—	6	4	—	—	—
5	6	5	6	K.6	6	6	5	6	6	—	5	2	—	3/A
5½ (m.g.)	—	—	—	—	—	—	6	5	—	7	6	—	7	—
6	—	6	7	—	6	6	6	—	6	—	—	1	—	2/A
6½	7	6½	7½	—	7	7	7	7	7	8	7	—	8	—
7	7½	7	8	7	—	—	8	8	8	—	—	0	—	1/A
8	8	8	9	8	8	8	9	9	—	—	—	—	9	1/0
9	9	9	10	9	9	9	—	—	9	9	9	00	10	2/0
10	10	10	—	10	10	10	—	—	10	—	10	000	11	—
11	—	11	—	11	—	—	11	11	—	11	11	—	—	3/0
12	11	12	—	12	11	11	—	14	—	—	12	—	12	—
Dust	12	—	—	—	12	12	14	—	—	—	—	—	13	—

SOFT SHOT

Eastern Size	Diameter in Inches	Diameter in Millimeters	Approx. No. of Pellets per Oz.	Weight in Grains
FF	.23	5.84	24	18.2
F	.22	5.59	27	16.2
TT	.21	5.33	31	14.1
T	.20	5.08	36	12.2
BBB	.19	4.83	42	10.4
BB	.18	4.57	50	8.8
Air Rifle	.17½	4.44	55	8.0
B	.17	4.32	59	7.4

A tabulation and comparison of all the various foreign sizes of shot have been given elsewhere.

The above tabulation gives the sizes of what are known as American Standard Shot, and these specifications are pretty generally adhered to. However, one may occasionally find a particular maker loading shot that will measure between these sizes; this is done to obtain a higher pellet count to the pattern, or something of the kind, with shot of approximately the designated size. It is the practice of some loading companies to vary the sizes slightly; thus they may actually have three separate dimensions of the larger sizes of shot, as, for example, Small Sixes, Sixes, and Large Sixes. The actual difference, however, will be so small as hardly to be noticeable, and can be told only by the slight variation in the number of pellets to the ounce.

In addition to familiarizing himself with the different sizes of shot, the investigators should make a close study of the wadding used by the different makers and learn to identify the various felt and cardboard gun wads. Many different materials, such as felt, linoleum, cork, paper, leather, etc., may be used in the manufacture of these wads, and sometimes certain patented shapes will be seen. Each maker has his own particular practice in assembling the different combinations with his own distinct colors and sizes, and a study of all this will help greatly in enabling the investigator to correctly identify any wadding which may be involved in an intentional homicide or a shooting accident with the shotgun.

It is, however, impossible to give all these data in a complete form for reference, as the practice of any maker is subject to change at any time, and what is correct today will perhaps be entirely different tomorrow. The investigator will do well to make a close study of shotshell loading practices and components and keep constantly abreast of current practices, with particular attention to improvements that are always being made in the art. It is an easy and entirely safe practice to cut open the paper case of a loaded shotgun cartridge and see accurately how it is assembled and how the different components are used. Cut a longitudinal strip from the paper case, dump the powder into a pill-vial, and preserve it and the wadding and shot for future reference and study, being careful to label it correctly as to maker, type and date.

Sometimes shotgun shells will be loaded to fire single spherical lead balls that pass through full choke guns. These balls, called punkin balls in some sections of the country, are of doubtful value. A shotgun is a very dangerous and powerful weapon; it will kill an enraged bull instantly if hit in a vital spot at a range of under 15 feet, even with a charge of bird shot. The spherical ball ammunition, however, greatly reduces this potential; it is less powerful in close. At ranges beyond which a charge of ordinary shot would be effective, the punkin ball begins to be so dreadfully inaccurate as to be of little value.

A German by the name of Brennecke invented a bullet to be fired from choke-bored shotguns, obtaining very accurate flight by means of a gravity adjustment, wads permanently attached to the bullets and veins about its sides to cause it to spin as a rifle bullet does. These rifle slugs are now sold by various American ammunition manufacturers who have discarded the attached wads but kept the old hollow base idea and the spiral grooves along the outside.

Sometimes, particularly in certain rural sections of the country, someone will cut through the entire cardboard shell casing behind the shot charge and load it into a gun. Such a "cut" cartridge is very dangerous for the safety of the gun in which it is fired. The entire front portion of the cartridge is usually forced in a single mass down the bore. Such a projectile is much too large for the barrel and can easily cause the barrel to burst. These cartridges have, however, been used in crime and, although irregular in their action, sometimes inflict the same sort of "rat hole" wound at up to 30 or 40 yards that a regular charge of shot will do at five feet.

Buckshot is also sometimes used in crime. Since there are relatively few of these pellets in a charge—there are nine 00 buckshot in a 12-gauge shell—the pattern and accuracy are poor at medium ranges. However, buckshot does extend the range at which a shotgun will kill human beings.

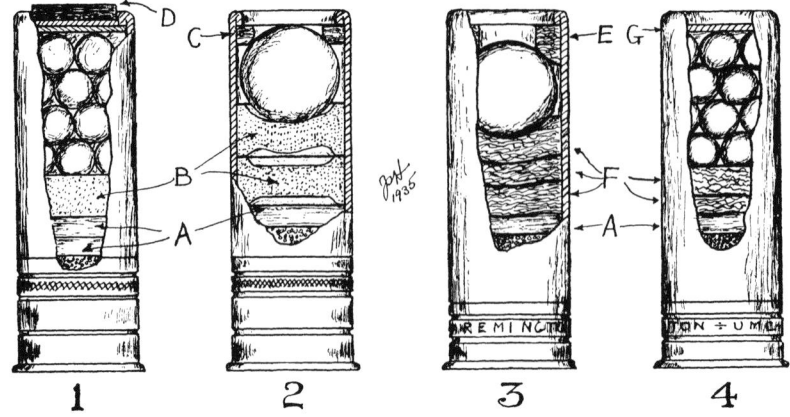

BUCKSHOT AND SINGLE-BALL SHOTGUN LOADS

1. Western Super-X load of No. 0 Buck. There are 12 pellets, in four layers. Over the powder there are two thick cardboard wads, then one cork composition wad, concaved on top and bottom. Notable is the top shot wad, which is of a patented construction as shown, in order to protect the crimp when the shell is carried in the magazine of a repeating gun for long periods, and taken out of the gun and replaced many times, as is the case when the shell is used by guards armed with riot type shotguns.

2. Western Super-X single-ball load.

3. Remington single ball load.

4. Remington load of No. 0 Buck.

A—Thick cardboard, as greaseproof wads. B—Cork composition wads, made with the top and bottom dish shaped, or concave, as shown in No. 2. C—Annular cardboard wad, or "Doughnut wad" used to hold the single ball in place. D—Patented top shot wad to protect the crimp on buckshot loads used by guards, which will usually wear out from handling before they are fired. E—Annular felt wad, used to hold the single ball in place. F—Felt wads. G—Cardboard top shot wad.

American Buckshot Sizes as Loaded

Years ago, there were two buckshot set-ups, an "Eastern" and a "Western," each with its own size standards. The Eastern finally won out and at this writing our American ammunition companies load buckshot cartridges only in the following combinations:

Gauge	Size	Number Pellets	Number Layers	Pellets per Layer
10	0	16	4	4
12	00	9	3	3
12	0	12	4	3
12	1	16	4	4
12	4	27	4	3 of 7, 1 of 6
16	1	12	4	3
20	3	20	5	4

The average hunter with the shotgun does not have much need for buckshot in his general hunting, but is likely to carry a couple of buckshot cartridges along for an emergency shot at big game. Such cartridges may lay around for years until used. Some years back, the following buckshot combinations were used and might possibly be encountered; hence we give these old loadings:

	Size		Number
Gauge	Eastern	Western	Pellets
10	000	2	9
	00	3	9
	0	4	12
	2	7	20
	3	8-9	27
12	2	7	15
	3	8-9	20
16	00	3	6
	2	7	12
	3	8-9	15
20	1	5-6	6
	2	7	12

BUCKSHOT

Eastern Size	Diameter in Inches	Diameter in Millimeters	Approx. No. of Balls per Lb.	Weight in Grains
000	.36	9.14	103	68.0
00	.34	8.64	122	57.5
0	.32	8.13	144	48.5
1	.30	7.62	175	40.0
2	.27	6.86	238	29.5
3	.25	6.35	299	23.5
4	.24	6.09	341	20.5

All the special shotgun ammunition such as spherical ball, rifled slugs, ringed shells and buckshot, are relatively rare. A crime, therefore, committed by means of one of these would probably have a shorter list of possible suspects. Not more than one hunter in 40 or 50 who purchases standard trap and field loads for sport will even know about rifled slugs. Their purchase can generally be traced if the crime warrants a large expenditure of time. Even buckshot is not very often found in the hands of sportsmen, save where state game laws preclude the use of rifles for deer hunting, or in the South where its use is almost universal among deer hunters.

Handloaded Ammunition

No discussion of ammuntion would be complete without some mention of the customloaders and the even more numerous handloaders. Before World War II, there was relatively little custom loading. A few individuals could handload ammunition more accurately than it could be loaded by any factory. Some of these men went in for gilt-edged handloads which always cost considerably more than their factory counterparts. However, since World War II, many who shoot a good deal have found it cheaper to buy handloaded ammunition from those making a profession of their old hobby and using semi-automatic machinery. Many police departments either reload their own ammunition or buy from one of those quantity customloaders. Further, a great many foreign weapons were brought back to this country at the end of the war. Some of these are superb sporting weapons for which ammunition was not available in this country. The professional handloaders and some amateurs have used great ingenuity in making up entirely reliable, satisfactory, and efficient ammunition for these weapons, either by necking down or blowing out already available American cases, or by combining parts of two or more American cartridge cases. As in all handloading, the cartridge case is the most expensive single component, so that once you have the cartridge case, additional reloading of it is much cheaper.

Cartridge cases and primers are almost always purchased from the ammunition companies. Frequently bullets also are purchased from them, although excellent metal jacketed bullets can be handmade in various swages and dies. Cast and alloy bullets are, however, easily and cheaply produced in the family kitchen or workshop. Most hand and custom cartridge loaders produce these themselves. This handloading and reloading is a business for a few but a hobby for thousands. Some of these folks just reload factory cartridge cases with cast or government bullets with a few dollars worth of equipment. Others have hundreds of dollars worth of equipment and do all manner of things both sane and otherwise. As soon as someone gets to be reasonably expert at such operations, he will probably leave characteristics of his tools and his personality on his ammunition. There is a good chance that a given mold or bullet die will leave its own individual characteristics on every bullet formed in it. However, those who do this work are for some reason very seldom criminally inclined. Their products, save as a result of hunting accidents or when they are also law enforcement officers and have to shoot for their lives, seldom come into the Ballistics Laboratories of the country.

MANUFACTURING PROCEDURES IN PRODUCING A ROUND OF .38 SPECIAL REVOLVER AMMUNITION

FOREWORD: The authors feel that a somewhat detailed description of the manufacture of this popular cartridge may be of value. Other types of ammunition, when not produced entirely automatically, are made in a similar manner.

Essentially, a round of ammunition is an assembly of four separate components. The propellant powder, which will be discussed in detail in Chapter 6, arrives at the ammunition loading plant in bulk containers. The primers are generally fabricated in an isolated area near the loading plant proper, with provisions for the safety not only of the operators, but also of the general public. We point out that the ammunition industry has a safety record that surpasses the average for non-hazardous industrial plants. The bullet is fabricated at the ammunition plant itself from raw materials as discussed hereafter. The cartridge case is the most involved and expensive of these components. We will discuss it below in considerable detail.

Cartridge Case

Strip brass of proper chemical composition and hardness is received in various thicknesses to conform with requirements for different cartridge cases. It is fabricated in accordance with requirements and procedures as indicated below:

1—The sheet brass is blanked and formed into a cup in one operation under a multiple punch press. We emphasize that a cup is formed in a single operation rather than two operations as was common a few years ago.

2—After cupping and preceding each drawing or forming operation, annealing is done automatically by a continuous process which takes advantage of modern controls. After annealing, these brass pieces are cleaned in a sulfuric acid solution and then neutralized, washed, and dried.

3—In the .38 Special cartridge case, three separate drawing operations, with anneals between each, are done in three different machines to produce the uniformly lengthened cartridge cases as shown in the accompanying photograph.

4—The completed deep-drawn case blank is trimmed to length.

Shown here is a picture of the .38 Special cartridge case in each stage of its production, from the small brass cup to the finished shell. The cup is stamped out on a double action press. The next three photographs show pictures of the cup after three successive drawing operations. The extra stock is then trimmed from the mouth of the case. Two operations are necessary in order to complete the exterior of the rim and undercut. The primer pocket is finally finished and primed. Finally, a cannelure is placed for the bullet to be seated against.

5—The base of this cartridge case is formed with primer pocket and head stamped.

6—On automatic machinery, the rim, together with the undercut, or extractor groove in front of the rim, is turned to size.

7—Perhaps the most interesting operation in the whole process is when, on one machine running automatically, the flash hole is pierced, the primer pocket broached to size, and the primer— suitably protected by waterproofing—pressed into its pocket. It is carefully inspected from all angles by the operator with a series of mirrors so as to prevent either an unprimed or unpierced round proceeding further.

8—An expanding operation is done to secure uniformity of mouth opening and, in certain cases, the waterproofing of this mouth where jacketed bullets are to be used. In general, this waterproofing is not necessary for lead bullets which are suitably lubricated.

9—A cannelure is impressed in the outside of the case to form a seat for the bullet. Sometimes for identification two may be rolled into the case; the upper sometimes forms the bullet seat, though sometimes these are spaced for identification only. Sometimes they are put in after the bullet is in place and form groove crimps to increase the initial pressure, particularly with light bullets designed for metal piercing.

The cartridge case is now finished and ready for the loading operations. It is in the condition known as a "primed empty" to thousands of handloaders. A few words are in order about the primer.

Primer

The individual primer cups are formed from thin strip brass or copper, which may or may not be nickel-plated, of the proper hardness and thickness for the weapons in which they are to be used. These cups are then filled with priming mixture to the proper depth. Both wet and dry methods of application are used. In general, this is the most dangerous of all loading operations. Even though Remington has not had a casualty of this type for many years, the individual operator is protected by a steel and shatterproof glass screen. In the wet process, the compound is relatively inert. In the dry process, a drop of shellac renders it temporarily inert.

A small paper or foil disk is pressed over the priming compound. The anvil is then introduced firmly, but gently, into the primer cup so that the pellet rests between the base of the cup and the anvil. After drying under controlled conditions, these primers are ready to be seated into cartridge cases or packed separately as desired. Almost all primers are now both non-corrosive and non-mercuric; they leave no residue to attack either steel or brass.

Bullet

Lead, antimony, tin, and, on occasion, certain other ingredients are received at an ammunition plant in the form of relatively pure commercial products. Sometimes the exact alloy required is purchased. More often the alloying is done in an efficient continuous proc-

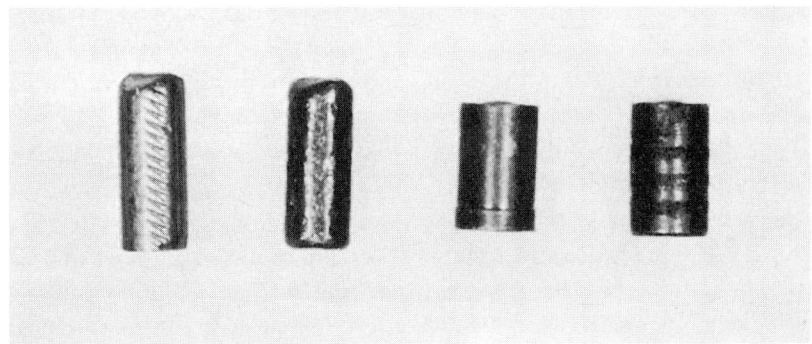

Four views of the four stages in the manufacture of a .38 Special wad cutter. First, there is the cut-off slug; second, after this slug has been tumbled; third, after the bullet has been swaged to size; and fourth, after the bullet has been grooved and lubricated.

ess furnace, and the alloy then so produced is cast in cylindrical billets of about 250 pounds each. These are fed automatically as required into a cold extrusion machine, where the lead-alloy, under the enormous pressure of a large hydraulic ram, is forced through dies of the proper size in the form of lead wire. This lead wire, in some instances, is coiled into barrels to be separately fed, for the formation of some bullets, into chopping machines. However, certain types of lead wire travel directly to machines which cut it into segments of the proper length for specific bullet slugs.

These slugs are tumbled in graphite to remove the sharp corners and slightly lubricate the lead for the next operation.

In the case of the .38 Special bullets, the next operation is the swaging of the lead slug into final bullet shape in a single operation.

The bullet is then grooved and lubricated in another operation, and is now ready for assembly in ammunition, or sale as a factory bullet.

Some high-velocity .38 Special bullets are made of solid zinc for metal piercing; others have a gilding metal nose swaged and crimped into place.

Assembly

The most complicated of all operations in ammunition loading is the assembly of primed case, powder and bullet. A group of operators and inspectors work together in line.

First, they shake the individual primed cartridge cases mouth-up into plates of a suitable size. The number of cartridge cases loaded into each plate depends, of course, on the size of the cartridge. In .38 Specials, about 130 are loaded at a time.

Second, these plates, complete with primed cartridge cases, are again inspected to determine blemishes, lack of piercing, primers, and the like.

Third, powder is introduced in exact quantities into each individual case by means of individual powder chambers exactly positioned over the cases. The operator of this particular machine stands with his back near an outside wall, and reaches through an opening in this wall for each container of powder that he uses. The individual powder containers have a capacity of only two (2) pounds. Not more than four (4) pounds of powder is ever inside the wall at any one time. A relatively few containers remain in a fragile magazine outside the wall. The empties are periodically collected from this outside magazine, and full containers deposited ready to be brought into the building and used in the loading of the hopper above the charging plates.

Fourth, every individual charge is now checked semi-automatically to assure a high degree of uniformity.

Fifth, the block containing the cartridge cases, with their appropriate charge of powder, is sent along a conveyor to meet the bullets that have separately been shaken into similar blocks. The two charged blocks are accurately positioned, one over the other and the bullets pressed into the mouths of the cases and crimped.

The ammunition is now ready for testing and packaging.

Ammunition Terms and Names

Ammunition for the various type firearms is spoken of in several different ways, and distinctive terms are used in various sections of the country, any of which may be correct. The word "ammunition" of course means any unfired assembly of primer, powder and ball which might be used in a firearm; but today the word is generally used when referring to a supply of assembled cartridges in bulk, as in boxes, cases or carloads. However, "ammunition" is often used to refer to the supply a person may be carrying in his pockets, although the term "cartridges" is about equally applied. The term "round" is often heard, especially by the military, it refers today to a single cartridge. The word is a relic of the earliest muzzle loading days and originally meant "A single charge of ammunition for a firearm"—powder and ball in flintlock days, and powder, ball and cap in percussion day. With the coming of the wrapped powder-ball assembly the word cartridge came into use.

Amongst the ammunition and firearms trade, shotgun cartridges are commonly referred to as "shotshells," while rifle ammunition is spoken of as "metallics;" further divided into "rimfires" and "centerfires," the latter comprising practically all calibers other than .22. When a trade representative used the word "cartridge" he invariably refers to some rifle or revolver metallic cartridge; while in speaking of any shotgun ammunition he will use the term "shells" or shotshells mentioned above.

The layman uses the abovementioned terms indiscriminately, although as a general rule he speaks of "cartridges" when referring to rifle or revolver ammunition, and "shells" when on the subject of shotguns. But a fired, empty, cartridge case is invariably spoken of as a "shell" also, although the backwoodsman might speak about "busted hulls" and still be in good social standing in his local court, even though "fired cases" is the strictly correct designation.

Amongst the uninitiated, the word "bullet" is most often misused, as it is commonly applied to any sort of an unfired cartridge. Actually, the bullet is only that solid portion of a loaded rifle or pistol cartridge which leaves the muzzle of the gun, and which does the striking or the killing. The word can really only properly be used in connection with rifle or revolver ammunition, but other common designations for the bullet are "projectile" and "ball" the latter a relic of old muzzle loading days when all rifle projectiles were round, lead balls.

Although there are available in special types, various patented bullets for use in modern shotguns, they are seldom seen in this country and the word "bullet" is never correct in connection with shotgun ammunition as generally manufactured. The cloud of small, round, lead projectiles which leave the muzzle of a fired shotgun is properly referred to as "shot," or as the "charge," or the "load;" while the individual, round projectiles are properly termed "pellet" or "shot." The entire charge, when moving through the air, can also be properly referred to as the "pattern," although strictly speaking the pattern is the registered strike of the entire load upon a paper target. Buckshot has its own peculiar and established appelations; the balls comprising the charge are often incorrectly referred to as "slugs," while in many backwoods sections, particularly in the South, they may speak of the larger sized buckshot as "blue whistlers" or "whistlers" while the smaller sizes are called "low moulds." The use of the round, single lead ball in shotguns is very greatly restricted in this country, but is popular to a degree in some of the mountain sections in the East where the use of buckshot is prohibited by State laws; such loads are often locally referred to as "punkins" or "punkin balls."

The Examination of Ammunition

A thorough and accurate knowledge of ammunition of all types is essential to the firearms expert and it often becomes necessary to break down cartridges in order to obtain needed information. The public, in general, has a greatly exaggerated idea of the dangers incurred in handling or taking apart any type of small arms ammunition; actually there is little chance of a mishap occurring if a bit of care and commonsense be used.

If necessary, such work can be done in an office and at one's desk. Just lay aside whatever you happen to be smoking and have any

BREAKING DOWN RIM FIRE CARTRIDGES

Watch out! This *can* be dangerous.

Rim fire ammunition is primed with a detonating priming mixture that, while wet, is spread evenly around the entire inner rim of the cartridge case. In modern ammunition manufacture this compound is put in place by automatic machinery that distributes the wet material only into the rimmed section of the case, but occasionally, somehow or other, spots of this priming compound are smeared on the inside of the cases; this fault used to be quite common in earlier rim fire ammunition. If one of these spots when dry, no matter how small, be torn apart or scraped, that speck of priming detonates and is almost certain to fire the powder charge and main priming. Even the little .22 short can "blow" with considerable effect.

The safest method of taking apart rim fire ammunition is to disregard any probable defacement of the projectile and to clamp the bullet firmly in the jaws of a vise but not engaging the case mouth in any manner. Then, wrap several folds of cloth around the cartridge case and twist or bend it loose from the bullet. Wearing glasses, and a thick glove or gauntlet may induce a greater feeling of safety. Once the bullet is out, the powder can be dumped; if a black powder cartridge of any age, the powder charge may have become firmly caked, in which case scrape it out, a little at a time, with a pointed wooden stick.

nervous persons go out of the room. Don't permit interruptions or allow yourself to be bothered. In case you are not accumulating a collection of cartridges and components, have at hand some small receptacles into which the discarded materials can be dumped. All discarded powder and primers, or emptied primed cases should be taken away and destroyed by the examiner himself; do not throw such into a wastebasket for disposal by a janitor or by some person who is not aware of what is being carried away.

Under proper conditions and with a bit of care, any sporting or target ammunition is perfectly safe to break down, but do not attempt to take apart military ammunition loaded with tracer bullets, or incendiary or explosive bullets of any sort. Such cartridges should

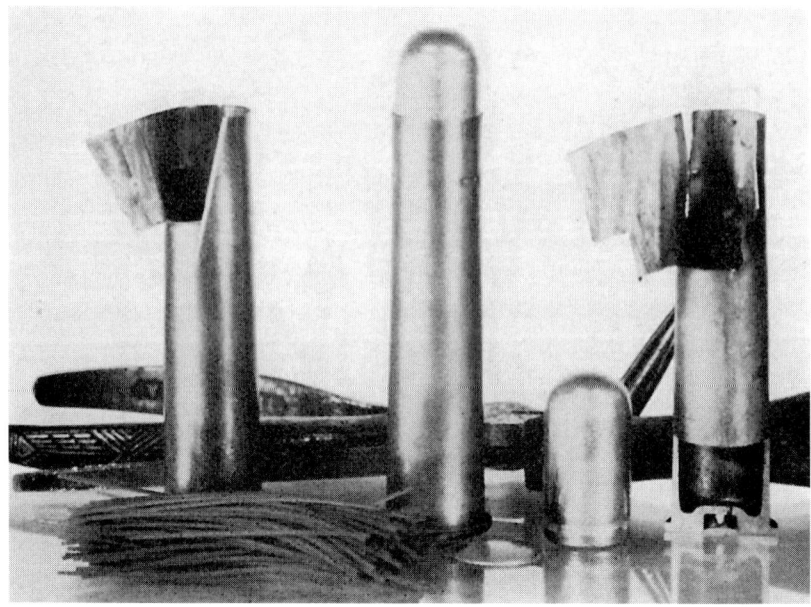

BREAKING DOWN CENTER FIRE CARTRIDGES

The simplest method of taking apart metallic ammunition is by means of a bullet puller, of which there are several different makes now on the market. However, with crimped ammunition of any type, this method scrapes off and tears the sides of the bullets so badly that they are useless for examination or comparison. Hence, most of the bullet pullers are out, although some of them work well on metal-cased bullets that are not crimped in any manner; these come out undamaged if held only friction-tight by the neck of the cartridge case.

Any modern, center fire, metallic cartridge can be opened safely by hand, if a few sensible precautions are observed. All that is needed is a three-square or knife-edge file and possibly a pair of pointed pliers. A small vise is often of assistance.

Take the file edge and cautiously file a slit lengthwise of the bullet and at a slight angle, as shown above to the left. File neatly through the neck of the cartridge, but not into the jacket or side of the bullet; with a bit of care the case mouth can be broken open and the bullet taken out unmarked. Break or twist out the bullet with the hands and dump the powder charge. The writer has opened a few thousand cartridges in this manner and has yet to have one flash or explode on him.

An equally good, and as safe, method is to clamp the neck of the cartridge in a vise with only the thickness of the case neck above the vise jaws. Take a flat, smooth-cut file and file six or seven strokes across the case neck, stopping *just* before the file breaks through into the bullet's side. Then take out the cartridge and, with the fingers, break through the film of brass remaining. The cartridge shown on the right was cut open in this manner.

The primer can be backed out with a small drift or flat-pointed punch. Once out and in the hand it should be treated with considerable respect. Do not play with it. It can explode with serious effect.

The three cartridges shown above are the British .600 Nitro, the largest of our modern big-game cartridges. The case to the right has been filed open to show the primer pocket, which takes the Berdan type of primer with the anvil as an integral part of the case. Those strips bunched together to the left show the powder charge, of British Cordite in its true "cord" form. (This term "cordite" is now much abused and misused by present-day news reporters and writers, who apply it to all types of smokeless powder). This .600 Nitro is the favorite cartridge with African elephant hunters, most of whom select it because they cannot get anything bigger.

SHOTGUN CARTRIDGE EXAMINATION

Easiest of all cartridges to open and examine are the modern shotgun shells. A pocket knife is all that is needed to do the job and the operation is perfectly safe.

The quickest and simplest method is to cut off the crimp at the mouth of the case, pour out the shot, then tear down the paper case past the wadding and dump the powder charge. The primer is not likely to be of interest but if so it can be punched out by a pin inserted through the flash hole in the base of the case.

Another method is to cut an open slot down the length of the case, as shown above in the cartridge to the right; this permits a close examination of the wadding, with the wads remaining in place. The shell so opened and shown above is a crude job, suitable only where the components are to be dumped out and later discarded.

A much cleaner job can be had by neatly cutting out this oblong opening so that no shot pellets or powder grains are lost, and then sealing it with a strip of transparent scotch tape. The specimen may then be saved and filed away for future reference and study; if this be done a complete record should also be made of all data available, including the Lot Number of that particular loading, as given on the box from which the cartridge was taken.

Notice that the wadding arrangement differs greatly in the two shells shown above, the loading to the right having almost twice the wadding as the shell on the left. All shotgun cartridges *must* be of a certain standard length; although the amounts of powder and shot that they contain may differ greatly this standard length is maintained by varying the length of the wad column between the powder and shot. In the many different loadings this wad column will invariably differ in length, furthermore some of the wads will vary in thickness and in composition. In some shotgun homicides this wadding and a few pellets of shot may be all the evidence available for investigative study.

only be broken down by experts who know fully the type, construction and component make-up of all such ammunition.

Where extensive study of ammunition is to be engaged in over a period of time, or by different individuals, it might be well to

Standard, wide mouth, glass jars, as shown above and which can be obtained in all grocery stores, are ideal for the collection of cartridges and components which every serious firearms expert will gradually accumulate during his studies and travels.

secure from a local used-car lot the entire safety-glass windshield from an old automobile and with it arrange some sort of a safety barricade on the workbench, from behind which most types of ammunition can be broken down in complete safety.

Now a few words about small arms primers. Primers *can* be dangerous, as that small assembly of fulminate and metal contains power beyond comprehension, considering its size. Our American type of primers, with their separate anvil contained in the primer cup can, if exploded in the open, develop power enough to propel that small bit of brass into the chest cavity of a person, or to destroy the sight of an eye. Treat primers with every consideration. *Do not* attempt to pry out that anvil with an awl or any pointed piece of metal of any sort. If the primer is to be taken apart, first soak it in water for a couple of hours and then separate the components by holding the cup with a pair of pliers and tipping out the anvil with the primer faced away from your body. Once wet, the priming compound (pellet) can safely be disturbed, upon drying it will again be sensitive beyond belief.

The serious investigator should make a point of picking up a few cartridges whenever possible, especially in calibers new to him, or of different makes from what he already has, or of old or odd ammunition of any description. Get three or four rounds if that many be available, but get two rounds by all means, one for breaking down and one for maintaining intact for future reference.

Such a collection should be filed away and treated as necessary equipment. Cartridge collecting is now quite a hobby and there are thousands of such collectors. Their methods of caring for the collections vary from the most elaborate of plush-lined drawer cabinets on down to a box shoved into a closet corner and into which anything in the line of ammunition is thrown.

The firearms expert can go as far as he wishes to in this matter of collection care, but at the start about as good a method as any, and one certainly simple enough, is to gather in a few score of these wide-top, low, clear-glass, refrigerator jars in which pickles, cottage cheese, peanut butter and the like are sold. One particular caliber of cartridge, components and all, can be dumped into such a jar, along with an index or reference card, where everything can readily be seen and handled. By the time you have accumulated a room full of such jars you should be quite an authority on ammunition—and can then begin to think about installing the plush-lined cabinet to hold the junk and show it off.

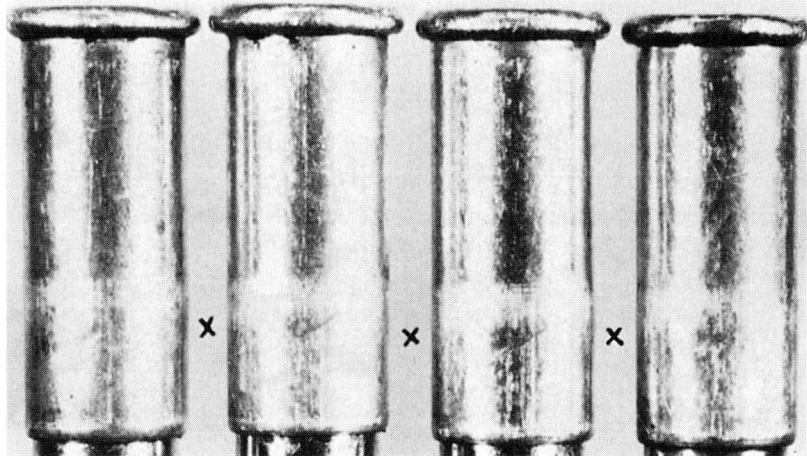

Here are shown four .22 cases fired in the same rifle—and this chap has two definite visible counts on himself without any reference checking in the comparison microscope.

These cartridges were fired in a rifle, the chamber of which has been noticeably eroded from the use of .22 short cartridges in the chamber, which was intended to be used with the long rifle cartridge. The degree of erosion shown above is such that the comparison microscope would give a positive match of any two cases fired in this style.

In addition, this chamber has become scratched on one side (at 3 o'clock, based upon position of the firing pin imprint) to the extent that it clearly marks every long rifle case fired—as indicated by X.

CHAPTER 5

MODERN FIREARMS MANUFACTURE

IN THE past, civilizations have been classified by the skill and relative advancement of their weapons makers. Firearms as made in industry today are superb; they are the finest the world has ever known. We are in the steel age. The principal components of small arms are made of steel.

The fabrication of precise objects from steel by forging and machining is a life-long study. Modern machine shop practice is detailed and intricate and, in essence, out of the realm of the firearms Expert. However, a brief description of the methods and procedures involved in firearms production may be of value. Where possible, an Expert should endeavor to visit the arms company factories in person. One of the favorite ways of reducing the value of an Expert's testimony is to cast doubt on his general knowledge of his subject. Frequently, an opposing attorney will ask questions in connection with manufacturing procedures used to make specific arms. A knowledge of these, while perhaps not really pertinent to the case at hand, increases both the Expert's confidence in himself and that of the jury in him.

Materials and Methods of Working

Although steel is the main element in all modern guns, other materials are employed to some extent. Bronze has been used for firing pins and the like. Aluminum alloy is sometimes now employed for frames of pistols and for other purposes. Stainless steel is used for gas cylinders. Stellite liners are sometimes placed in the throats of machine gun barrels. Beryllium copper is being used for floating chambers in experimental shotguns. Stocks are made of wood, plastics and other materials. Sometimes small quantities of precious metals, bone and ivory are used for sights and decorations.

106

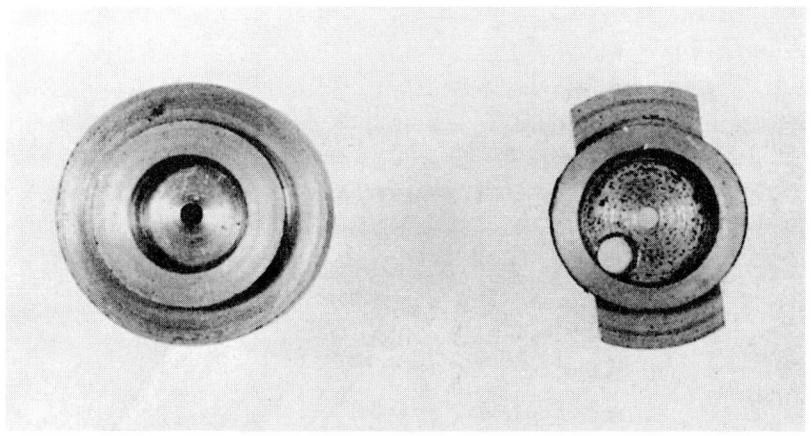

These are end views of the bolthead for the Remington Model 721 in two different stages of manufacture. The fully round piece has just come off the automatic screw machine and still has several more operations to be performed on it. The flatter piece is now completed and is ready to be brazed into the bolt sleeve in accordance with the procedure describing the manufacture of this weapon.

In this country in the commercial manufacture of firearms most steel arrives at the arms plants in the form of bar stock and special forgings. In foreign counties, particularly Spain, castings are still used, even for small parts. We formerly fabricated firearms almost exclusively by machining operations, either directly on bar stock of various sections or on forgings usually made from bar stock. These forgings are generally made by drop hammers with special top and bottom forming dies. Forgings are particularly desirable for such irregular shaped components as frames for revolvers, receivers of rifles and large parts of shotguns where machining from the solid would be wasteful and expensive.

The use of special high quality castings may increase in this country in the future. At this writing, 1954, some companies are using them for revolver frames and shotgun actions. Recently, particularly in war production, stampings from flat stock have been used. This practice continues in the cheaper grades of commercial guns, although firearms made up in this way are somewhat unattractive. A stamped floor plate and trigger guard are cheaper to produce than corresponding parts machined from forgings, they probably work equally well, but they certainly do not have the same appeal to the eye or to the hand.

Steel itself can be of many different types. These are generally known by the American Society of Mechanical Engineers' designations in which the first two digits indicate in general the alloy composition of the steel, and the second two digits the approximate carbon content. For instance, ASME 1010 is a mild steel having the usual silicon and manganese and .10% carbon. ASME 4340 on the other hand, is a chromium-molybdenum-nickel steel containing .40% carbon. Steels are a life study in themselves and are beyond the realm of the usual firearms Expert.

Heat Treatment

One of the most involved subjects in firearms manufacture is that of heat treatment of steel. Many .22 caliber rimfire barrels and the barrels for some revolvers as well as many other gun parts are not heat treated at all. In fact, the superb precision of certain .22 rimfire weapons is undoubtedly aided by the extremely easy machinability of the barrel steel. It is always a little more difficult to finish .22 caliber bores than larger bores because of the small clearance in the bore for supporting the rifling tools. It is easier to rifle precisely a soft steel barrel than a hard one. Target .22 rimfire barrels show no appreciable wear after firing thousands of rounds of their mild ammunition.

Many components of firearms are, however, heat treated either to develop greater strength in the metal or to produce a hardness for wear resistance or smoothness. Steel containing about .20 of one per cent of carbon or more, if heated above a certain temperature, depending upon its composition—usually about 1350° F.—and quenched, becomes both stronger and harder but brittle and less resistant to shock. By tempering, which is heating to a great deal lower temperature, much of the brittleness can be removed and the strength and hardness adjusted to the desired amount. By varying the chemical composition, temperatures, times and methods of quenching and tempering an astonishing variety of properties can be obtained.

Sometimes a low carbon steel part will have its surface only carburized; later heat treatment will affect this outer skin or case but not the underlying portions of the component. Some .22 automatics have a small area at the rear of the barrel heat treated to prevent the peening action of the slide from deforming the breechface. On the other hand, almost all high-powered rifles are machined from heat treated barrel blanks by the use of high-speed steel cutting tools. Sometimes carbide tools are used. Industry has for

Views, respectively, of the blank bolthead, the finished bolthead, and the finished bolt of the Remington Model 721 sporting rifle.

some time been able to machine reasonably cheap heat treated steel with a hardness approaching 40 Rockwell C—about the hardness of a good knife blade.

Even a few years ago, engineers believed that heat treating had to be done before the final machining of barrels. It was next to impossible to prevent pieces of steel of irregular sections from slightly changing shape or warping during the process. However, the Colt Company is now (spring, 1954) heat treating ASME 4050 finished barrels with only a very minor outside operation done afterwards and no operations whatever inside. Additional surface hardness has been obtained after finish machining in connection with .50 caliber machine gun barrels. One of the writers did considerable experimenting during World War II with surface hardening by a nitriding process on barrels after they were finished. We wanted to produce an internal surface more resistant to erosion than the barrel steel untreated. A thin film of hard chrome plate is a better answer to this problem. This is done now commercially on special order by various companies. These processes have as their object the lengthening of the life of a barrel. The throat area right in front of the chamber in high-powered weapons is eroded or burned out quite quickly; anything that will harden this surface and increase its resistance to heat and abrasion would, of course, be desirable. The stellite inserts used in the .50 caliber barrels were perhaps the best answer.

In low and medium-powered small arms, the necessary strength can be obtained without heat treatment. In many instances, resistance to physical damage is more of a determining factor than actual stresses set up in firing. For instance, the muzzle ends of cheap shotgun barrels are now made thicker than they need to be in order to withstand the internal pressure; they are easier to make and resist dents better this way. Target revolvers are made in general far stronger and heavier than they have to be to withstand factory loads. The ultimate in factors of safety undoubtedly is embodied in the fine target revolvers handling .22 rimfire ammunition.

However, in certain rifles, notably the old Springfleld, heat treatment played a very large part in the history of the weapon. Early in the manufacture of the Springfield the receiver was made of carbon steel case-hardened; they were sometimes hard all the way through which gave a high tensile strength accompanied, however, by little capacity to withstand external blows. A few of the carbon steel receivers did rupture. They were reasonably easy to break by blows of a hammer or iron bar. A double heat treated alloy steel was then employed which did not have the smoothness of the old carbon steel action but was considerably stronger. Finally, a nickel steel was employed which is undoubtedly strong but unpleasant to manipulate. The bolt seems to stick in the action.

Barrel Manufacture

Most rifled barrels made in this country today are drilled from the solid bar. This applies not only in small arms but also in large ordnance, save in connection with certain seamless tubing used in mortars and rocket launchers. Some shotgun barrels are made from tubing. Mass produced sub-machine gun barrels have been made from the same type of stock. However, drilling from the solid is far more common. The layman would be astonished to see the precision and speed of a deep hole drilling operation, particularly involving relatively long bores in comparison to the diameter. In general, the drill is stationary and is mounted on a long hollow spindle through which oil is forced under high pressure. The barrel blank is revolved, so that the drill tends to keep cutting along the axis of revolution and therefore comes out almost exactly centered at the other end. The points on these drills vary somewhat from factory to factory; in general, they have only one cutting edge in the smaller sizes but two for artillery. The barrels, once drilled, usually have several external operations done on them, positioning on the rough-drilled bores. They are then ready for reaming. Several different methods are used for this reaming, depending upon the factory and the size and length of the bore. Centerfire pistol barrels are generally brought up to size internally by use of the type of reamer known as hand reamers. In the Smith & Wesson factory this operation is actually done by hand. In the larger bores the packed-bit boring principle is used but still usually called reaming. There are at least two and possibly three or four reaming operations in order to produce an extremely smooth bore. Honing is

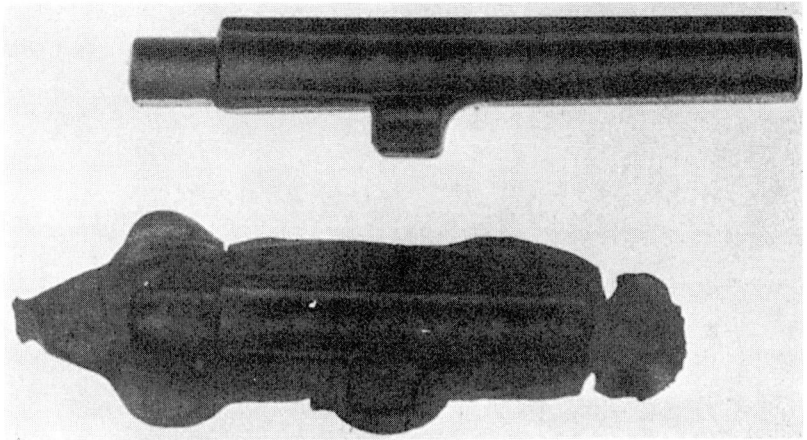

Two barrel forgings, one before and one after the flashing has been sheared off. This process is used in the making of Smith & Wesson barrels and saves very considerably in manufacturing time over producing the same part from solid stock.

sometimes employed to finish the bore exactly to size, particularly in artillery. In small arms a fine reamed surface is usually considered sufficient.

When the bore is finished, the chamber is frequently completed, although it is also done after rifling. In general, this chambering may be done on automatic machinery, a turret lathe, or by the use of reamers by hand. A turret lathe by all odds is the most usual, commercially. It is possible to contour grind chambers; however, boring or reaming them is more desirable at the present time in small arms. Most bolt action rifle barrels are rough chambered before assembly to the receiver, and finished chambered to an individual bolt.

Rifling is the most important single process in barrel manufacture from the standpoint of the identification expert. The scrape and hook methods of rifling were used for many years and were continued in most plants to World War II. In these methods each

Component parts of a Smith & Wesson Revolver—those four parts at top are drop forgings before being cleaned off.

groove was cut individually, or two grooves diametrically opposite each other were cut at the same time. The rifling cutter in one instance had a hooked cross section; in the other, the cross section was that of a pyramid. This negative rake on the scrape cutter produced a smooth finish but was slow in operation. The Springfield Armory preferred the scrape method; most commercial concerns used hook cutters.

At the beginning of World War II, artillery barrels were being rifled on broaching machines; a series of broaches were pushed through the bores with high-pressure cutting oil clearing the chips out ahead of these broaches. This arrangement was far quicker and did a more uniform job than the old rifling head used for so many years. Further, it transferred the skill necessary for the operation from the man handling the machine into the tool room to the fellow who was sharpening the broaches. If the broaches—a series of from 20 to 50 for any individual artillery barrel—were kept sharp and in order, this operation was almost foolproof.

During World War II, with the vast quantity of small arms required, a great deal of experimenting was done to speed up the rifling operation. For instance, the M3 sub-machine gun had a mild steel tube swedged into a hardened tool steel or carbide

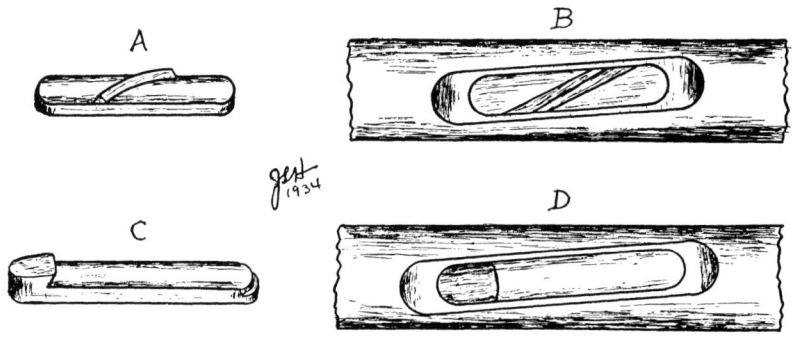

RIFLING CUTTERS

These are the tools which cut the grooves into the bore of the barrel of a rifled firearm. These particular types are not used today to any great extent in the mass production of service firearms, but are still used by private gunmakers and skilled amateurs for the making of extremely accurate target and hunting rifles and pistols.

A. Blade of a "scrape" rifling cutter.

B. A scrape cutter in place in the rifling head at the end of the rifling bar. There is another cutter exactly like this one on the opposite side of the rifling head, as with the "scrape" system, two opposite grooves are rifled at the same time. The cutters are arranged to operate in both directions and take a shaving off the metal of the bore on both the "push" and the "pull" strokes of the rifling bar. The scrape system is used only when an even number of grooves are being cut in the bore.

C. Blade of a "hook" cutter.

D. Hook cutter in place in the rifling head. There is no cutter opposite this one, because with the "hook" system but one groove can be rifled at a time. The cutter only works in one direction, as it is being pulled through the bore. This is the system which must be used when an odd number of grooves is being used.

Both these systems or rifling have today been supplanted by more modern and rapid systems of rifling where a greater output is necessary.

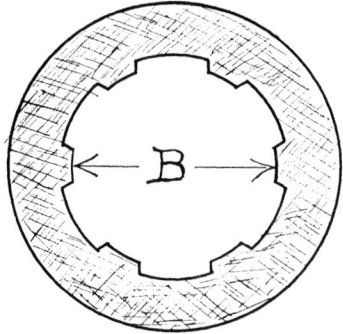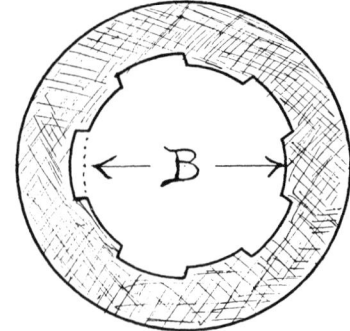

FIREARM BORE DESIGNS

Diagrammatic sketches, to show how the barrels of firearms appear in cross section. These are revolver bores; the left sketch shows the Colt type of rifling, with six grooves, each approximately twice as wide as the lands. The right shows Smith & Wesson type, with five grooves, each approximately the same width as the lands.

The diameter of the bore, as shown by dimension "B" in each sketch, would be the caliber of the barrel.

mandrel between heavy rollers working on the centerless grinder principle. These swedged barrels were produced quickly and with a minimum of skill save, of course, in the making of the mandrels themselves. It might be of interest to the Expert to know that test bullets from these barrels are rather easily identified with crime bullets from the same weapon. Perhaps the swedging process removes most individual marks in the bores; however, the pushing out of the mandrels seems to introduce plenty of new ones.

This form of swedge rifling is possible only with soft steel barrels. However, an adaptation of the broaching system used in large gun tubes has tremendously reduced the operational time for the rifling of small arms barrels made from heat-treated stock. Several broaches are mounted at intervals on a central stem. The entire assembly— actually it is all made from one piece of steel—is pulled through a barrel which is properly revolved during the passage of the gang of broaches through it. Cutting oil is supplied under high pressure. In general, one pass of a series of about 25 broaches is sufficient to complete the rifling of a pistol bore. Barrels made in this manner have quite uniform grooves. However, they give no trouble in identifying bullets with individual barrels. Actually, most experienced identification men look mainly at the striae in the grooves of the bullets which are, of course, made by the lands of the barrel. Regardless of how smooth the lands are made when the bore is finished, every rifling process will mark them considerably. This is particularly true of the gang broaching operation even though no metal

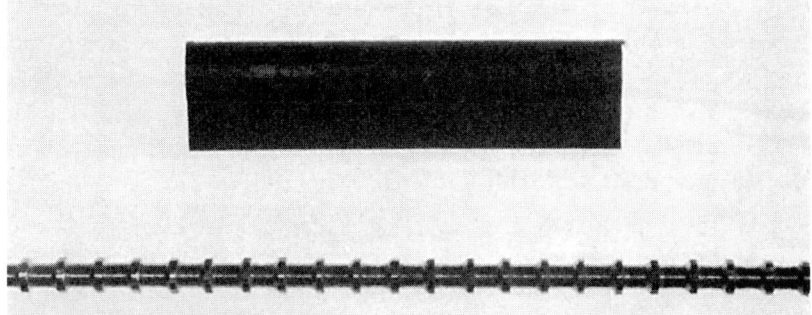

The mid-section of a broach used to rifle .45 ACP barrels, and a piece of the keyhole section stock from which these barrels are made.

is taken from the lands. Examination under magnification will show very definite scoring of the lands by the riding surfaces of the broaches or by chips.

In the broaching of high-powered rifle barrels, one pass with each of two gang broaches is generally required to finish the bores complete. Proper indexing is necessary to have the second gang broach follow in the path of the first with, of course, the same lead or pitch. This method of rifling both pistols and rifles is much faster than the old single or double groove cutters. In a large factory today there will be only one or two single spindle rifle broaching machines; a few years ago, there were banks of the old type machines which had in total far less potential output.

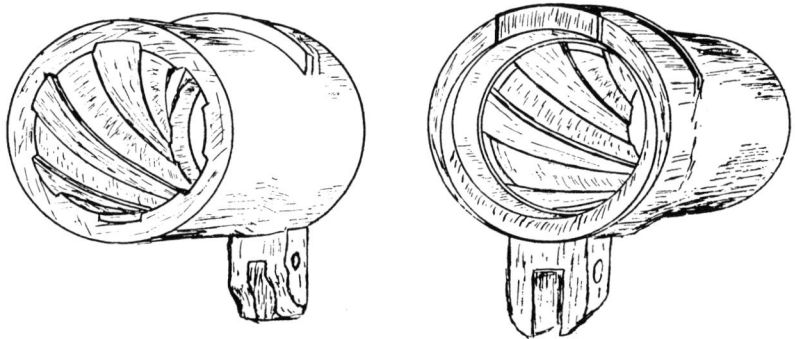

LOOKING DOWN THE BARREL

Sketched to show how a rifled automatic pistol barrel looks when viewed through either end. The left represents a view into the muzzle end of a .45 automatic pistol barrel and shows an exaggerated appearance of the rifling. The right sketch shows the same barrel from the breech end—note the chamber shoulder, beveled ramp and leads to each land.

Notice that the left-hand rifling appears to slope in the same direction whether it is viewed from the muzzle or breech ends—don't allow yourself to be fooled on this fact at any time.

A third method of rifling is presently in use in this country. A tungsten carbide "button," with the rifling contour negatively impressed on it, is forced through the finished reamed or burnished barrel. This method of rifling was introduced into production early in World War II both here and abroad. An appendix attached to this chapter deals with this in detail.

Breech Face Manufacture

Next to the finishing of the inside of the barrel, the most important portion of a gun for identification purposes is the section of the breech which supports the cartridge case at the time of discharge. Almost every machine tool has at one time or another been used to produce this surface on one or more models of small arms. However, in general, in bolt action rifles and similar weapons, some form of end mill is employed which will always leave characteristic marks which vary slightly from weapon to weapon of the same make and model. Broaching of breech faces is also done. As dis-

A PRONOUNCED BOLT FACE SIGNATURE

This is the bolt face of an Enfield Model 1914 rifle, originally chambered for the .303 British cartridge—later rebuilt to shoot one of the hot .22 wildcat cartridges based upon the .30/40 Krag cartridge case.

Note the deep, curved scorings remaining on this bolt face, caused by finishing it off with some form of revolving tool, such as an end mill.

To the right is shown the base of a cartridge case fired in this rifle, the primer of which has clearly embossed itself into those tool markings left upon this bolt face. This particular cartridge develops a breech pressure of around 50,000 pounds and every cartridge fired in this rifle has markings clear enough to be plainly visible to the eye.

The firing pin hole through this bolt face has a clean, square edge, resulting is a sharp firing pin imprint into the primer, with no protruding rim around it. Under the microscope this imprint would show distinctive firing pin markings not visible to the eye.

A 'HOME MADE' POSITIVE COMPARISON

This presents an accurate and positive comparison between the bolt face and fired cartridge shown on opposite page. It was made with an 8 x 10 view camera having a 30″ bellows extension and fitted with an Ektar 4.7 lens of 127mm focal length. A lens of shorter focal length would have given even more magnification than this lens gave—and of course, the magnification shown above is much less than that actually obtained on the larger 8 x 10 print.

Comparisons of this nature can be made by any investigator possessing a somewhat similar outfit or having access to the average professional photographic studio.

Such a view camera is fitted with some suitable short focus lens and the subject focused *sharply* to the full capacity of the film. *Cut film must be used.* Both bolt face and cartridge case base must be photographed together *in accurate alignment* and at right angles to the lens. Focus sharply and then stop down the lens to its limit, so as to get as much depth of focus as possible. The lighting should come in from above, or from one side, and both bolt face and case be positioned so that all illumination be thrown *across* both tool and primer markings.

Develop for contrast. Two prints are made from the same negative, the bolt face print being made in the usual manner with emulsion side of the negative against the paper. Then the cartridge base section of the negative is printed, but *with the negative turned over,* printing from the back—this reverses the image and permits a direct line-to-line comparison by trimming away the primer print and imposing it over the bolt face photograph. Try different grades of printing paper until suitable contrast is obtained.

In making the prints, a slight enlargement may possibly be used to advantage, but not too great an increase should be attempted. An increase of about one diameter is generally all that can be gained with any advantage. Get as much magnification on the negative as is possible; get it as sharp as possible; then enlarge very little, if any. Get just enough magnification to convince that the comparison is a positive one or otherwise. Then, if further proof is needed, go to the comparison microscope with both evidence and test cases.

In the above illustration the primer area of the cartridge base section of the print has been trimmed away sufficiently to permit an accurate register of all the continuous markings necessary to prove a positive match.

Component parts of the Colt Official Police Revolver.

cussed in the appendix, Colt broaches the breech surface of the .45 caliber automatic pistol and then finishes a portion of this surface with a rotary file, by hand. At Smith & Wesson the cylinder opening in the frame of revolvers is broached but a recoil plate is staked in place later and hand finished. In other words, in both these weapons the last breech face marks are from hand operations, so that under magnification these surfaces are quite different from one weapon to another of the some model.

The breech faces that give the most trouble are those of cheap shotguns which are made in enormous quantities from soft steel

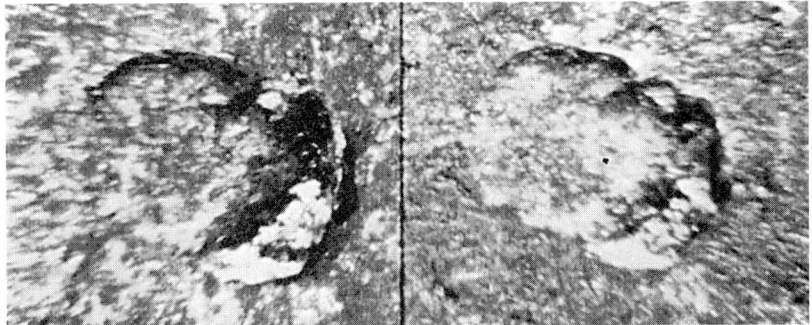

An example of a distinctive breech signature—the ejector marking on the base of a fired .45 case from an M-3 Submachine Gun.

by standard cutters without any hand finishing at all. The relatively low pressure of the ammunition is also a factor. However, the abuse or lack of care given by the average owner of these cheap shotguns, together with the rapid wear of the soft steel used, invariably imparts highly individual nicks, scratchings, pittings and

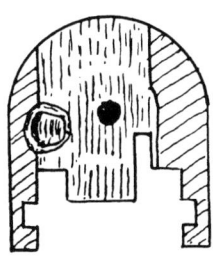

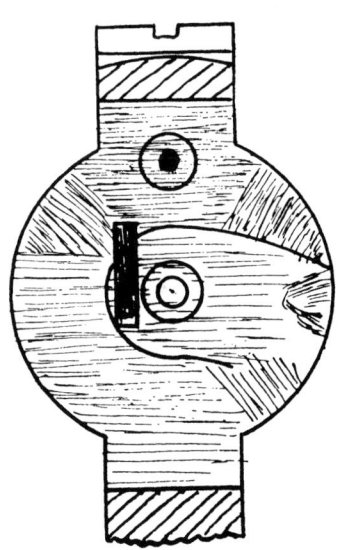

VARIATIONS IN TOOL MARKINGS

Showing the direction of file-marks or tool marks on revolvers and automatic pistols.

Left, the breech face of a .45 caliber Government Model Automatic. Note that the tool marks are vertical, as the stroke of the tool had to be up and down through the ejection opening at the top of the slide. Cartridges fired in such a gun will have vertical striations impressed in the primer.

Right, the breech face of a .45 Caliber Revolver, Model of 1917. Tool-marks or file marks cannot run up and down, as they do in the pistol, as the top and bottom of the frame prevent an up and down motion. The breech-face was finished by filing straight across, and the fired cartridge will have horizontal marks impressed on the primer.

Of all types of firearms, the single barrelled shotgun is undoubtedly the most per-
sistent producer of individual markings on the fired cases. This is largely due to the
cheaper metals used and the crude and hasty methods of assembly, but in many cases
can be due to repairs made by the owner or by some rural jackleg who takes on gun
repairing as a side line to his regular garage work.

The manufacturer is not entirely to blame, however. Generally the care, or utter
lack of it, which these shotguns receive from their owners is more likely to account
for the manners in which they mark their fired cases. They are seldom cleaned and
never oiled or greased. In the rural sections such guns may stand in a corner or in
back of a door for months at a time, gathering dust and rust and permitting particles
of almost anything to fall down the barrel and lodge against the breech face, where
it is flattened and pounded-in by the back thrust of the fired cartridge.

However, the most drastic and permanent markings are those put vertically into
the standing breech and across the profile of the firing pin by the action of the extractor as it drags across when the gun is opened and closed. Present day single barrel
shotguns are fitted with a simple type of extractor actuated by a stiff, always loaded
spring which forces the extractor back against the breechface at all times and wipes
across the firing pin, which latter is not automatically retracted and often is so rusted
that it sticks out continually unless the gun is loaded. This wiping action of the extractor soon scores the breechface and affects the profile of the firing pin, either
blunting its nose or putting more of a point to it—which facts explain the differences
in the various firing pin and primer imprints shown on the 12 shell bases appearing
on the opposite page.

The above illustration shows the breechfaces and firing pin profiles of three typical,
much-used single barrel shotguns—including the rust and other debris which their
extractors were unable to keep brushed off. The two breeches at either end show a
pronounced difference in firing pin profiles. The breech shown in the middle caused
its owner quite a bit of trouble—the center of the standing breech blew out but the
local garageman promptly welded in a new 'recoil plate,' faced it off quickly (as can
readily be seen) and for further insurance pinned-in the weld from the left side of the
action. The new firing pin was made from a wire nail and the gun has been used
in this condition for the past ten years.

the like that are readily recognizeable by any skilled investigator.

In digression, it should be remarked that many weapons used in
crime have been so badly abused that nicks, scratches, rust pits,
and the like are more important, both in the bores or on the breech
faces, than the original machining marks. This is less likely to be
the case where the breech face has been hardened, as in the old
carbon steel Springfield bolts.

Sometimes a breech face finish is so characteristic of a weapon that
the marks on the base of a fired cartridge case will identify the weapon
used even if no suspect arm has been picked up. The M3 sub-

SHOTGUN BREECHFACE SIGNATURES

These show the bases of an even dozen shotgun cases which were fired in 12 different single barrel shotguns. Each of the above fired cases clearly showed, to the eye alone, distinctive markings differing from all the others and each fired case could probably be matched by eye. Negative matches need be matched no further.

In preparation for the above illustration the writer went around in his immediate neighborhood, borrowed 15 single barrel shotguns and fired two shells in each gun.

Out of these 15 shotguns, the fired cases from 12 showed similarity enough in their firing pin imprint and primer set-back to enable any experienced investigator to reach a postive check and opinion with his eyes alone. In most instances the comparison was far more obvious than can be seen in this illustration, as the camera has no means shown the many shadings, scrapings, abrasions and scoring caused by the up-and-down passage of the extractor across the face of the standing breech—which markings are individual to each gun and which are imprinted or embossed to a more-or-less degree upon the base of every cartridge case fired in that gun.

Any skilled investigator should be able to make a rapid and accurate preliminary check, by eye, between evidence and test cases marked such as the above, and then verify his findings with the comparison microscope, if possible. In many instances, an accurate and positive check can be made by means of photomicrographs, made with a large camera, sharply focused to more magnification and using full negative capacity to take in all of the evidence and test cases.

machine gun frequently had a sand blasted breech face which marked both primers and the heads of cartridges fired in it in an extremely unusual manner.

The portion of firearms that set off cartridges, although not integral with breech faces proper save in certain fixed firing pin full automatics firing from an open bolt, leave their impressions on primers as part of the general breech face pattern. There are several different types of firing pins in use today. In crime weapons perhaps

Here are five of the fired cases appearing in the preceding illustrations but showing also identification faults on the sides of the brass base or in the rim edges.

On the left is a case fired in a shotgun having a bulged chamber, a faulty condition which is not any too well shown in the illustration although it is clearly apparent to the eye. The fault appears as shallow bulges of greater or less extent, which bulges bear a fixed relation or position to the other base markings.

The next two are cases fired in guns the actions of which have become very 'loose'; due to long usage; a worn hinge-pin; or excessive set-back of the standing breech. This fault was very pronounced in the case of the middle cartridge shown, resulting in what is known as a 'split rim,' which is a dangerous condition.

The two cases shown at the right were fired in guns having 'expanded chambers,' due to the metal in the barrels gradually stretching and thus enlarging the chambers, which is also an unsafe condition, especially with modern high-velocity ammunition.

All the above examples were fired in cheap, single barrel shotguns, and there are many single barrel guns in similarly poor condition. An old and worn double gun may be in similar condition. The faults shown here are constant and will show on almost every shell fired in the gun in question. They, combined with the markings on the base of the case, will be of much assistance to the experienced investigator.

the most common are the firing pins of automatic pistols and the noses of revolver hammers.

Firing pins, not only from automatic pistols but also for other weapons using them, are now usually axial for centerfire weapons

MAUSER H S C PISTOL FIRED CASES

The two cases were fired in a 7.65mm (.32 A. C. P.) Mauser H S C automatic pistol, now a rather common and very popular pistol here in the U. S. A.

The action of this pistol is a violent one and its breech face marks up the cases upon insertion into and ejection from the chamber of the pistol. Note the battered areas of the bases of these two cases shown above—which actually were much more noticeable upon examination than the photograph showed.

or move parallel to the axis of the bore in rimfire arms. Further, they are usually simple in construction as opposed to some of the extremely complicated early designs which were necessary to utilize the motion and power of the old outside lock mechanisms. Firing pins today are usually made upon automatic screw machines and heat-treated. No special effort is ever made to eliminate completely the tool marks from these surfaces since they are far too small to interfere with the action of the primer. A given firing pin is hard to identify with a primer; however, two primers fired by the same firing pin are usually extremely similar, and easily identified.

Certain revolvers have hammer noses forged integral with the hammer itself, although the more usual procedure is for an inserted

DOUBLE DERRINGER IMPRINTS

A double barrelled pistol, like a double barrelled shotgun, will imprint its fired primers differently. Above are shown two cases from .38 S. & W. cartridges fired in the different barrels of a Great Western Double Derringer.

To the left is the case fired in the upper barrel, which shows a normal firing pin imprint. To the right is the case fired in the lower barrel; note the distinctive imprint from the flattened tip of its firing pin. This lower firing pin became flattened simply through sloppy finishing during manufacture.

non-rigid nosepiece to be employed. The flexible nosepiece produces an identification in the primer similar to that of an axial firing pin, although differing from it in certain respects. Many hammerless weapons, particularly shotguns, employ strikers which are essentially inside hammers; these generally have integral noses. The integral hammer nose produces a characteristic indentation with striae formed in part by generation rather than merely negatively impressed as is usual. Any irregularities in either of these noses will be reproduced in each and every primer so fired. A few revolvers, mostly of foreign origin, have separate firing pins mounted in the frame and actuated by actual blows of flat, exposed hammers.

Frames, Actions and Large Parts

The major parts of firearms are usually made from forgings; these are so extremely diverse in shape and design that even to give general principles is difficult. However, these will be worked on by several types of machines in special fixtures according to a fixed procedure involving as much as 50 or even more operations. Revolver and automatic pistol frames, bolt action rifle receivers and shotgun actions are in this category. The ingenuity of the process engineers in some firms is astonishing. These extremely complicated pieces of hard steel are produced interchangeably, with no great skill required of the individual operators, on almost semi-automatic machines. Not only machining but also handling and inspection are considered and worked out.

Since these large components show externally, their appearance is important commercially. It is important psychologically to the owner. Many different machines are used to produce different types of contours. The final finish, however, is usually produced by hand on a buffing wheel. If a smooth machined surface has first been obtained a skillful female operator can change it, in only a few seconds, into the glossy satin finish required for proper blueing.

Smaller Components

Many of the production headaches of past years have been eliminated in modern drafting rooms; small arms design has been simplified to an astonishing extent. Anyone who thinks modern weapons are hard to make should disassemble and contemplate the mass of miscellaneous parts, each complicated in itself, which form, for instance, the early British Soper military rifle or with World War I SMLE. Even a big Mauser pistol and the Thompson sub-machine gun are too complicated for efficient modern production.

Every individual component of a military or commercial firearm has been carefully considered for modification in design, if by so doing time and money can be saved. Wherever possible standard commercially available products such as pins, screws, and springs are used. Any pieces that can be adapted to automatic screw machine production are made on them. Stampings and welded assemblies are used only after due consideration has been given to their inherent unattractiveness.

There will always remain, however, a good many special and reasonably complicated small parts in every firearm. These are produced by routines similar to but usually not so lengthy as those used for the larger components. Various machines with special fixtures, sometimes holding dozens of parts at a time, are used. Even the inside parts are well finished and sometimes hardened and blued.

Final Assembly and Testing

Years ago firearms were slowly and painstakingly assembled from groups of parts, by skilled men. A good deal of fitting was necessary. To a larger extent than some people realize this is still the case with some weapons. A fine modern revolver has usually received the skilled attention of a master fitter and had a few hand operations performed on it, such as spinning in the recoil plate and fitting the cylinder. At the other extreme are the cheap shotguns which are slapped together with little or no fitting; their parts are loose enough to slip into place without it.

Every firearm made in America is test fired at the factory with proof loads which are considerably more powerful than any factory ammunition. In a cheap firearm, operation and safety only are tested. Some of the finer weapons are fired many times to check the sight alignment, the smoothness of operation, and even the patterns of shot fired from shotguns.

Firearms Factories

A few words might be in order in connection with the factories producing firearms in this country. There are so relatively few of them now. In 1900 there were perhaps twenty times as many manufacturers as today; now almost all the small producers have been eliminated.

There are only three large manufacturers active in this country producing pistols of above .22 rimfire caliber at the present time; these are, of course, the Colt's Manufacturing Company in Hartford, Connecticut, Smith & Wesson in Springfield, Massachusetts, and Harrington & Richardson in Worcester, Massachusetts.

A number of other concerns produce .22 light rimfire revolvers, notably Iver Johnson, Harrington & Richardson, and Sturm, Ruger & Company. Automatic pistols of .22 rimfire caliber are produced by High Standard Manufacturing Corporation and Ruger, as well as Colt.

Centerfire rifles of medium and high power are produced by Remington, Winchester, Marlin, and Savage-Stevens for sale under their own names. Of these, only the Winchester Company can be said to produce a wide range of models and calibers. Remington is a large producer but puts out only a relatively few models. The High Standard Company barrels a great many imported FN actions and completes these rifles, which are sold nationally under another trademark.

Most of the above firms, with the exception of Colt and Smith & Wesson, produce shotguns of varying quality and type. These same firms produce tens of thousands of .22 rimfire rifles each year of several types, from singleshot bolt actions for boys through the beautiful heavy barreled target weapons. The Mossberg .22 rifles are also well known. There are in addition a few other makers of shot-guns and .22 rifles. All commercial factories of any size producing standard model commercial firearms are located in New York and New England.

There are, however, literally hundreds of small gun shops through-out the country who make up special weapons to order, including very occasionally even the fabrication of special actions. Some of these like Weatherby, Johnson Automatics, and Dunlap have developed semi-standard models. However, they usually take an action already produced either in this country or in Europe, mount a special barrel in it, chamber the barrel, make small action changes, and finally stock and sight the weapon to the customer's specifications. Most of these shops do only rifle work, although some also work on pistols and shotguns. There are literally dozens of barrel makers through-out the country producing highly accurate, carefully made rifle bar-rels. These small barrel makers have a flexibility in regard to bore size, twist, and form of rifling impossible in a large factory. Further, their equipment can be set up to take most any size barrel blank.

Custom rifles are seldom used in crime. These special weapons, produced after much thought by the purchaser and usually exhibit-ing great skill by the gunsmith who made them, receive years of loving care from their owners. One of these gunsmiths or even a skilled amateur could make a weapon not conforming in any way to regular firearms. However, the very rarity of it would greatly aid in the solution of any crime in which it was used.

Actual Production Routines

In the course of preparing this volume, the authors have been courteously allowed to visit the firearms manufacturing plants of the country. Whereas, we do not want to write a treatise on production techniques, we do feel that the four special appendices to this chapter may be of value to the Expert in indicating along general lines that which is done in fabricating firearms today.

APPENDIX 1

Procedure for Manufacture of Commercial Model .45 Caliber Automatic Pistol Barrel by Colt

Hot rolled steel bars approximately 12′ long are purchased commercially; they are of an irregular, or "key-hole," section. This steel is ASME 4050. It is a chrome molybdenum nickel steel containing approximately one-half of 1% carbon. It is received in an annealed condition with stresses relieved after rolling.

The steel bars are sawed into blanks of an appropriate length slightly longer than the finished barrel. Most of the "key-hole" section is milled away leaving just enough to eventually form the link lug at the breech of the finished barrel. These blanks are then deep-hole drilled with the usual single lip gun drill made of high-speed steel. A horizontal two spindle machine is used with the drills stationary and the work revolving at high speed. Each drill is hollow and carries cutting oil under a pressure of 400 to 500 p.s.i. to the tip. The speeds and feeds used are those common with high-speed drills. A hole is drilled in the center of the bar of .427″—.430″ diameter. The finish obtained in this operation is good.

These blanks are now turned externally on profile lathes to a few thousandths above finished diameter and reamed. This reaming operation is unusual in machine shop practice. The equipment used requires a good deal of hand labor. The machines are small, simple, and could be relics of the Nineteenth Century. The barrels are supported in a kind of floating jig and are pulled onto straight flute hand reamers by a hand adjusted spring. They are successively reamed in this manner three

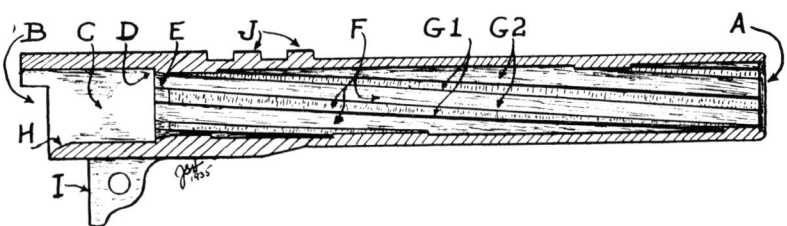

AUTOMATIC PISTOL BARREL CONSTRUCTION

A cross section of the barrel of the .45 Automatic Pistol. The entire barrel and chamber are one integral part, as is the case with practically all automatic pistols. A is the muzzle, B—breech, C—chamber, D—shoulder (against which the mouth of the case abuts and is positioned for firing), E—lead (pronounced leed) or throat, or bullet seat, or forcing cone (into which the bullet rests before firing), F—bore, H—ramp, I—barrel lug, and J—locking lugs. G1 designates the lands and G2 the grooves of the bore.

CARTRIDGE CHAMBERING IN AN AUTOMATIC PISTOL

Here is the manner in which the .45 Automatic pistol cartridge chambers in the barrel. Note that bullet is seated up into the throat of the bore, bearing right against the rifling before the cartridge is fired. The twist of this rifling is left handed.

All automatic pistol cartridges are so seated before being fired. But not all chamber exactly as the above cartridge, which is positioned into place by the square edge of the case rim (this cartridge never being crimped but its bullet held friction tight); many are loaded into a semi-rimless case and are positioned by the protruding rim. Others load into bottle neck, rimless cases and position on the shoulder of the case.

times with the removal of a total of about .015″ stock. The amount of metal removed declines and the finish improves with each operation.

These barrels are then finally reamed in a delicate hand operation with the removal of less than .001″ of metal on a square burnishing reamer with wood backing. The bore is now completely finished.

The barrels receive a milling operation for the removal of metal around the link slot and for the cutting of the locking lug slots in the top of the barrel. The breech irregularities are finished in minor operations. Chambering is done on a small turret lathe with floating tools, largely with hand feeding to positive stops.

The barrels are semi-finished ground externally and then rifled. The rifling process is done in a single pass with a pull broach containing 25 individual buttons made integral on a central stem. The broaching machine does a fast, accurate job. The barrel is revolved to produce a definite pitch by means of gearing; the broaching head and the work are always in a positive relationship to each other. The broach is pulled through a barrel, the barrel removed, and the broach returned to the ready position. There are two vertical spindles; the machine can rifle 900 barrels in eight hours.

There are now milling operations to form the clevis in the link lug and various finishing operations.

The complete barrel is heat-treated by immersion to obtain a hardness of Rockwell C 40-44. The barrel then returns and is ground externally on a centerless grinder to finish diameter. After proof firing, polishing, and blueing, it is ready for final assembly.

Other important factors of Colt manufacture are as follows:

First, finishing of the breech of the Government .45 automatic slide is done by a pull broach in a fixture set at the appropriate angle. However, a certain amount of hand finishing is necessary to remove burrs and a shoulder that cannot be broached; this is done with a rotary file. This process is certain to leave unmistakable marks of individuality on every single breech face.

Second, Colt revolvers have the recoil plate installed in them before the breech face is finished. The aluminum frame revolvers follow a slightly different procedure. The breech faces in both of these may, at final assembly, be slightly refinished by hand.

Third, all Colt hand gun barrels are now broach rifled. The old rifling machines with hook rifling cutters were superseded about 1948. The scrape method of rifling was not used for several years before that.

Fourth, all barrels for Colt commercial firearms are lead lapped with about fourteen strokes per barrel. The lead lap is cast in each individual barrel and mechanically lapped in a floating horizontal position.

APPENDIX 2

Procedure for Manufacture of the Smith & Wesson .357 Magnum Revolvers

We will discuss in some detail the production of these weapons, particularly in regard to barrel, cylinder and recoil plate.

The Smith & Wesson Company in its new plant has introduced a good many efficient modern procedures in regard to their product not commonly followed by some other concerns engaged in firearms manufacture. This is particularly true in the initial drop-forging operation in connection with barrels, frames and some minor parts. These pieces are machined from forgings, which are produced in the plant. Steel is received in suitable lengths and sizes for forming into these forgings; the bar-stock is heated in ovens immediately adjacent to the drop forges themselves.

Oil furnaces heat bar-stock to 1800°-2000° Fahrenheit, depending upon the steel. It is then grasped with tongs by an operator and placed between suitable dies. A manually operated trip mechanism allows the individual operator to forge the bar-stock into proper shape in about six strokes of the drop hammer. The last stroke cuts off the forging proper and allows the remainder of the bar-stock to be returned to the furnace for reheating.

The forgings fall onto the sand floor in a pile and cool there. They are then gathered into heavy wire baskets and passed through stress relieving furnaces. The excess metal, in the form of flashing, is removed on punch presses. The forgings and pieces of bar-stock for cylinders are then heat treated.

We will now follow briefly the barrels through their various operations:

1—All forgings are placed in fixtures on engine lathes and the breech ends faced and turned.

2—All forgings are deep-hole drilled in horizontal spindle machines with the usual type of gun drill. For the magnum barrels, these drills have carbide tips; they are of the usual single lip type and are hollow, with cutting oil at a pressure of approximately 600 pounds fed to the tip.

3—The barrels are rough reamed by hand on vertical spindle machines.

The operators place a barrel in what amounts to a glorified tap wrench fixture and then, with their forearms supported in leather stirrups, force the barrels onto the vertically mounted revolving reamers. Four reamers remove among them about .009″ of metal from the bores.

4—Positioning on the semi-finished bore, the outside of the barrel is suitably profile milled. These formed milling cutters do a first-class job over all. The fitting of the sights and polishing only need to be done later.

5—The barrels are now finish reamed internally with only a small amount of stock removed. This last reamer really burnishes rather than cuts.

6—The barrels are broach rifled on a horizontal spindle machine with as many as 30 buttons to each individual gang broach. One pass of the broach is sufficient to complete the rifling. The barrels are held in place one at a time and the tool shank passed through the barrel by hand and secured; it is then pulled through the bore with the proper lead.

7—The breech threading is cut by milling which insures absolute uniformity of threads so that each barrel will screw up tight into the frame with the front sight and rib in the proper position.

8—All bores are lapped with lead laps cast in each individual barrel on a vertical spindle lapping head.

Similar operations are done to produce the magnum cylinders from the same heat-treated stock.

In addition to the barrels and cylinders, extremely important operations are done in connection with the recoil plate installed in the frame. These are done in the area which marks each cartridge case and primer fired in the individual weapon. The major opening in the frame is broached to exact dimensions to receive the cylinder. However, the small recoil plate is not installed until assembly of the revolver. This recoil plate is forced into the frame on a small arbor press and staked in place. The recoil plate protrudes a couple of thousandths above the surface of the breech face. Each is finished flush by hand filing. Although no appreciable amount of stock is removed from the breech, file marks remain on both the recoil plate and breech face. This method of finishing means that there will always be individuality in primer and cartridge case base markings in different weapons.

APPENDIX 3

Button Reaming and Rifling of Firearms at High Standard Manufacturing Corporation

H. H. Sefried of the Research and Development Department of the company as well as others has been of great help in, to some extent, unraveling the conflicting stories about this process and its evolution. It is widely used, yet a great deal about it is secret.

The basic idea of the button or swage process is to press down the bore of a firearm, or other deep hole, something larger in diameter than the inside of the bore. It is possible in this manner to change radically the internal surface. This basic idea is not new. Actually, it has been used for years in cases such as burnishing of holes in soft metals by pushing an oversized ball bearing through them. This is being widely applied today with at least one company supplying spherical ball bearings and engineering details for any job. One or two passes of such a ball—now called Ballizing—of appropriate interference was known to tremendously improve the finish of certain manufactured products when one of the authors was a young machinist.

This same process can be used with steel, sometimes of great tensile strength. The object passed through the bore is frequently not a ball bearing, but a so-called button. In firearms this resembles a torpedo, of slightly more than bore diameter in the center. The contacting surfaces which swage the bore comprise a leading angle, a flat burnishing surface, and a trailing angle. The leading angle is about 3° to 5° included. The flat varies from about .025″ to .150″ in width. The trailing angle is not important. Since this type of button is round in cross-section at any point, a cylindrical bore is produced with much improved finish. Actually, this operation is equivalent to a burnish ream but much faster and productive of a better finish. If a fairly smooth reamed barrel of .2165″ in diameter is operated on by a cylindrical button .2199″ in diameter a bore of about .2185″ will result with a surface in the neighborhood of 15 to 25 micro-inches—very smooth indeed! The bore has permanently opened up .002″ in diameter (but closed in behind the button almost .0015″ in diameter).

Originally, however, button reaming was not experimented with nearly so much as button rifling. If the interference area of the button, instead of being circular, is formed with a negative impression of the desired completed condition of the bore, a rifled barrel will result. In other words, negative grooves cut in the button will press positive grooves in the bore and leave a portion of the old bore as lands. A rifling button with negative grooves .225″ in diameter will leave grooves about .223″ in diameter. If the button is properly made and applied, the rifling in the finished barrel will be perfect—superior to that produced in any other way.

Just when and by whom this buttoning process was applied to firearms is not clear. The Remington Company patented a process for producing high speed steel buttons in 1939. Inland Steel is thought to be the first producer in the United States to use the process successfully. Certainly .30 caliber barrels were so produced during World War II, probably by pushing a free button through the bores mechanically. Both hydraulic and powder pressure were used experimentally. The Germans also used buttons to rifle small arms in World War II. The fundamental idea is undoubtedly not patentable.

To appreciate the buttoning method of finishing rifle barrels, it is necessary to consider the standard rifling and reaming methods and the theories which have established present designs. The final finish of the bore, as obtained with fluted reamers, will show cutter marks at approximately right angles to the bore. These minute serrations, which remain on the lands of conventionally rifled bores, are at right angles to the path of the bullet through the barrel and have a tendency to wipe the outer surface of the bullet with resultant high friction and the probability of metal fouling of the barrel. The conventional finishing of the grooves, however, is accomplished by a cutter which travels in the same direction as bullets do later. Any striae are in the direction of motion and do not produce nearly so much friction or metal fouling. It is obvious that the optimum inner surface of a conventional bore, as far as friction is concerned, would be the widest groove and narrowest land obtainable. The narrow land, however, is prone to wear excessively and erode at the throat. This was the old Metford system that would not stand cordite. Our present designs, like most other things in firearms, resulted from a compromise between these two factors.

Button reaming, however, has changed this. When the button is forced through the bore, the high spots of the reamed surface are swaged over the low spots. This burnishing action vastly improves the surface that will later be the lands. A larger land area can, therefore, be used while still reducing friction and metal fouling. Button reaming definitely aids both performance and design, but is a recent development.

Button rifling, at first, met with limited success because of inherent difficulties in connection with the system of rifling that was to be impressed. For instance, the old four-groove Springfield Rifle barrel had a good deal of the inside of the bore removed during the rifling process. As pointed out above, by removing as much as possible of the rough bore, less friction was created in the passage of a bullet down the rifle barrel and less fouling met with in service. Further, it was far easier to put four fairly wide grooves in a barrel than six or eight narrower ones. The old rifling cutters could be made more substantial in the widest acceptable width. Actually, at the end of World War II, many barrels were being made with only two grooves; however, this was a manufacturing consideration rather than a ballistic one. Six and eight-groove barrels are considered better ballistically.

It was difficult to pull or push a four-groove button through a .30 caliber chrome molybdenum steel barrel. The interference was too great to be accomplished with entire satisfaction by attaching a rod to the front of the button and pulling it through the bore. A free button can give troubles, although we pushed them mechanically with a rod of almost bore size acting by hydraulic pressure in a tightly fitting channel. The Germans used a free button and direct acting hydraulic pressure with less than perfect results because of chatter of the button in the bore. Experimentally, an American company shot hardened buttons through smooth bored barrels with proof loads, buttoning both lands and grooves. Sometimes a perfect barrel would be produced; more often the button would strike a hard spot and tear out portions of the bore and the buttons were seldom usable when recovered.

With the end of the war, however, this buttoning method was found by the High Standard Manufacturing Company to have immediate and extremely successful application in the multi-grooving of relatively soft steel .22 caliber barrels. It was possible to pull-button these entirely satisfactorily, with a finish on both lands and grooves smoother than any obtained in production heretofore. At first, this buttoning was confined to the impressing in the soft steel barrels of the rifling grooves. However, in search for the ultimate in smoothness and accuracy from .22 barrels, button reaming was also done. Extremely good results were obtained in this way, whereas all former finished ream bores, although appearing mirror-smooth when viewed from the ends of a barrel, are really quite rough, if the barrels are cut apart and the surface of the bore examined at right angles.

A multi-groove bore which has been button reamed twice will present a more serious identification problem than conventionally produced barrels do, so long as the bore of the firearm is kept in good condition. Whether this extreme smoothness is necessary for accuracy is a debatable point; however, it is certainly a problem for the identification Expert.

At the present time—the spring of 1954—at least one manufacturer of heavy caliber rifles is using button rifling. It is easily identified since a buttoned surface has a characteristic appearance. We list below some experimental data from the High Standard Company. They do not yet produce their centerfire barrels, however, by this process.

First, it is possible to reduce the finish on a smoothly reamed .30 caliber ASME 4140 heat treated barrel with a smooth button, with an interference area of approximately .075″ in length and .003″ in diameter, to below 10 micro-inches by two or three passes of a bore button. Such a finish is simply unheard of in firearms. It has never been obtained before in small arms and had only been approached in artillery when honing machines using a series of stones finishing with extremely fine grit—440 for instance—were used.

Second, a .30 caliber barrel with a bore diameter of .301″ was buttoned experimentally with a rifling button with a land diameter of .312″, and six lands .078″ wide. The lands, of course, form the grooves in the bore. The groove diameter of the barrel after rifling was .3105″, with a bore diameter of .3016″. In other words, the bore was going slightly oversize, while the groove diameter was closing in .0015″.

Third, with the same type of barrel with a .301″ bore, a six-groove button was then passed through with lands of .310″ diameter and .098″ wide. This button produced a bore diameter again of .3016″, but with a groove diameter of .3085″, which was about perfect for .30 caliber barrels chambered for the standard .30-06 or .300 H & H Magnum cartridges.

Various experimenting has been done with the mechanical means of buttoning. For instance, a free button can be used under hydraulic pressure alone. This button will, of course, produce its own lead as it is forced through the bore. However, certain irregularities in pressure, perhaps caused by air in the hydraulic medium, seem to create a chatter, as is noticeable in some German military barrels. Further, a pulling of a button through a hard steel barrel is still extremely difficult in production. However, it is thought that a combination of the two, with a hydraulic pressure at about 75% of that required to start the button in the bore, will work out very well.

A few words in connection with the making of buttons and troubles encountered in their use may be in order. They are generally made from a high tensile strength tungsten carbide and last for thousands upon thousands of barrels if not broken through carelessness. They are ground to size with diamond wheels on universal cutter grinders. The leading angle, flat burnishing surface, and trailing angle are produced as in regular cylindrical grinding. The bore reaming button requires no other attention.

The rifling buttons require extreme care, however, so as to function properly. For instance, in a six-groove button the included angle on the wheel forming the groove must be no more than 60° so that the sides of the impressed grooves in the bore are not undercut. The plane of revolution of the wheel must be precisely perpendicular to the axis of the button at all points along the helix of the groove in the button.

Lead, once impressed upon a button, must be matched by the sine bar setting on the rifling machine. A few test barrels are run with trial and error adjustments so that the natural lead on the button is matched by that of the machine. This means variations in lead when buttons are changed. In theory, a free pulling-head would be best; it is not practical.

APPENDIX 4

Marlin Micro-Groove Rifling

FOREWORD: Through the courtesy of the Marlin Company, we are quoting below from their original release in connection with this new type of rifling. At the end of the quotation itself, the authors will have a bit to say in connection with this rifling and experience with it, both at the Marlin Company ranges in New Haven and at their own ranges and laboratories.

"After several years of careful design and testing, a new and revolutionary type of rifle barrel has been perfected. Based on results of many general firing tests at 100 feet, and more intensive careful firing at 100 yards, the new barrels have shown a net gain in accuracy improvement of about 20% as compared to conventional barrels.

"The new Micro-Groove barrel is revolutionary both in its exceptional accuracy, its generally high bullet velocities and its low chamber pressure. It is also striking in the ease with which it may be cleaned and in its durability. The unique design of the Micro-Groove barrel, with many more rifling lands than conventional barrels have, results in the engravings on the bullet, made gently but firmly, giving a secure spinning grip for stabilization.

"The relatively shallow grooves coupled with a bore of greater than standard size result in a bullet or bullet jacket not deeply grooved or distorted. This happens in the case of conventional barrels in their action on bullets.

"The Micro-Groove barrel develops a light multiple grip on the bullet or bullet jacket, with more than adequate driving area in the lands to assure reliable bullet spinning. With this accomplished there is no objectionable finning or squeezing-out-of-shape the bullet fired which occurs in other barrels.

"Marlin Micro-Groove barrel has been deliberately designed to produce required rotation in the bullet with an absolute minimum or plastic deformation, i.e., disturbance of the original cylindrical shape of the bullet. By use of a special land design, the barrel more effectively grasps the bullet than can be done with the standard rifle barrel. Gas leakage by bullet—between bullet and bore—is practically eliminated. Leading and other types of bore fouling are minimized by shape and surface condition of the bore.

"Ultimate verification of the Micro-Groove rifle compared with conventional barrels was commenced in January, 1953, following many thousands of rounds fired experimentally with various modifications of the new rifling. The conclusive test was made with final standardized Micro-Groove barrels.

"One hundred Model 88 and 81 micro-groove barrels were made. All 100 barrels were assembled to guns and shot on the Marlin 100 foot targeting range with iron sights. Average performance of all 100 guns compared to regular barreled guns of these two types proved the definite superiority of the micro-groove system."

The authors wish to point out that a large number of grooves have been used before in the history of firearms. However, the polygroove type of rifling was used for the original round ball rifles which went out of style 100 years ago. For instance, 12 or more grooves were used in most of Hall's military breechloading rifles of the early Nineteenth Century. Many other muzzleloading rifles firing round balls used an even larger number of grooves. Even the early American rifles, sometimes called Kentucky rifles, had more grooves than are common today. The Marlin Company is the first modern arms manufacturer to design rifling of this type for cylindro-conoidal bullets. Whereas, it is difficult to say that something did not take place, the authors know of no weapon of any considerable importance using a large number of grooves for a modern shaped bullet, with the possible exception of the paradox rifling for elongated bullets fired in weapons smoothbore until a short distance before the muzzle. A large number of shallow grooves were used in this instance to disturb as little as possible the shot charges regularly fired from the weapons. The Marlin's rifling is an achievement in precision. They have worked out beautifully the actual grip on the outside of the bullet so as to reduce to a minimum all disturbance and distortion.

The authors have fired a number of light Marlin barrels both in rimfire and centerfire chamberings, and found the accuracy extremely good. It is possible to fire a five-shot group at 100 yards with a standard Marlin autoloading .22 rimfire rifle measuring less than two inches center to center. For weapons selling in this price range, this is truly remarkable. However, because of weight and other considerations, a bit different hold is necessary. Once you are set in your bench rest position, do not move a muscle. Since the weapon is an autoloader, this is relatively easy to do.

The centerfire micro-groove barrels are equally remarkable. Bolt action rifles of moderate weight produced experimentally shoot very well indeed, even with sporting sights. However, at this time (Spring, 1954), experimentation of this type has not proceeded far enough to predict what may be expected from bench rest weapons using micro-groove heavy barrels.

Identification-wise, the Marlin micro-groove rifling is going to be an extreme headache. The number of grooves presents a problem in itself. Since each groove must be compared with every other groove, it is probable that a negative identification will take an Expert about two days rather than an average of two hours as at present. Further, positive identification will have to be checked extremely carefully since the work that the authors have done in connection with test bullets fired through these Marlin barrels indicates that when the weapons are new, there are closer family resemblances than are usually present in consecutive barrels of other designs.

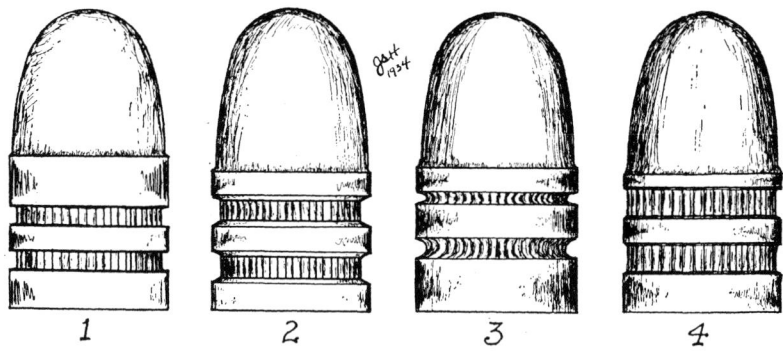

1   2   3   4

CANNELURES ON LEAD BULLETS

The way in which differences in the form, appearance and location of the cannelures can assist in identifying the make of a bullet is quickly apparent from the above sketch. This shows four bullets of the same caliber and weight, and approximately the same date of manufacture, from the four leading American cartridge factories.

CHAPTER 6

POWDERS

THE explosion within a firearm is the basis of all ballistics Interior, Exterior, and Forensic. An Expert should know as much as possible about the explosion itself and have at least a general knowledge of the various substances which are used to produce it. Few terms in firearms are so misleading as that of powder. The word originally meant and still means fine dust. However, today, in firearms, powder means something quite different; it refers to any of the many different substances which are burned in the chamber of a weapon in order to give velocity to the projectile or projectiles. Powder in its original sense of fine dust has not been used for about two centuries; some of the present propellants called powder and used in large naval guns are composed of pieces several inches long and two inches or more in diameter. Even larger individual grains have been produced experimentally for certain special purposes. It is all still called "powder."

Black or Gunpowder

In the discussion of the first firearms, we mentioned that gunpowder was certainly known in Europe by the middle of the Thirteenth Century. This early black powder was a combination of charcoal, sulphur, and niter or potassium nitrate. The percentages of each seem to have varied considerably; however, an explosive mixture can be formed with widely different proportions. At first, these three elements were finely divided and then merely mixed together in their dustlike condition. Early in the Eighteenth Century, however, it was found possible to greatly increase the power and quickness of the explosion by combining these three materials into a solid cake and then breaking up the cake into grains. The air space between the individual grains meant a much more rapid burning, particularly inside a confined space such as the chamber of a firearm. Black powder is with us today and is superior to all other explosives for some purposes—though it is seldom used as a propellant in modern cartridges.

137

The reason that this mixture explodes when ignited is that it contains two fuels, charcoal, and sulphur, which are very easily inflammable; in addition it contains another solid substance which when heated gives off large quantities of oxygen. If charcoal and sulphur are mixed together without the niter, the result is a black substance that looks in all respects like the primitive gunpowder described above, and if it is ignited, it will burn, provided that plenty of air is supplied to support the combustion. The oxygen in the air enters into combustion with the charcoal to form carbon dioxide gas, and with the sulphur forms sulphur dioxide gas. However, there is no explosion because these gases can be formed only as fast as air can be brought in contact with the burning powder to supply the necessary oxygen for the combustion. In other words, this mixture of sulphur and charcoal would be a non-explosive powder.

Now when we mix powdered niter with it, we change its nature so that it becomes an explosive powder. The instant it is lighted it burns rapidly without waiting for any air to come in contact with it. This is because the niter is very rich in oxygen which it gives up when heated. Gunpowder supplies its own oxygen for combustion from within and can burn without any outside air.

As nearly half the products of combustion of gunpowder are gases, which occupy a great deal more space than the solids from which they were evolved, it is evident that the combustion of gunpowder results in a sudden expansion from the size of the original powder to the much larger size occupied by the gases which are given off. It is this sudden expansion which causes the gunpowder to exert force when it is ignited. The heat of combustion adds to this effect because hot gases occupy more room than cold ones.

Black powder relies for its explosive properties on three qualities which are typical of all explosives. First, when ignited it will burn by itself without aid from the outside air; this burning is very rapid. Second, in burning, it gives off a large amount of gas. Third, a considerable amount of heat is also evolved.

To be successful for use as an explosive, a substance must possess all three of these qualities. We have seen above that there are substances which will burn and will give off a large amount of gas and heat, such as the mixture of charcoal and sulphur, but they do not burn without the aid of the outside air and consequently they do not burn rapidly enough.

We do have substances that will burn without the aid of outside air and burn rather rapidly, but they must also give off a large amount of gas or else there is no explosion. Thermite, which is a mixture of oxide of iron with aluminum powder, will burn rapidly without air when ignited. It gives off great heat but it does not give off any gas; therefore, there is no explosion. When this powder is burned it forms melted iron and slag which is aluminum oxide.

Even though a substance can change from a solid to a gas without the addition of oxygen from the outside air, there will be no explosion unless the change from a solid to a gas involves the liberation of heat. Some substances change into gas rather rapidly, but in doing so, instead of giving off heat they take up heat. Substances such as these do not cause an explosion when they decompose because as soon as they have started to decompose, they chill the surrounding air so much that the action comes to a stop until more heat can be supplied. A very familiar example is the "dry-ice" in which ice cream is often packed. This is carbon dioxide, CO_2, and it can exist in solid form only at an extremely low temperature. The minute dry-ice is exposed to heat, it begins to give off great quantities of gas, but in giving off this gas it absorbs all the heat around it so that the gas can only be evolved as fast as the heat is supplied. You can lay a big chunk of dry-ice on a hot stove; it will sizzle, hiss and jump around for a long time, but there will be no explosion.

The basis of gun powder is the substance which gives off oxygen; that is the niter or saltpeter. No doubt the original invention of gun powder followed the discovery that when saltpeter is thrown into a fire, it crackles, hisses and makes the fire burn very much more brightly because of the oxygen it gives off. By mixing this saltpeter with incendiary materials, it was found that extremely hot fires could be made when the mixture was ignited. Finally, no doubt the proper combination was obtained so that when one of these mixtures was ignited it went off with a real explosion.

The average composition of black powder is saltpeter 75 parts by weight, sulphur 10 parts, and charcoal 15 parts. There is, however, quite a wide variation in compositions that have been used at different times. Powders used for blasting, for example, are more effective if they do not operate quite as quickly as the ordinary gunpowder. This is for the reason that in blasting it is desired to rend the rocks and tear them out in chunks, rather than to break them up in fine pieces and toss them about, which a stronger powder would do. The average blasting powder has about 70 parts saltpeter, 14 sulphur and 16 charcoal, but some powders of this kind have been made with compositions as low in saltpeter as 40 parts, with 30 sulphur and 30 charcoal. Frequently, sodium nitrate is used instead of potassium nitrate.

The speed of burning of black powder, and therefore to a certain degree its strength, is controlled by the size of granulation. Powder with extremely large grains burns more slowly than a fine grain powder and, therefore, is less sudden in its action. Moreover, large grain powder is somewhat harder to ignite.

As the size of the granulation decreases, the strength of the powder increases up to a certain point. However, when a powder is made very fine, like dust, the speed of burning is again reduced, because all the spaces between the grains are filled up; there is no way for flame to communicate itself rapidly through the spaces and ignite the whole charge at one time, as would occur with powder where there is somewhat larger granulation. Powder with very small grains is the easiest to ignite, this kind of powder was often used for priming charges in the old muzzleloaders.

Black powder as at present made and sold to the trade comes in irregular shiny metallic looking black grains designated as to size by the letters FG, FFG, FFFG, and FFFFG; the more Fs, the finer the grain.

The powders are measured by screening them through a coarse screen which catches all the very large lumps, letting the rest fall onto a fine screen which allows all the very finest powder to pass through. The screens through which the different grades of powder must pass and on which they must be retained are given in the table, the figures representing the number of meshes per inch in the screen. The du Pont Company's system of values is given in the table below.

Size	Must Pass	Must Be Retained On
Grade A-1	6	10
FG	14	16
FFG	16	24
FFFG	24	46
FFFFG	46	60

Nitro Explosives and Smokeless Powders

Smokeless powders so common today were unknown as propellants 100 years ago. Our Civil War was fought throughout with black powder, both for use inside the guns and for bursting charges in shells and mines. However, as far back as 1846, a chemist by the name of Schoenbein interested the Austrian government in the possibility of using nitrated cellulose fibers in firearms. This was the original nitrocellulose which began to be experimented with to some extent both in Austria and in England after that time. The great trouble was that the various laboratories and manufactories were continually blowing up. About this time, the explosive nature of nitroglycerine was also discovered. It was proposed as a firearms propellant; however, the substance, when set off, would always blow the gun to bits.

These two substances differ from gunpowder in that they are not mixtures of fuels and oxidizing agents. Instead, they are chemical compounds containing a large amount of oxygen. These chemical compounds are capable of instantly rearranging themselves into more stable compounds which are gases. A sharp shock will cause this rearranging process to start, and when once started it spreads almost instantly throughout the entire mass, thus resulting in a violent explosion.

Nitroglycerine is inconvenient to use by itself, as it is a liquid. Films of nitroglycerine leaking to the outside of containers are subject to explosion by friction. Moreover, pure nitroglycerine is dangerously sensitive. It was soon found that unless something could be done to overcome these disadvantages, nitroglycerine would never attain great popularity as an explosive.

Perhaps the most dramatic of the several discoveries made about this time was the fact that when nitroglycerine was absorbed into a porous earth, called Kieselguhr, it became insensitive and, further, had its potential for explosive work increased. This discovery was made by Nobel and was the basis of dynamite. This substance when reasonably fresh can be burned in an open fire, carried around in one's pocket, and kept safely for at least several months. However, when it is set off by a cap containing originally a small quantity of fulminate of mercury, it will detonate with tremendous power.

This discovery of Nobel's, however, did not have any application immediately to propellants since, of course, detonation will wreck any gun. A murder was committed in this manner in Des Moines, Iowa, in 1937. A man persuaded his wife to shoot a shotgun which he had loaded with a quarter-stick of dynamite tamped into place with a blasting cap where the shot charge was in the original shotgun shell. The primer in the shot gun shell and the regular charge

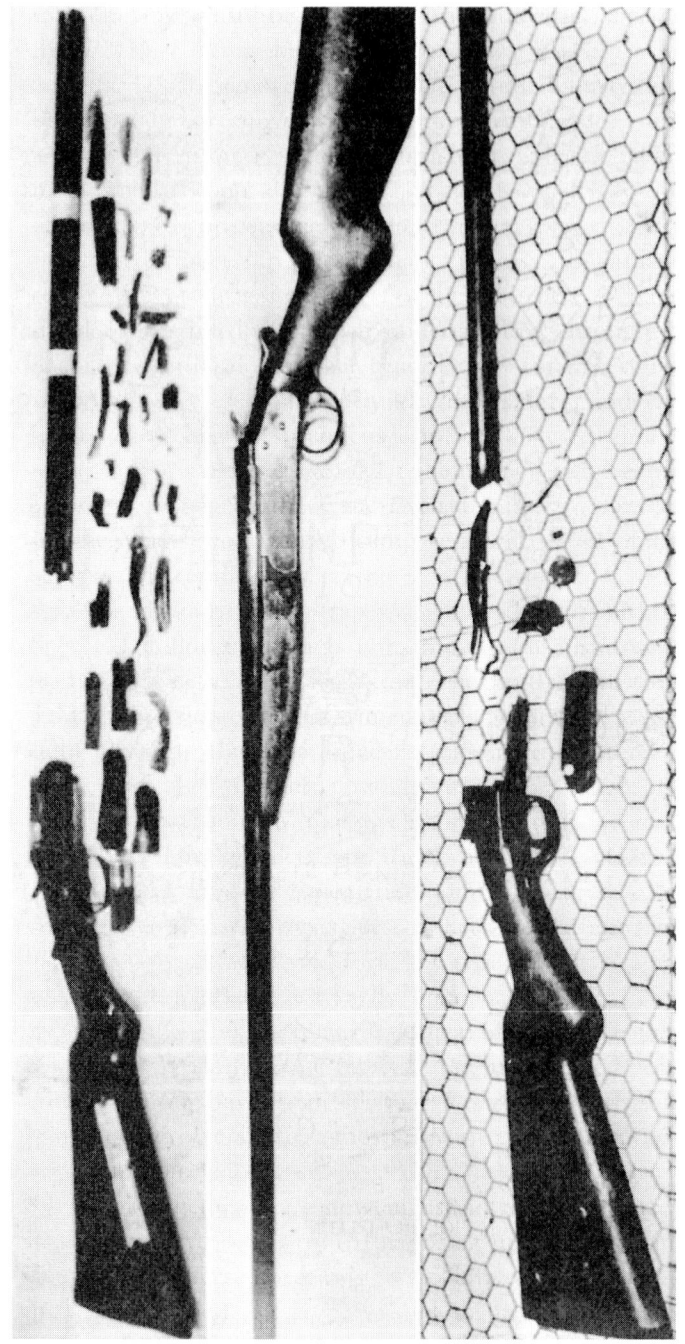

A BLOWN-UP SHOTGUN

When a firearm is damaged or wrecked by a high-pressure cartridge; by an obstruction in the barrel; by double or lodged bullets; or from practically *any* cause, the user invariably reports that the gun was "blown up." Actually, such a statement is very seldom the correct one to make; a true account would generally be that the barrel burst; a case-head blew out; a bullet lodged and split the barrel on succeeding shot; or some similarly accurate explanation.

(Continued at bottom of opposite page.)

of powder was fired in the normal way. This set off the blasting cap and dynamite. The woman was practically decapitated. The husband's explanation of gun failure was not accepted by the firearms investigator. In fact, he duplicated the occurrence, as illustrated, before the husband confessed.

Gradually scientists began to appreciate the difference between deflagration, sometimes called low order explosion, and detonation, which is high order explosion. Both nitroglycerine and nitrocellulose compounds are capable under certain conditions of each of these two reactions. Because of the great potential for military and commercial use of the extreme power of these two compounds, many people did work in connection with them.

It will be remembered that gunpowder is merely a mixture of two fuels and one oxidizing agent, whereas nitrocellulose and nitroglycerine are single chemical compounds. Such compounds in general explode more violently than simple black powder and are known as high explosives. Moreover, there is a big difference in the manner of explosion of high explosive as compared with black powder. When black powder is ignited it burns with reasonably great rapidity, giving off a large volume of gas. However, this burning is the only explosion that occurs. On the other hand, many high explosives such as guncotton, nitroglycerine and tri-nitro-toluol or TNT can be burned without causing any explosion at all. A stick of dynamite for example, can be cut up and thrown into a fire bit by bit with very little danger of exploding, though each piece will burn rather intensely. In the same way, TNT can be burned in small quantities without any explosion.

What these high explosives need to cause them to exert their full power is a shock which starts a chemical rearrangement of

---

However, the opposite illustration shows a shotgun that actually did "blow up." The upper section shows the death weapon in a murder case some years ago in Iowa. The husband in this case had loaded a shotshell with dynamite and then tricked his wife into shooting the weapon. The resulting explosion practically decapitated her, tore off her left hand, mutilated the right hand and arm and lodged a piece of metal in her chest. The tragedy occurred in a basement of the home and thus most of the pieces of the blown-apart gun were found, as shown here. Firearms and explosive experts testified in the case. The husband was found guilty of murder in the first degree with recommendation of death and later signed a written confession.

In checking up on this case, the sheriff's office obtained and blew up a similar weapon, which is shown in the central section.

The lower section shows this test-gun as it appeared after it had been fired with a cartridge containing a quarter-stick of dynamite—the shot being removed from the shell and a No. 10 blasting cap inserted amongst the powder with its open end towards the primer. Only the seven pieces shown here could be found afterwards, as the gun was discharged in open country. Observe how well this check-test duplicated the results of the original crime.

the molecules into gas instead of solids. This chemical rearrangement is called a "detonation." Once a detonation is started in a high explosive it spreads through the entire mass, with a sort of wave action, with great rapidity. The speed of detonations varies in different explosives, but in some it is as high as 7000 yards a second. A detonating fuse consisting of a long thread of TNT encased in a lead or fabric covering, will thus set off a number of charges of explosive almost instantly if a detonating cap is exploded at one part of the fuse. Ordinarily the detonation of high explosives is started by exploding a blasting cap in contact with them. These blasting caps, which originally contained fulminate of mercury, now generally have a charge of lead azide. PETN is an HE used in detonating cord and requires a blasting cap to initiate. Tetryl, PETN, etc., are "boosters" of main charges in small HE shells like 20mm.

Owing to its porous nature, dry nitrocellulose burns with extreme rapidity; it detonates with violence from a moderate shock. However, when it is moistened with 25% of its own weight of water, it becomes safe to handle, for it can be burned only with extreme difficulty, and it can be detonated in this state only by exploding a priming charge of dry nitrocellulose in contact with the wet mass. When it does detonate, however, it is just as powerful as the dry nitrocellulose. For many years, wet nitrocellulose was used as the explosive charge in submarine mines and torpedoes. As long as the dry nitrocellulose priming charges were kept out of contact with the wet mass, it was perfectly safe to store or handle.

After the discovery of nitro explosives, an attempt was made to use them for propellants in guns, but they detonated with such instantaneous violence that they would simply blow the gun all to pieces before the bullet had time to move. To make them suitable for use in guns, it was necessary to find some way to prevent them from detonating and to slow down the burning to a reasonable speed. This was finally accomplished in the case of nitrocellulose, which now forms the basis of most of our smokeless powders. Instead of leaving the nitrocellulose in a light fluffy form, it is partially dissolved in some solvent, and then evaporated into a glue-like mass, or colloid, which detonates with difficulty or not at all, and which burns with a reasonable speed.

One of the earliest successful smokeless powders was produced by making a colloidal solution of nitrocellulose and nitroglycerine and adding vaseline. This soft mass was extruded into long cords and

the free solvent evaporated; this is the British propellant called
Cordite. Actually the word "Cordite" refers to a whole series of
powders made in this way. Several different proportions are still
in use. The most common today contains about 60 parts nitrocellulose, 35 parts nitroglycerine, and 5 parts vaseline.

Many of our most popular rifle and pistol powders are made by
colloiding nitrocellulose and nitroglycerine, and then forming the
resultant mass into small grains or flakes. Examples are the well
known rifle powder Hi-Vel, and the pistol powder Bullseye. These
powders, made with both nitrocellulose and nitroglycerine, are called
Double Base powders.

Another method of making nitrocellulose into smokeless powder
is colloiding it in a mixture of ether and alcohol, thus forming a
sticky mass having very much the same consistency as melted glue.
This is pressed in various ways. First, large blocks are formed. These
are then shredded and the mass extruded through intermediate
presses in which a reasonably homogeneous consistency is attained.
Finally, this colloid is extruded into little tubes like fine macaroni,
which later are cut into short lengths. The remaining ether and
alcohol are evaporated off, leaving the hard grains well known to
all handloaders. This dried out colloid of nitrocellulose is called
a Single Base powder. Examples are the old Government Pyro
powder, used in the '06 Springfield Rifle cartridges of before World
War I and the du Pont Improved Military Rifle Powders. The du
Pont Pistol Powders are also of this type.

Perhaps in digression, we point out that nitrocellulose as used in
propellants is not a fixed chemical compound. The nitrating of
cotton linters or wood fibers can be done completely in the laboratory so that the final product contains 14.3% nitrogen. However,
commercially, this is not attempted, as the cost would be extreme.
Guncotton, one type of nitrocellulose, contains 13.4% nitrogen. Pyro
cellulose has 12.6%. Most powders are made from blends of different degrees of nitrated celluloses. The maximum nitrogen content of commercially produced nitrocellulose is about 13.42%.

Originally, most American smokeless powders were in the form
of flat flakes, small porous balls, and the like. However, these were
expensive to produce and sometimes involved reworking up to 40%
of production runs as the grains were either too large or too small.
The cylindrical grain with a hole down the middle is now most
common. In fact, some of the new powders are in essence only old

ones produced in a new, cheaper to manufacture, shape.

The object of having the hole through the center of the powder is to control the rate of burning so as to hold up the pressure inside the barrel as long as possible. The great trouble with getting high velocity in a gun is the fact that pressures fall off with bullet movement. When the powder in the cartridge is ignited, it turns into gas. This gas confined in the small volume of the cartridge case creates a very high pressure which pushes the bullet along the bore of the gun. As soon as the bullet starts to move along the bore of the gun, a larger space is created for the gas to occupy, hence there is less pressure. The pressure of the powder gases would immediately fall off to nothing unless special means were taken to keep it up.

One of these special means is the perforation of the powder grain, which causes it to have a larger burning surface as the combustion proceeds. This is because the primer flash ignites the inside of the tube as well as the outside. As the grain burns, the outside surface gets smaller, hence the rate of evolution of gas would tend to decrease; however, this is balanced off by the fact that as the inside of the grain burns, the inside surface gets bigger. Therefore, there is a larger surface on the inside for the evolution of gas as the combustion proceeds.

The balance between these two surfaces can be controlled by the ratio of the inside diameter to the outside diameter in the finished grain of powder. In cannon powders, there are usually up to seven perforations instead of one. More than two dozen have been used in some of the exceptionally large grains.

Another method of controlling the burning of the powder and making it more progressive, that is, making it hold up its pressure longer during the travel of the bullet, is by coating powder with a substance which makes it burn slowly at first. As this coating burns off the outside of the powder, the speed of combustion increases. These progressive burning powders tend to give a more uniformly distributed pressure, which is sustained longer during the travel of the bullet. Moreover, the maximum pressure is not so high, because instead of being exerted all at once, the pressure is spread out more evenly during the entire travel of the bullet. The du Pont "Improved Military Rifle" powders have all been progressive burning.

Like black powder, smokeless powders are controlled as to their speed of burning by the grain size also. Powders with very fine grains burn up in a hurry, and therefore are suited for short barrel weapons. Powders with very large grains take longer to burn up; therefore they are suited to long barrel weapons.

For any given size of gun such as the .30 caliber rifle, there is a granulation powder that is suitable, and a larger or smaller size will either give too high pressures or will not burn up completely before leaving the barrel. One of the authors has attempted with only limited success to use powder of large grain size designed for .50 caliber machine guns in smaller cartridges. In the .30-06 cartridge grains may be either long and slender or short and fat and still be about the same weight. In general, better results are obtained if the grains are long and slender because this leaves bigger air spaces between the grain for the priming flame to penetrate and better ignition is obtained.

In most powders today—all those made by extrusion processes—the grains are uniform in size for any given type of powder. The .50 machine gun cartridge is loaded with larger grains than that for the .30 cartridges. However, the Germans in World War II used an extremely ingenious arrangement in their propellants for high angle fire weapons. There would be a basic minimum charge of relatively fine grained powder for short range use. The powder increments used to increase the range were composed, however, of progressively larger grains which burned more slowly giving a greater average pressure without any corresponding increase in the early maximum pressure.

The rate of burning of the powders also depends on the resistance that the bullet offers to being moved. If any of these powders are ignited in the open air they will burn slowly, the same as a pile of grains of celluloid; there would be no explosion. On the other hand, when they are ignited in a gun the first grains burn and give off hot gases and these hot gases in turn accelerate the burning of the other grains. The more the gases are confined the hotter they are and the faster the combustion of the remaining powder will be. With a heavy bullet, or with the cartridge case full of powder, the combustion is more rapid than it is with a light bullet or with a cartridge case only partly filled with powder.

There are some powders which must burn with a big air space in the case and a light bullet. These are powders used for reduced loads. These powders are extremely dangerous if improperly used, as it is easy to overload with them. Investigation of blown-up weapons can usually be traced to too heavy charges of these powders. Dozens of revolvers are blown up each year by the detonation of too much powder of this type in a given case.

There are also powders which must burn with no resistance at all in the form of a bullet to hold back the gases. This is the case with powder used in blank cartridges where there is no bullet, but only a paper wad. The best known of these is Hercules E. C. Blank Fire Powder which is sold only to regular customers having a legitimate use for it. It is actually a high explosive; with proper initiators it forms one of the best fillers for hand grenades, where it creates almost perfect fragmentation. Under low pressure, however, as in a blank cartridge, it burns progressively. It is also used as a fuel to absorb liquified tear gas. These blank powders are generally colored pink or red and should *never* be used behind a bullet; the additional resistance will probably cause the powder to detonate.

Powders on occasion can be of considerable importance in Firearms Investigations. The powders available commercially for handloading purposes are discussed in detail in an Appendix. In special circumstances other powders may be encountered. For instance, the .50 machine gun cartridge powder already mentioned was left over from experiments during World War II in the Princeton University laboratories.

In general, powders not produced for handloading should be viewed with caution. In the first place, identification is reasonably difficult. Secondly, they vary from lot to lot, even in the same type. The powder companies make many more powders than those they sell to handloaders. In general, only certain lots of a given type of powder will be released for sale in this manner since uniformity of weight and volume to pressure is absolutely essential for the handloader but of no great importance to the ammunition factory. They can vary their components to get exactly the pressure and velocity they require. Much commercial ammunition is loaded with types of powders not available to handloaders. For instance, high velocity pistol cartridges are often loaded with a type of Hercules Unique never sold in small quantities.

The Appearance of Powders

Most types of powders can be identified by appearance. The Firearms Laboratory should be fully provided with samples of powders of all popular brands, together with as many obsolete and foreign types as possible. On occasion, the identification of un-burned and semiburned particles of both black and smokeless powder have solved cases and provided proof. For instance, in shotgun crimes, there may be recovered from the clothing of the victim, or even from the wound, particles of powder that matched the ammunition carried by only one of several different suspects. The grain comparison is sometimes done microscopically.

The appearance of black powder is quite characteristic, and once it is seen and carefully examined, it will always be easy to tell it from smokeless powder at a glance. Black powder is composed of irregular grains of a fairly uniform size, with a glossy black sur-face. There are various sizes, or granulations, of black powder used, but in any one cartridge the grain sizes will be fairly uniform, for the larger and smaller sizes are removed by screening, as was described earlier in this chapter.

Some varieties of smokeless powder also have a glossy black sur-face, but they look very different from black powder, in spite of the similarity of color. Almost all smokeless grains have a per-fectly definite shape, such as small squares, discs, or cylinders, with or without a hole in the center; while the grains of black powder are of the irregular shape caused by forming the damp mass into a cake, and then, when this cake has hardened, breaking it and crushing it into grains of the desired size, then polishing the grains by tumbling them in a barrel, and then removing the large grains as well as the dust and very small grains by differential screening.

No smokeless powder is naturally black; the reason why some of it in the finished state has a black surface is because it is coated with graphite. This graphite coating is put on because the powder is a hard amber-like substance which is a non-conductor of electricity, and is very strongly susceptible to becoming electrified by friction, especially the friction of one grain on another in handling and load-ing. In cold dry weather the grains have a tendency to stick to-gether and become difficult to load in even charges. Moreover, the occurrence of sparks of static electricity is a very serious fire hazard, especially when the air may contain highly inflammable volatile vapors of the solvent used in making the powder. Coating the sur-face with graphite makes it a conductor of electricity, and prevents trouble from this cause, also helps it to flow better through the loading tubes.

SMALL ARMS PROPELLANT POWDERS

Shown here are small samples of various types of rifle, shotgun and revolver powders such as have been in common use for the past 75 years and which are still being loaded into ammunition.

Nos. 1 and 2. These are early types of British smokeless powders. No. 1 is 'Cordite,' a powder developed around 1890 by the British as their official military rifle powder and used as such until the period of the First World War. No. 2 is 'Axite' a powder of the same type and composition but developed for use in sporting cartridges. Both are high nitroglycerin content powders, very erosive to the bore of the rifle; they are used today only in a few of the heavier caliber, big game, sporting rifle cartridges where a smokeless propellant developing a low chamber pressure is essential.

(Continued at bottom of opposite page.)

Rifle powders are most usually in the form of small cylinders, with a single perforation through the center, and with the surface graphited, so that the color is a shiny black. Small flat squares are often seen.

Dense pistol powders are usually of a greenish-brown color, and on close examination are seen to consist of small flat chips.

Bulk powders for both rifle, pistol, and shotgun are mealy in appearance, like fine sawdust, and the color may be grayish white, light yellow, cream color, gray or pink. Other colors are sometimes seen and are due to certain dyes or coloring substances used by some makers.

Ball Powder

This new powder is frequently met with both in the M1 carbine ammunition and in some of the .45 ACP government loaded ammunition as well as commercial loads. It is fairly important today and may prove to be even more important in the future. Some .30 and .50 cartridges were loaded with it during World War II.

Nos. 3 and 4. Both of these are what is known here in the United States as 'flake grain' powders of the 'Ballistite' type. Such smokeless powder is made in sheet form and then rolled off into small squares, as shown above. The form originated in the '90s and for years was the most commonly used type of powder amongst European nations; here in the United States it was made only as a shotgun powder under the trade name of 'Ballistite.' The Germans were once the great proponents of such powder and used it for years in their military and sporting cartridges. However, during World War II much of their military and sporting ammunition was loaded with the American type tubular grain form of powder. Both 3 and 4 are high power rifle powders, 3 is from a 7mm German sporting cartridge and 4 is from the 7.5mm French military rifle cartridge.

Nos. 5 and 6 show American types of smokeless shotgun powders—the new and the old. 5 is one of the present day dense, sporting powders while 6 shows the old and famous duPont Shotgun Smokeless, a bulk, porous grain powder developed about 1895, which held its own for the two following generations.

No. 7 is a present day revolver and automatic pistol powder of fine granulation necessary for use in all modern handgun cartridges.

No. 8 is a dense, high power rifle powder of the conventional tubular grain, single perforation type. This shape of powder grain is of American origin and has been in general use here in the United States for almost three generations.

Nos. 9 and 10 show the earliest and the latest in small arms propellants. No. 9 is 'gunpowder,' as black powder has been called since its earliest days. The sample shown is FFFg granulation, suitable for use in revolver and small rifle cartridges. Very little black powder is used in the United States today, other than in obsolete muzzle loading types of firearms.

No. 10. Along with the earliest firearms propellant known, is shown the latest in smokeless rifle powder—the recently developed type known as 'Ball Grain' powder. Although not yet in general use in all types of firearms it shows great promise and with further development may possibly become our main type of powder.

Smokeless powders started to come into general use about 1895 and by 1921 they had entirely supplanted black powder for use in modern small arms here in the United States. The various smokeless sporting powders, especially the shotgun powders, have been made in every conceivable shape of grain and in all the colors of the rainbow; red, orange, pink, yellow, brown, blue, gray, purple, green and in-betweens, as well as in black and white. Therefore, don't be surprised if something looking especially gaudy comes out of an old shotgun cartridge.

Today, it is being used by Olin Industries Subsidiaries to load some of their commercial rifle pistol and shotgun cartridges. However, it has never been available to the handloader.

This powder was originally made by the Western Cartridge Company Division of Olin Industries from over-aged nitrocellulose cannon powder with both new nitrocellulose and nitroglycerine added. The powder itself was originally made spherical by hardening globules of nitrocellulose while in a colloidal solution. The process is involved but revolutionary. It is largely automatic; handling is by means of pipelines conveying the powder, either in the form of a mud-like mass or spherical grains in a water slurry, from one operation to another. The individual grains, when finally completed, are spherical and composed of both nitrocellulose and nitroglycerine. This Double Base globule is coated to a certain depth with an inhibitor so that it burns progressively in spite of its shape. It is a very good powder. Sometimes the grains are flattened rather than spherical. This was done intentionally in manufacture before the grains were hard. Larger globules were made to perform ballistically like smaller globules in this manner.

Powder Safety

A firearms Expert is often asked about safety in connection with powder. Many people confuse powder with dynamite and other high explosives. Unfortunately, however, black powder is definitely dangerous. It explodes unpredictably in reasonably large quantities and does so with extreme violence whether or not it is confined. Extreme care has always been taken in its manufacture, transportation and storage, yet accidents have happened both here and abroad. During the Civil War the Confederate loading plant at Richmond went up from an unknown cause with considerable loss of life. More recently, extreme safety precautions rigidly adhered to have kept explosions of black powder relatively few and small.

Smokeless powder on the other hand, when unconfined, burns with a flaring yellow flame that resembles that of burning celluloid; in small quantities in the open air, it burns at about the same speed. A large grain of cannon powder can be lighted with a match and then blown out. General Julian Hatcher used to tell of a friend who would take a grain of cannon powder in his hand, ignite it with a match, use it to light a cigarette, and then blow out the half-burned grain. But you cannot do this with black powder, or even with all kinds of smokeless powder.

Smokeless powder, particularly of the Single Base, nitrocellulose variety, is not sensitive to shock or friction, and is not dangerously easy to ignite, and moreover, when ignited in an unconfined condition, it merely burns with extreme rapidity and great heat, and usually does not explode. For these reasons, it is much safer than black powder to store or to handle. One disadvantage that it had in the past was that, in some varieties, chemical changes were likely to occur with age. The powder deteriorates.

Some years back, a magazine at the Curtis Bay Ordnance Depot containing 175,000 pounds of smokeless powder was destroyed by spontaneous combustion, which was a definite hazard in the early powders. Two colored men and a team of mules were standing by an adjoining magazine, 400 feet away, when they heard a hissing sound, suddenly increasing to a noise like a safety valve on a locomotive. Looking around they saw flames begin to spurt out of the doomed building. They started to run, but the terrific heat overtook them, and so blistered the backs of their necks that they fell to the ground. Fortunately, they fell into a slight depression; the worst of the heat then passed over them. They were not seriously hurt. The mules simply trotted around the corner of the building, so that they were in a shadow which cut off the intense heat. The two men had placed their lunch in paper bags, on the ground, and had hung their coats on a couple of spades leaning against the side of the building where they were standing, which, as mentioned above, was 400 feet from the fire. The lunch bags and the coats were burned up, and the shovel handles were charred.

An aviator, flying across the nearby town at an altitude of 2000 feet, saw a yellow tower of flame rise to a height that he estimated as 2000 feet higher than he was flying, then slowly die down. The metal siding and roof sheets of the magazine whirled high in the yellow flame, like leaves in an autumn wind; a fountain of powder containers sky-rocketed around. All this lasted a few brief but exciting seconds; then everything was over. There was no explosion and no noise except a loud roaring sound, like escaping steam. There were no casualties. People in the nearby city did not know that there had been a fire until they read about it in the papers.

That was a typical smokeless powder fire, or "flare"; there is no explosion, simply a fire, of extremely short duration, and accompanied by almost unbelievably intense heat.

But if that had been black powder, it would have gone off with a bang, and the blast would have levelled every building within hundreds of feet in all directions. For that reason, black powder is always stored in very small lots, in individual huts, well separated. No such explosion with black powder can happen simply because it is so dangerous that no large quantity of it is ever collected together at one time. Great care is exercised in the protection of the public in such matters; neither high explosives nor black powder are ever stored in dangerous quantities in the vicinity of a city.

In other words, smokeless powder is safer to manufacture, store, and handle than black powder. The former is no more dangerous than celluloid and many other plastics and less dangerous than gasoline. However, when one gets into ammunition and the firing of it, smokeless powder is far more dangerous. This is because the smokeless powders are potentially much stronger than the black powder. If a strong vessel should be filled entirely full of black powder, and the powder should then be ignited, the maximum pressure that would result, even if the vessel were strong enough to confine all the gas into the original space occupied by the powder charge, is about 86,000 pounds per square inch. This has been tried experimentally. Nitrocellulose, nitroglycerine, and other high explosives, as well as smokeless powder, are all too strong for such an experiment, but calculations indicate that they would, under such conditions, give pressures of around 360,000 pounds per square inch. Fulminate of mercury, which is even stronger, would give nearly five times as much pressure as black powder, or about 412,000 pounds per square inch.

This explains very clearly why smokeless powder is potentially more dangerous in guns than black powder. Greater care must be used in loading smokeless than in loading black powder. A small change in weight of charge may reduce the airspace in a cartridge and change the whole character of the burning of the powder to something like a detonation. About the general limit of safe loading with any powder is 90% of the chamber volume. If this is exceeded even with a normally slow burning powder, detonation can occur. This 10% of free space, however, includes the space between the grains and inside the central hole. Nothing like this can occur with black powder, which burns in the same way whether it is compressed or loose. In a muzzleloader a double charge can be used, or two bullets be seated in the gun, without disaster.

It is well for the legal authorities to realize that ordinary smokeless rifle powder, such as is used by rifle club members and others for reloading cartridges, is not the type of explosive that can be used to blow things up. A misapprehension on this score has caused much unnecessary commotion among the police and others at times.

Unknown ammunition loaded with smokeless powder is always open to a certain amount of distrust, particularly if it may have been handloaded. However, age plays no part in this. The authors have fired smokeless powder ammunition loaded more than half a century ago without any ill effects whatever. The aging process does not, save to a very small extent due to the initial drying out of the powder, add to chamber pressure. Even decomposition and spontaneous combustion are now extremely remote; certain inhibitors prevent the formation of the old nitrous acids. There is no such thing as smokeless or black powder cartridges detonating because of the long period of storage. In at least one case where a revolver was ruined by firing a round of stored ammunition, several other similar loaded rounds were opened. Someone had inadvertently loaded all these rounds with at least a double charge of Bullseye powder, which would wreck a revolver even more surely when it was originally loaded than later.

Perhaps a word should be said at this time in connection with the many stories current of death due to the exploding of a cartridge while in the hand, after being thrown into the fire, or while in a hunting coat in front of a fire. These are without question mere old wives' tales. Both experiments and personal experience with .30-06 ammunition inadvertently tossed into an open fire leads to the absolute conclusion that mortal injury is impossible. A cartridge case of the usual type will rupture long before pressure sufficient to expel the bullets with any considerable velocity is reached. One of the authors was hit on the cheek by a .30-06 bullet from one of the old M1 cartridges, at a range of about 3 feet, while bending over a campfire, with no more than a slight bruise. A blow of this type in the eye might have led to loss of sight; however, there is no possibility whatever of any real penetration from any standard round of ammunition safe to shoot in a small arm.

To demonstrate a point in an actual case recently, a member of the Firearms Identification Section of the Massachusetts State Police put a rifle cartridge primer downward on a hot plate and held his hand, protected only by a light glove, over it until the bullet was expelled from the case by the primer going off. The blow was hardly perceptible; there was insufficient restriction inside the cartridge case to develop pressure to burn the powder.

EXPLODED CARTRIDGES

Above are shown the remains of a few cartridges of different type and caliber that have been exploded "out of the gun" and the fragments then gathered up and photographed. Nos. 1, 2 and 3 were fired by being heated in a lead pot with a corrugated cardboard box covering the top—No. 4 is the result of a thoughtless small boy setting off cartridges in a vise.

No. 1 is a .38 S. & W. smokeless powder, brass case. The bullet merely scraped an inch or so along the bottom of the melting pot. The case broke a hole through the corrugated board, which happened to be two thicknesses at the point where struck, it then fell back into the pot. The edge that is bent back over the interior of the case is what struck and this piece of flying brass would have made a nasty flesh wound at that 2″ distance.

No. 2 is a black powder .41 rim fire—an old cartridge. It was stood on its base in the melting pot; upon firing the bullet turned over in its two inches of flight and struck the corrugated board base first. It merely dented the board and is still in perfect shape—this bullet could be used over again.

No. 3 is a Remington copper-case .22 long rifle cartridge which was stood on its base in the pot and fired. The explosion blew the corrugated board off, but bullet and case remained in the pot, the bullet merely marking the corrugated board lying two inches above it.

No. 4 is a sadder story. This small piece of brass is a fragment from a .22 long rifle cartridge case taken from the eye of a 12-year old boy after the eye had been removed from his head. He had been amusing himself by clamping these .22 Hi Speed cartridges in a vise and exploding them with the point of a screwdriver struck with a hammer. His eyes were 18 inches above the vise and the fragment shown was blown into his left eye.

The authors are aware of stories that have appeared in the press from time to time about fatal accidents of this type. However, in every case where the details could be verified, the circumstances were not as originally stated. An English game keeper was indeed killed by shot entering his body from a bag containing loaded shotgun shells. The shot charge had come entirely through the bag from a gun accidentally fired by an old friend three feet away. The shooting had been intentionally misrepresented in the press, for it was beyond question accidental. Vague rumors were heard of two men blown up by a single case of .22 rimfire ammunition when a single bullet hit it. Several cases of this were subjected to dozens of shots at Marlin. A very few primed rims went off and no explosion at all could be caused in this way.

A loaded small arms ammunition collection is no more dangerous in a home or elsewhere than a group of old celluloid photographic negatives. Stored smokeless powder is no more dangerous in a home than various petroleum products, paint, and the like in equal volume. We point out, however, that we are speaking only of propellant powders and small arms ammunition and not in connection with the various high explosives and artillery shells to be discussed in the next chapter.

APPENDIX 1

The Manufacture of Double Base Powders

FOREWORD: Through the courtesy of Hercules Powder Company we are able to give an account of the manufacture of double base powders as seen at their Kenvil, New Jersey plant. The authors wish to point out that the descriptions are their own and the Hercules Powder Company is in no way responsible for the shortcomings that may be evident to those well versed in the industry.

Upon entering the plant one is immediately struck by the wide dispersal of the buildings. The Kenvil Plant resembles good open bird-shooting country with pine trees and thickets separating the buildings. The manufacturing structures are intentionally of light construction. Doors are held by friction fastenings and the walls, except those protecting the operators, are intended to stand only a reasonable internal pressure. In some instances the roof is made so as to lift away with a very moderate pressure. Since explosive operations can be hazardous unless safety rules and devices are constantly enforced, every effort is made to protect the employees. It is interesting to note that an enviable safety record has been compiled by the Industry. Only the Ladies Garment Industry has fewer occupational injuries.

Smokeless powders are either the single or double base types, that is the principal ingredients are nitrocellulose only or nitrocellulose and nitroglycerine. Those made by Hercules for commercial use are of the latter type because they are non-hygroscopic and generally more efficient. The proportions of nitroglycerine to nitrocellulose are varied from 10-90 to 40-60. The higher the nitroglycerine content, the hotter and more powerful the powder, other things being equal.

Besides the main ingredients, nitrocellulose and nitroglycerine, certain deterrents, stabilizers, certain metallic salts of nitrates, graphite, and small quantities of other chemical materials are used in making double base powders.

Wood pulp and cotton linters are nitrated at the Hercules Parlin, New Jersey plant. Usually a blend of nitrations is used, the blend being being made from 13% and 12.6% nitrocellulose, depending on the type of powder to be made. The nitrated cellulose is transported to Kenvil in a water wet condition. As long as it is correctly packed in heavy steel drums and carried in an enclosed truck, it is comparatively insensitive and safe to handle. Nitroglycerine is made at the Kenvil

Plant and transported to the point of use dissolved in acetone, which makes it relatively insensitive.

The water in the nitrocellulose is removed by displacing it with alcohol in a hydraulic press. Weighed quantities of the dehydrated nitrocellulose and nitroglycerine are added to a mixing bowl in relatively small quantities, and the mass is mechanically mixed with the chemical ingredients at a controlled temperature for about three hours. The temperature is controlled by a water jacket surrounding the bowl. The solvents, alcohol and acetone, are used only in sufficient quantity to form a dough rather than a solution. This has been popularly known as a colloid. It is then transferred to a blocking press where it is compacted into cylindrical blocks resembling the old "store cheeses" in both consistency and shape.

The blocks are then put into extrusion presses and the material forced through dies by means of hydraulic pressure of 4000-6000 pounds per square inch. The extruded powder comes out in the form of long spaghetti-like strands about the size of fishing lines, which are collected by coiling into a fibre drum. In the case of Bullseye, the strands are cut into thin wafer-like discs by feeding into a specially constructed cutting machine with a revolving head containing many knives. By varying the speed of the strand feed and the speed of the revolving knife, the thickness of the flakes can be controlled. In Bullseye, the thickness, which is the web in the case of flake powders, is very thin, 1/275". Water is sprayed on the revolving knife head during the cutting operation to keep it cool.

Essentially the same procedure is used for producing the long hollow grains used in rifle ammunition. The grains usually have a single hole running through the center of the cylinder. Cannon grains have up to seven such perforations. The web in these cases is the thinnest dimension, that is, the distance from the outer diameter to the nearest edge of the perforation. In powders which are made of the same ingredients the web controls the rate of burning.

Drying of smokeless powder is accomplished in one manner by spreading it out in shallow trays and blowing heated air over it. The drying time depends on the web of the powder and varies from a couple of days to more than a week for large grains.

After drying, the powder may be coated with a deterrent material to control the initial rate of burning. Some powders are left uncoated. The intended use dictates the procedure. Shotgun, .22 caliber and pistol powders are usually fine cut flakes, because the entire powder charge must be burned rapidly in order to propel the bullet as desired. This type of powder would be less likely to need coating than a coarser rifle granulation. However, this is a generalized statement; it is often necessary to coat even these powders to counteract the effect of over-ignition. While on the subject, it might well be mentioned that the proper ignition of powder is too often ignored. The primer bears the same relation to correct performance in the gun as does the spark plug in a gasoline engine. The best gas performs poorly if the spark plug is fouled or the spark plug incorrectly set. Powder is a fuel that must be ignited and burned properly and exactly if it is to propel the bullet as uniformly as is required.

The coating is accomplished by tumbling the powder in a temperature-controlled copper barrel. The coating material is sometimes added dry, sometimes with water, or maybe dissolved in a solvent. Again the choice of procedure is dictated by the effect to be accomplished; it is originally determined only by experimentation by the research group.

After coating, the powder is redried and then glazed with graphite by again tumbling in a barrel. The purpose is to make it conductive so that static electricity will not build up, and to make it free flowing.

The glazed powder is then screened to remove partial grains (tailings from the cutting procedure) or pieces that may have accidentally been broken from grains during processing. Samples are then sent to the Ballistic Laboratory for exhaustive physical and ballistic tests. The burning speed is determined and various batches are blended together to achieve the correct ballistic characteristics. Some of the tests made are:

(a) Determination of the moisture and volatiles content.

(b) Physical check on grain size, flowing characteristics and the volume occupied by specified quantities.

(c) Burning speed, velocity, pressure, recoil and unburned powder as compared with a standard or reference powder.

(d) An accurate check on the ignition and burning characteristics, peak chamber pressure, barrel time and muzzle pressure are obtained by Piezo pressure time curves which portray these characteristics on an oscillograph screen. The phenomena are then photographed and measured.

Chemical analyses are also run, as well as checks on the stability, by short term laboratory tests. Samples are stored in a surveillance oven at 150°F with controlled light and humidity for the life of the powder which is, on the average, well over a year under these extreme conditions. A similar sample is also stored in an unheated magazine for five years with annual checks on its stability. Every 10th lot is kept for the life of the powder, which is apparently more than 50 years.

Double base powders normally have a longer life than single base powders. Both are considerably more stable and long lasting than generally realized. Samples of the first Infallible made by Hercules in 1912 are still on hand and show no sign of deterioration. Powder made in 1899 is kept in the laboratory stored under water, it is tested periodically; the last ballistic tests were made in 1949 after 50 years of such storage. The powder was dried for five minutes in a convection oven and gave ballistic results only slightly under those of the reference powder.

APPENDIX 2

The Hercules Smokeless Powder Types

FOREWORD: Through the courtesy of the Hercules Powder Company, we are reprinting below information made available to the general public by this company in connection with some of the powders that they make. We point out that the Hercules Powder Company makes several powders upon which data is generally not released publicly. These powders may or may not be available for sale in large containers. For instance, the Hercules EC Blank Fire Powder is not available. However, certain grades of Unique not advertised may sometimes be bought. We would like to point out that among the Hercules powders described below, there are frequently several different grades, only one grade of which is put up in small containers. For instance, there are several grades of 2400, only one of which—a relatively slow burning type—is available in small containers. All Hercules powders sold under their trade mark are of the double base type, although they have made for the United States Government and for other governments a great deal of single base powder.

A series of the more popular brands of Hercules smokeless powders are now packed in small containers for the convenience of reloaders. The five grades selected for sale to individuals include at least one powder suitable for loading any pistol, rifle, or shotgun cartridge. The standardization on these few grades insures the user that he is always buying fresh material produced to meet the high Hercules standards of quality and uniformity.

Hercules HiVel No. 2 Rifle Powder

HiVel No. 2 Smokeless Powder is the slowest burning grade of the various rifle powders sold to the retail trade. It is developed primarily for loading the .30/'06 Springfield, Krag, and other types of large capacity cartridge cases and can be used in most of the bottleneck cartridges. It gives a high degree of accuracy with full loads in these cartridges, as well as with lower charges which may develop pressures around 50% of the full load. It owes its remarkable accuracy to ease of ignition and uniform burning. It is made in small cylinders with a single perforation.

HiVel No. 2 is generally accepted as the standard powder for the famous International free-rifle loads and for long-range prone rifle shooting.

The powder has been on the market since 1908 with an unusual record of storage stability. There has not been a single case of decomposition under normal storage conditions.

HiVel No. 2 Powder requires a relatively lower weight of charge for standard velocity with lower breech pressure than most other powders used in this class of cartridges. Because of its rather large grain size, much more uniform charges can be obtained by weighing than by measuring. HiVel No. 2 Powder is eminently adapted for use with metal case factory bullets; it has also been found entirely satisfactory with gas check cast bullets.

Available in 20-lb. kegs and 1-lb. cans.

Hercules 2400 Rifle Powder

Hercules 2400 is a progressive-burning, smokeless rifle powder of a rather fine granulation. It was developed especially for such centerfire rifle cartridges as the .22 Hornet, the .25/20, .38/40, .44/40, and others. It introduces to this class of ammunition the ease of ignition, uniform burning, low hygroscopicity, outstanding accuracy, and low weight of charge which characterize HiVel No. 2 in the large volume rifle cartridges.

Because of its grain form, Hercules 2400 Powder gives exceptionally uniform charges when thrown by any of the adjustable commercial measures.

Hercules 2400 is quite flexible in its use. It gives outstanding accuracy and satisfactory results in loads somewhat below maximum in many of the larger cartridges, such as .25/35, .30/30, and .30/'06, and is well adapted for use with the lighter weight bullets available for many of these large volume cartridges; but it is not intended for maximum loads in any cartridge except for the small volume cartridge cases of the .22 Hornet class.

Hercules 2400 can be loaded in practically the entire list of rifle cartridges from the .22 Hornet up to and including medium-power loads in the .405 W.C.F. and in cartridges of the Self-Loading Class, such as the .32 W.S.L., .35 W.S.L., and .351 W.S.L, in which few other powders will give satisfactory performance

Available in 20-lb. kegs and 1-lb. cans.

Hercules Unique Rifle or Pistol Powder

Unique was first manufactured about the year 1898, and has been continuously on the American market ever since. Reloaders find it a stable and dependable powder for low-power or gallery loads for all caliber centerfire rifle cartridges.

Hercules Unique gives very uniform and accurate short-range target and hunting loads. It is especially well adapted for use with lightweight bullets of either the lead or jacketed type, for mid-range work in the military line of cartridges or the so-called sporting rifle type. One of the greatest difficulties with light charges for the larger bottleneck, high-power rifle cartridges is to obtain ignition and uniform burning of the few grains comprising the charge in the large cartridge case. Unique fills this requirement very satisfactorily, and no matter whether the rifle is fired horizontally or vertically, the results will still be uniform. Owners of heavy-frame centerfire pistols will find that Unique will give them the high velocities for which those guns are designed without unduly high pressures.

Some Unique powder stored in a laboratory for almost 50 years still retains its original strength. One sample has been under water for that period and when the surface water is removed by means of blotting paper, the powder still gives normal ballistics.

Hercules Unique should not be used for maximum loads or full charges in any rifle cartridge. It is intended primarily for use as a mid-range or reduced power load. Uniform charges of Unique can be thrown by any adjustable measure, although it is desirable to check the weight of charges on an accurate scale or apothecary's balance.

Available in 15-lb. kegs and 13 oz. cans.

Hercules Bullseye Pistol Powder

Hercules Bullseye revolver and pistol powder is a very fast-burning, dense, non-hygroscopic smokeless powder which is well adapted for use in light and standard loads for all revolver and pistol cartridges. Since its manufacture in 1900, it has been loaded by the United States Government and by commercial loading companies for revolver cartridges as well as for automatic pistols.

Bullseye ignites easily with any regular smokeless powder primer. Its ease of ignition means that the powder will perform uniformly regardless of the position in the cartridge case. This is important because charges of Bullseye occupy only a small portion of the cartridge case, and these few grains must ignite easily and burn evenly regardless of the position of the gun when fired. Variation in the ignition of a powder may affect the rate of burning and change the ballistics of any given load. The low weights of charge used for standard ballistics make Bullseye especially economical to load. In reloading revolver or pistol ammunition, it is recommended that the mouth of the cartridge case be crimped tightly to the bullet so that there is no danger of the bullet moving back in the cartridge, thus reducing air space and causing high pressures.

It is not necessary to clean the fired cartridge cases before reloading with Bullseye powder. Because of its homogeneous, dense, colloidal composition, it does not absorb water and it is not affected by the residue which may remain in the fired shell.

Available in 15-lb. kegs and 11-oz. cans.

Hercules Red Dot Shotgun Powder

Hercules Red Dot Smokeless Shotgun powder is one of the most popular modern powders on the American market. It is intended primarily for use with light or medium-heavy loads in all gauge shotguns and is especially adapted for use in skeet and trap loads.

Hercules Red Dot Shotgun powder has been extensively loaded by all cartridge companies since its introduction. It is also popular with the handloader who desires to prepare his own ammunition. Any modern smokeless powder primer furnishes sufficient flame and heat to ignite it properly. Its composition and regular-sized grains give it uniform burning so necessary for consistently high percentage patterns.

Hercules Red Dot powder contains a small percentage of red grains blended with the regular black grains. The red grains are of the same composition and ballistics and are used only to provide positive identification of Hercules Red Dot powder.

Hercules Red Dot is intended for use in flatbase shells adapted for bulk powders. Its low weight of charge needed to give standard ballistics makes it economical to load. It is recommended for use with light or medium loads in any gauge.

CHAPTER 7

BOMBS AND EXPLOSIVES

HOMEMADE infernal machines, grenades to be thrown by hand, and high-explosive war souvenirs certainly are not firearms; however, in most police departments there will be no other division so well equipped to handle cases of this type as the Firearms Investigation Laboratory. Unfortunately, there are not many bomb disposal experts; it is not an attractive occupation. Some of these men have been killed in the course of their work.

An explosive hand-thrown missile is probably as old as firearms. In fact, there is reason to believe that gunpowder was used for such purposes long before it was used for discharging projectiles from tubes. The word "grenadier," so common in European armies, was derived originally from the fact that these companies or regiments were supposed to throw grenades. This method of making war fell into disuse in the Nineteenth Century save in connection with siege operations but was, of course, revived during World War I and has continued common since then.

The early grenades were filled with black powder. However, since in most military explosive work a maximum of pressure is desired instantly the new nitro explosives are far more efficient. The basic military explosive before World War II was T.N.T., which was highly stable, safe, and extremely shock-resistant. Dynamite was more suitable for commercial application since it was cheaper, usually more powerful, and just about as safe when fresh. In order to deliver their potential power these two explosives must be caused to detonate. Either may be safely burned in small quantities in an open fire.

Blasting Caps and Detonators

The means used to cause this detonation was and is some form of detonating cap which is usually nothing more than a copper capsule filled with fulminate of mercury, one of the most sudden or "brusque" of explosions. These caps are extremely dangerous, for they will explode from any moderate abuse, and when they do explode, they go off with extreme violence. A common item in the daily papers is, "boy pounds blasting cap, loses fingers." Blasting caps are cylindrical in shape, somewhat smaller in diameter than a lead pencil, and from an inch to two inches, more or less, in length. If you find any such object as this, act with extreme caution. Unless you know exactly what it is, do not handle it anymore than absolutely necessary to put it where it can do no future harm, as at the bottom of a deep river. Even the man who knows the danger is not immune. On May 1, 1934, Captain Clarence O'Leary, Ordnance Department, U.S.A., an expert on explosives, and the author of the Safety Manual for handling them, was investigating the theft of some small-arms

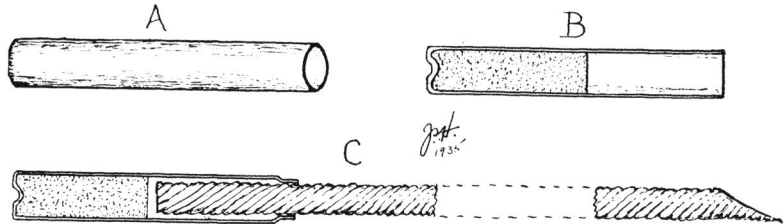

A. Number 6 Blasting Cap, full size. B. Cross section of same. C. Blasting cap crimped onto a length of safety fuze. Note that inner end of fuze must be cut square across, while distant end may be split or cut on a slant as shown for ease of lighting if desired. Fuze burns usually at either 30 seconds or 45 seconds per foot.

ammunition and, alone at the time, was searching in a magazine area on a large Army Post. Chancing upon a culvert, he noticed a piece of metal protruding from the earth. He pulled it out with a stick, and started to rub the mud off of it preparatory to examining it, when it exploded, tearing away his fingers, and inflicting severe wounds on his chest, abdomen and face. With iron nerve, he walked 100 yards to his parked automobile, managed to start it, and alone and injured as he was, drove approximately a mile to the Post Hospital. On arriving there, he walked to the door, and was carried to the operating room. He died four days later. The object that caused this accident was a bomb fuse detonator, that had lain in the mud since the time, ten years previously, when some defective ammunition had been destroyed there.

A vital point and one which cannot be emphasized too greatly in regard to blasting caps and detonators is that they are intended solely to be used to set off a much larger explosion of some substance which detonates. The caps themselves are very much more sensitive than these explosives that they are intended to detonate. Therefore, it is one of the universal rules of safety in handling and storing explosives, never to keep the detonators or blasting caps near the high explosives, until the moment they are put together to produce the explosion. They should not be stored in the same room, nor should they be handled or transported together in quantity. The particular lesson in this statement is intended for those members of the police who may not have had an opportunity to learn about the safety precautions required with such materials, but who very naturally might, in connection with an arrest of bomb makers, or seizure of an illegal stock of explosives very innocently carry all the components involved together in one conveyance to one place of storage.

Wartime Souvenirs

Throughout the land today there seem to be hundreds of wartime explosives in private hands. Blocks of T.N.T. were recently floating about an eastern harbor after military maneuvers. Frequently the Firearms Laboratory will receive for identification both artillery shells and hand grenades brought home by returning servicemen as souvenirs. Usually the hand grenades will have been emptied of explosives, although not always. A patrolman recently found some children in a New Jersey town playing with what he thought was a hand grenade. He picked it up and questioned the boys about it. No one seemed to know anything definite. He took it into the station where it was sent along to the Firearms Investigation Laboratory. It turned out to be a fragmentation grenade completely charged and capped but the safety pin was still in place.

High explosive shells are numerous and likely to be charged. They get more dangerous with age. High explosive shells from World War I have had the original explosive rendered unstable by long storage and may detonate with the slightest movement. Explosives of this type should be disposed of with as little handling as possible as common sense dictates. One method is to detonate them in a pit with a new block of TNT fired electrically. Another arrangement is to drop them into really deep water.

Whereas war souvenirs will probably be with us for a long time, they do not constitute any large-scale problem. In general, most people unacquainted with them have a dread of them that far exceeds their power. After all a 75mm high-explosive shell, whereas it can kill a number of people, cannot wreck a city block.

Homemade Bombs

These have always been used to some extent by criminally inclined persons. The notorious Englishman, Guy Fawkes, planned to blow up the Houses of Parliament with gunpowder on November 5, 1605. However, beginning in the 1920s, handmade bombs of various sorts used by gangsters began to be a very serious problem. They

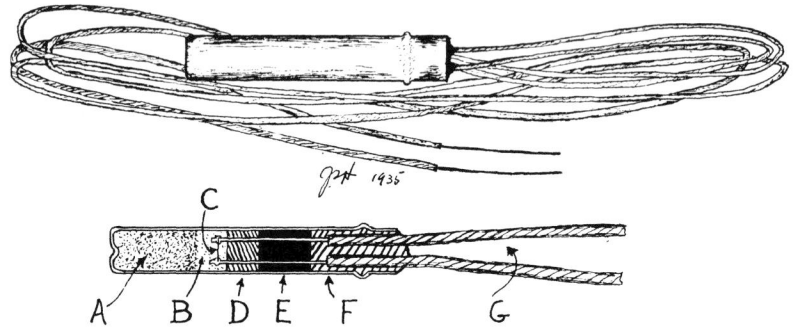

Top. Electric blasting cap, complete. Bottom. Section of electric blasting cap. A. Compressed explosive. B. Loose explosive. C. Bridge wire which heats when current is applied and so fires the cap. D. Sulphur and asphalt. E. Asphalt. F. Sulphur. G. Lead-in wires.

can be dangerous and are fairly frequently employed in some sections of the country even today.

The simplest of all of these bombs is a section of pipe from 1½″ to 4″ in diameter, commonly known as a nipple, with pipe caps screwed on each end. Into the enclosed space a quantity of black or blasting powder can be placed and a piece of safety fuse run into it through a hole drilled in one of the caps. This fuse can be lit from a cigarette or a match and the bomb tossed where desired by the criminal. Bombs of this type were considerably used to intimidate small store owners in the various protection rackets which flourished before World War II. The bombs are so simple that they can be made by anyone. The ingredients can be obtained without trouble. No skill is required for their combination and use.

Occasionally, this same form of bomb has been made from smokeless powder and even from match heads. If the pipe nipple in question is of high quality, with first-rate pipe caps, sufficient pressure can be built up within the bomb itself before it bursts to get a fair percentage of the energy from the smokeless powder. The igniting heads from common "book" safety matches, carefully taken off with a knife, produce an explosive even more powerful than black powder, although it is more dangerous to handle.

However, the damage potential of none of these fillings approaches that of dynamite, which can usually be obtained by criminals without too much trouble simply by burglarizing a quarry or construction storage shanty over some weekend. Dynamite requires a little more knowledge and training to handle but is almost as easy to use. A number of sticks can be taped together into the form of a bomb and a blasting cap with fuse attached placed in one of them. The fuse can be lit and the bundle thrown in the same manner as with a black powder pineapple.

A few sentences in regard to dynamite in its various forms and its use may be in order. As already pointed out, dynamite is nitroglycerine absorbed in some material such as sawdust, fuller's earth, kieselguhr or the like. It is sold in this country according to percentage of nitroglycerine contained in the entire stick. For instance, 40% dynamite is less strong than 60% dynamite, since the latter contains 20% more nitroglycerine. Dynamite usually used by criminals is the standard half pound stick eight inches long by 1¼" in diameter wrapped in paraffin-covered heavy brown paper. The quarter pound stick is also fairly common.

The detonators used in bombs designed to be thrown are of the type operating from burning fuse. This fuse has a center composed of mealed black powder which burns in the standard American safety type fuse at a rate of approximately 32 seconds per lineal foot before it explodes the cap and detonates the dynamite. This safety fuse once ignited will burn under water and is quite impervious to shock and the like so long as it is taped firmly in place.

Another type is fired electrically from a distance. The passage of current through the circuit sets off the fulminate which is in contact with a fine wire suddenly made red hot.

Infernal Machines

Criminals do not always throw their bombs. In fact, to a certain type of mentality, the sending of a bomb through the mail or the placing of it in the home of an unsuspecting individual seems to have a tremendous appeal. Many bombs have been sent or placed by cranks in the past—some of them with fatal results. However, not all of these bombs are used by cranks. In the twenties, a number of gangsters learned to place several sticks of dynamite under the seat or floorboards of a rival's automobile with an electric blasting cap in place wired so as to explode when the ignition was turned on, when the starter was pressed, or the like. Further, these same individuals learned to plant similar bombs in the homes of their enemies which would go off with the turning on of a light or almost any appliance which could be utilized to close a circuit.

Political criminals have used these contrivances abroad. A bomb of unknown origin exploded when being carried from the British Pavillion at the New York World's Fair in 1939, killing and wounding policemen. Investigation of these infernal machines are without question about the hardest task that any police officer will ever encounter. Some of them are made so that there is nothing whatever that anyone can do to prevent them from exploding. They are truly infernal.

Perhaps the simplest of these was the early type of delivered bomb masquerading as something else which would explode upon some-

A stick of dynamite or "cartirdge," as it is called in explosives parlance. A common size is 1¼ inches in diameter and 8 inches long.

one opening the package. It would not explode if the package was not opened. If a bomb has been delivered by mail or express a short time before being reported to the police it is undoubtedly of this type and can probably be safely removed from the premises and placed where it will do no harm if it does blow up.

A slightly more complicated type of bomb is one containing a timing device. It will blow at a predetermined fixed time if not opened. It may and usually does have also an arrangement for detonation if it is opened before the time decided upon. Mail and express deliveries are not usually so predictable as to make this type of firing adaptable to their use. A more advantageous arrangement is to hide a time bomb near where an intended victim will be at a predictable future moment. The assassination of Hitler by the German Generals was attempted with a bomb of this type.

The most difficult type of bomb or infernal machine to combat will be contained in a parcel, suitcase, overnight bag, typewriter case, or the like, which mysteriously appears where it should not be—as, for instance, on the front porch of a judge's home, on the desk of a municipal official, or in the lobby of a bank. Anyone finding such an unexplained object regardless of the circumstances should investigate fully without allowing it to be moved, tilted, or in any way disturbed. Police officers will not generally find these themselves, but will be called upon to investigate them by others from time to time when the circumstances are sufficiently suspicious to arouse the recipient or finders to action.

The first step that a police officer should take in a case like this is to clear the area of all persons, place a responsible individual to make sure that it remains clear and summon help. Save in the very large cities no one will know very much about bombs, but someone has to take over.

A great deal of common sense is necessary in this. For instance,

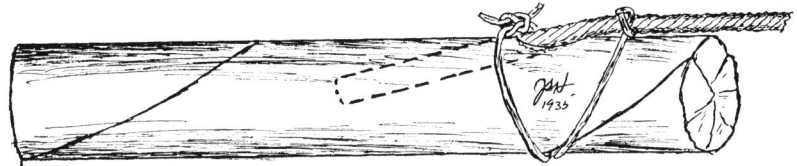

Blasting cap, with fuse attached, imbedded in a stick of dynamite ready for inserting in position for firing. When several sticks are to be fired at once, only one has a cap in it, and this one is called the"Primer."

if a bomb must be moved, an expert will probably figure out some method of doing it without getting himself killed; however, please remember that many of these individuals have finally been outwitted by the insane cranks that make these machines and have been blown up. Military bomb-disposal work is relatively safe in comparison. Three policemen skilled in this work were injured recently in a large Eastern city while testing homemade grenades seized in a raid.

A policeman's life and the life of innocent civilians are far more important than damage to property. After all unnecessary personnel have been removed from the immediate area of a suspicious package and perhaps the water, gas and electricity turned off, a bomb expert can get to work. Save in most exceptional cases, protective devices of various types such as mats used by contractors for rock busting and shields that will direct the force of the explosion where it will do the least harm can be employed.

The most important attribute of a bomb expert next to common sense and caution is a thorough knowledge of what may be inside an infernal machine. There is no foolproof way of finding out what the bomb contains although a fluoroscopic examination with portable X-ray equipment may yield a great deal of information. Such X-ray equipment can give photographs showing the same details.

It is important to remember that the usual infernal machine cannot possibly contain any tremendous amount of explosive. People sometimes believe that three or four sticks of dynamite can blow down blocks and blocks of buildings. This is, of course, not the case. A few sticks of dynamite could easily kill a number of people if they are concentrated around the bomb when it explodes; however, remember how much in the way of aerial bombing was necessary to destroy by blast the various structures destroyed in Europe during World War II. A 25-pound bomb would in general contain from 10 to 15 pounds of explosive. Such a bomb was considered very small but will usually be a lot larger and more powerful than the largest infernal machine.

Intuitively or by the help of the fluoroscope an expert may guess the construction of a bomb and how to keep it from going off. If at all possible he should personally stand at a non-fatal distance and do his manipulating with grapnels operated by cords and pulleys, preferably behind a shield. No definite procedures are possible.

One of the authors made a bomb for demonstration purposes that has never been successfully moved or tilted. When set but charged only with a small firecracker several experts as well as countless recruits have carefully approached the box. Some of the skillful have in one way or another counteracted some of the firing devices but not all. It has invariably gone off as soon as actually moved. A bomb so made could not possibly be handled; it would have to be exploded where it was discovered. A blasting mat of wirerope will smother any portable infernal machine.

A great deal that is said about bombs is sheer nonsense. For instance, immersion in oil or water may or may not hurt them. It can cause them to go off. If a bomb can be moved at all it should be placed where it can be disposed of without danger. If a bomb is known to contain a clockwork firing mechanism—determined by an external listening device—oil immersion may stop this device, if the oil reaches the mechanism.

A word of caution may be in order in connection with the definite identification of a bomb before it is destroyed. A police department in New Jersey still has a somewhat red face in connection with a suspicious package found on the front porch of the dwelling of a judge who was at the time presiding at the trial of some well-known criminals. The judge immediately reported the package to the police who came out with their portable fluoroscopic equipment. The equipment unfortunately was not working quite right. Their diagnosis that the package contained a bomb was based on the picking up of a few springs, levers, arms, and pins made of steel. The package was moved most gingerly into a trailer behind a police car and escorted by motorcycle policemen several miles to a suitable range. The package was then subjected to rifle fire to set off the explosion where it would do no harm. The entire procedure was undoubtedly completely in order, save that after five rounds, the local Chief of Police came running out to say that the judge's daughter had bought him a new wire recorder which she had left on his front porch, since no one was at home. A wire recorder with five .30-06 bullets through its middle is not of much account.

The potential of the criminal and crank has been greatly increased by various techniques and new materials developed during World War II. Many different types of high explosives have been developed in addition to TNT and dynamite. Such things as PETN, pentolite, RDX, C, C2, C3, and other new military explosives could easily have been stolen from various military establishments and used by the criminal. However, these explosives are, for the criminal's purposes, no better than the various forms of dynamite which have been available with certain restrictions for many years. "More powerful" is a relative term with explosives. The very maximum possible power per pound is not in excess of the potential of two pounds of dynamite. Atomic bombs will not be made by petty criminals in the near future. Further, many of the military explosives are harder to set off. They may require not only blasting caps, but also booster charges of other explosives in order to get them started.

Perhaps the greatest potential danger is in the use of techniques of booby traps and tension and compression firing devices made for them to government specifications. Some of these are in storage and they turn up from time to time but usually not in the hands of criminals. Apparently the men who fought for their country, especially those with sufficient intelligence to be picked for special explosives training, do not usually become criminals. A change in economic conditions could perhaps lead to a partial change in this condition.

Primacord

There is one thing that all police officers should be on the lookout for, since it resembles another similar but quite different product. Safety fuse, which has been called, erroneously, in the past at times "primer cord," is the stuff that burns at the rate of 32 seconds to the foot. Occasionally instantaneous cord or instantaneous fuse is found, which burns at a much higher rate, although it is far from instantaneous, in really long lengths. This is red in color and rough in texture; it burns at 120 feet per second. Both these products are used commercially in blasting and in other operations. They should, of course, not be confused since someone believing that he had safety fuse, when he had actually instantaneous fuse, might easily blow himself up after lighting it. However, this is a mere nothing compared to so-called Primacord which is actually a detonating cord yellow and green in color and pebbled in outside texture. This cord has many uses in warfare and in demolition work. For instance, a single detonating high explosive may be set off either electrically or by time fuse connected by Primacord looped about it and then looped about other explosives. The detonation will be communicated from the original explosive to all other explosives connected to it by the Primacord instantaneously. Actually, this detonation wave passes along the cord at the rate of about 7,000 meters per second. By means of this, a bridge can be blown up at both ends simultaneously and dropped precisely into a stream. Rails can have good long chunks blown out of them in this manner without the aid of other explosives. A couple of turns of prima cord wrapped around a telephone pole will usually take off the pole neatly at that point. Many ingenious uses of this cord were thought of in connection with defense of isolated positions and booby traps.

Chemicals and Incendiaries

Various chemicals have been used in the past for both explosive and incendiary purposes. One of the oldest applications is in connection with the action of various acids on separators made of cork, rubber and the like. Acids will eat through these other substances in a reasonably-accurately determined time interval. In other words, chemical action can be substituted for the more usual cheap clock in time bombs. The acid can then cause an explosion or fire, either directly or by mechanical means. Sulphuric acid and chlorate tablets used for sore throats produce a rather violent explosion when they come together. Acid, once through a partition, can release a pre-loaded spring firing device.

Incendiary bombs have, of course, been used for a long time, particularly by arsonists making fraudulent claims on the insurance companies. Whereas these can be ingenious and highly complicated, they are beyond the limits of this discussion. Fortunately, however, most of this type of bomb can be detected after it has done its work which, of course, defeats the purpose of the crime.

Not all bombs used in the protectionist rackets were explosive. Some of them were designed to create odors of extreme potency that would be hard to eradicate from merchandise. Others merely threw particles of acid or dyes about over a large radius to ruin stock in retail stores of various types.

Miscellaneous

Perhaps a few words as regards explosives used by safe burglars in the past might be in order. These men were at times real artists in what they were able to do. They used a liquid explosive which was almost pure nitroglycerine; they boiled it out of dynamite on a kitchen stove. Nitroglycerine used to be carried around by these men in glass bottles. It was, of course, extremely dangerous. More than one of them was killed when his supply inadvertently went off while in his pocket. A clever man was, however, capable of introducing this viscous liquid between the door of a safe and the safe proper and then exploding it in such a way as to open the safe neatly and easily. These men occasionally used office carpet, furniture and the like to deaden the explosion and to minimize the shock, so that people in the near vicinity of the explosion were completely unconscious of it. A superblaster is said to have once opened a safe without disturbing the office cat on its top.

Identifying Explosions

Occasionally, a Firearms Identification man doubling in connection with explosion work will be called upon to identify the source of an explosion from evidence left by the explosion itself. Frequently there is a question as to whether the explosion came from natural causes as, for instance, in connection with leaking cooking gas, or gasoline vapor, or was caused criminally. In general, explosions of accumulations of gas will be identified by a fairly gentle push over a large area. For instance, where the whole side of a house has been pushed out some form of gas explosion should be suspected. However, where an explosive occupying a relatively small volume, such as a high explosive bomb, has done the work, there will in general be a focal point that can be fairly definitely determined. A bomb explosion will not generally push out the side of a building but may easily send fragments of the bomb itself or of material in the immedite vicinity of the bomb quite long distances. A diligent search of the area surrounding a bomb explosion will generally yield one or more fragments of the bomb itself. Amazing as it may seem, in many cases the actual fingerprints of the crank who made the bomb can still be brought out on these fragments.

Fire may or may not ensue after a gas explosion; it is rare in the case of ordinary explosive bombs. Incendiary bombs are, of course, intended primarily to cause fires.

There are three cartridges that will fit into the chamber of any firearm chambered for the .22 long rifle cartridge, and these are the .22 short, .22 long and .22 long rifle. Years ago, the .22 short was the most popular, but it has entirely lost out to the more highly developed and more powerful and accurate .22 long rifle. The .22 long, which has no advantages over the .22 short and is not nearly as good as the .22 long rifle, is running a poor third in popularity. In fact, it has no real reason for existence, but it still sells in some of the less enlightened parts of the country, simply because some types of intellect will always take 'long' in preference to 'short' just because to them it sounds bigger and better. In 1934, the standardization committee of the Sporting Arms & Ammunition Manufacturer's Institute decided to drop the .22 long, but one company declined to go along with the committee's recommendation and the others all followed suit, so that this practically obsolete .22 long cartridge is still made, though the chances are it will not be very much longer.

All modern .22 caliber hunting and target rifles are now chambered for the .22 long rifle cartridge but the .22 short is used to a great extent in settled communities and by boys and others of limited means.

When the .22 short cartridge is used solely, or to excess, in the longer chamber it erodes or 'burns out' the forward portion of the long rifle chamber. With continued use, it may become difficult or even impossible to extract the longer fired cases by normal means.

The above illustration shows .22 long rifle cartridge cases which were fired in a rifle in which shorts had been used to a considerable extent. Note how the forward third of these cases have been scraped by the eroded front of the chamber. This is a mild condition shown here, but the degree of erosion is permanent and will increase with further use of the short cartridge.

CHAPTER 8

MARKINGS ON FIREARMS AND CARTRIDGES

THE Firearms Investigator will come in contact with many different markings on evidence presented to him. Unfortunately, firearms and their ammunitions have now been made for so long that even in modern investigations occasionally markings will turn up that confuse completely the most experienced men. However, all recognizable marks on firearms and ammunition should be immediately recorded.

Marks on Arms

Every handgun made in America today and for several years past is stamped with the maker's name and address, the caliber, and most important of all a number referring to that particular specimen and no other, called a serial number. This serial number is by far the most important marking on a firearm, although there are several qualifications in particular makes and models.

Some automatics and revolvers as well as many shoulder weapons will be encountered completely devoid of all serial numbers and even distinguishing marks. Sometimes the marks on them will be meaningless or obviously fraudulent. For instance, awhile ago, a crime was committed in New Jersey with a weapon marked "Joseph Manton, London Fine Twist." This was a damascus double-barreled shotgun made in Belgium; however, the only indication of its Belgian manufacture was in the proof marks, which are not visible until the gun is disassembled.

Further, some Spanish-made revolvers outwardly resemble fine American revolvers but are devoid of markings. Other similar weapons will have deliberately misleading markings such as ".38 Smith & Wesson cartridge" or "Frontier Six-shooter." Some of these weapons are not marked at all.

The Expert will receive into his laboratory in general a far greater proportion of weapons that are reasonably clearly marked and whose

175

general place of origin is immediately apparent. For instance, the most numerous class will contain the maker's name, address, and probably a serial number. These should all be recorded and checked with any papers which are received with the weapon when it arrives in the laboratory. A simple transposition of two digits of a serial number greatly handicapped the prosecution in the famous Sacco-Vanzetti case.

In some instances a suspect or crime weapon may have not only manufacturers marks but also identifications placed on the weapon by the detective who discovered it or by others at various times. Although these extra markings are largely unnecessary on weapons with bonafide serial numbers, they should also be recorded. The prosecution must establish identity between a weapon introduced as evidence and that taken from a suspect or connected with him in some way. Recorded marks of various types on the weapon are the simplest means of proving this.

Handgun Serial Numbers

There is a second use of marks on firearms and particularly serial numbers. In theory a police laboratory should be able to trace a weapon from the factory through its various owners to the man who used it in a crime. This is almost never the case. In fact all the time, money, and effort spent in recording each transfer is largely wasted.

The system is supposed to work in connection with handguns about as follows: The manufacturer stamps each weapon with a different serial number and records the name and address of the jobber to whom it is sold. The jobber in turn should record the name of the dealer who buys it from him. The dealer then records the name of the purchaser. Each time the weapon changes hands thereafter a record should be made of the transaction and it should be recorded with appropriate authorities.

No one with any knowledge of firearms thinks that such a system would work. It is not even legally required. The manufacturer must make a record of his sales; dealers are required to do the same in most states. However, jobbers are not usually required to keep records. People will not bother reporting sales or gifts or small items of personal property, even if the items happen to shoot. Smith & Wesson has records going back to 1870; Colt and Stevens go back almost as far. But all these will tell is the name of the jobber, who is extremely unlikely to be in business today. Even if he is, he is not likely to keep records which cannot possibly do him any good but may cause him to waste valuable time checking them and possibly take up space needed for other purposes.

All this theory is based upon serial numbers. A few words in connection with them, therefore, are in order. For instance, most revolvers have their serial numbers stamped in more than one position. At least one well known make of revolvers has this number in five different places, the positions of which cannot be given in any generally available publication; however, these positions are revealed to any Police Department upon application to the company. No single revolver of this make received over a long period of years in the New Jersey laboratory has had every one of these five numbers obliterated.

There are quite a number of conditions encountered reasonably often by the various Police Departments that have led to delay and erroneous information. We quote below in its entirety a letter from the Colt company in connection with this subject:

"Following out your request to outline such information as we require to positively identify the model of a Revolver on which you want our record of disposition—on our present line of Revolvers known by Models such as New Service, New Service Target and Shooting Master—these three models are in the same series of serial numbers. The Official Police .32/20 caliber, .38 and .41, and Officers Model Target .32 caliber and .38 are in the same series of serial numbers. The Official Police and Officers Model in .22 caliber have their own series of serial numbers; they are not in the same series as are the other calibers in this model. It would, therefore, be necessary to specify caliber when requesting information on these two models.

"It will be necessary to specify caliber as well as model and serial number when requesting information on Police Positive Revolvers. The reason for this is that below 300,000, we ran the .32 caliber arms on a smaller frame, and with its own series of numbers. At 300,000, we combined the .32 and .38 caliber Police Positive Revolvers into the same series of numbers, and discontinued the small frame Revolver.

"Our greatest problem is when Police Officials request information on .38 Police Positive Revolver, and what they really want is a record on .38 Police Positive Special Revolver. We give the information requested, and after checking and finding that .38 Police Positive Revolver is still in the hands of the purchaser, we receive a call to please re-check our records. We then know what the trouble is, and give information on Police Positive *Special* Revolver, advising the one sending in the inquiry of the difference between the two models—as each of these models has its own series of numbers, it is necessary to clearly identify the model when asking for a record.

"The model roll on the barrels read as follows:

"On the Police Positive model—Colt Police Positive .38, and the Police Positive Special model—Police Positive Special. Where the caliber of the Police Positive Special is .38, this numeral is placed before the words "Special," and when the caliber is .32/20, the markings follow the word "Special."

"The serial numbers on all Revolvers manufactured by us with cylinders that swing out (with one exception the New Army and New Navy models, which were marked on the butt) are found on the frame and crane where the cylinder swings out.

"The barrels of our Double Action and New Army model Revolvers are marked the same—that is, Colt D.A. .38. The Double Action model Revolver has a grip that is shaped and very often called "bird's bill grip," while the New Army model Revolver has the square butt identical with our Official Police Model Revolver. The serial number on the Double Action Revolver was placed at three points; the butt, front of trigger guard, and bottom front of frame. The New Army Model Revolver has its serial number on the butt, there is a number on the crane and frame, but it is of no assistance to us in looking up records—we must have the butt number.

"When requesting information on .45 caliber Automatic—when the Pistol is the Commercial model always mention the letter "C" before the number. If you fail to mention this—the information you will receive will be for the model supplied to the United States Government without the letter "C" before it.

"We also manufactured a .45 caliber Automatic in a model called the Military Model which was patterned after our .38 Automatic Pistol, Military Model; this model should be defined as the Old Military Model.

"When requesting information on .38 caliber Automatic—care should be exercised in specifying the model; this is because we have the Super .38 with its series of numbers; the discontinued models of .38 caliber Pocket Model and its series of numbers, and the Military Model with its series of numbers, and do not drop the "O" when requesting information on our .380 caliber as is very often done.

"Our New Service Revolvers manufactured for the United States

Government marked on butt Model of 1909, and Model of 1917—please give if possible both the Colt serial number found on the frame and crane, and the Government serial number found on the butt of the gun. This will minimize the time required in looking up our records.

"We trust you will find some information that will be of assistance in the above, and please feel free to ask any questions on any models that may be a puzzle to you, and we will do our best to assist."

Tracing from a manufacturer's records is impossible for weapons made by defunct firms, most foreign factories, and even by manufacturers in business today, if the firearm was made sometime ago. Harrington and Richardson informs us that they can give no information on weapons with serial numbers lower than those listed below:

Model Name	Caliber	Serial No.
American D. A.	.32	168251
American D. A.	.38	168351
Safety Hammer	.32	175651
Safety Hammer	.38	180751
H & R Bull Dog	.38	178651
Victor Large Frame	.32	22701
Victor Large Frame	.38	22601
Model 4	.32	111301
Model 4	.38	111251
Model 6	.22	98001
Model 5	.32	85101
Young America	.22	446151
Young America	.32	444651
Automatic	.32	474101
Automatic	.38	473671
Victor Small Frame	.22	98751
Victor Small Frame	.32	99751
Premier	.22	452301
Premier	.32	452501
Hammerless	.22	18851
Hammerless	.32	257101
Hammerless	.38	173251

A great deal could be written about the positions of the various numbers on weapons manufactured in America alone; however, in general, these numbers are visible without disassembly in at least one place. The individual investigator will do well to look over with care every weapon that comes into his hands to familiarize himself with the probable position of numbers on similar weapons.

The U. S. Model 1911A1 .45 automatic is frequently used in crime. However trying to trace a crime weapon of this type is almost always useless. The answer from the Ordnance Department after an appropriate delay is usually, "this pistol has been traced to such and such a combat area where accountability ceased."

Based upon the experience of Eastern Police Departments known to the authors not more than one handgun in 10 used in crime is of a type that tracing by serial number is possible. More than 90% are over age or foreign. Even where tracing is possible it seldom is conclusive. Criminals seldom use a weapon that can be traced to them.

Firearms Registration

Many states including New Jersey have purchase permit laws. A citizen must get formal approval for his purchase of a handgun from the local Chief of Police in advance. Reputable dealers abide by these laws; however, in most towns there are exceptions. Further, more than half the handguns in any given state were not purchased from dealers at all. Finally reference to a collected purchase permit file, as in the central office of each state, will reveal the most astonishing irregularities. Patent dates are frequently given as serial numbers. Impossible calibers are listed. Few people with experience have any confidence in the information contained on some of the permits.

Even if one state should manage to accomplish the really tremendous task of getting all its purchase permits in order and have its citizens register all their handguns properly, that state would merely disarm its honest citizens. Only a tiny percentage of honest people are willing to go through the red tape, trials, and tribulations in order to keep a firearm they may need desperately later. Such regulation would accomplish little since the average honest citizen does not go about committing crimes of violence with his gun. The criminal who does commit such crimes is not going to bother to register his weapon or purchase it legitimately from a dealer with a purchase permit obtained from the local Chief of Police. Disarming the citizen aids crime. Few criminals want to meet on an equal footing a man ready and able to defend his home, his wife, and his property. Criminals want stringent firearms regulation.

The task of disarming the criminal is impossible. A revolver represents high value for relatively small bulk. If the government could not stop the importation and distribution of alcohol during Prohibition, it has no chance to stamp out traffic in arms. Only an honest citizen worries about a fine or the extremely unlikely short prison term for possessing or carrying an illegal arm. The gangster, robber, and murderer have much more serious penalties to worry about.

Shoulder Weapon Serial Numbers

Firearms for shoulder firing are not legally required to have serial numbers. However, most high power rifles made in this country in the past years and at present do have them. There are pertinent exceptions to this, however. For instance, the Remington Model 721 and 722 do not have serial numbers; neither do certain of the Stevens-Savage centerfire rifles, although the Model 99s do. Many fine sporting weapons built on foreign actions will be unnumbered or have had the numbers unintentionally removed during refinishing and the like. Most fine American shotguns and many shotguns made abroad will have serial numbers. Many of the foreign rifles of the upper price brackets also will have serial numbers.

The largest single group of shoulder weapons which come into the laboratory for identification—the cheap modern low power rifles and shotguns—will have a maker's name or a familiar trademark but no serial numbers at all. The rifles and shotguns mass-produced for the large mail order houses will be marked with trade names of the house but not the manufacturer's name and will have no serial numbers. These weapons are so numerous around the country that most investigators are familiar with most of the variations. Any one having a legitimate interest in the names of the makers could obtain this information. Sometimes there is more than one source of manufacture for the same model. Further, in some instances, a given maker will produce a model for sale under his own name that differs little from that sold through the mail order house marked with the trade name of the retailers.

Reliability of Foreign Serial Numbers

Identification of evidence by recorded marks and serial numbers can be done with any weapon having them. However, positive tracing of foreign weapons falls down in many instances because several different weapons undoubtedly bore the same numbers. For instance, most of the war souvenirs fall into this classification, although not all of them. Most German automatic pistols made for the army bear serial numbers; however, these are obviously too low to mean what they do in this country. There must have been hundreds of thousands of Lugers, yet a serial number over 40,000 is unknown. Several different firms made these weapons. The German Ordnance Corps had a system of identification as to plant and number; their codes, however, are not completely known to us and would be of little use if they were. The records are gone. By giving all the numbers, prefix and suffix letters, data and other lettering stamped on these pistols, guns bearing the same serial number can be distinguished from others made in different months or years, or at different arsenals. For example, a Luger may bear the serial 2468a on the front end of receiver, the number 2468 but no suffix letter on slide; 1941 on top of slide, byf on top of bolt. Another may have 2468a, but 1939 or 1940 on top of slide, and/or DWM on top of bolt. These would show up in the records by serial numbers as the same gun, when in fact they are obviously not.

Even more confusing are the assembly numbers frequently used on several different parts of foreign military weapons. The last two digits of the real serial number will be repeated several different times on different pieces of the weapon as it originally went out of the factory. However, in service or since being brought to this country, many of these weapons have been disassembled and reassembled in combination with other similar arms, or had broken pieces replaced, so that two or more numbers appear on parts eventually assembled into a single arm.

Some commercial arms made in Europe apparently have serial numbers referring to an original order only. When the quantity made for one exporter was completed and another order started for a second firm, the serial numbers began again at 1.

Foreign military rifles may or may not have non-repeating serial numbers. Most Mauser actions will be numbered; however, during World War II several dozen different plants were manufacturing the Model 1898. Each of these different plants started off apparently with Serial No. 1. Further, some military weapons were not marked at all with serial numbers. Only a relatively few British service rifles are numbered. Apparently most of the numbered ones were made in this country, although the vast quantity of No. 4 SMLEs (sometimes called just Rifle #4) made by Stevens were unnumbered.

Restoration of Serial Numbers

Sometimes a weapon which should have a serial number does not. Usually this means that someone has deliberately removed it. Restora-

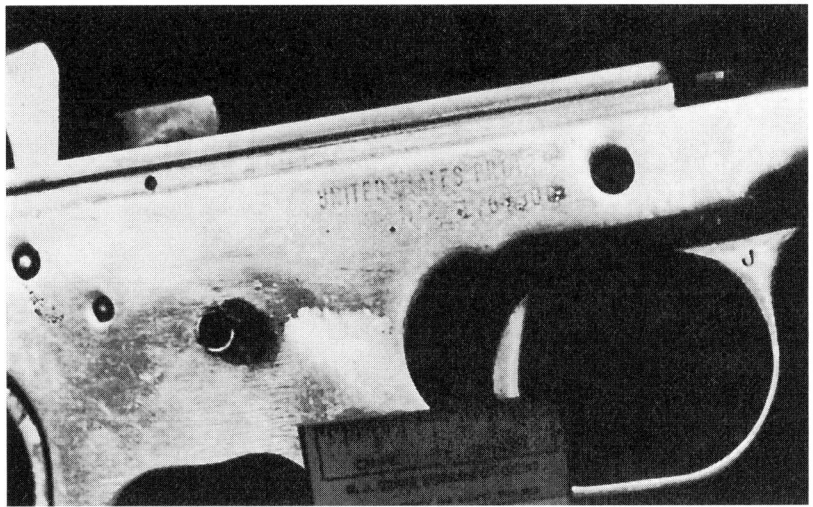

Illustrating a "recovered" serial number (1,767,300) on a .45 service automatic pistol. Serial number had been ground off and the gun inexpertly polished for reblueing—note the elongated holes, a sure sign of the untrained hand on a polishing wheel.

tions of serial numbers by either chemical or magnetic means are now well known procedures in every laboratory. Perhaps a few words as to their efficiency are in order. In general, any markings removed from steel by surface filing or grinding can be restored by the repeated application of various chemicals. The actual solutions depend somewhat on the metal to be treated. Those actually used in the New Jersey State Police laboratory are given below, though there are many others.

For first quality relatively modern weapons:

Concentrated Hydrochloric Acid	40 cc
Distilled Water	30 cc
Alcohol	25 cc
Copper Chloride Crystals	5 grams

For older and more abused weapons:

Concentrated Hydrochloric Acid	120 cc
Distilled Water	100 cc
Copper Chloride Crystals	90 grams

If these do not work, use a 4% Nitric Acid Solution in amyl alcohol. If this does not work, it will have so hurt the steel that there is very little likelihood of anything else working either.

The surface is polished to a high luster initially and is then repeatedly swabbed with the chemical, allowed to dry, rubbed bright, and then swabbed again. Sometimes numbers or words will not be legible until after several hours of such treatment; however, in general, the numbers will be restored sufficiently to be read within a far shorter time. The restoration should be photographed, for the numbers may fade out after a short time.

Another method for the restoration of numbers is by polishing bright and then subjecting the surface to so-called black light inspection. When the numbers were originally stamped in the metal, sufficient derangement of molecules was done below the actual stamping depth so that there will be a magnetic concentration of force in the outline of the original numbers, visible with this special equipment.

Roughly, about 90% of the serial numbers that are removed can be restored; however, the remaining 10% in general cannot be restored because of the method of removal. For instance, numbers that are center-punched out evenly and cleanly cannot be restored by any method now available. Further, we recently had in the laboratory a rifle made about 90 years ago. Certain marks were stamped on the barrel when accepted. We were completely unable to restore these marks even temporarily. Perhaps we were not using the proper chemical reagents for this old, mild iron. An electromagnetic inspection revealed so many flaws in the original longitudinal weld which coincided with the known position of the original marks that nothing whatever could be found by this method either.

The restoration of serial numbers is dramatic. The layman might believe that it aids tremendously in the solution of major crime. It seldom does; generally it merely proves that the firearm is stolen. Tracing a weapon will often reveal that an honest citizen lost it a few days before the crime, sometimes without missing it until the police call. Few men who possess a pistol legitimately commit a crime. Fewer still will file off the serial numbers of an arm that can be traced to them and use it in premeditated law breaking.

Government Ownership

There are quite a number of other markings on weapons. For instance, prominently displayed on many United States service pistols and revolvers is "United States Property." The actual ownership of these weapons is a debatable point. For instance, the authors have in their possession several Model 1917 Colt and Smith & Wesson revolvers which were purchased through the Director of Civilian

RESTORATION OF OBLITERATED NUMBERS

Upper panel shows portion of an automobile block with spurious serial number thereon.

Second panel from top shows same engine block with this spurious serial number ground off, preparatory to restoration of the original number—and this photo was taken before treatment to restore the correct number.

Following panel is a photo of engine block after treatment to restore the original number. The spurious number (CS646277) is plainly visible, and the original and correct number (AF393144) is faintly visible.

Bottom panel shows a different original number clearly recovered on another block. These engines were two of a number stolen by a gang specializing in stealing nearly-new automobiles, repainting them, re-numbering the engine and serial numbers and forging new bills of sale.

Marksmanship with a bill of sale, before World War II. There cannot be any question that these particular weapons were not stolen from the Government and were legitimately sold by the Government to the original owners; however, in only one case can we actually prove this with original orders and bills of sale. However, the mere absence of proof could not legitimately be used as a reason for instituting proceedings against us or any other of the very many owners of these weapons who may have legitimately purchased them from the Government and lost their papers or have obtained them legitimately elsewhere. All orders to the Director of Civilian Marksmanship had to be accompanied by a statement that the materiel purchased was for the individual use of the purchaser; however, the purchaser did not agree not to sell it or trade it after a period of several years. Further, many weapons have been inherited from

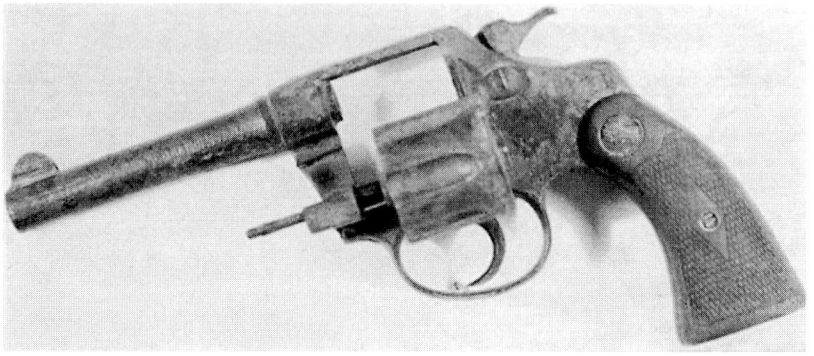

SERIAL NUMBER RESTORATION
This Colt revolver was so badly rusted away that its serial number was obliterated, yet proper cleaning and treatment restored it legibly enough to permit correct reading.

the original owners or legitimately sold by their estates. The American law presumes a man to be innocent until proven guilty. Even though one of these revolvers or even a Model 1911A1 Automatic —some of these were sold to regular and reserve officers—in a citizen's possession can be proven to have been stolen, the possessor is presumed innocent until it is proved that he was connected with the theft or knew that the weapon was stolen.

The same holds true for many other service weapons to a greater or lesser degree. The M1 carbine of World War II has an appeal to many people. Apparently many thousands of these were taken home, with or without permission, by servicemen at the end of their fighting. Technically, however, until quite recently none of these weapons was ever sold by the Government, nor were certain component parts of them ever sold. However, inadvertently, or perhaps because the Government no longer cared, quite a large number of receivers—the last component not sold as surplus—were legitimately sold to at least one dealer. He can sell you a complete M1 carbine

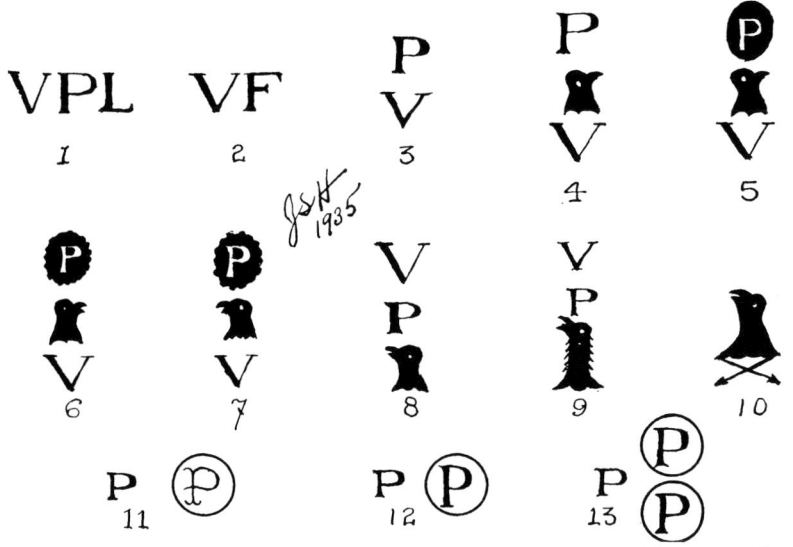

Various proof and view marks found on U. S. Government rifles or muskets. Their approximate periods are:—

1.	1799	5.	1819	10.	1869
2.	1802	6.	1824	11.	1878
3.	1803	7.	1837	12.	1903
4.	1806	8.	1842	13.	1934
		9.	1857		

with a bill of sale to accompany it which will show that the weapon is actually yours. This transaction eliminates the validity of statements formerly made that no one could possess a carbine legitimately.

Apparently the only time that the M1 Garand was ever sold by the Government was on Okinawa during the last days of World War II and immediately following; however, bills of sale for these weapons are now in the hands of civilian owners together with the weapons themselves. The mere losing of a bill of sale does not constitute a crime; however, of course, the Government may easily be able to trace a particular serial number and prove a conclusive case against an individual in possession of a Garand. Not many of them are private property, although they are to be sold in limited numbers to the contestants in the National Matches.

In the British service, a broad arrow has been the equivalent of "United States Property" for more than two centuries. In theory, at least, when a gun is legitimately sold by the British Government, a second broad arrow is placed directly opposed to the first; however, neither the British Government nor our own has ever been noted for following a single procedure in all its different divisions over a long period of years.

Proof Marks

Besides the marks already described, there is another class of markings on firearms that are very interesting and important to the Identification expert. These are the "proof marks" that are always found on arms made or sold in any of several European countries. The proof marks and their use are stringently regulated by law in the countries where they are used, so that they form a sure method of determining the country of origin of many foreign arms. This is likely to be a matter of considerable importance in the case of cheap firearms of uncertain make, as the proof mark may be the only real clue to the origin of the arm. Moreover, American arms sold in the countries requiring proof marks must bear a mark show-

AMERICAN PROOF MARKS

In the U. S. there is no central proof house, and no Federal law on the subject. Some makers regularly proof fire all guns, and others proof fire only those intended for export to countries requiring proof before sales. Each maker uses his own mark to indicate that the arm has been proved.

1. Colt Patent Fire Arms Co. proves all arms and uses the mark shown. 2. Harrington & Richardson proves arms intended for export, and marks those going to Australia and New Zealand. 3. Hunter Arms Co. prove all barrels and apply the mark shown. Iver-Johnson use the letter "P" in an oval for the provisional proof, 4 and the monogram IJP 5, for the definitive proof, 6. Marlin Arms Co. uses the letters JM in an oval. 7. O. F. Mossberg proves barrels intended for export, and stamps them with the letter "P," or with "P" in a circle, as shown. 8. Winchester Repeating Arms Co. subject all arms to a provisional proof, and later to a definitive proof, and use the monogram WP in an oval to indicate that the arm has received both proofs. Parker Brothers give all arms a definitive proof, 9. and stamp the barrels as shown. R. F. Sedgeley fires all rifles with a 50 percent overload, and marks them 10, with the letter "S" in an oval. 11. Remington Arms Co. prove all arms intended for export to countries requiring proof, and mark guns so proved with the letters R.E.P. (Remington English Proof) in an enclosed space, as shown, Savage Arms Co. give their arms both a provisional and a definitive proof, 12. and mark the rifles so proved with the letters "SP" in an oval. The shotguns are marked with the letters SAP 13, in an oval. Proof marks used in common by the Ithaca Gun Co., 14-19, the Lefever Arms Co., and the Western Gun Corp., all of whom are located in Ithaca, and conduct their tests jointly. Guns are fired with an overload giving about 40 percent excess pressure.

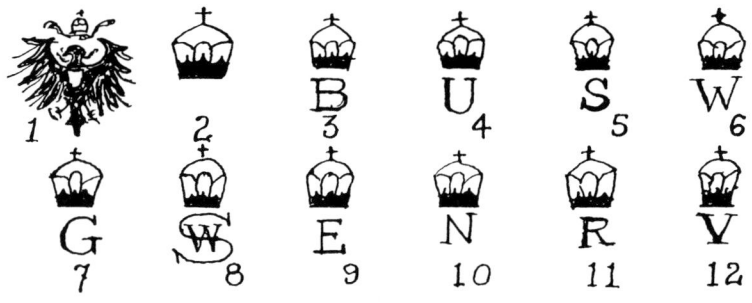

GERMAN

1. Provisional Proof of barrel.
2. Single proof for pistols and revolvers.
3. Single proof of rifles and shotguns.
4. Together with No. 1, Definitive Proof of barrels and actions.
5. Shotgun mark.
6. This mark appears on choke-bored shotgun barrels.
7. Mark for Rifled Shoulder Arms.
8. Shot guns with a rifled portion in the choke.
9. Mark for "Express" rifles.
10. Nitro powder proof.
11. Arm re-proved after modification.
12. Mark used on guns already completed, but unsold, on passage of the proof law of 1891.

ing that it is an arm of foreign make which has been tested and found safe for sale. Thus, for example, a Colt Revolver bearing British proof marks was probably sold in Great Britain. One of the authors has had in his possession several such arms.

Before proceeding any further with this subject of proof marks, it is in order to say a word about the proof tests themselves, and the reasons for them. It is almost the universal practice to test all firearms during or following the manufacture, to be sure that they are sufficiently strong to prevent their blowing up in the hands of the user.

As the part of a rifle or shotgun most likely to burst is the barrel, many manufacturers make it a practice to test the barrels in the rough, before assembling them into finished guns, by firing a very heavy charge in them, which subjects them to a much greater pressure than they will ever be called on to stand in ordinary service. Thus, if the barrel is weak, or the metal is seamy, the rough-bored barrel will split open or burst, and the maker will be saved the time and expense of finishing it up. Moreover, if the barrel should be assembled into a finished gun, and then should burst on the final proof, it would most likely ruin the other expensive parts of the gun, and would thus result in a considerable loss to the maker.

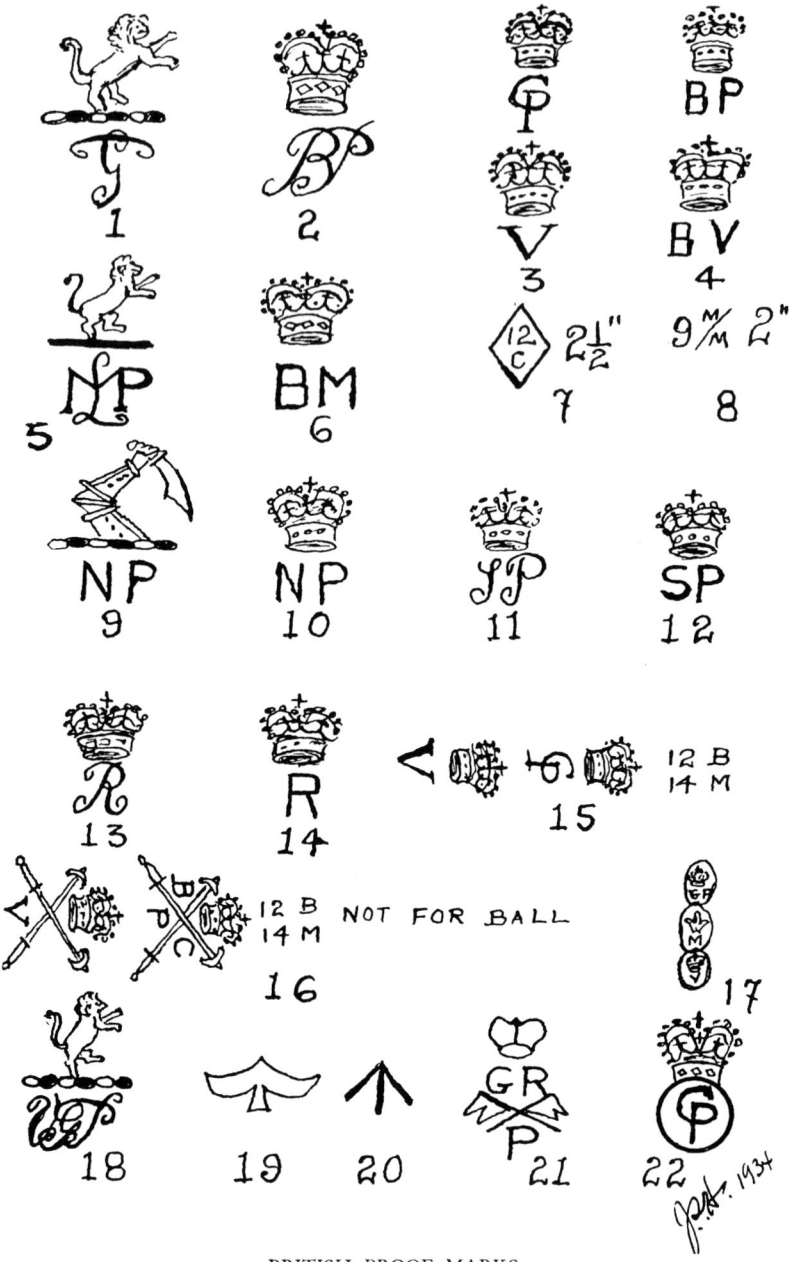

BRITISH PROOF MARKS

1. London Provisional Proof.
2. Birmingham Provisional Proof.
3. London Definitive Proof.
4. Birmingham Definitive Proof.
5. London Mark for special definitive proof of Rifle barrels, Class 10.

(Continued at bottom of opposite page.)

ITALIAN

1. National Small Arms Testing Bureau. Identifying mark of the Brescia Branch.
2. Identifying mark of the Gardone Branch.
3. Italian Proof House. Provisional Proof of barrel.
4. Italian Proof House. Mark on Rifled Arms.
5. Smokeless Powder Test. (Polvere Sensa Fumo).
6. Black Powder Test. (Polvere Nero).
7. Mark is added to the ones described above on finished arms.

Therefore, this first test is very desirable in cases where experience has shown that a certain percentage of the barrels may be expected to fail. Naturally, it is not to be recommended in the case of guns which never have to stand much pressure, such as a .22 rimfire, where the barrel does not ever have to shoot a high powered cartridge; and where the quality of the steel used in the barrels is so uniformly high that failures very rarely occur, it is cheaper to put off the testing until the entire gun is finished, for all high powered arms must be tested after completion, regardless of any previous test of the barrel. This test of the partly finished barrel is what is called a "Provisional Proof."

When the entire gun has been finished, it is inspected to see that it is properly and safely made, which is called "Viewing," and it is then fired with one or more high pressure charges, and some-

6. Birmingham Mark for special definitive proof of Rifle barrels, Class 10.
7. Chamber mark, breech loading shotguns. (Bore and length of cartridge.)
8. Chamber marks, breech loading rifled arms. (Caliber and length of cartridge.)
9. Nitro Proof, London.
10. Nitro Proof, Birmingham.
11. Special Proof, London.
12. Special Proof, Birmingham.
13. Re-proved Arms, London.
14. Re-proved Arms, Birmingham.
15. Choke Bore Mark, London, now obsolete. Upper figure, "12 B" shows gage at breech, lower figure, "14 M" shows gage at muzzle.
16. Choke Bore Mark, Birmingham.
17. An old London Mark, now long obsolete.
18. London Mark employed prior to new rules adopted in 1925.
19. The "Broad Arrow," sign of Government ownership.
20. Another form of the Broad Arrow.
21. Military Proof of rifles and automatic pistols.
22. London Definitive Proof for arms of foreign make. Followed by the words "Not English Make." Marks numbers 1, 2, 3, 4, 5, 6, 9, 10, 11 and 12 are treated in a similar way when applied to arms of foreign make after British proof; consequently all these marks may be seen with the letter or letters enclosed in a circle, and followed by "Not English Make."

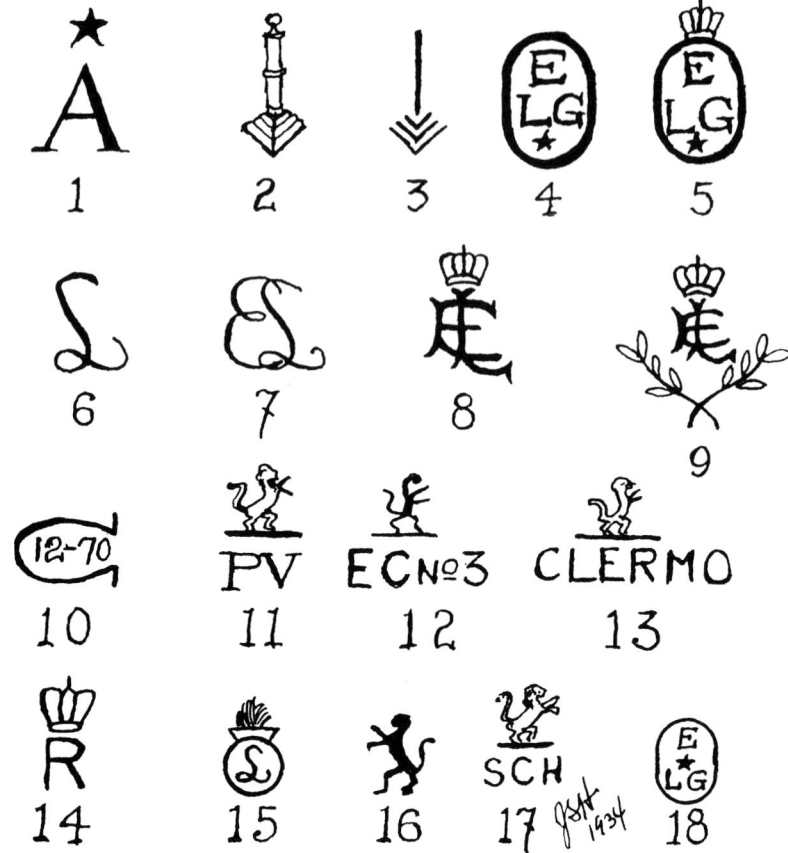

BELGIAN PROOF MARKS

1. Countermark of the Inspector. Any letter may be seen thus surmounted by a star, as the initial of the inspector is used.

2. The Monument, or "Perron" of Liege, indicating inspection after assembly.

3. A simplified form of the last described mark. It is more often seen as shown here.

4. The mark of acceptance.

5. Acceptance after proof with excessive charge.

6. Optional Provisional Proof of barrel for muzzle loader.

7. Standard Provisional Proof.

8. Double Proof.

9. Triple Proof.

10. Chamber stamp for shot guns. Shows bore dimension and chamber length.

11. Mark indicating arms of standard dimensions.

12. Special test with smokeless powder not officially recognized. This mark indicates test with E. C. Powder No. 3.

13. Same as last described mark, showing marking for another type of powder, in this case Clermonite.

14. Proof of barrel of low power target arm.

15. Proof of foreign arms.

16. Unfinished arms to be completed after exportation.

17. Mark indicating proof with a particular kind of smokeless powder.

18. An old Definitive Proof Mark, now obsolete.

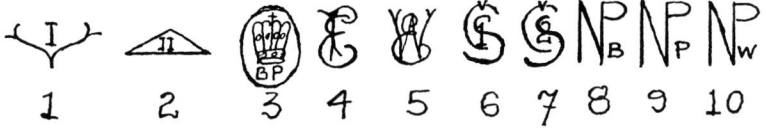

HUNGARIAN AND CZECHO-SLOVAKIAN

1. Buda-Pesth, 1st proof, single barreled shotguns.
2. Buda-Pesth, 2nd proof, double barreled shotguns.
3. Buda-Pesth, 3rd test on multi-barreled guns, the second on double-barreled shotguns, the one and only test on pistols, revolvers and small-bore arms.
4. Prague, 1st proof.
5. Weipert, 1st proof.
6. Prague & Weipert, 1st (Voluntary) proof of shotgun barrels.
7. Prague & Weipert, 2nd, (Obligatory) proof of shotgun barrels.
8. Nitro-proof, Buda-Pesth.
9. Nitro-proof, Prague.
10. Nitro-proof, Weipert.

times with regular charges also, which is called the final or "Definitive" proof.

In many countries, there are government rules or laws which forbid any gun to be placed on sale which has not passed certain prescribed inspections and proofs. In some countries these tests are made by a government proof house, and in others they are made by a proof establishment operated by a guild or co-operative association of gunmakers under government regulation. Some of the most notable proof establishments are those of Liege, in Belgium; the Gunmaker's Company, in London, and the Guardians in Birmingham, which are arms makers' guilds, operating under government sanction; and the proof houses of Eibar and of Barcelona in Spain.

In Great Britain, for example, no gun can be placed on sale which has not been proved according to regulations recognized by these establishments. Some countries forbid the importation of unproved arms. For this reason, some American makers submit their arms to a proof test corresponding to those required by the British regu-

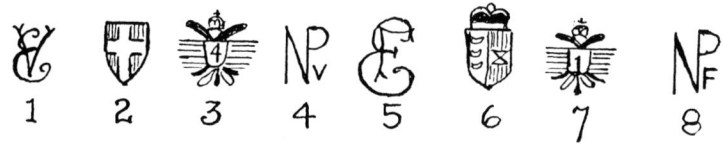

AUSTRIAN

The first four are Vienna marks, as follows:
1. The first proof on single barreled shotguns.
2. The second proof on double barreled shotguns.
3. Third proof on multi-barreled guns, the second on single barreled guns, and the one and only proof required on pistols, revolvers, and small-bore arms.
4. Nitro-powder proof.
The remaining four marks are the same as above, except for the proof house at Ferlach.

SPANISH PROOF MARKS

1. Present Eibar Proof.
2. Old Eibar Proof, under the monarchy.
3. Present Barcelona Proof.
4. Old Barcelona Proof.
5. Definitive Proof for smooth bore muzzle loading arms.
6. Provisional Proof for smooth bore breech loading or automatic guns.
7. Definitive Proof for the smooth bore breech loading or automatic guns.
8. Single Definitive Proof for rifled carbines or muskets.
9. Single Definitive Proof of low power pistols and carbines.
10. Single Definitive Proof of non-automatic pistols.
11. Old form of the mark just described.
12. Definite Proof of revolvers.
13. Definitive Proof of automatic pistols.
14. Definitive Proof of foreign made arms.

(Continued at bottom of opposite page.)

SPURIOUS PROOF MARKS

These are spurious markings found on unproved firearms. Such spurious proof marks, or markings intended to look like proof marks, are used by makers of cheap arms for the purpose of misleading the uninformed or unwary customers.

lations, and place a proof mark of their own on all guns so proved; but the practice is by no means universal, as there are no rules whatever requiring arms sold in the United States to be proved, or to bear a proof mark. However, it has always been the practice of our Ordnance Department to proof fire all guns made in the Government Arsenals, and to stamp the finished arm to indicate that the proof has been accomplished. Formerly there was a provisional proof as well as the definitive proof, but in late years the provisional proof of Government small arms has been omitted.

By far the most comprehensive work that has ever been published on the subject of proof regulations, proof practices in various nations, and proof marks, is the series of articles by Col. Calvin H. Goddard, Ordnance Reserve, in *Army Ordnance* (magazine) on "Proof Tests and Proof Marks," beginning in November-December 1933, and running in five parts to finish in July-August, 1934. This excellent paper has, we are glad to report, been reprinted, and at this writing is available from The American Ordnance Association, Mills Building, Washington, D. C. This monograph is profusely illustrated with cuts of practically every proof mark of which a record or description could be found, and it should be in the library of everyone interested in firearms.

15. Definitive Proof of rifled muzzle loaders.
16. Indication that arms have been inspected and meet the required tolerances.
17. Old form of "View Mark" just described.
18. Supplementary Proof of breech-loading and automatic shotguns.
19. Old form of last described mark.
20. Second supplementary proof of breech-loading automatic shotguns.
21. Single Definitive Proof of Breech loading shotguns.
22. Special Proof with excessive overcharge.
23-29. These are old marks now superseded.

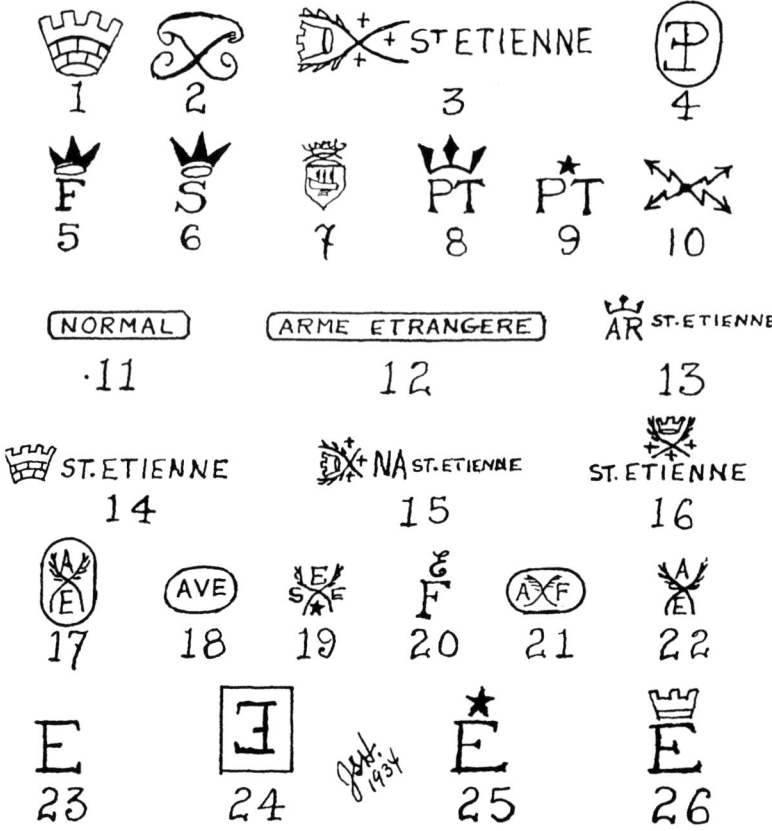

FRENCH PROOF MARKS

1. The St. Etienne Mark for provisional proof of separate shotgun barrels.
2. The Paris Mark for the same proof.
3. The St. Etienne Mark for definite proof of barrels finished and assembled.
4. The Paris Mark for the same proof.
5. St. Etienne Black Powder Proof of finished arms.
6. St. Etienne Special Black Powder Proof.
7. Paris Mark for standard Black Powder Proof.
8. St. Etienne Mark for Standard Test with Smokeless Powder, "Poudre T."
9. Paris Mark for this test.
10. Mark placed on arms when ready for sale and delivery.
11. Mark indicating arms of standard dimensions.
12. Mark placed on arms of foreign make.
13. Proof of Rifled Arms.
14. Proof of hand weapons.
15. Proof of barrels finished but not assembled.
16. An old St. Etienne Mark.
17-20. Old marks formerly used at St. Etienne.
21. Mark used on French Government made arms.
22. A mark placed on foreign made arms proved at St. Etienne. Special marks at one time placed on arms made at other French towns.
23. Chattellerault mark, the letter E.
24. Sailleville, reversed E in rectangle.
25. Mutzig, E surmounted by a star.
26. Tulle, E surmounted by a crown.

Cartridge Case Markings

The markings on cartridges and their components are of infinite variety. For instance, occasionally in the United States and fairly frequently in some other countries, obsolete weapons firing cartridges that are really cartridge collector's items are used. The shape, construction and make up of some cartridges are unusual. Their markings, of course, are as varied as the times and places of their manufacture. For instance, a pinfire revolver was recently taken from criminals on the New Jersey Turnpike. It was loaded and ready to be used. Actually, one chamber had been fired fairly recently. The remaining cartridges went off either on the first or second blow from the hammer. Pinfire ammunition is extremely rare and has not been used in any quantity over here for half a century.

In general, however, reasonably modern ammunition is used in more than 95% of all firearms crime. It is of types already discussed. The .22 rimfire is about the only frequently used rimfire ammunition in the country today. Centerfire cartridges will be found in many sizes for shotguns and rifles as well as pistols.

In general, the rimfire ammunition will be unmarked save for a symbol or letter stamped at the time the cartridge case is made on the head of the case itself. The different symbols and letters on recently manufactured ammunition in this country are limited; however, a good many variations can be found in cartridge collections.

Shotgun shells are almost always head stamped either with indented letters which are common in American brands, or with raised letters found in shells of European manufacture. Occasionally, shotgun shells are unmarked and may have been of private manufacture.

Rifle ammunition is almost always head stamped in the same way that shotgun shells are. Many varieties of ammunition of this type will be found in a large cartridge collection or even in one of the big police cartridge collections. However, in general, the usual centerfire rifle crime, which in itself represents only a small percentage of the investigation of any department, will be confined either to military ammunition or to a relatively small amount of recently manufactured commercial ammunition.

Finally, pistol cartridges follow pretty much the same line as rifle and shotgun ammunition, although more variety is met with. American commercial ammunition is stamped clearly, not only

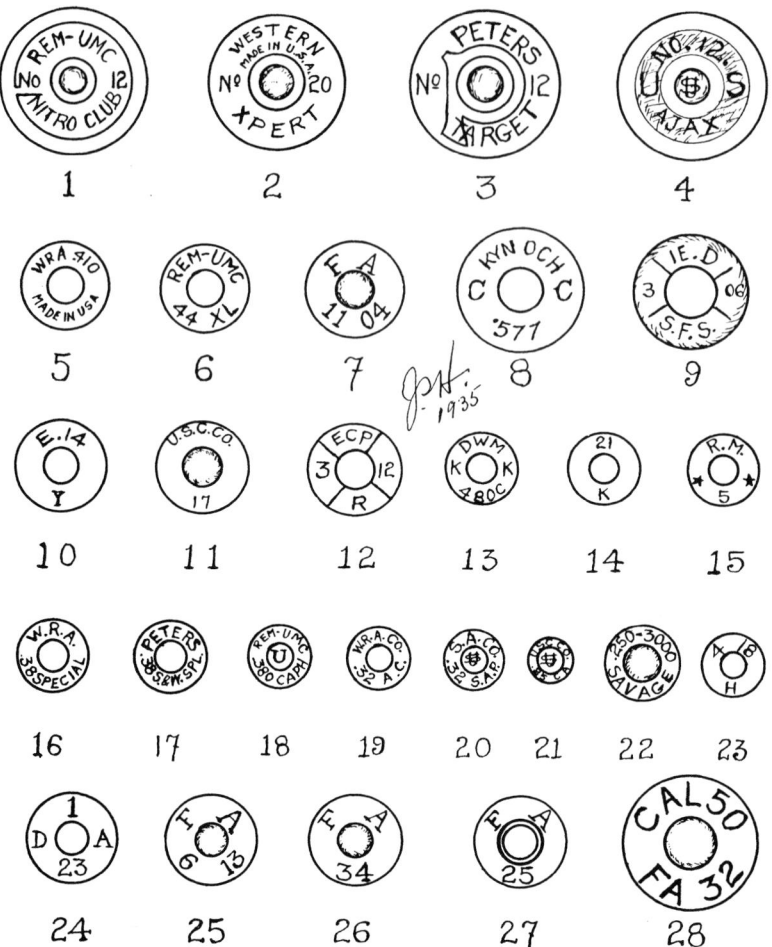

TYPICAL MARKINGS ON THE HEADS OF CARTRIDGES

These sketches were carefully made from actual samples in the author's possession. Both foreign and domestic ones are shown, and some rather old ones have purposely been included, for very often the cartridge that may figure in a shooting case has been lying forgotten for years in the chamber of some old gun in a household where arms are seldom or never used.

    1. Remington 12 gage shotgun shell.

    2. 20 gage shotgun shell made by Western Cartridge Company.

    3. 12 gage shotgun shell by Peters Cartridge Company.

    4. 12 gage shotgun shell by U. S. Cartridge Co. Note the monogram on the primer. Formerly it was the practice to place monogram or initial on primers used for smokeless powder, while those intended for black powder were left plain. At present, all standard primers are intended for use with smokeless powder, and will work equally well, with black, so that the initial on the primer is not often used.

    5. .410 gage shotgun shell by Winchester Repeating Arms Co.

    6. .44 Metallic shot cartridge, for use in rifles.

    7. .30 caliber Krag cartridge made at Frankford Arsenal, November, 1904.

    8. Cartridge for .577 Elephant gun, with 750 grain bullet and 110 grains of Cordite. Made by Kynoch, Ltd., in Great Britain.

(Continued at bottom of opposite page.)

with the maker's identification, but also the cartridge size. Large assortments of pistol and revolver cartridges are kept by all Firearms Laboratories. We will discuss this matter in greater detail later.

The firms loading ammunition in the United States today are even more limited in number than those making firearms. Two parent firms make almost all commercial ammunition sold: First, du Pont controls the Remington Arms Company, which makes cartridges both under Remington's own name and under the Peters trademark; second, Olin Industries, Inc., which sells ammunition under both the Winchester and Western trademarks. The Federal Cartridge Company produces only the large volume types of ammunition such as .22 rimfire, shotgun shells and primers. The United States Cartridge Company was bought out by Winchester years ago. The Union Metallic Cartridge Company has long been recognized as Remington and survives only in the REM-UMC head stamp.

No one familiar with service ammunition can have failed to wonder at the code letters found with the last two digits of the year of manufacture on these cartridge case bases. By special permission we quote as follows from *Hatcher's Notebook*.

9. 8 MM French service cartridge for the Lebel Rifle.
10. .455 Webley Self Loading Pistol Cartridge, made by Eley, London. Purchased in 1932.
11. Wartime .45 Automatic pistol cartridge, made by U. S. Ctg. Co.
12. Foreign 11 MM revolver cartridge.
13. 9 MM Luger Pistol Cartridge, made by Deutche Waffen und Munition Fabriken at Karlsruhe. The number 480C marked on the base is the factory pattern number of this particular cartridge. Other D. W. M. Pattern numbers are as follows:

403	7.63	MM Mauser	480D	9		MM Luger carbine
413A	6.5	MM Bergmann No. 3.	484	5		MM Clement
416A	5	MM Bergmann No. 2	487	9		MM Mauser
456B	9	MM Bergmann "Mars"	488	8		MM Bergmann Simplex
466	7.65	MM Mannlicher	497	7.65		MM Mannlicher carbine or carbine pistol
471	7.65	MM Luger				
497A	7.65	MM Mauser (.32 A. C. P.)	508A	6.35		MM Automatic (.25 A. C. P.)
480C	9	MM Luger Pistol	540	9		MM Short

14. 9 MM Luger pistol Cartridge, German make.
15. 9 MM Cartridge for Bayard Army Pistol.
16. .38 Smith & Wesson Special made by Winchester Repeating Arms Co.
17. .38 Smith & Wesson Special made by Peters Cartridge Co.
18. .380 made by Remington with the old marking ".380 CAPH." (Colt Auto Pistol Hammerless).
19. .32 A. C. P. made by Winchester.
20. .32 A. C. P. made by Savage, marked .32 S. A. P. (Savage Automatic Pistol).
21. .25 A. C. P. made by U. S. Ctg Co. (Old).
22. .250-3000 cartridge made by Savage Arms Co.
23. 9 MM Browning Short Cartridge, German made.
24. 8 MM Swiss service cartridge made at Government factory, Altdorf.
25. .45 caliber pistol cartridge made at Frankford Arsenal, showing old marking.
26. Same, showing present method of marking.
27. .30 caliber M 1 rifle cartridge, made at Frankford. Note that primer is crimped in.
28. .50 caliber ball cartridge, made at Frankford.

HEAD STAMPS ON U. S. SERVICE CARTRIDGES

D A .... Dominion Arsenal ....
D-18 .... Dominion Arsenal .... 1918 tracer
D I .... Defense Industries, Canada, Ltd. ...
D E N .... Denver Ordnance Plant ....
E C .... Evansville-Chrysler .... .45 Caliber only
E C S .... Evansville-Chrysler-Sunbeam .... .45 Caliber only
E W .... Eau Claire Ordnance Plant ....
F A .... Frankford Arsenal ....
F M .... Fabrica Nacional de Municiones
              (Mexico) None issued. .... Experimental.
G E .... General Electric Company ....
K S .... Allegheny Ordnance Plant .... .50 Caliber only
L C .... Lake City Ordnance Plant ....
L M .... Lowell Ordnance Plant .... .50 Caliber only
M .... Milwaukee Ordnance Plant .... .50 Caliber only
P C .... Kings Mills Ordnance Plant ....
P C C .... Peters Cartridge Company ....
Q A .... Dominion Arsenal (Quebec) ......
R A .... Remington Arms Company ....
T R .... Three Rivers (Quebec) ....
T W .... Twin Cities Ordnance Plant ....
U, U Y .... Utah Ordnance Plant ....
V C .... Verdun, Canada .... Non-corrosive
W, W 18 ...... Western Cartridge Company ....
W C W C C .. Western Cartridge Company ....
W R A .... Winchester Repeating Arms Company

Markings on Primers and Bullets

At one time, other components of ammunition were frequently marked. In some cases, companies would use primers bearing an identification symbol or letter. For instance, for many years, a "U" was placed on primers manufactured by the Union Metallic Cartridge Company which became a part of Remington Arms before that company was bought by du Pont. Similarly, Winchester used to stamp all its jacketed bullets with a small "w" on the side. The old United States Cartridge Company used to stamp a small "s" in the same place. Some jacketed bullets were stamped with a symbol or letter in the exposed lead at the base. Remington used a large "U." Today, however, the authors know of no bullets intentionally marked during manufacture. Some British primers are still stamped with the Imperial Chemical Industries symbol.

Besides the intentional stamps formerly placed on bullets, a great

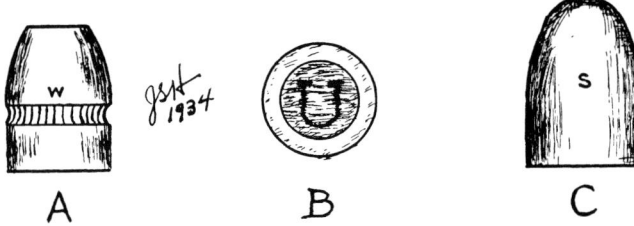

MARKINGS ON BULLETS

Years back, it was the custom of some manufacturers to stamp or otherwise designate the bullets of their make with initials or marks. Occasionally old ammunition will be found with bullets so marked. The examples shown above were sketches from actual samples.

A—Metal cased bullets with "W" stamped lightly into the side. Made by Winchester Repeating Arms Company. B—Base of .45 Automatic Pistol bullet, Remington make, with "U" stamped into the lead core exposed at the base. C—Another .45 Automatic pistol bullet, with an "S" stamped into its side, made by the old United States Cartridge Co.

deal of work has been done in the past on identifying each individual criminal bullet brought into laboratories with its original maker. This work has sometimes ended in utter confusion. An ammunition company itself, upon being queried as to the depth and serrations as well as the positions of its cannelures, has given information entirely in error. For instance, one large ammunition manufacturer finally admitted that the engineering department could prescribe the exact width, depth, and constructions of the knurling tools to make the bullet cannelures without getting much cooperation from the production department. If a toolmaker made a small error in the production of a certain manufacturing contrivance for impressing these grooves in each bullet, no one would scrap the tool. Further, even in one particular lot of ammunition, these tools would be changed so that at a certain point the bullets would suddenly change very considerably in appearance, although not much in over-all shooting performance since weight, diameter, and general shape would be the same.

There are certain peculiarities of loading that will mark bullets. For instance, they may be permanently indented by staking or crimping operations. This was particularly true of .45 pistol ammunition of World War I, where three point-stakes secured the bullet in the cartridge case.

In essence, the Firearms Investigator should be extremely hesitant about identifying a bullet from its physical characteristics alone unless it is obviously of a general type made by only one ammuni-

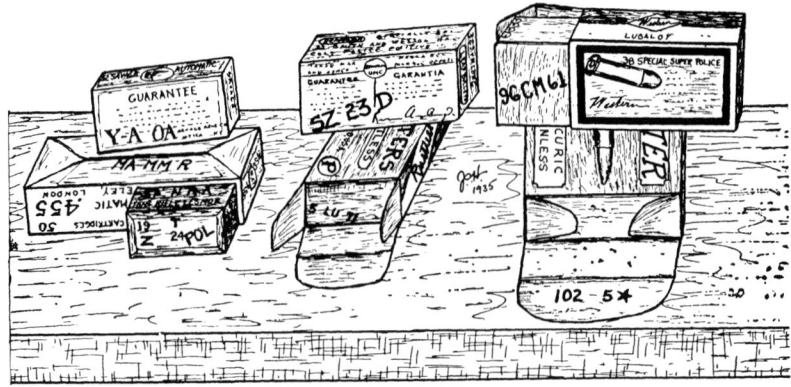

MANUFACTURER'S CODE MARKINGS AND LOT NUMBERS

If the customer will closely examine the box of cartridges or shotgun shells he has purchased, he will find, on the bottom of the box or tucked away on some odd place, a combination of numbers and letters, generally printed with a rubber stamp or numbering machine. These are the manufacturer's code recordings to indicate the lot number of the batch of ammunition from which particular box came. This information can, at times, be of value to the investigator; by means of these numbers he should be able to duplicate this loading of ammunition.

In the above sketch are shown a few examples of these code markings. They are:

Left, a box of 50 cartridges for the .455 Webley Self Loading Pistol, made by Eley brothers, London. The lot symbol MA MM R is found on the bottom of the wrapping which encloses the pasteboard box. On top of this is a box of .32 A. C. P. cartridges made by The Savage Arms Co. The lot symbol Y-A OA is stamped on the bottom of the box. Lying on its side in front of the other two is a box of U. S. Cartridge Company's .22 long rifle cartridges, which has a number of symbols stamped on the bottom.

Middle, A box of .32 S. & W. Longs of Remington make, lying on its side with bottom visible, showing the lot number SZ 23 D stamped over the guarantee. Under it is a box of Peters .38 Police Positive cartridges, showing the lot number 5 LU 71 stamped on the end of the inner sliding portion of the box.

Right, bottom view of a box of Western .38 Special Super Police, with lot number 96CM61 stamped on bottom of inner sliding part. It is lying on a box of Winchester Staynless .30 cal. U. S. Gov't rifle ammunition, with lot number 102 5* stamped inside the flap of carton.

tion company. An identification of this type is not essential or even of any particular significance. All that is required is to choose for test purposes an ammunition reasonably similar. A mistaken identification of the company that loaded a cartridge used in a crime is an unimportant detail, but could be used to considerably embarrass the Expert. Caution and common sense must be used here as in every other place in Firearms Identification. What the Firearms Investigator in general wants to determine is what sort of weapon may have been used and whether a given bullet did or did not come out of a suspect weapon.

It is rarely necessary, or even desirable, for a report to give any more data on evidence ammunition than is obvious. For instance, you will record the make and caliber of the cartridge cases recovered; you will rarely say more than that the bullet is of a given weight and caliber and whether lead or jacketed. What the Expert is really to determine is whether there is a positive or negative match between a suspect weapon and a criminal bullet. He need not, and frequently cannot, identify the criminal bullet as to maker.

APPENDIX

New York State Police on Serial Numbers

FOREWORD: We have received permission to reprint in its entirety from the bimonthly Bulletin of the Bureau of Criminal Investigation, New York State Police, Volume 18, No. 2, a most interesting monograph on serial numbers and their uses, including an actual case of duplicate Luger serial numbers. Everyone has known that duplication of numbers in this make of weapon is frequent; however, few departments ever have these duplicates. The New York State Police did have these, and have explained various possibilities lucidly and well.

FIREARMS NOTES—DUPLICATE SERIAL NUMBERS

The purpose of serial numbers placed on firearms is for the identification of that firearm by the manufacturers as well as distributors, law enforcement agencies, and owners of the firearm. Law enforcement agencies are particularly concerned with serial numbers of firearms in order to keep accurate records of firearms seized in crimes so that they may be checked against the records of the lost and stolen firearms.

The New York State Police Pistol Permit Bureau has at the present time approximately 250,000 pistol permits registered. This means that these persons possess at least one firearm. The routine of investigation of the Pistol Permit Bureau is that whenever a firearm is taken in by a police agency, the firearm is fully described in a teletype message directed to the Pistol Permit Bureau for check against their files. This description should include manufacture, caliber, serial number, model and type of firearm. When these facts are properly presented the Pistol Permit Bureau checks the lost and stolen firearms file and then checks the legally registered firearms file. This file is maintained first by serial number, then by manufacture and caliber. During the past, many duplicate serial numbers have been found. In most of these instances the firearms involved have been found to be of different manufacture or different caliber or different models.

However, very recently a 9mm Luger semi-automatic pistol was recovered from a burglar. This firearm was described as having the serial number 9165. A check with the Pistol Permit Bureau revealed that a 9mm Luger, serial number 9165, was carried on the pistol permit

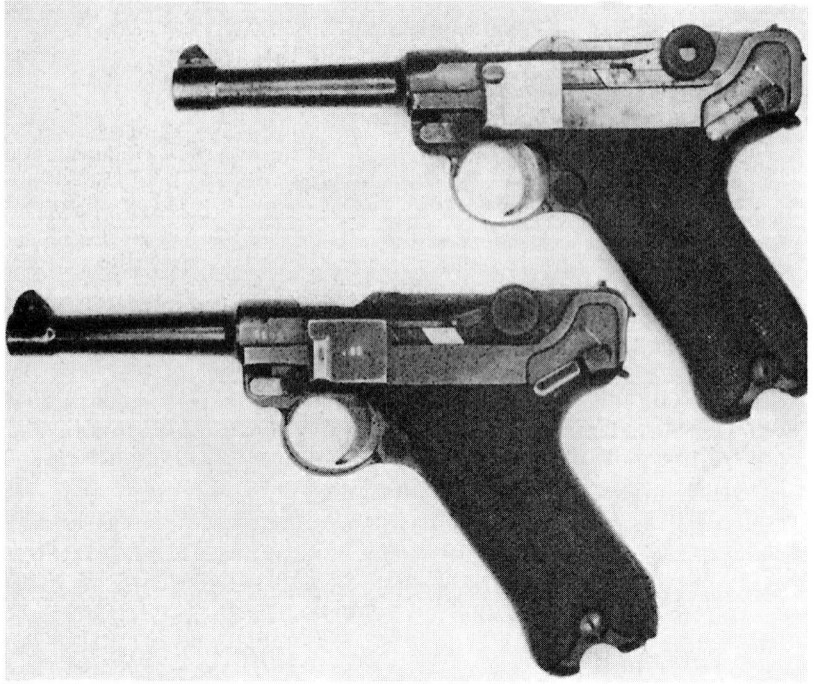

Two Luger pistols bearing the same serial number, and apparently identical when viewed from the side.

of a person located in the opposite end of the state. An investigation by the State Police of that area indicated that this firearm was sold to a friend in the same community. A check at that location revealed that the person had in his possession and displayed to the police officer, a 9mm Luger automatic pistol, serial number 9165. This fact was reported back to the Pistol Permit Bureau and State Police Laboratory.

In order to investigate scientifically this duplicate serial number problem, both firearms were secured and were examined.

Examination indicated that it was true that both firearms possessed the same serial number—9165. However, there were a great many differences in the other markings on the firearm which, if properly recorded in the first instance, would have easily distinguished between the two firearms.

Figure No. 1 is a side view of the two pistols; it will be observed that the serial number 9165 is located on the left side of the weapon. If a person was satisfied with that as a complete description of identifying marks it would definitely indicate duplication. However, Figure No. 2 represents the top of the slide receivers of the two firearms. It will be observed that the upper firearm shows the numerals 1913 and the numerals 65, which represent the last two digits of the serial number located in three different positions on the gun. In addition, the letters

Top and bottom views of these same Lugers show further markings which definitely distinguish between the two pistols.

DWM are present on the top of the toggle bolt. The firearm in the bottom portion shows the presence of the numerals 1940 and the same digits 65 located in exactly the same three positions. In addition, the number 42 is present. These numbers are very definite descriptive and identifying marks of the firearms. The number 1913 represents the year of manufacture so that it is obvious that one gun was manufactured in 1913 and the second gun in 1940. The DWM indicates manufacture by the Deutsche Waffen und Munitionsfabriken, which was an organization formed by a combination of the arm factories of Ludwig Loewe & Sons of Berlin with the cartridge manufacture plant of the Deutsche Metallpatronenfabrik. It will therefore be seen that information contained on the top of the slide receiver, even though the serial number is not included therein, nevertheless gives information which definitely distinguishes the two firearms by distinguishing between year of manufacture and place of manufacture.

The third illustration is of the bottom of the barrel and the front end of the frame. It will be observed that the upper firearm shows the letter "C" beneath the serial number on the frame, while the lower firearm shows the letter "A" beneath the serial number. The letters "C" and "A" indicate different months of manufacture of the firearms. Here it is readily seen that the inclusion of the letter with the serial number present on the frame of the firearm will give rise to another means of distinguishing between the two firearms.

It will also be observed that nowhere on the Luger pistols is the word "Luger" to designate that it is that type of firearm. The word "Gesichert" which is present on the left side of the frame near the safety lever is the German word for safety. It does not designate the name of a manufacturer.

It is noted that careful examination of two firearms, same caliber, same manufacture, same model, same serial number, reveals the presence of markings that clearly distinguish between the two firearms. This emphasizes the importance of being complete, exact and accurate in describing firearms when the description is to be transmitted over teletype systems or by correspondence. Only by submitting a complete, exact and accurate description, can a thorough check be made with any pistol permit bureau or even with the manufacturers.

These four .22 W R F cases were fired in a rifle which has been snapped on an empty chamber to an excessive extent. The 'anvil' of its chamber (that portion of the rim counterbore lying directly under the point of the firing pin, and against which the priming must be snapped and suddenly crushed) has accordingly been beaten down and peened into the chamber.

This condition is a common one in .22 firearms and it often offers a means of negative identification. As will be seen above, it crushes inward that portion of the case rim underneath the firing pin and roughens to a greater or less extent the adjoining section of the chamber. The result is that this slight disfigurement of the chamber edge distinctively embosses the fired cases; in addition each case is scribed or scratched its entire length as it is extracted from the chamber.

CHAPTER 9

PHOTOGRAPHY IN INVESTIGATION OF FIREARMS CRIME

L ITERALLY hundreds of pictures are taken by the police in the course of the average murder. A good rule to follow in some large police departments is to photograph everything and put aside 99% of it later on as useless. In general these photographs may be divided into three types.

First, there are the pictures of the crime scene, the victim, and any large objects of importance. These photographs are taken with a general purpose camera.

Second, there are the pictures of small objects taken with the camera lens quite close to the object being photographed. Finger prints, crime bullets, and miscellaneous "clues" are of this type. These relatively short focal length cameras are special equipment.

Third, there are the pictures taken through microscopes, called photomicrographs. These may be specifically of firearms evidence usually taken through the comparison microscope or other substances and materials viewed through a single instrument. Similar pictures can be taken with other equipment.

General Scene Photographs

Pictures have come to be standard procedure in almost all forms of police investigation. They are extremely important in serious firearms crimes. The first photographs will usually be taken at the scene within a few minutes of the discovery of the crime.

In medium and large cities police photographers are generally summoned to the scene of a homicide or to where the victim is found if transported away from the place where killed. Photographs should be taken before anything is moved. Some of these will be designed to explain as far as possible, visually, the whole surroundings of the crime. These photographs, in combination with a chart or plan,

207

will show other investigators, and eventually the jury, what the conditions were when the police arrived.

Other pictures will be taken to show the victim, wounds and the like. Photographs of this type are taken at the time of the post mortem medical examination, for specific purposes. A picture of the victim with wounds may aid to establish the crime—the corpus delecti. Probes can be used to show the path of the crime bullet or bullets into or through the body of the victim. Such photographs may be of considerable importance later.

Large Police Departments have photographers working for them, full time, who are professionals. It is not the purpose of this chapter to write a manual for professional photographers; they are far more capable in their field than the authors. However, certain tips may be of use to the smaller departments, sheriff's offices and the like, who do not have a professional photographer working directly for them and sometimes cannot arrange for the services of such a man in an emergency. Wherever possible, however, it is desirable to have photographs taken by someone not presiding at the investigation since, even if he is a photographer, his time is better spent in general supervision. However, occasionally, an investigator from even a large police department may have to take some photographs for himself.

Every police officer, sheriff and the like who may find himself in such a position should know a bit about photography and have a general purpose camera available. If at all possible, one of the press cameras, such as a Speed Graphic complete with flash equipment, should be obtained. Unfortunately, crime is seldom committed in a location where photographs can be taken using daylight only. Frequently, flash equipment can be used by itself, although for general lighting of the scene of crimes photo floods, in appropriate portable stands, can be used to advantage as supplementary lighting.

Photography demands special techniques which cannot be learned entirely from a manual. Anyone in an official capacity taking pictures at the scene of a crime should have enough experience to know what he is doing and enough equipment to accomplish his purposes.

In general, the larger the negative, the more flexible will be the use to which it can be put. For instance, a 4 x 5 negative from a Speed Graphic allows much more latitude than a 35-millimeter negative from a Leica or its equivalent. The larger negative always means, however, a heavier, bulkier camera which cannot always be taken around. In general, a good photographer with a small camera will

This is the kind of evidence photograph where a wide angle lens is necessary for proper scene coverage when the exposures are taken in the average sized rooms or apartments.

The photograph above was taken from the doorway to the living room in which this tragedy occurred. The bodies of three homicide victims are shown, together with the body of their killer, who committed suicide. Nothing had been moved or disturbed when this photograph was taken.

It is now the practice of all the larger or better equipped police departments to photograph every and all possible objects or bits of evidence that may be of use or value in the examinations, hearings and court trial to come later. They photograph just about everything and keep it for the record.

In the murder-suicide shown here, the authorities took, in all, 42 photographs of different aspects of their investigations: bodies (distance and close-ups), blood-stained areas, ejected cartridge cases, bullet holes in walls and in doors or floor, ricochet markings on walls, position of weapon used, postmortem views before and during examinations, entrance and exit wounds with and without probes in place so as to show path and effect of bullet. In addition, overall views of interior and exterior of house and living quarters are taken to show jury, if necessary, the conditions under which these people lived.

produce far better results than a poor photographer with a big camera.

The speed of lenses for police photography is not important. Extremely few photographs will ever be taken of criminals in motion. Almost all negatives exposed will be of still life, which can be satisfactorily artificially lighted. The camera can be placed on a tripod when necessary.

There is one desirable feature of a lens, however, which should not be overlooked. The widest possible field will be desirable for all area photographs. Lenses giving this wide angle are expensive and slow, but necessary. Space inside a room to move the camera back and increase the field of view is usually not available. Bodies can not be moved around to better suit photographing.

Small Object Photographs

Police investigation often involves many pictures of small objects. Special cameras are used for finger prints. Extreme care must be used to get the very best pictures possible of tool marks, foot prints, tire tread marks and the like. Sometimes casts will be taken also, but the photographs are seldom dispensed with. Wherever possible small objects should be taken to the photographic laboratory and worked on at leisure, but this is obviously not always possible.

Cameras with lenses of short focal length are used by professional photographers both at the scene of a crime and in their laboratory for photographing small objects. They want a negative that shows the object photographed at least actual size; sometimes the image will be considerably larger. With cameras of this type the speed of the lens is of no importance since the camera is always supported rigidly and the object carefully illuminated and brought into exact focus by means of the ground glass panel before the film holder is put in place. An exposure of several minutes can be used if necessary.

For the small department with no full time photographer a camera capable of taking portrait photographs is better than nothing, though whenever possible the services of a professional and a bellows camera should be secured. Most objects of this type can be brought back to the office and later sent to the nearest photographer's shop.

Quick Development Cameras

Many Police Departments have used in recent years, particularly in minor crime, the Land Polaroid camera which has the tremendous advantage of producing a picture in a matter of about one minute after taking it. The present film is capable of black and white enlargements of a surprisingly good quality, considering the circumstances. A camera equipped with this film and the new preservative which allows the snapshots to be handled almost immediately after being produced will be of great value in emergencies, particularly to small departments. It can be equipped with quite satisfactory flash equipment and can, of course, also be used with floodlights.

In general, however, when possible, the Polaroid camera should only supplement the use of the conventional type camera in a serious crime, although the former does have many advantages. For instance, the part-time cameraman will never be positive with a conventional camera that he has a record of what he has photographed until the films are developed, whereas the Polaroid camera will ease his mind. If he missed, he can try again immediately.

Evidence Photographs

Photographing of firearms evidence is, of course, more particularly the province of the Firearms Investigator than scene and victim photographs. These evidence pictures are made in the laboratory rather than at the scene of the crime. Normally, they are taken by professional photographers since a department having a Firearms Investigation Laboratory usually has also a Photographic Laboratory. The Special Agents in the FBI Laboratory, however, take photomicrographs of their own comparisons.

The photographs taken will be of several different types. It will be advisable in certain cases to take pictures of a suspect weapon, even though the weapon itself will later be introduced as evidence into the court. Wads, unburned powder grains and the like also are frequently photographed as well as bullets and cartridge cases.

N. J. STATE BUREAU
OF IDENTIFICATION
DIV. OF STATE POLICE

An example of infra-red photography applied to firearms investigation. Above is the top shot wad from a buckshot cartridge. This wad was removed from the body of the victim in a murder case. It was impossible to identify fully because of blood, but infra-red film brought out the lettering, as shown above.

A particularly important field of photography has been developed recently in connection with powder residue, nitrates and the like on the clothing of the victim or even on the body of the victim itself. By the use of different types of lights, far more accurate evidence can be found of such residue than can be seen by the naked eye. In some cases, a type of photograph will be taken in which there is no camera used at all. For instance, a prepared photographic paper will be placed in contact with the cloth surrounding the wound and have

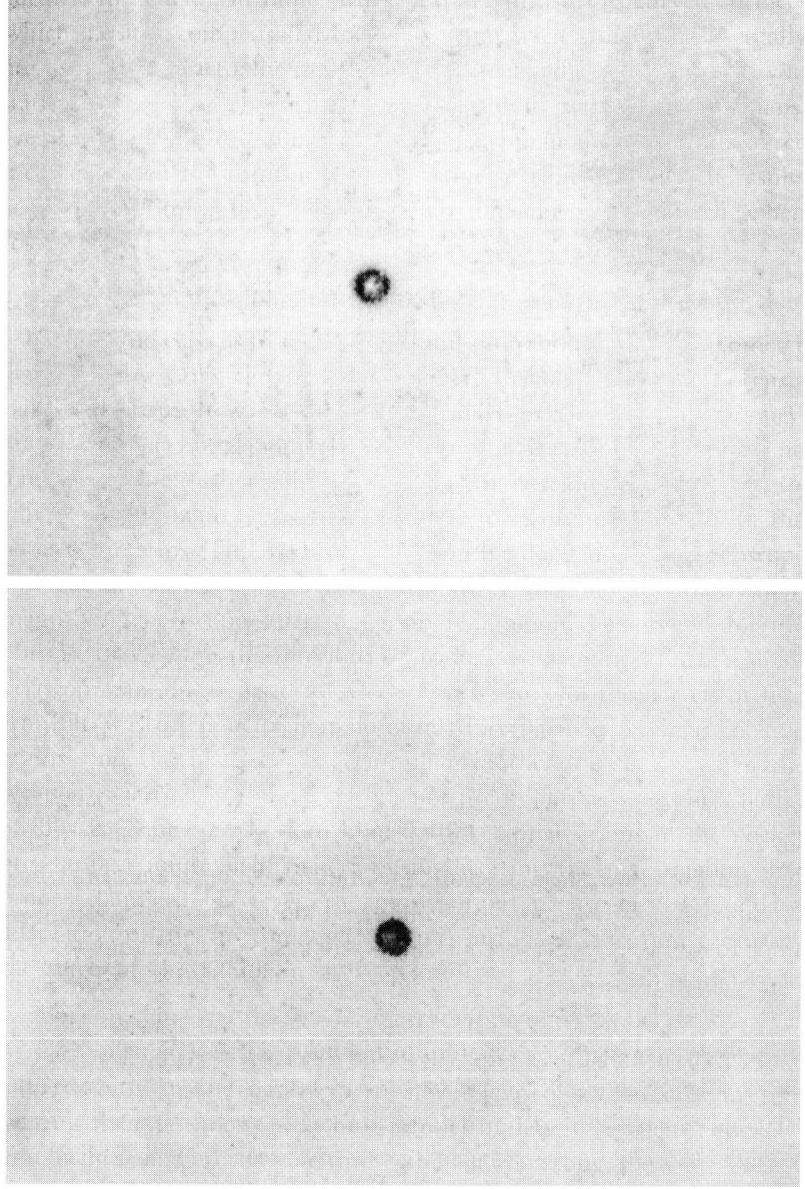

Showing the value of infra-red photography in powder pattern work. The above two photographs show definite powder patterns, on woolen cloth of horizon blue color.

No pattern was visible on the cloth when viewed through the microscope under ordinary light, at 10 inches distance.

The infra-red photographs showed indentifiable patterns of powder burns at 21 inches (top) and 30 inches (bottom).

These photographs were taken to refute a claim of self-defense. The coat of the victim showed no trace whatever of even a single granule, furthermore the entrance hole was in the middle of the back.

the cloth pressed down against it with a hot iron. Any particles of remaining nitrates will show up on such paper as black spots of various sizes.

The use of infra-red light and film in Firearms Identification is particularly valuable in powder stains and burns, and tattooing on dark colored cloth, such as the usual dark blue service uniforms. By naked eye and even under a microscope, it is very difficult to detect grains of partially burned powder and blackening of the cloth at the edge of the bullet hole. With infra-red light and film, the dark surface of the cloth appears white, and the tattooing, powder specks, and soot appear black. From these photos, it is comparatively easy to fire test shots with similar ammunition and gun and determine the range at which the crime shot was fired. Recently, two policemen, who felt neglected and wanted to be in the limelight, shot through their own clothing and invented stories highly flattering to themselves. Infra-red examination showed the range to be between 4″ and 6″, whereas the gun battles had occurred according to their stories at considerable ranges. Confronted with the evidence, they admitted their rather ridiculous conduct and were dropped from their departments.

Photographs Taken Through Microscopes

At one time by far the most important part of Firearms Investigation photography was micro-comparisons of cartridge cases and bullets taken under considerable magnification. With the comparison microscope the investigator may or may not find that the fatal bullet matches the test bullet. If it does, he has visual proof of it, but this proof will no longer be available after he takes the bullets out from under the lens, unless he takes a photograph of what he has seen through the microscope. This matter of photographing through a microscope is common laboratory practice, and is called Photomicrography. All that it requires is a camera with a holder to retain it in a vertical position over the eye-piece.

When a camera is used in this manner, for photographing through a microscope, the camera lens may be removed. The lenses of the microscope take the place of the camera lens. With any microscope an image of the object viewed can be projected on to any white surface such as, for example, the ceiling over the microscope. Of course, in such a case as this, where an image is projected on to a screen a very strong illumination of the object under the microscope must be used; this would be rather hard to do satisfactorily in the case of dull

Showing the effect of unequal light on the bullets being compared and micro-photographed. The section of bullet shown on the right has much too much light, its striae being consequently lost in the glare. The duplication of this same comparison shown in the left half of the illustration is a different and longer print exposure, made in an attempt to "save" the negative which was poorly lighted to begin with.

colored objects such as bullets. However, the image projected on to the plate of the camera is plenty bright enough to photograph when even moderate illuminations are used, because if the illumination is dull a longer exposure can be given to compensate for the lack of brilliance in the picture. Thus photography can be accomplished even when the projected image is not brightly enough illuminated to be satisfactory for visual examination.

All microscope makers such as the Bausch & Lomb Optical Company, The American Optical Company and others make cameras specially mounted for use with the microscope. These cameras are mounted on a vertical post attached to a base plate for holding the microscope. The camera is mounted in a vertical position and is so arranged that it can be placed directly over the microscope or can be swung to one side to allow the microscope to be focused by eye. In addition some cameras furnished for this purpose have an eye-piece of their own sticking out at right angles from the shutter housing so that the image in the microscope can be viewed by eye without removing the camera from its position over the microscope.

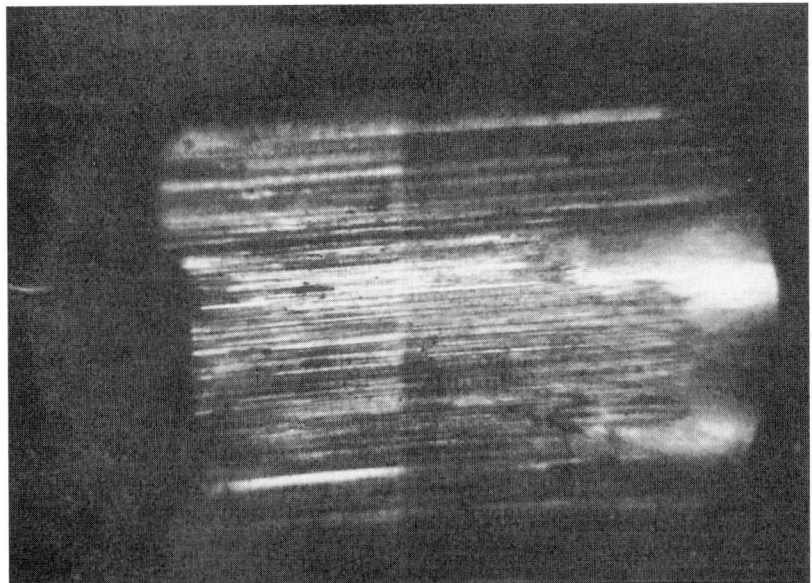

Two excellent microcomparisons, under high magnification, of positive matches.

In order to take a picture through the microscope it is only necessary to set up the object under the microscope, adjust the illumination, and then focus the instrument until the field of view and the appearance of the object are satisfactory. Then the camera is swung over the microscope and the shutter is opened. An image will now be

An example of the base and primer markings as shown by photomicrographs of two different shells fired in the same pistol. An examination of these two shells, made while in the microscope, would show their similarities much more clearly than is possible by means of these separate photographs.

The above shells were fired in the same pistol, but the photographs do not show things as clearly as could be desired. A trained and experienced man can see and evaluate in the microscope many identical marks to prove they came from the same gun. However, the photographs could allow some doubts to settle in the minds of some jurors.

It is sometimes possible to get more prominent markings on the head of a fired cartridge case by various means, this in an attempt to duplicate the markings on an evidence shell from the same gun. Since all brass manufactured is not always of the same degree of hardness. even in the same make of ammunition, the test ammunition may come from a lot of brass much harder than that of the evidence shell. It is then worth while to fire additional shots; or to oil the chamber of the gun, which will allow a greater thrust of the head of the shell against the breech or bolt face. This in no way changes the evidence shell or gun, but it does serve to remove any doubt from the mind of the Firearms Identification Expert.

These pictures, and others shown in this chapter, were part of a series used in class instruction, to illustrate the necessity of careful focusing and lighting when it is deemed desirable to make photomicrographs; also to point out the fact that a photomicrograph will show differences as well as likenesses. We feel that a thorough knowledge of photography and photomicrography are necessary to round out the education of a Firearms Identification Officer.

projected on to the ground glass plate in the back of the camera; this image may be viewed by looking at the ground glass plate the same as is done in taking an ordinary photograph. Usually the image will be very dim; it will be necessary for the operator to place a black cloth over his head in order to shut out extraneous light and render the subject clearly visible. The object should then be brought into sharp focus by moving the stage of the microscope up or down. It is very useful indeed to have at hand a focusing magnifier, which is merely a small magnifying glass in a metal tube which is of the right length to bring the ground glass into focus when the tube is set on top of the ground glass plate. A linen tester is ideal for critical focusing.

After the focus has been correctly obtained, the shutter is closed,

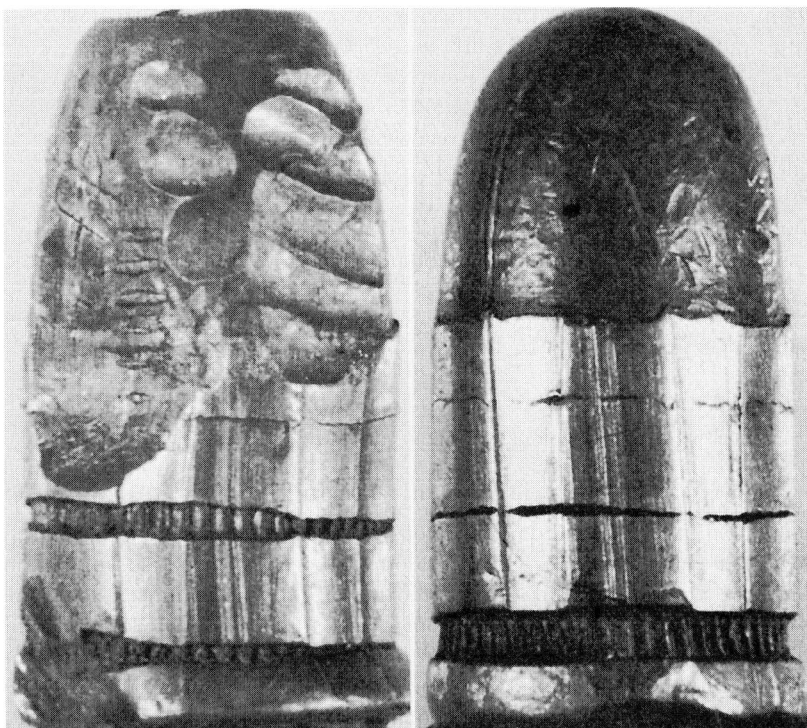

Excellent photographs of a crime and a test bullet from a homicide. The deformed crime bullet, shown on the left, shows marks left by forceps used in removing it from the body. The test bullet, shown on the right, shows typical skid marks, frequently found on .38 Special bullets fired in old, loose fitting revolvers—and in some good ones.

the plate holder inserted into the camera, and the slide withdrawn. Then the shutter is opened for the required length of time to make the exposure. The length of exposure required is learned only by experience, as it varies considerably, depending on the brightness or darkness of the object, the amount of illumination and the magnification used. The higher the magnification, the more the available light must be spread out and the greater must be the exposure time.

With a camera having an extension bellows the magnification depends not only on the power of the microscope objectives and oculars that are used, but also on the extension of the bellows. The further the plate is from the eye-piece the more the image will be spread out and the greater will be the resulting magnification of the picture.

A type of camera with a box of fixed length instead of a bellows becomes increasingly popular for micrographic work. Such a type of camera has several very considerable advantages. In the first place

A poor example of photomicrograph, showing the effect of uneven lighting of shells. This can also be caused by photographing a nickel-plated shell with a brass shell. The use of filters is an aid in avoiding negatives such as this.

it is made of a fixed length of 250mm, which is the usual distance for clear vision with the unaided human eye, and a picture taken with this camera gives precisely the same magnification as is obtained when the specimen is viewed through the microscope alone. Thus

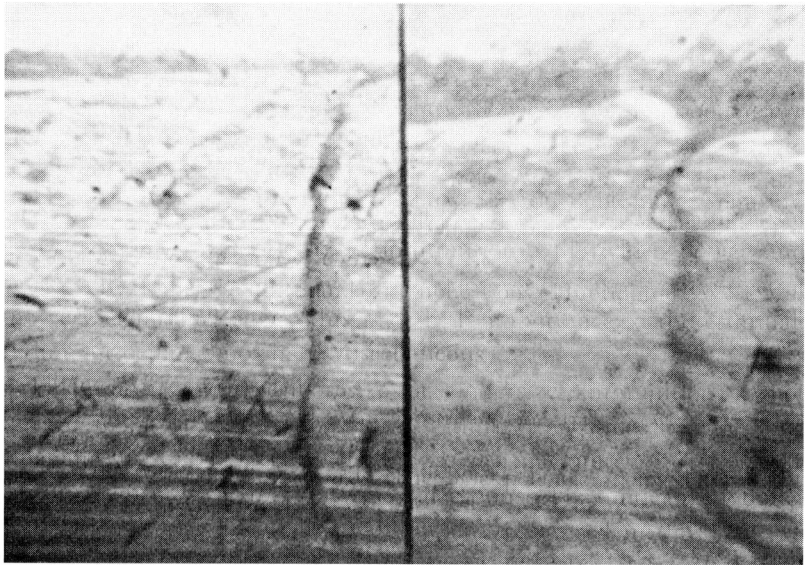

A photomicrograph made to show the effect of having one bullet out of focus.

An excellent microcomparison under low power of a crime and test bullet. It is unusual for a low-power photograph through the comparison miscroscope to show such obvious similarity throughout an entire groove on each bullet.

the magnification obtained in the picture can be determined closely enough for all practical purposes merely by multiplying the magnification of the objective by that of the eye-piece used in the microscope. As all modern objectives and eye-pieces are plainly marked with the magnification that they give, this process is extremely simple. An even more important advantage of this camera is the fact that no focusing on the ground glass plate is necessary. Just below the camera shutter there is an eye-piece projecting at right angles to the body of the instrument, which enables the specimen to be accurately focused on the plate merely by looking through this eye-piece. The view seen through this eye-piece is just the same as would be seen when looking down the tube of the microscope, hence it is not necessary to move the camera from above the microscope when mak-

This illustration was made from a contact print of a negative taken in an 11 x 14 Levy Process camera, using a Wollensak Velostigmat Enlarging 4¼″ focus f4.5 lens. It is NOT an enlargement of a negative or taken through a microscope.

Here is shown a .45 Colt metal cased pistol bullet after being fired in Colt M1911A1 pistol. It represents an experiment to illustrate the fact that most of the valid striations on a bullet are the land signatures in the groove of the bullet, and not the groove signatures. Thus, the reaming operation and not the rifling operation has more influence on the identification of a bullet with a barrel; a fact that is particularly evident in metal jacketed bullets, which more often fail to "bottom" in the grooves of the barrel than do the softer lead bullets.

ing visual examinations. An object under the microscope can be viewed and examined as much as desired. When it appears correct through the horizontal eye-piece the mirror can be shifted out of the path of line, the slide removed from the film holder and the picture can be taken without any further focusing.

All this sounds easy. Actually, however, for the Expert with a full schedule of work who has not made a real study of photography either professionally or as a hobby, it can be well nigh impossible. There just is not enough time. Even the best men will be disappointed with most of their efforts along this line. The very best equipment will not work without skill and patience. Many laboratories do not have the best. The whole routine of the laboratory may be disrupted by the tying-up of equipment needed for other purposes.

Magnification By Camera Alone

While we have been referring above to the taking of enlarged photographs through a microscope, the reader should clearly understand that a microscope is by no means necessary for taking enlarged photographs, for pictures with all the magnification necessary for bullet comparison work can be made with no other apparatus than a long bellows camera and a short focus lens. If the lens is placed close up to the object and the camera bellows is extended to the proper point the image will be much larger than the object. Using a lens of 48mm (approximately 2″) focal length, it is possible to take photographs magnified up to as much as 20 diameters.

Years ago, after it had been determined that the test bullet and the

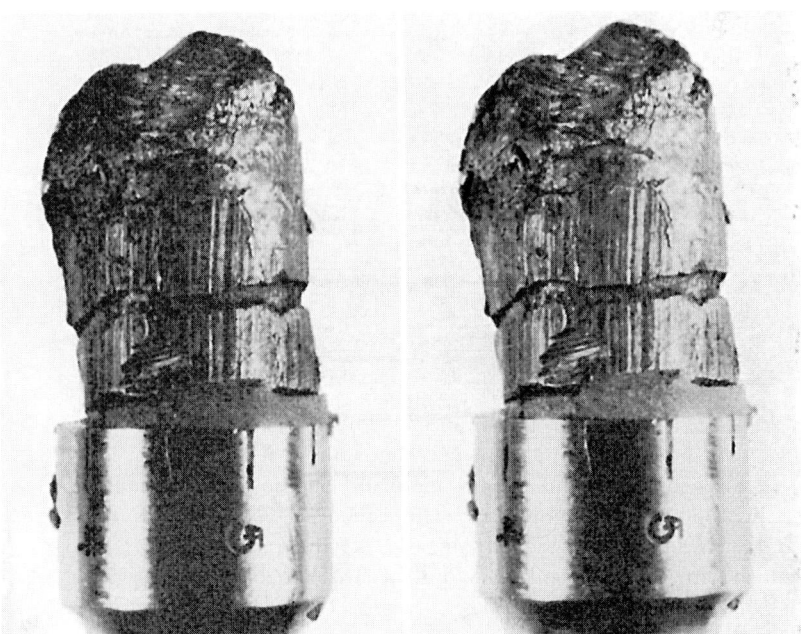

Two examples of a fatal bullet, with contact prints made from the same negative. Taken with a Levy 11 x 14 Process camera having a short focus lens of 4¼″ focus @ f 4.5.

The example shown on the right is much better for visual comparison of the photographic print and shows the striae much clearer than the example to the left. Shown to illustrate the importance of using different grades of printing paper and varying the exposure until the best possible print is obtained.

AN EXPERIMENTAL COMPARISON

This picture is the result of an experiment in making a positive comparison of an evidence (and test) .22 caliber lead bullet removed from the body of a valuable hunting dog.

The pictures of both bullets were made, as an experiment and for the information of all parties concerned, with an 11 x 14″ Levy Process camera fitted with a 3¼″ focal length lens taken from a small $39.00 enlarger. Lighting was with two carbon arc lamps and one photo flood light. Exposure was for 40 seconds. Bellows extension of about 3½ feet.

The result was a very good positive match, entirely suitable for the purpose intended. The unusual feature is the amazing depth of focus obtained, so much so that three land's impressions on the bullet are in good focus. This evidence bullet was in almost perfect condition, as it did not upset or deform in the tissues of the dog's body.

These pictures, while accurate and interesting, are not practical for most court cases. They are, however, suitable for study and comparison on the part of the investigators —and they can be obtained without the use of expensive and possibly inaccessible laboratory equipment.

fatal bullet match, the corresponding lands and grooves of the fatal bullet and the test bullet were sometimes photographed for comparison. This photographic work was usually done by an ordinary camera without the use of the microscope.

For this purpose an investigator will need a special lens known as the Spencer Micro-Teleplat or the Bausch & Lomb Micro-Tessar. These lenses are made in various focal lengths, but 48mm is the most suitable for bullet work. They are made to fit on the tube of the microscope, but an adapter can be obtained to enable them to be used directly in the camera shutter. This adapter is regularly furnished by the makers of the lens for a small additional price.

These short focus photographic lenses may be used on the microscope in place of the microscope objective for all the bullet work that has been described. They are more expensive than the ordinary microscope objective, but they have the important advantages of being more finely corrected and of having a diaphragm between the two elements of the lens. A diaphragm is absolutely necessary for good results in bullet photography, because when the diaphragm is closed down to a small opening the depth of field is increased and the parts of the bullet which are slightly above or below the point of focus can be clearly seen.

Interesting results have been obtained by the New Jersey State Police Laboratory in recent experiments in photomicrography using an ordinary low--priced 3½″ enlarging lens in the big Levy Process Camera. Three land impressions on a .22 caliber bullet were in good focus with the bullet filling a 5 x 7 negative.

Enlarged Photographs

Everyone who has had any experience with picture taking knows that a small photograph may be enlarged to almost any size desired. It is therefore possible to take a small picture of a bullet and enlarge it to the same size that a micro-photograph of it would have. However, the results are far from being the same in this case as they are when a real micro-photograph is taken with either a bellows camera and a short focus lens or with a camera and a compound microscope, as anyone who has studied the subject of microscopy will fully realize. When a micro-photograph is taken, the lens is close enough to the object to separate points which lie very close together, and thus give resolution of detail. When an ordinary photograph is taken the lens is too far away from the object to separate very close points, and en-

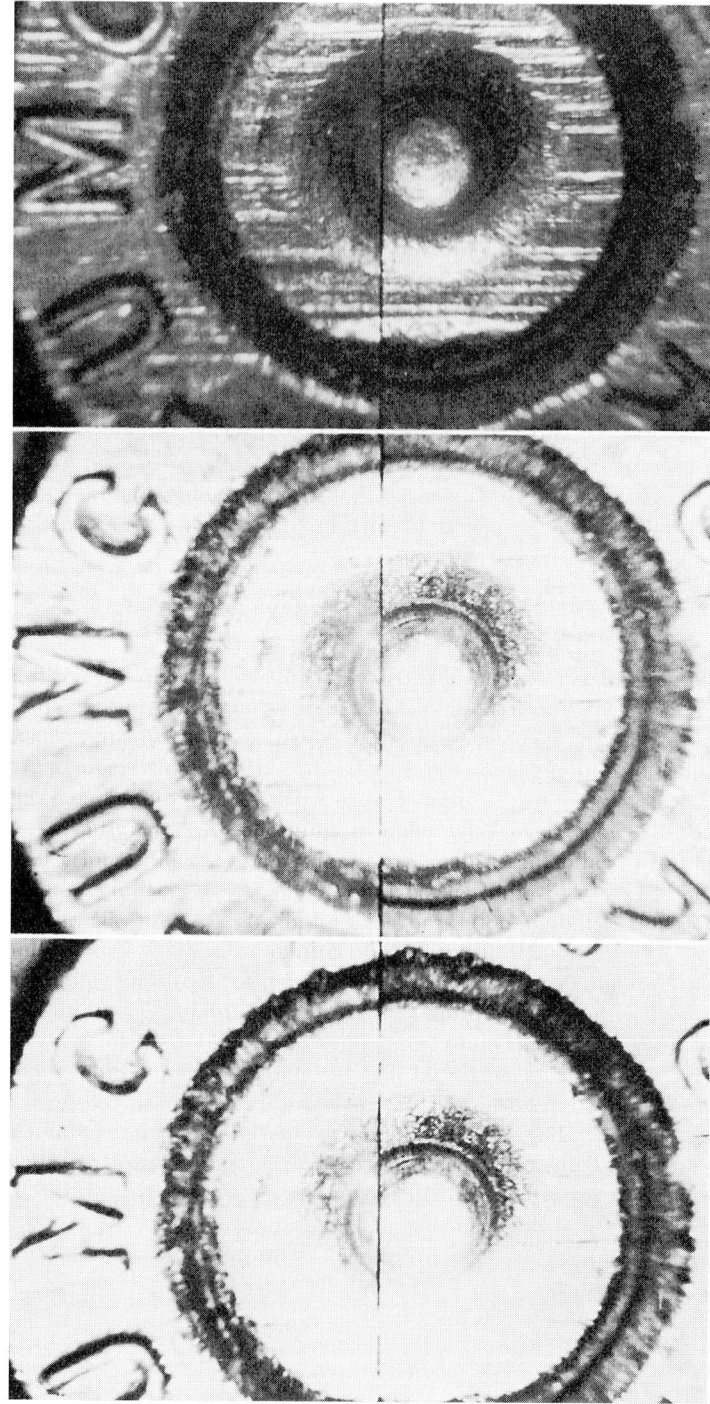

Photomicrographs of primer comparisons, as printed on different grades of photo printing paper. Made to show the differences in the effect of various papers used in printing. Section to left is F-4 grade; center section is F-2, and right section is F-0 grade. As can readily be seen, F-0 paper gave the most satisfactory results.

larging the resulting photograph will not add any new detail. It merely spreads out what is already in the picture, and gives what the microscopist calls "empty magnification." Suppose, as an illustration of how this works, we take a picture of a ship, showing ropes. We may enlarge this picture until the ropes appear as thick as a man's wrist, but we will not see any indication that the rope is composed of fine fibers twised together. The line representing the rope merely gets wider and wider as we increase the enlargement, but it does not gain detail. On the other hand, if we take a close-up picture of the rope with a lens of, say, 2″ focus, we can show the individual fibers with great clearness. It may be advisable at times to use enlarged photographs for demonstrations and the like, but the difference between an enlargement and a photo micrograph should be clearly understood.

As pointed out in our discussion of the History of Firearms Identification, photographs of bullets taken under high magnification and at various angles were at one time used to prove identity between a crime and test bullet. This was of very considerable importance early in the history of this science. It was even developed as a method to take the place of the comparison microscope in several large and complicated machines, which actually amounted to two cameras projecting images first onto a ground glass plate which could be examined. After the match had been obtained, a film holder could be substituted for the ground glass.

However, photographs taken directly through the comparison microscope largely supplanted all these separate photographs. Without question, the comparison microscope is at the present development of the science the one and only method at all practical for bullet examination. In order to make a visible record of what is seen in a comparison microscope, a photograph must be taken right through the instrument.

Decline in Use of Micro Comparison Photographs

Throughout the country, however, comparison microscope photographs are now rarely used. There are several reasons for this. Undoubtedly, the most important of these is that Firearms Identification is now accepted in courts far more readily than it was even a few years ago. Further, the findings of the Expert are accepted by the judge and jury without the visual proof required when Colonel Goddard and others were first presenting their cases. The psychologist, the doctor, and scientist, when they testify in court are not required

to present photographs to prove their findings. Neither should the
Firearms Expert when he is testifying in his specialty.

Perhaps the second most important reason for the decline in
photography is the vast increase in the number of cases handled. If
photographs were taken in one case, they would have to be taken in
every case. The photographic expense alone of the New York City
Police Department would probably be on the order of $750,000 a year
which would, of course, be far too great, particularly since the luxury
of such presentation is not really necessary.

There are also photographic reasons for the almost entire abandon-

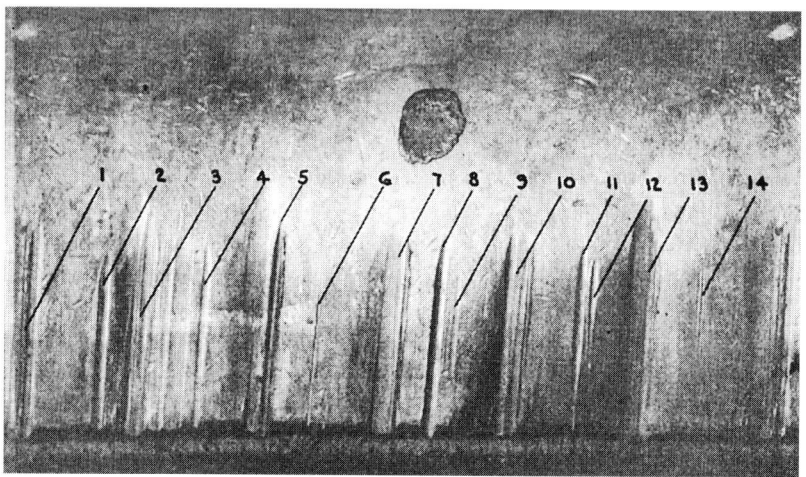

Pantascopic picture of a .32 Auto, full metal patched bullet which killed a man
in a public washroom. The murderer was never found and this shooting is still an
open case in the police files. The above bullet and the one shown on opposite page
have both been photographed and are on record in case the murder gun will one
day be discovered.

ment of this method of presentation. Unfortunately, the lenses of a
camera do not adjust in the way that the human eye adjusts. The
depth of field is extremely limited. Unlike the human eye, a photographic film has far less tolerance to variations in lighting. Photographs are almost always unsatisfactory to the Expert who has made a
positive comparison through the microscope. You photograph only
what you can see at one single time. The camera cannot move along
the surface of bullets to pick up identity after identity.

Further, these pictures were not understood by juries. A good deal
of knowledge and experience are necessary to evaluate them. Some
men after years of working in Firearms Identification refuse to make a
positive identification from pictures alone. Clever defense attorneys

by stressing lines in photographs that do not match create a doubt in the minds of jurymen that would not have been there at all if the Expert had merely testified without producing any pictures. Very few photographs have been presented in court recently. Actually, some departments have not taken pictures through their comparison microscopes in years.

Other Methods of Photographic Comparison

Several unusual methods of photographic comparison have been used in the past. Even though these are almost completely discarded at the present time, a few words in connection with them might be

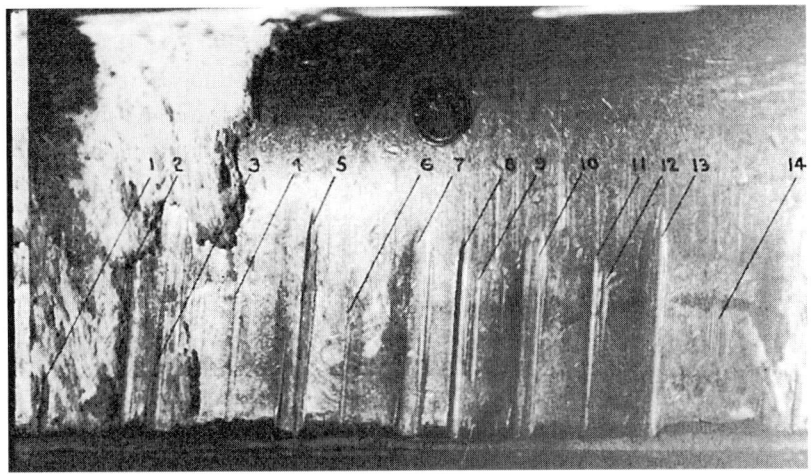

Pantascopic picture of the second, spent bullet which was recovered from the floor of the washroom, where it had fallen after striking plaster wall. Note the many visible points of comparison between these two bullets. These two photographs may prove a positive tie-in to the perpetrator of this crime if the murder gun is ever located.

in order. In addition to the still photographs taken in several different positions around the circumference of a bullet, a continuous picture was taken with a geared camera already discussed. This gave a strip photograph showing the exact bullet surface as it would have appeared if it had been possible to unroll this in a single plane. These photographs may have been of value in the hands of Experts used to dealing with them. They had the disadvantages, however, of all strip methods of comparison and further were completely unintelligible to Experts not dealing regularly with them. These cameras are still present in some of the larger Police Laboratories but have fallen into disuse. One of the advantages for this arrangement was that photographs of crime bullets could be sent around to other departments for possible identification with test bullets from suspect arms. This would work only if the other departments also had such a camera and knew how to use it. The best man the authors know with one of these geared cameras may need two or three days to get an acceptable picture of a mutilated crime bullet. The rig is not really practical.

Another method of photographing crime and test bullets was by positioning them individually in the middle of a conical mirror and taking a picture of the mirror and bullet from directly overhead. This method was thought to have certain advantages but depended upon mounting the bullet axis exactly coincidental with the axis of the mirror. The mirror itself was difficult to produce. After considerable experimenting, the Federal Bureau of Investigation found not only that the normal crime bullet could not be mounted satisfactorily, but even undamaged test bullets were difficult to place a second time in this arrangement so as to give an exact coincidence between the first and second photographs.

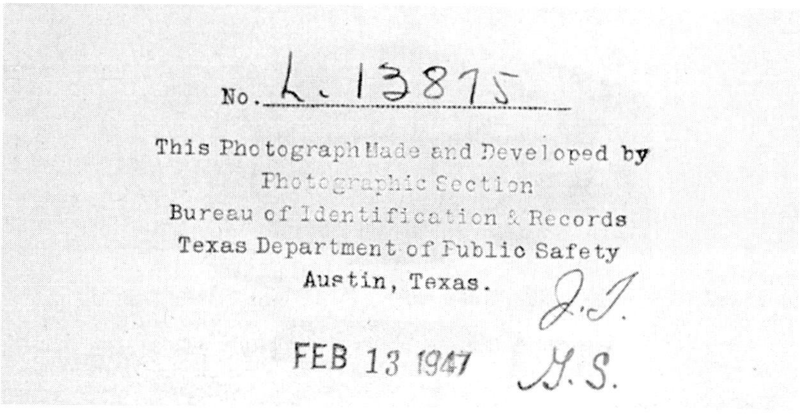

No. L. 13875

This Photograph Made and Developed by
Photographic Section
Bureau of Identification & Records
Texas Department of Public Safety
Austin, Texas.

FEB 13 1947

Any and all evidence photographs taken, which later might be used in the court-room, should be authenticated, dated and initialed by two or more investigators, as shown above. File number and date of all such photographs should also be entered in the log.

CHAPTER 10

THE FIREARMS IDENTIFICATION LABORATORY AND ITS EQUIPMENT

L ABORATORIES started in the past have often "just growed" like Topsy; most departments are not by any means as efficient as they could be if not handicapped by poorly arranged and cramped quarters. Much equipment now in use is not the best. Some highly skilled laboratory staffs are working without things they could use to advantage. The authors visited many Firearms Indentification Laboratories in the Eastern states and two in Great Britain in their preparations to write this book. Only the FBI Laboratory and the H. P. White Laboratory approach perfection.

Physical Laboratory Layout

As in all cases of efficient design, the basic requirements of the finished product should be clearly understood before starting to consider the details. The whole laboratory will eventually be used for certain activities and to house materials, equipment, evidence, and files. The most important activities, not necessarily in the order of their importance, are listed below with a few words in connection with each.

First, receiving the business, professional or friendly callers who may be coming into the laboratory for anything from bringing firearms evidence just taken in on an atrocious crime to merely paying a friendly call. People being what they are, a fairly large number of them will visit a laboratory in the course of a week. In most instances, it would not be good policy to clamp down too tightly on the fellows who in essence just want to "shoot the breeze" about some case or other. It is essential, however, to prevent them as far as possible from disrupting laboratory procedure and completely from endangering evidence.

229

Second, doing the actual basic Firearms Indentification work on evidence and test bullets and cartridge cases. This is the heart of the laboratory activity. It should be done in absolute safety and privacy with as much freedom from interruptions and discomfort as is possible. Police departments in general do not go in for these things.

Third, doing the thousand and one small jobs like soaking revolvers, restoring serial numbers, and patching up weapons so they can be fired. This is really shop activity and requires the facilities of a shop to be done efficiently. This includes firing of suspect weapons for test bullets, cases and patterns.

Fourth, doing the paper and desk work that is always required.

The things that must be kept in a laboratory include the following:

1. Pleasant and comfortable furniture for receiving visitors.
2. The ultimate equipment such as comparison microscope, stereoscopic microscope, balance, microscope camera, with their auxiliaries, attachments and the like.
3. Shop equipment including machine and hand tools, supplies, and materials.
4. The store of ammunition and the bullet recovery equipment.
5. The office equipment including desks, typewriters, paperwork files and the like.
6. The evidence files, drawers and racks for crime and test bullets and cartridge cases and suspect firearms.
7. The ammunition collection.
8. The library.
9. The firearms collection for reference and replacement parts.
10. Any special unusual equipment which the laboratory may be in charge of such as permanent magnets, portable fluoroscopes for bomb work, and the like.

With the above in mind, the individual expert will have to adapt what he has towards the perfect setup. Even if an expert has the good fortune to be setting up a laboratory from scratch, he will not be able usually to plan his activities and his equipment in the most efficient arrangement and then build a building around it. Most police departments occupy old quarters. All too frequently the Firearms Laboratory gets what is left after others have had their choice. However, let us just dream for a few moments. Here is what we would like for a laboratory handling between 200 and 600 cases a year and having an Expert, an assistant, and a secretary-clerk. It

This photograph shows, in the opinion of the authors, the very best and most efficient Firearms Identification Laboratory receiving station and work area. A Special Agent of the Federal Bureau of Investigation Firearms Division of their Physical Laboratory has just opened evidence received on a clear table, save for equipment, envelopes, and report sheets, which will be made out in ink with the pen shown, as the examination progresses. First, the evidence will be examined under the stereoscopic microscope mounted on the universal arm of a special support. Any comparison microscopic examination will be conducted on the instrument, independently mounted to the side. Note the orderly condition of everything—the desk drawers for small supplies which may be required, the excellent general and individual lighting, and the standard camera for photographing microcomparisons. This set-up can hardly be improved upon.

will be large enough for reasonable expansion without overcrowding.

It would be a rectangular area about 20′ x 60′ with daylight at least down one side, well heated and ventilated, with a strong, vibrationless floor. A permanent partition with a single door would divide this area in half. The front portion would be used as an office and receiving room; it will be divided again in half by a grill down the center over the middle of a counter including a self-locking, electrically released door.

The front portion of the front room—that outside the grill—would have the waiting room furniture and perhaps the arms collection but no evidence firearms. It would be as attractive as possible to keep visitors satisfied to stay here. An attempt should be made by

Two views of the Federal Bureau of Investigation firearms collection. These weapons are kept not only as examples of what might be encountered in crime, but also for purposes of tentative identification of a bullet when no firearm is submitted with it, and in order that weapons submitted incomplete may be test fired. It is also important to have complete weapons in order to identify components—sometimes very small parts of components in cases in which a criminal endeavors to destroy a crime weapon.

the Expert and his assistant to come outside the grill rather than inviting all and sundry inside.

The second half of the front room would contain the office equipment and the desks of all three of the personnel. It would also have the paperwork files, and the locked drawers and racks for active evidence firearms. Police officers calling professionally and others would undoubtedly have to be admitted to this area

THE F. B. I. HANDGUN COLLECTION

Showing a portion of the many pistols and revolvers on hand at the Federal Bureau of Investigation laboratories. Each handgun is in perfect working condition and is instantly available for test purposes, checking with evidence guns, disassembly for examination of small parts or for other ballistic purposes.

to discuss matters with the Expert at his desk. Even authorized persons, however, should not be allowed inside the grill alone.

The rear half of the whole area would be separated from the front or office-reception half by a solid wall and door. This rear half would be the shop and laboratory proper. Every effort would be made to keep this free from all visitors.

About 60% of this rear area would be a general shop containing the hand and machine tools supplies and rough equipment. About

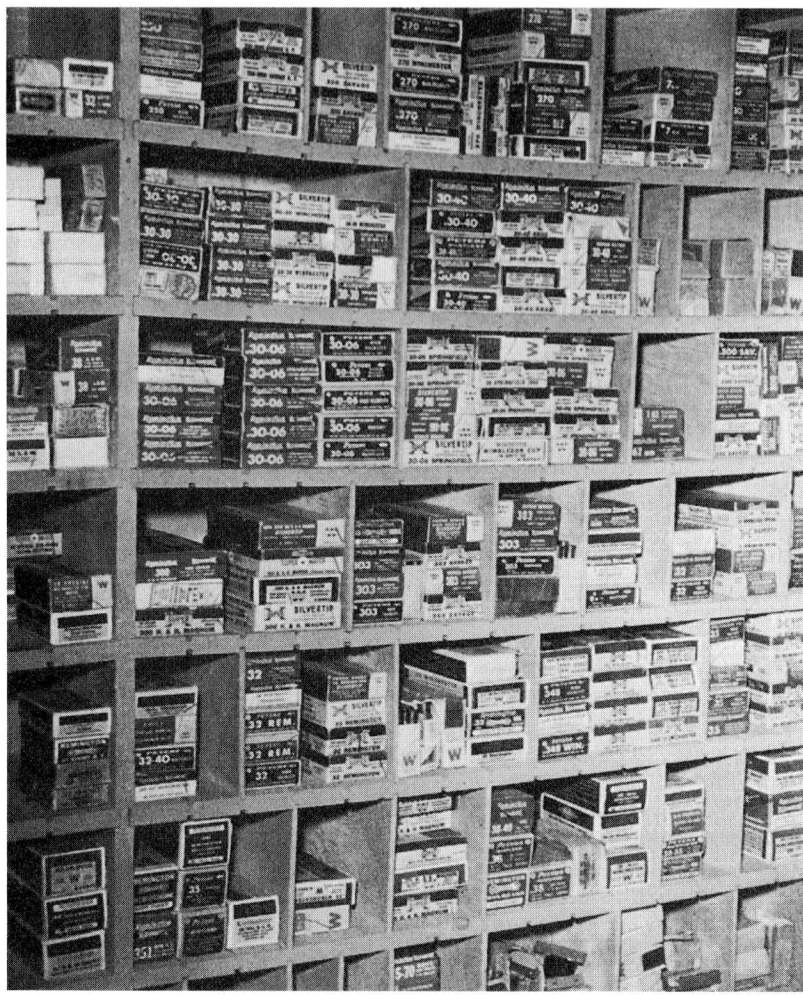

A part of the very extensive ammunition inventory of the Firearms Division of the Physical Laboratory of the Federal Bureau of Investigation. This is located within the soundproof room for test firing, but outside the bulletproof door. An investigator has immediately at his fingertips, in an orderly workmanlike arrangement, anything that he may require in the nature of standard ammunition. Note that each caliber is represented by all four brands of American-made ammunition.

20% would be a long narrow sound-proofed rectangle containing the active ammunition supply, the ammunition collections, and two types of bullet recovery equipment—the cotton waste box and a water tank, preferably sunk in the floor.

The last 20% of the rear area would be the most private of all. It would contain the precision instruments already mentioned. The lighting would be perfect from both natural and artificial sources.

The room should be air-conditioned, or at least dehumidified. Here is where the Experts would spend the long hours of drudgery necessary on difficult comparisons. The Experts should have everything that will aid them in continuing at near top efficiency.

Of course, our perfect layout might not please others. Further, it is unlikely that any of us will ever have it. However, the basic principles are the same in adapting a much smaller and differently shaped area for your uses.

General Firearms Equipment

Now let us think about the equipment of the laboratory; we skip the furniture and office equipment and supplies. The two collections, one of firearms and the other of ammunition are of considerable importance. Every laboratory should have a weapons collection containing at least the more common firearms in use in that vicinity by criminals. In general, this collection will build itself up in a few years without too much specific effort from anyone. Prosecutors and police departments are usually delighted to turn over to the laboratory for "exhibition, experimentation, or destruction" crime weapons after the conclusion of a case.

The ammunition collection, which is not the same as the supply of cartridges for test firing, will give more trouble. As a starter one can take two rounds from each type, caliber, and make of test ammunition that you have on hand, separate one into its components, and then put the bullet as well as the unfired cartridge case and a second complete round in a small glass or plastic jar, labeled with data as to make, caliber, dimensions and weights. File cards are made up on each one. Some departments add a fired bullet and cartridge case. All bottles are then arranged in some orderly manner. Whenever possible, specimens of foreign or obsolete cartridges should be added. In the case of shotgun shells, the powder and wads also should be displayed.

A file should be maintained also of test bullets and cartridge cases fired with various types of weapons. These will be extremely useful in identifying the probable make and model of an arm used in a crime when only the bullet, or cartridge case, is available. Such identification will seldom be conclusive, but a tentative identification is desirable.

A reference library is absolutely essential. This will include not only books on Firearms Identification and Criminology, but also books and catalogues on firearms and ammunition in general. Old catalogues are especially desirable. The writers include an appendix of titles and authors of those volumes actually in their combined libraries. There are probably additional books giving important information; however, we have far more than are actually necessary.

The shop equipment can vary from a polishing head, a bench grinder, a few hand tools, supplies of oils, steel wool and abrasives to a small machine shop such as that of the H. P. White Laboratory. In an actual case recently, one of the authors made a replacement part necessary to fire a suspect weapon on a lathe in less time than it would have taken to order it. We ourselves have a surface plate and a complete set of precision measuring instruments up to and including Johanson blocks, but they are seldom used. We have also machine and hand tools, welding and heat treating apparatus, an arbor press, as well as a lot of other things.

In general, though often convenient, a firearms Expert does not have to have a completely equipped gunsmith's shop. These things, however, will be found almost indispensable:

1. A really good strong vise with several different sets of jaws made of copper, lead, leather and felt lined hardwood, including one set, preferably of brass, for grasping tapered objects.
2. Mechanical polishing equipment including abrasives.
3. Soaking trays, solvents, and oils.
4. A good bench grinder.
5. Wrenches, screw drivers, scribers, hammers, punches, a brace and screw driver bits, hacksaws, and other hand tools.
6. Cleaning rods and other cleaning supplies and wipers.
7. A stock of drill rods and brass rods of various sizes.
8. Equipment for pulling bullets from unfired cartridge cases.
9. A mounted illuminated magnifier; it will be used for many things from a preliminary examination of test bullets to reading restored serial numbers.
10. A good drill press.

Bullet Recovery Equipment

Nothing in the science is so extremely variable as the different methods that have been successfully used to recover bullets fired from suspect and crime weapons as well as from the laboratory firearms collection specimens. These bullet traps can be made in several different ways and can be filled with various substances. For low

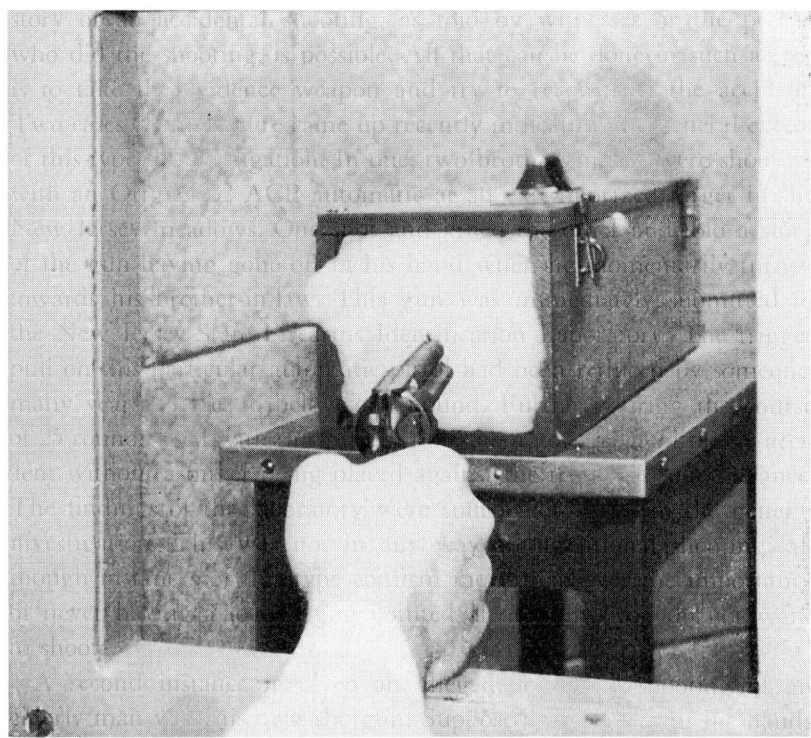

This is the ultimate in efficiency and safety in connection with catching undamaged test bullets from pistols and revolvers. A Special Agent of the Firearms Division of the Physical Laboratory of the Federal Bureau of Investigation has prepared the bullet trap with rolled absorbent cotton of tightness to correspond to the bullet being caught, and has then retired and closed a bulletproof door in which there is a 12-inch by 12-inch opening. Note also the sound-absorbing wall covering which allows firing to be done adjacent to other laboratory workers without unduly inconveniencing them.

velocity bullets, undoubtedly the simplest and easiest to use is a metal box of some description—a wood box will serve almost equally well—filled with rolls of absorbent cotton, or waste, or a combination of the two. In the past, oiled sawdust, loam, flour, corn meal, rags, snow, gelatin, and perhaps other substances have been used in various kinds of traps. One famous Expert used no trap at all but fired into a bundle of absorbent cotton placed on his desk. He caught thousands of bullets without an accident, though most people would not have wanted to share his office with him. One Eastern laboratory actually catches a lot of their test bullets in a common bushel basket filled with absorbent cotton.

In the New Jersey laboratory we use a steel trap of our own design. It is only 30" deep, but when properly set up, will stop the metal piercing .38 Specials and .45 ACPs, the .357 Magnum and the hottest .44 Special handloads. It is usually used with absorbent cotton in front and waste behind. If a lead bullet of low velocity such as the .32 S & W is being caught, new surgical cotton is unrolled and the paper removed. It is then re-rolled fairly loosely. The test bullet, upon striking the cotton, will immediately pick up a wad of it which protects it from injury. It will probably be recovered in its small cocoon after having penetrated halfway through one roll of this absorbent cotton.

Jacketed bullets from automatic pistols are easily handled in a cotton or cotton-waste trap. For best results, the absorbent cotton should be rerolled after each bullet is fired through it. The rolling operation can be done either quite loosely for .25 ACP, .32 ACP, and .380 ACP; or extremely tightly for 9mm Lugers and 7.63mm Mausers. In general, jacketed specimens can be recovered almost undamaged; in other words, you can look at them as they were when they came out of the bore of the gun. Even though a crime bullet may be smashed badly, you want to compare it with a test bullet that is in the best possible condition.

Although our bullet trap can stop the .357 Magnum and .44 Special handloads, the job is extremely unsatisfactory. The velocity of these bullets is so high that the cotton must be rolled very tightly and waste packed in under pressure. Even so, they tear right through the first layers of absorbent cotton before picking up a wad of it on the nose. Even cotton wipes off some striae; waste can put on new marks all its own if the cotton cocoon does not form. These bullets are badly deformed when recovered, frequently past any use. Several are sometimes fired before one usable one is recovered. However, test bullets from these weapons do not have to be fired at full velocities. We reduce the charge for test bullets, and fire full loads for test shells.

Our regular bullet recovery trap falls down in another respect also. The .22 rimfire test bullets fired into it are very likely to be damaged. We have resorted to water to stop these; whereas the powerful revolvers are almost never used in crime, the .22s are very common.

Two arrangements are now in use in the New Jersey State Police Laboratory. An oil drum full of water is one answer, but is clumsy since the room ceiling is too low. The barracks indoor swimming pool is better. Recovery used to be done and still is done in summer by diving. In winter, however, a blob of modeling clay on the end of a pole is more pleasant. With a little practice most bullets can be recovered on the first poke with no clay at all adhering to them.

One of the authors has set up a tank made of two oil drums, with all save one end removed and welded together. When full of water and resting on the cellar floor the top is level with the first floor. With a removable cover in place having a small hole in the center, test bullets can be fired easily and comfortably. Recovery is also easy with the stick and clay idea, if the inside of the tank has been painted white and the light above good.

In some localities the water supply is chemically treated, and the bullet should be washed with alcohol to prevent eventual destruction of the striae by action of the chemicals in the water.

The Royal Canadian Mounted Police fire into a tank of water horizontally using an extremely clever automatic plastic window arrangement to shoot through. Recovery is also easy—a kind of funnel arrangement apparently, so no searching through water is necessary.

For rifle ammunition, special arrangements are necessary. In general, the test bullets should be fired at reduced velocity; the regular traps will then suffice. One of the authors prefers to remove bullets from test cartridges—usually by hand with standard lathe collets— pour out all powder and replace with six to eight grains of Bullseye. This gives uniform ignition which is frequently not obtained with, say, a half charge of rifle powder. The bullets are readily caught in the usual manner and are entirely satisfactory for use in identification.

Solid point jacketed rifle bullets can be caught at full velocity, but with unnecessary difficulty. Striae created by passage through a bore with a muzzle velocity of 600 to 1,000 feet per second are fully identical with those obtained at full velocity. Reduced velocity is absolutely necessary if expanding bullets are to be caught in any condition fit for examination. It is also the best method of getting perfect test bullets from the .357 Magnum and other revolvers of this type.

Another method is the use of oil instead of water. Various grades have been tried, from No. 2 fuel oil to the motor oils up to SAE 20. While the oil does not damage the striae on the bullet, it must be cleaned off and is generally unsatisfactory.

The Inner Laboratory Equipment

Here is the heart of our science. The area in which actual examination of evidence and comparison is done should be kept free of

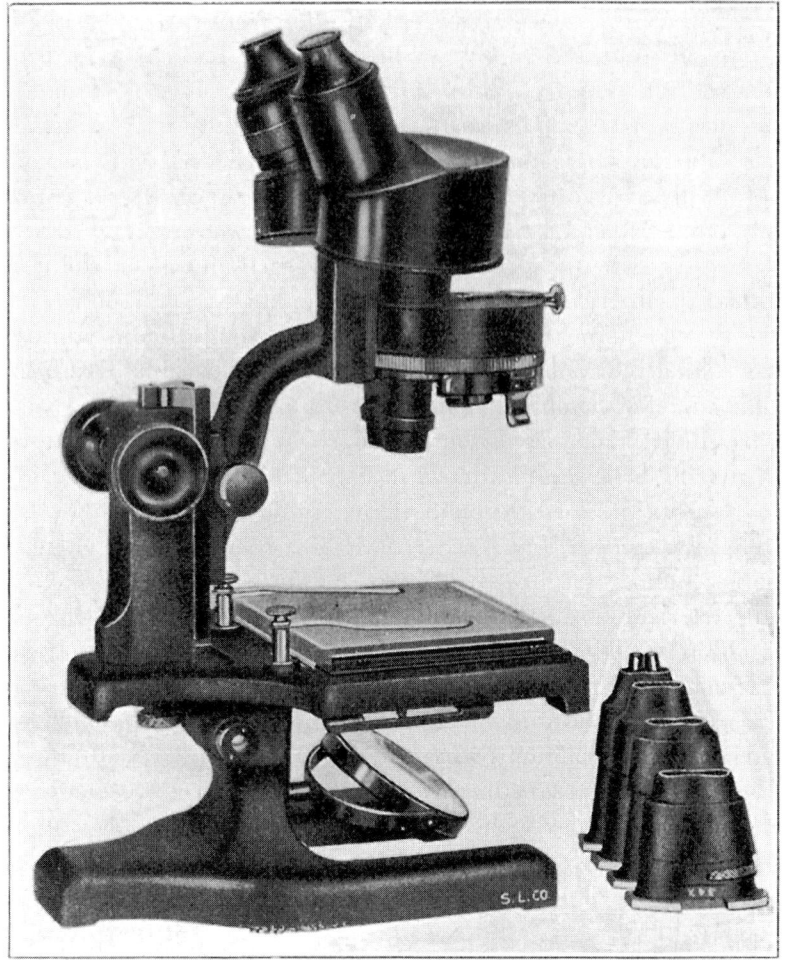

Wide-field binocular microscope, which is extremely useful for the examination of bullets, powders, cartridge cases, handwriting, etc. This instrument not only allows the use of both eyes, but it gives a stereoscopic effect which enables the depth of depressions and scratches to be seen, and gives specimens under examination a very pleasing appearance of reality, which is of great assistance in judging what is seen.

all unnecessary equipment, visitors, small movable objects and the like. Here you are dealing with human lives; even if your inner laboratory is only a desk in a large room, you owe it to yourself to clear off everything that could possibly lead to distraction. Actually, very little equipment is used in this work. However, the very best quality of what you do use is essential.

First, a good balance is required, capable of weighing to 0.1 grain. This should be to the side of the examination area, but accessible to the Expert without moving from his chair.

Second, some form of magnifier of a rigid type and a wide field is needed. The FBI and the authors use a widefield stereoscopic microscope capable of being swung on an arm over the evidence to examine it and then being pushed back out of the way. The illuminated stand magnifier is also used for this purpose.

Third and, of course, most important of all, a comparison microscope is needed. This is the one and only instrument used for actual practical bullet comparisons. Many experts use it also for their

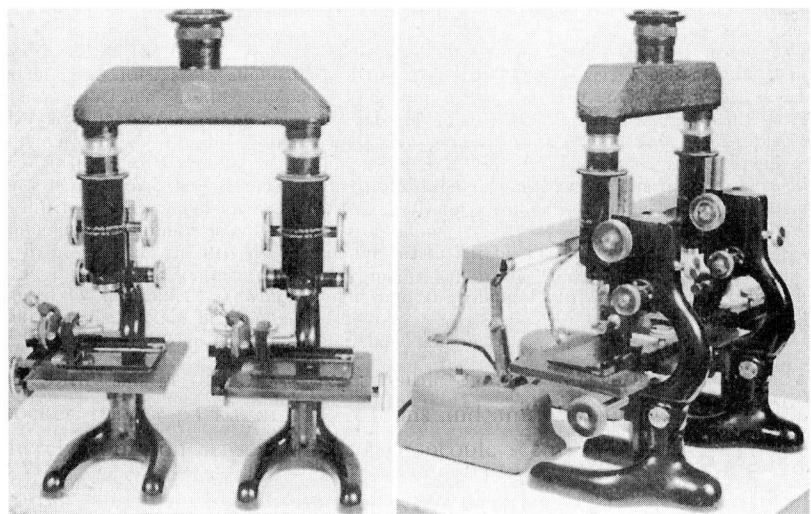

As shown elsewhere in this text, a comparison microscope may be a special instrument constructed for the purpose and for no other. However, equally good results can be obtained from two separate but matching instruments, known as metallurgical microscopes, with suitable bullet mounts and a comparison eyepiece to join the two instruments together in one unit. This particular instrument is from the Firearms Identification Laboratory of the New York State Police and is first class in every way. It has the additional virtue of the disassembly and use as separate instruments, if this should be required due to unforeseen circumstances. The fluorescent lighting through individual transformers is excellent, although other laboratories may prefer slightly different means of directing the illumination.

cartridge case work, although some men, including Sir Gerald Burrard and one of the authors, prefer the stereoscopic microscope for this work.

As already pointed out, a comparison microscope is basically two microscopes joined optically by means of a bridge so that one-half of the image seen through the bridge eye-piece shows one bullet and the other half the other bullet. Any two things to be compared can be placed on the two stages; to mean anything, bullets must, of course, be pointing in the same direction.

The comparison microscope should be equipped with a number

of different oculars or eye-pieces and perhaps two sets of objectives, although the latter is not necessary. The lenses should be of the wide angle type. Examination under about eight magnifications is usually sufficient for preliminary work. The magnification then can be doubled and finally, if necessary, tripled. However, any magnification beyond about 30 diameters is not only unnecessary, but may be a source of grave error. At a really high magnification the field is so small that an apparent match at a single spot is conceivably possible on bullets from different barrels.

There are two makes of American comparison microscopes: that put out by American Optical Co. and the Bausch & Lomb instrument. A few foreign makes, notably the Leitz, are also used. Quite a number also have been made up from parts by dealers, or even the laboratories themselves. The finished instruments from manufacturers cost in the neighborhood of $2,000 and involve a delay of several months. A perfectly satisfactory instrument can be made from two metallurgical microscopes with a comparison bridge and special bullet mounts.

In general, the Expert will prefer the instrument that he is most familiar with. The authors are familiar with several instruments. All allow first class work as soon as a man is familiar with them. The Spencer, since it can be mounted on a regular table and tilted towards the Expert, seems to the authors to be entirely satisfactory. The Bausch & Lomb is excellent but uncomfortable for many people to use, since it must be worked on between one's knees.

The new Leitz instrument seems to be the last word in convenience and optical excellence; however, we have not used one of these. The Watson instrument made for Major Sir Gerald Burrard more than twenty years ago and now in the personal possession of one of the authors is still a first rate comparison microscope. It has the advantage of being set up so that the axis of the tubes makes an angle of only about 30° with the horizontal. This position is very restful.

In America we demand mechanical stages. It is hard for our Experts to cope with the clamps and screws securing test and crime bullets in place under the older foreign instruments. The bullets themselves are fixed with some plastic such as beeswax, modeling clay, or plasticine to a base of appropriate size which is then mounted on a manual or mechanical stage. In the case of the latter, the entire bullet assembly can be moved precisely and evenly in both directions. The universal joint can be adjusted for changing the axis of the bullet and a separate control used for turning the bullet about that axis.

An Expert seated in front of his comparison microscope should have in easy reach but out of the way in drawers all the small supplies he will need, as well as the extra objectives and occulars for the instrument. Beeswax, cleaning fluid, small brushes, lens cloth, evidence wrapping paper, marking fluid, and wipers are only a few of the small objects that will accumulate. Where possible these should be kept in separate compartments in the drawers. The lenses should be placed in their cases in blocks of wood cut to receive them.

Safety of Evidence Files

From time to time, the evidence contained in the laboratory will be of tremendous importance to various individuals, many of whom are the most unprincipled type. Further, it is the definite duty of an Expert to maintain the evidence that may mean a man's life in the very best of condition, order, and efficiency. Undoubtedly, during the course of the usual day, many people visit the laboratory for one reason or another. All evidence should be preserved not only safely, but completely beyond all chance of tampering. If a defense attorney was ever able to insinuate that there was a possibility of substitution, most jurors would lose confidence in Firearms Identification. If evidence files are in the same room with even official visitors, they should be kept locked. The best place for them is in the inner laboratory which is entered only by the Expert and his assistant and where the actual examinations and comparisons are done.

The entire laboratory should, of course, never be left open unattended at any time. Even though police departments are seldom broken into or held up, such possibilities should not be disregarded. A man whose life is endangered by the striae on a single bullet could either take a very long chance himself to get a crime bullet back or persuade friends to do it for him. Actually, something of this type happened in the famous Sacco-Venzetti case.

Log or Laboratory Diary

An important record is the laboratory book which should be bound (not loose-leaf) and kept chronologically, showing in detail exactly what is done each day in connection with every bit of evidence officially examined. This laboratory book is perhaps the most important single aid to an Expert when he is called upon to testify as to exactly what he did in a case handled several months previously. We cannot emphasize its importance too highly. It should always be kept in ink and have full and complete entries made by the Expert or his assistant, rather than by a secretary.

Special Equipment

The Firearms Identification Laboratory should be able to conduct infra-red and ultra-violet examinations and photography. It is far better if this general equipment is not attached to the Firearms Laboratory proper, but is available for the use of all divisions of the Identification Department when they require it. In some organizations, the Firearms Laboratory technician sticks rigidly and completely to firearms; in others, he also does other scientific investigation in connection with firearms and other crimes. The nonvisible lights will frequently be of use in connection with the charred areas surrounding bullet holes. Photographs taken with these illuminations will reveal conditions invisible under ordinary light to the naked eye. Frequently in this same connection polarizing microscopes will be found helpful. They permit examination of cloth, skin, and the like for particles of certain elements such as nitrates and lead dust.

The most recent large addition to a Criminal Investigation Laboratory is the spectroscope and photography in connection with it. Any non-organic material whatever can be placed in the hollow of the cathode of a carbon arc and the resulting light broken prismatically into its component wave lengths. This pattern can be photographed and very precise determination made of the elements in the material placed in the center of the cathode. Since only tiny morsels of material are necessary, an extremely accurate chemical determination is possible with only a small amount of evidence. This instrument has tremendous potentialities. It is sometimes capable of determining precisely, in connection with lead shot, what lot and batch of ammunition was used.

THE STRIAGRAPH

Of the various firearms identification procedures developed within recent years, probably none is more novel or significant than that originated by John E. Davis, Chief Criminologist of the Oakland, California, Police Department. This method (first described in the FINGERPRINT AND IDENTIFICATION MAGAZINE, July, 1951) involves the application of a specially designed contour analyz-

STRIAGRAPH FOR BULLET ANALYSIS

Illuminating system and reflecting prisms not shown. Recording disc not visible from this angle. Tracer point rests on .45 caliber bullet on centering head. Note lever bracket, and chains for balance. Tail-blade of lever, and projection microscope to left.

ing instrument known as the 'Striagraph," which permits the operator to obtain a photographic chart or recording of the surface contour of a specimen bullet or bore.

The Striagraph is a mechanical and optical device capable of detecting minute irregularities in surface contour, and recording them in greatly amplified form on photographic paper. The instrument permits considerably greater magnification of heights than of widths so that the smallest of features become clearly discernible on the

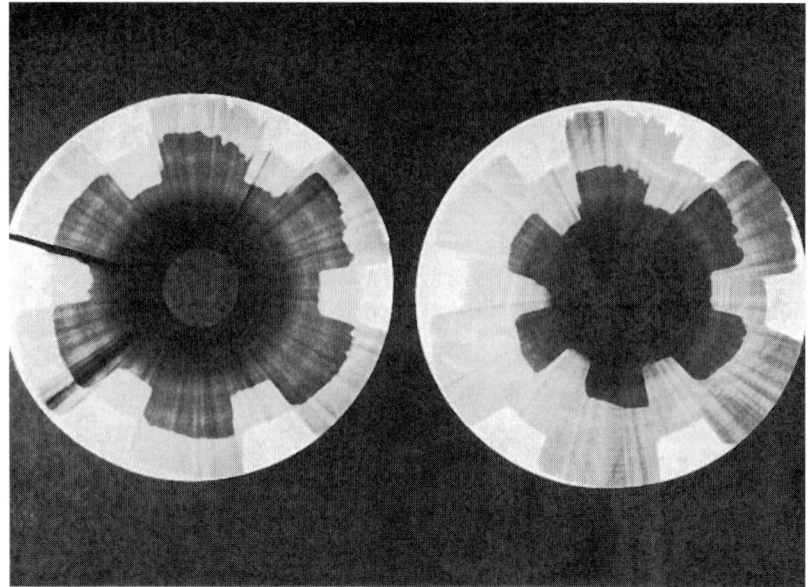

.38 CALIBER REVOLVER BULLETS—FIRED FROM COLT REVOLVER

Two separate bullets, two analyses of each. Center run on one offset lengthwise for an angle-of-twist determination. (No previous microscopic selection of areas for analysis.)

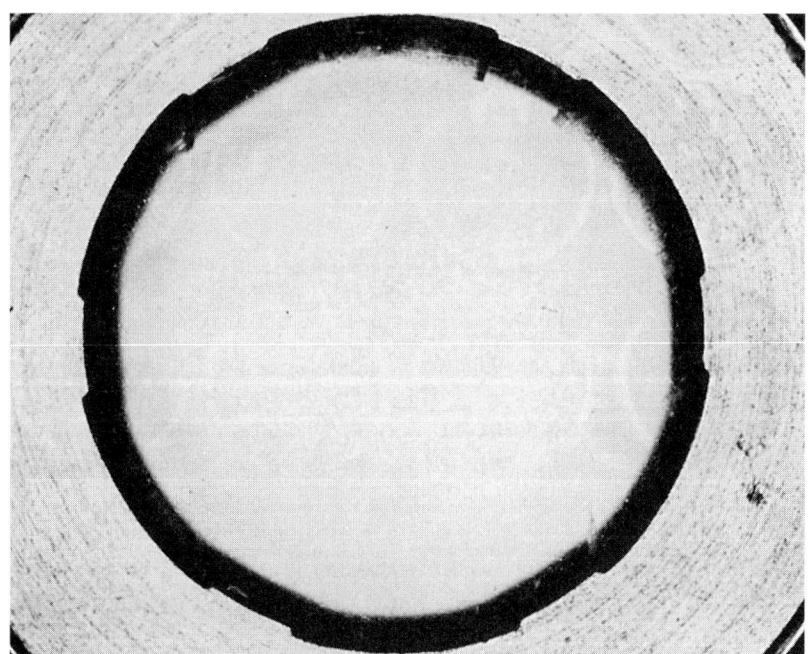

PLASTIC IMPRESSION OF SECTION OF COLT .38 BORE. Enlarged.

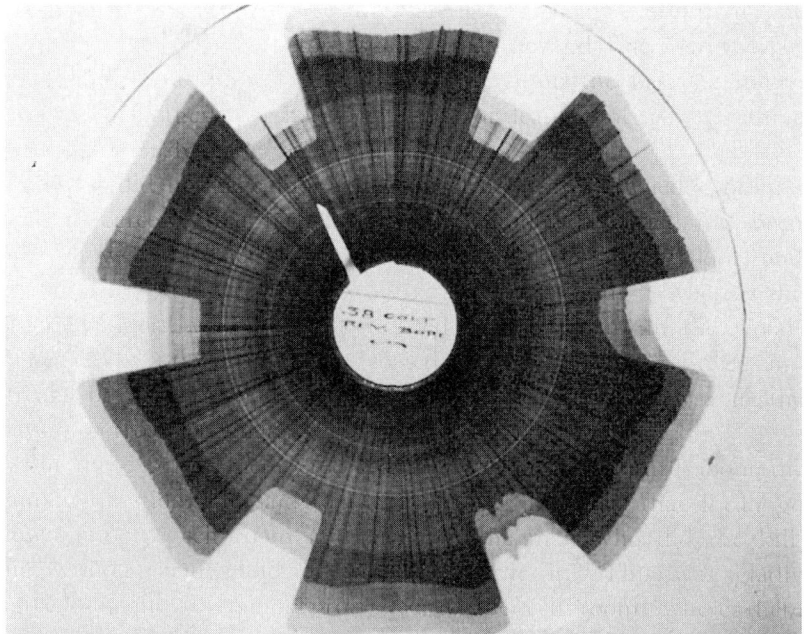

STRIAGRAM OF BORE SHOWN IN ILLUSTRATION ABOVE
Gross radial size shown here approximately the same as that of the bore illustration, as determined by bore groove levels. Note that groove bottoms are not concentric with bore at that diameter. (All interpretations are made with reference to a circular line when analyzing cylindrical specimens.) This chart shows three analyses at the same position inside the muzzle, but with different gross amplifications. Chart not calibrated but bore lands approximately 0.004 inches in height. Land and groove widths and depths easily measured on the chart as compared with the bore itself.

final chart. Application of this instrument permits the detection of slight "waviness" irregularities not previously detectable (either on bullets or within the bore) and a demonstration of land and groove dimensions, bullet diameters, bore diameters, etc., with an accuracy not possible by other means. Tests with this instrument have shown that where an identification is demonstrable with the comparison microscope, it may also be illustrated by contour analysis. Its ability to detect general irregularities of a non-striated character enables the operator to form a more definite opinion as to "identity" or "non-identity" in those instances wherein striae are present in insufficient numbers for microscopic analysis.

Although it is similar to certain industrially used surface-analyzers, the Striagraph was specifically designed for firearms and tool mark analyses on soft materials. It utilizes as a tracing mechanism, a very light weight lever, balanced on needle-point pivots, and provided with

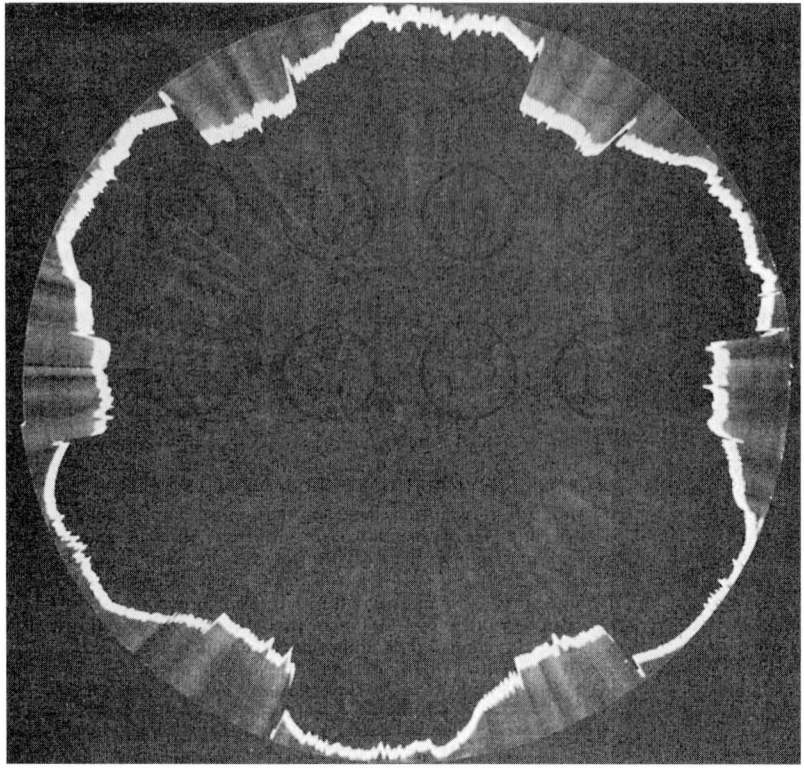

.45 CALIBER BULLET FIRED FROM COLT AUTOMATIC

Two analyses of the same specimen at one setting, but with different gross radial amplification. Same detail amplification in both. One run "positive" and the other "negative." Observe fidelity of detail as indicated by identity in the two runs.

an extremely fine tracing stylus. The tip-radius of the tracer stylus is approximately 0.0001 (one ten-thousandth) inch. This stylus rests lightly upon the rotating specimen bullet (or within the bore at the muzzle) in such a position that it is moved up and down as the striated specimen rotates beneath it on a centering head. The opposite end of this lever is positioned in front of a horizontally disposed microscope objective, which projects an image of the "tail blade" edge of the lever outward to reflecting prisms placed approximately three feet away. Light striking the prisms is reflected back toward the instrument where the blade image is focused on photo-sensitive paper affixed to a large disc contained within a housing. The housing is provided with a fine radial slot, through which the light passes to expose the paper as it rotates with the disc. The specimen bullet or bore being analyzed is affixed to the same shaft as holds the record-

ENLARGED BULLET LAND AREA FROM CHART
Another chart of the .45 caliber bullet enlarged to show consistency and extent of detail revealed by the tracing process. Radial lines due to binding of drive wheel, have no effect on contour tracing. Circular lines due to dust and microscopic irregularities on the slot edge.

ing disc, so that perfect relative motion is obtained between specimen and chart. Horizontal (gross radial) amplification is controlled by the position at which recording is begun relative to the center of the disc, while "detail" amplification depends upon the length of the light-lever used, the optical system, etc. For the average bullet specimen, typical amplification would be 20x "minimum gross radial" (chart twenty times as large as the specimen) and 400x detail, wherein all striae and surface irregularities are shown twenty times as high as they would appear in a simple 20x enlargement of a true cross-sectional cut of the specimen.

The chart obtained from a Striagraphic analysis is a sort of "shadowgraph" of the surface contour of a specimen. Two or three tracings may be recorded on one chart, at different gross radial amplifications, to permit verification of characteristics shown on each. The tracing lever is sufficiently light in weight that no cutting or scratching is visible on the analyzed specimen, and lever balance is so controlled (by fine chains) that no difficulty is encountered in the usual

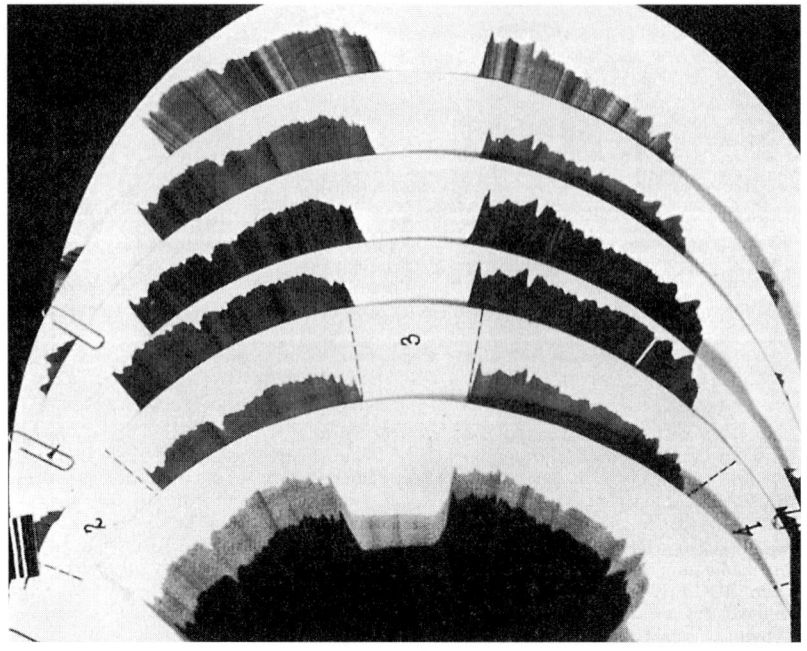

.30 CALIBER RIFLE BULLETS

Four charts of one bullet, and two of another fired from the same weapon. Compared as to two bullet land areas. (For proper appreciation of these identities the entire chart must be compared; this cannot be satisfactorily done by text illustrations.)

operation. The method may be applied to selected areas of damaged bullets, or can be utilized to give a complete peripheral tracing of undamaged specimens on a single chart. While the method is less rapid than microscopic comparison, it provides a permanent record of each analysis within the few minutes it takes to trace and record. Once the specimen has been properly centered and all adjustments made, it takes approximately ten minutes for the analysis to be conducted—after which the chart is photographically developed.

The analysis and comparison of Striagraphic charts is generally relatively simple—particularly as regards mere striae form and position. Determinations involving such features as bullet diameter, bore dimensions, and groove widths, depths, etc., require a more thorough understanding of the technique than might be true of other measuring means, but is claimed to be considerably more accurate than the usual methods. It appears that charts illustrating identities obtained through Striagraphic analyses are more easily understood by the lay person than are photo-micrographs of matching striae, so that

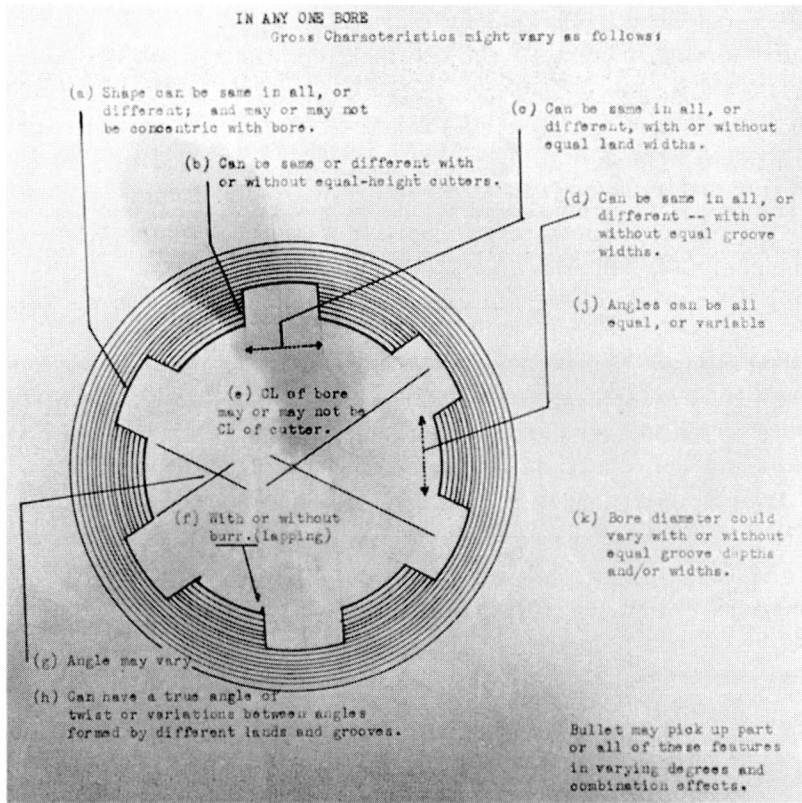

IN ANY ONE BORE
Gross Characteristics might vary as follows:

(a) Shape can be same in all, or
    different; and may or may not
    be concentric with bore.

(b) Can be same or different with
    or without equal-height cutters.

(c) Can be same in all, or
    different, with or without
    equal land widths.

(d) Can be same in all, or
    different -- with or
    without equal groove
    widths.

(j) Angles can be all
    equal, or variable

(e) CL of bore
    may or may not be
    CL of cutter.

(f) With or without
    burr.(lapping)

(k) Bore diameter could
    vary with or without
    equal groove depths
    and/or widths.

(g) Angle may vary.

(h) Can have a true angle of
    twist or variations between angles
    formed by different lands and grooves.

Bullet may pick up part
or all of these features
in varying degrees and
combination effects.

Possible bore defects and/or characteristics which may be detected with the Striagraph, and which may in part be detected on fired bullets by this technique.

the method seems well suited to court demonstrations. At the time of this writing, no attempt has been made to introduce these analyses in court, though there is no reason why they would not be acceptable as a valid technique.

The contour-analyzing method permits a completely new approach to the investigation of firearms evidence. It does not duplicate nor replace the familiar comparison microscope, but instead supplements it by providing a more specific analysis of minute and gross irregularities than is obtainable through microscopic or other means. The development of the Striagraph, and research conducted with it, have consumed a considerable amount of time over the past five years. As this work continues it is likely that still further determinations will be made possible through its application.

THE COMPARISON CAMERA

J. H. Mathews and Lee K. Henke

Dr. Mathews is Chairman of the Department of Chemistry and Director of the Four Year Chemistry Course at the University of Wisconsin. His interest in scientific methods of criminal identification dates back to the Magnuson bomb case in 1923—a case in which an identification was made by metallographic analysis. After this experience he was soon called upon to identify a rifle from a fired murder bullet. Working with meager equipment he was successful in building up a case that resulted in a confession during the course of the trial. Since that time he has been called on for assistance by the law enforcement agencies of Wisconsin on hundreds of occasions and has built up one of the best scientific crime laboratories in the country. A considerable number of new and improved instruments have been built in collaboration with Mr. Henke who is Senior Mechanician in the Chemistry Department. Mr. Henke has had many years of experience as an instrument designer and maker.—EDITOR.

Those of us who for years have had occasion to use a comparison microscope for the examination of bullets and other objects know how very tiring such examinations may become. Not only are they tiring but they are productive of eye-strain.

Both of these difficulties disappear with the use of the comparison camera—the basic idea of which was suggested by May and by Lewis. The operator sits in a comfortable position looking directly ahead at images on a large ground glass screen. The images are large and clear, are seen by both eyes and are viewed at a normal and comfortable distance, thus materially reducing eye-strain.

FIGURE 1.

The Comparison Camera and Rheostats for Light Control.

A series of knurled control knobs, placed in line a few inches below the ground glass, and one at either side of the ground glass, enable the operator to move the specimens in any desired direction and rotate them. After all adjustments of the two juxtaposed images have been made, one sees on the ground glass exactly what he may expect to see in the finished photograph. A time switch enables him to expose for a predetermined length of time.

Both plates and cut films may be used. Eastman Panatomic film is very satisfactory because of its lack of grain and its sensitivity to all colors.

The complete assembly is shown in Fig. 1. For the sake of clarity in description it is probably best to divide the instrument into several parts and describe each separately.

The Carriage

This part of the instrument (Fig. 2) has in its assembly the mechanism for transmitting vertical, horizontal, longitudinal and circular motion to specimens being compared and photographed. It is so arranged that the image of one object may be juxtaposed above the other, exactly. The carriage travels on a pair of $\frac{1}{2}''$ square tracks, $8''$ in length, secured to the base of the instrument and is actuated by $\frac{1}{4}''$ lead screw (G in Fig. 2) and having 28 threads to the inch. All lead screws are of the same diameter and pitch throughout. Rising from the carriage are two supports capable of being moved to the right or left independently of each other by means of two inclined planes and actuated by individual lead screws (H_1 and H_2 in Fig. 2). Each of these supports carries a bracket to which is attached a small worm and worm gear so arranged that there is virtually no back lash. These worm gear assemblies impart circular motion to the specimens they carry, by turning knobs R_1 and R_2 in Fig 2. The brackets are moved vertically by means of inclined planes built into the supports. The planes are moved by lead screws (V_1 and V_2 in Fig. 2). The shaft in the center of each worm gear is a small split stud to which different types of specimen mounts may be quickly and easily attached. All lead screws are coupled, by means of small Hooke's joints to the control rods which terminate in a series of control knobs (Fig. 2) at the lower edge of the plate end of the camera; thus the operator may remain seated while making all adjustments, once the bullets are aligned vertically.

The Camera

The camera is not unlike the conventional 8″ x 10″ view camera. It is cone shaped, being about 5 inches square at the lens end, about 12 inches square at the plate end and is 36 inches long. The small end has 2 lens rings mounted one above the other. Actually the comparison camera consists of two rigid camera enclosures mounted one above the other with an adjustable horizontal dividing septum to be described later. The lens rings carry a matched pair of Bausch and Lomb lenses. Various focal length lenses are available, a pair of 112 mm. Tessar Ic lenses are used for low magnifications, a pair of 70 mm. focal length Micro Tessars for greater magnification. The unique feature of the camera body is the fact that it contains an adjustable thin metal partition dividing the camera horizontally into two separate parts, one for each lens. This partition is surfaced with sateen cloth and sprayed with instrument black (a non-glossing lacquer) to prevent halo and ghost images, as the angle of incidence is very low, less than 15 degrees at times. There is no reflection from this mental partition. It is slip hinged at the lens end and supported at the plate end by two bearing blocks through which run two finely threaded vertical rods, one at each side of the plate holder but out of line with the light path. A single control knob (D in Fig. 2) at the top of the rear end of the camera, permits the adjustment of this dividing partition up or down, the two threaded rods being rotated in synchronism by means of a connecting chain and sprocket drive. The movement of the partition permits the making of comparisons of specimens at any point in their length. The rear of the camera is fitted with a conventional 8″ x 10″ plate back with ground glass permitting the use of standard 8″ x 10″ cut film or plate holders.

The Base

The base of the instrument is a 1″ thick American black walnut board edged with a modified "OG" bead all around. Inset on the under side of this base are two pieces of cold rolled steel ¼″ x 1″ through which are threaded the three adjusting screws whose ends terminate in knurled knobs, thus giving a three point support for the instrument. In order that the operator may be seated, in a normal position at the instrument, the entire assembly is set on three pyramidal wood blocks of the correct height for the operator's eye level.

Materials used throughout are stainless steel and brass except the inclined planes whose bearing surfaces are faced with cold rolled

FIGURE 2.
Rear End of Camera Showing Controls.

steel. All exposed brass surfaces are finished with black Hilo crystal-lizing varnish ("crackle finish"). The camera body is covered with Keratol (simulated leather) and the base is piano finished (several coats of clear Valspar varnish, hand pumiced and polished).

Lighting Equipment

At the lens end of the instrument and back of the carriage is a diffusing screen of opal glass illuminated by an ordinary 75 watt light bulb in a desk type reflector, thus providing a white background for the objects being photographed.

There are two adjustable microscope type object illuminators (Spencer. which are so positioned that they give oblique lighting to the specimens. These lights are controlled by two rheostats (shown in Fig. 2) so their intensity can be matched or balanced.

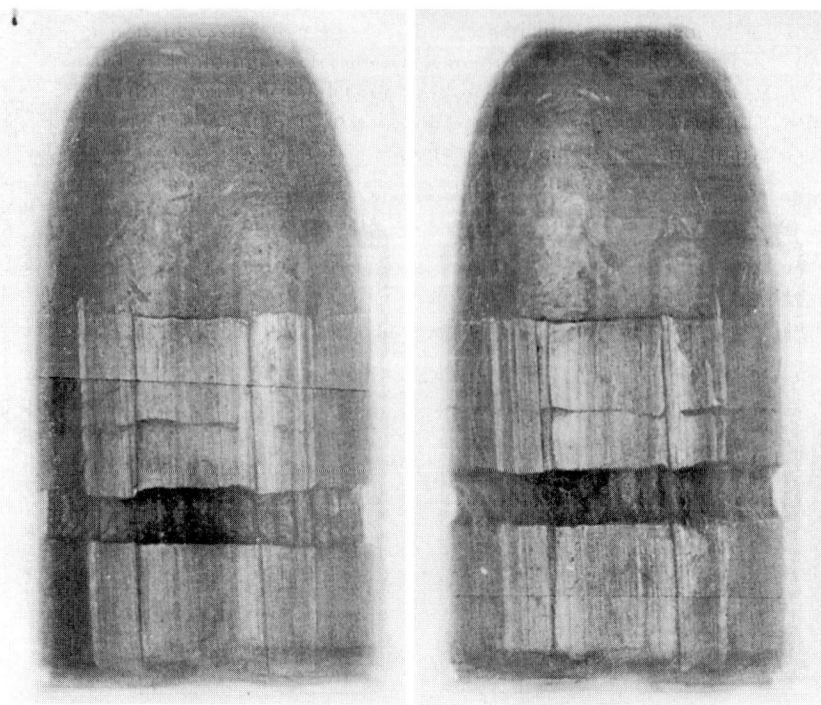

Comparison camera photographs of two different sides of the same bullet, one which featured in an actual murder case. Bullets were not in same position and the dividing line was moved—base end is the evidence, nose end the test bullet.

This is important as a test bullet is usually brighter than the evidence bullet.

Another similar pair of lights (not shown in the accompanying illustrations) is also used on the opposite side on occasion, or both sets can be used, to get the clarity and definition that is so necessary to good photography. Figure 3 shows a typical example of what can be accomplished with a pair of bullets that show good matching. Below the horizontal dividing line is shown the base end of the evidence bullet, above the line is shown the nose end of a test bullet fired from the suspected gun. Figure 4 shows what can be accomplished with the comparison camera in matching fired cartridge case and primer evidence.

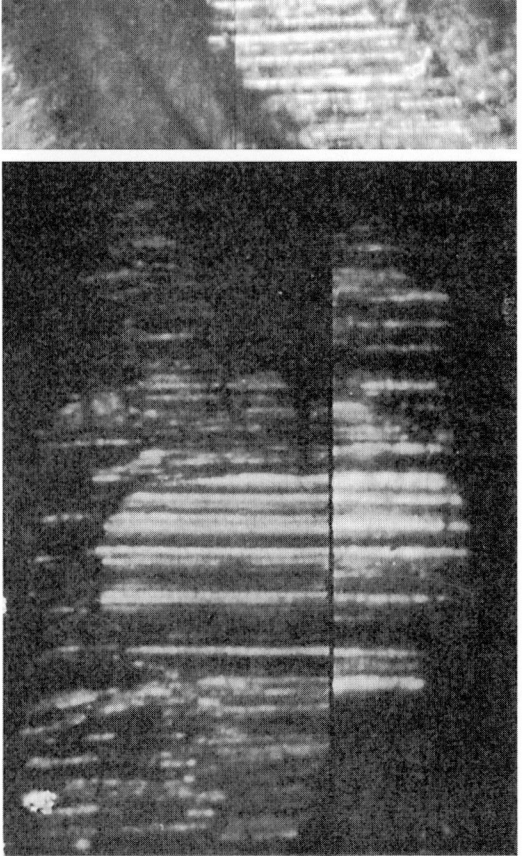

These are two photographs of fired case evidence, as taken and magnified by the comparison camera operated by Dr. Mathews. The left view shows the positive comparison of the breech block markings on a fired primer—a most excellent example of such evidence. The right view shows extractor marking comparisons on same fired case—also a very positive check. In both views the test shell is at top, evidence at bottom.

Reprinted from the POLICE SCIENCE incorporated in the JOURNAL OF CRIMINAL LAW AND CRIMINOLOGY

Obsolete and Experimental Equipment

No discussion of laboratory equipment would be complete without mentioning the instruments once quite popular for bore examination; there were also various measuring devices. All such work, however, is more or less unusual at the present time. It does little good in most instances to know why a bullet is marked in the way it is marked; it is quite sufficient to know that a given gun does so mark its bullets. Therefore, most of the comparison work today is done between a crime bullet and a test bullet with no effort to determine the whys and wherefores of the individual characteristics on either. However, in the past, a barrel scope was used.

Star gauges for measuring diameters at various places in the bores were also used. Machines were perfected to make lead slugs for passing slowly down bores. Casts were taken in sulphur or bismuth of bores and chambers.

One thing that was much talked of in the early literature was the question of telling from a crime bullet what make and type of gun fired it. This would involve extremely accurate and complete data as to the standards of every maker, and in addition, extremely accurate and complete measurements of the crime bullet, including such difficult things as determining accurately the pitch of the rifling, and so forth.

This theory is very attractive, but the actual hard facts of Firearms Identification place so many difficulties in the way of working out this identification that most of our best men do not even try to do it. They have practically abandoned the idea of attempting anything more by this method than eliminating the most unlikely guns, and even this they usually do by comparing the evidence bullet under the microscope with test samples fired from guns of known make, of which every good identification man will collect as complete a series as possible.

For this reason, the instruments that were designed and sold for such purposes as making very accurate measurements of groove widths, or of the twist of rifling in sample arms, are seldom used, if at all, by the research worker. Among these instruments may be mentioned the following:

Filar Micrometer

An extremely useful device for making measurements of the greatest precision is the Filar Micrometer eye-piece for the microscope. This is an eye-piece fitted with a movable thread or crosshair in the field of view; hence the name from the Latin filum, meaning thread. The crosshair is moved by means of a graduated knob on the outside of the eye-piece. Each rotation of the drum moves the wire 1mm across the field of view; as the drum is divided into 100 parts each division corresponds to a motion of .01mm or .00039″.

To measure an object with this micrometer, the object is placed in the field of view of the microscope and focused; then the amount that the crosshair must be moved to travel from one side of the image to the other is recorded. This is the actual size of the magnified image, and to get the actual size of the object itself this reading must be divided by the magnification.

For precision work it is necessary to measure some object of known length under the microscope; this shows you how much each division of the micrometer corresponds to for the magnification in question. It is usual to purchase with the Filar Micrometer a stage micrometer which is merely a little glass slip ruled with lines .001″ apart or with lines .01mm apart. Before making the measurement with the Filar Micrometer, the stage micrometer is laid on the stage or platform of the microscope and brought to focus; then the number of divisions that the drum must be turned to move the wire across one of the known divisions on the stage micrometer is noted. Thus if for example it takes 10 divisions of the drum to move the crosshair over one of the .001″ divisions on the stage micrometer then each division on the drum will be .0001 of an inch in measuring objects at the same magnification.

The type of Filar Micrometer eye-piece made by American Optical Company does not have a crosswire, but instead, has a finely ruled glass scale which is moved by the rotation of the micrometer head. Each division of the scale is equal to one revolution of the drum. The use of a ruled scale instead of a wire does not change the principle of the instrument, but is said to have the advantage that the index does not have to be traversed the entire distance of the object; the fraction of one division of the scale is all that must be read. In measuring the length of an object in the field of view, the scale is moved until one of the millimeter lines coincides with the margin of the object, then the total number of divisions of the glass scale lying on the object is noted, and the fractional part of the last division determined by reading it to hundredths by means of the drum, is added.

If the above has not confused you completely, let us continue a bit further. In general, in bullet work you are measuring curved surfaces which must be compensated for in defining what your results actually are. When you actually make measurements, the points where you start and stop are not really definite. Seldom will two technicians, even two college professors, making independent measurements agree precisely. Even the same man may arrive at slightly different results at different times. It is a tedious job and not worth much when completed.

The Goddard Helixometer

In the same class as to its necessity and desirability is the helixometer developed by Colonel Goddard and Mr. Fisher of the old Forensic Ballistic Bureau. It was a fine thing to have in a complete laboratory, but like other fine laboratory apparatus, it was expensive. It is no longer available; where already installed, it is seldom used.

It was thought to be useful to be able to examine the inside of the bores of firearms. The medical profession has long used the cystoscope and naso-pharyngoscope. These are nothing more or less than small magnifying periscopes carrying a small light on the far end for the purpose of viewing the interior of various body cavities. Such a small periscope can also be used for viewing the inside of the barrels of firearms. It was often helpful in locating dust particles such as tobacco grains, and the like, which accumulate in the gun barrel when the weapon is carried in the pocket but not fired. The presence in a barrel of tobacco grains of a certain kind and of similar grains in the pocket of a suspect might help materially in directing the investigation into proper channels. However, the most important use of the Goddard adaptation of this instrument is for measuring with extreme accuracy the pitch of the rifling in a weapon.

Court Equipment

Frequently in the past, experts have devised several aids in presenting their testimony to a jury. Model bullets have been made several inches in diameter and more than a foot long showing greatly exaggerated striae. These bullets have usually been cut in half and fixed loosely together to that the base sections would interchange. They were sold in pairs. This gave a good indication of the picture shown in the comparison microscope. The halves which matched did so very obviously; when improperly paired they, of course, did not. An even more ambitious arrangement was a double eye-piece comparison microscope which allowed an expert to point out through an extra eye-piece and movable pointer to a juror who was using the regular eye-piece the fine points of his identification. However, these aids, like photography, are not used to any great extent anymore. The testimony of the Expert is accepted without confusing visual demonstration.

Portable Equipment

The equipment for a complete laboratory such as described above is costly, difficult to obtain, and extremely bulky. It is not portable, save with considerable trouble. However, much in Firearms Identification can be accomplished with a good store of knowledge and a jeweler's loupe. For instance, in general, one knowing arms and having a bit of common sense can probably identify the caliber, make and model of a weapon by a projectile or a cartridge case only. Further, an examination of the crime bullet held in the palm of the hand will frequently eliminate many suspects, or at least suspected weapons

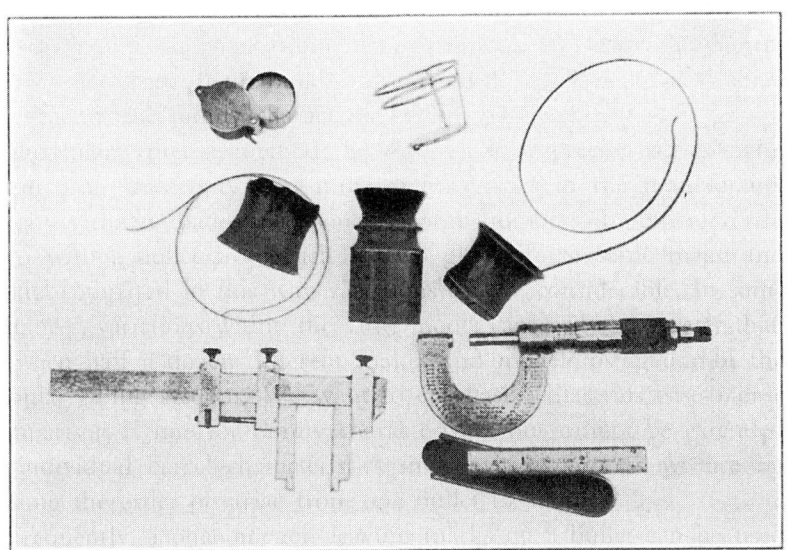

TOOL KIT FOR PRELIMINARY EXAMINATION

Some of the tools found useful in preliminary bullet identification. Top, left, Bausch & Lomb aplanatic magnifier, 7.5 power. Next, watchmaker's glass fitted to slip over the lens of a pair of spectacles. Middle of picture, Bausch & Lomb Focusing Magnifier, for use in getting the image sharply focused on the ground glass. On the right, Bausch & Lomb watchmaker's glass, 3 inch focus, with removable spring for holding it in the eye. The spring goes around the back of the head. Left, a similar glass with spring removed. Bottom, left, vernier calipers, useful for measuring the overall length of cartridges, etc. Right, 1 inch micrometer calipers, and Starrett 3 inch flexible steel rule No. 321.

which may have been in their possession. For instance, obviously a crime committed with a 6-groove left-hand twist weapon cannot be laid at the door of a man possessing only a weapon of the same size but having 6 grooves right-hand twist. A .38 Special crime bullet cannot have come from any .32 caliber weapon.

Some Firearms Identification Experts will, in the course of their regular duties, go to the scenes of firearms crimes. At least one member of the Ballistics Squad of the New York City Police Department investigates on the spot every serious firearms crime. Elsewhere, Experts are sometimes on call by the detective bureau when needed. Sometimes a detective has the basic knowledge to do the work himself.

In any case, a small kit of equipment will be useful. One of the authors has a pocket size kit that seems to contain most of what is usually needed. It has three loupes of $2\frac{1}{2}$x, 5x, and 10x magnifications, a six inch steel scale, a six foot carpenter's rule, a caliber gauge, a scriber, a micrometer, a screw driver set, a pair of pliers, a fishing line and some calipers. With a notebook, pen, evidence envelopes, flashlight, and a Land Polaroid Camera with flash equipment, reasonably good rough work can be done.

Personal Qualities Needed by the Firearms Expert

By far the most important part of the laboratory is the Expert. Perhaps a few words in regard to the attributes he will need most are in order. The first qualification must always be complete and uncompromising integrity. The lack of this even to a small degree—laziness for example—can do immeasurable harm to the Expert himself and his department. In general, ambition and hard work are far more important than academic training. Experience and gun knowledge are certainly valuable but are not absolutely essential. Most Firearms Identification Experts were gun cranks before their appointment. Natural intelligence and cleverness are, however, imperative. Common sense will do more in the long run than a Doctor of Philosophy Degree. A certain knowledge of microscopy is essential, but can be picked up as one goes along. The writers have no patience with Experts who state their knowledge to have come naturally. It takes no more natural flair to produce a firearms Expert than it does to produce a real toolmaker. In both cases, you cannot succeed if you work carelessly or hurriedly or without a true sense of the extreme importance of what you are doing.

APPENDIX 1

The Combined Library of the Authors in Connection with Firearms Identification

FOREWORD: There are two reasons why a library is needed by every Firearms Identification Expert. The first is for direct reference in connection with specific questions in connection with actual laboratory work. Reference books of one type or another will be of importance; however, of even more importance are the various charts, ammunition collections, and the like, that the laboratory will have built up for itself. These are in large measure compiled from various books and to some extent replace them.

The second purpose of a library is in connection with general firearms knowledge. Books of all descriptions will aid the Expert in gaining the complete familiarity with the whole subject so useful to him in his profession. However, other things besides books are important too. Extensive shooting with many different weapons under varying conditions, as well as gunsmithing and experimenting, play their part. We have included in books of general knowledge only those that are fairly pertinent. There are many others in connection with ordnance history, the use of sporting weapons, hunting, and the like, which may be valuable in some lesser degree. We point out that the two purposes mentioned above are not separate. Frequently, the same book will be of importance both for actual reference and for general knowledge.

PERIODICALS AND ANNUALS

We list below certain magazines, catalogs, and the like, which are important. We point out that the importance of these is not limited to the current issues. We maintain files of all these publications over a period of several years, and frequently refer more often to, for instance, Stoeger's 1935 Catalog, than to the current issue.

All Available Catalogues of the Major Ammunition Makers

The Journal of Criminal Law, Criminology and Police Science, Quarterly
>   Published by Northwestern University School of Law, Chicago, Illinois.

The American Rifleman, Monthly
>   Published by The National Rifle Association, Washington, D. C.

Ordnance, Bi-monthly
>   Published by American Ordnance Association, Washington, D. C.

The Ideal Handbook, Annually
>   Published by The Lyman Gun Sight Corporation, Middlefield,

Stoeger's Catalogue, Annually
>   Published by Stoeger Arms Corporation, New York, New York.

The Gun Digest, Annually
>   Published by The Gun Digest Company, Chicago, Illinois.

FORENSIC BALLISTICS AND CRIME

These books cover both actual Firearms Identification and other criminal investigation important to the Expert in ratio of his participation in actual criminal investigations. For instance, some departments occasionally call on laboratory personnel for special crime scene consultations. As referred to many times before, the New York City Ballistics Squad investigates every firearm crime.

Textbook of Firearms Investigation, Identification and Evidence
> By Julian S. Hatcher, Samworth, Georgetown, S. C., 1935.

The Identification of Firearms and Forensic Ballistics
> By Sir Gerald Burrard, Jenkins, London, 1951.

Forensic Chemistry and Scientific Criminal Investigation
> By A. Lucas, Arnold & Company, London, 1948.

The Identification of Firearms
> By Jack Disbrow Gunther and Charles O. Gunther, Wiley, New York, 1935.

Criminal Investigation
> By Dr. Hans Gross, John A. Adam and J. Collyer Adam, Sweet & Maxwell, London, 1949.

Modern Criminal Investigation
> By Dr. Harry Soderman and John J. O'Connell, Funk & Wagnalls, New York, 1952.

Homicide Investigation
> By Dr. Le Moyne Snyder, Thomas, Springfield, Illinois, 1950.

Encyclopedia of Criminology
> By Dr. Vernon C. Branham and Dr. Samuel B. Kutash, Philosophical Library, New York, 1949.

Questioned Documents
> By Albert S. Osborn, Boyd, Albany, New York, 1950.

Lie Detection and Criminal Interrogation
> By Fred E. Inhav and John E. Reid, Williams and Wilkins, Baltimore, Maryland, 1953.

GUNSMITHING, AMMUNITION AND HANDLOADING

This list of books is self-explanatory. Even though not of direct use to the Identification Expert, these are of extreme importance in some of his restorations and the like. They also add to his general familiarity with weapons, and more important still, ammunition. The book by White and Munhall is particularly recommended in its small field.

Gunsmithing
> By Roy Dunlap, Samworth, Georgetown, South Carolina, 1950.

Advanced Gunsmithing
> By W. F. Vickery, Samworth, Georgetown, South Carolina, 1940.

American Machinists Handbook
> By Fred H. Calvin and Frank A. Stanley, McGraw-Hill, New York, 1940.

Mechanical Engineers Handbook
> By Lionel S. Marks, McGraw-Hill, New York, 1941.

The Chemistry of Powder and Explosives
> By Dr. T. L. Davis, Wiley, New York, 1941.

Centerfire Metric Pistol and Revolver Cartridges
> By H. P. White and B. D. Munhall, Sportsmans Press, Washington, D. C., 1948.

Cartridges
> By Herschel C. Logan, Standard, Huntington, West Virginia, 1948.

Principles and Practice of Loading Ammunition
> By Earl Naramore, Samworth, Georgetown, South Carolina, 1953.

Complete Guide to Handloading
> By Philip B. Sharpe, Funk & Wagnalls, New York, 1942.

Firearms In General

These volumes should be read slowly, carefully, and outlined. One of the authors spent a period of more than 24 months on the first two, and took examinations in these from appropriate professors. They aid immeasurably in broadening the general firearms outlook, as well as, in some cases, being extremely interesting reading.

Naval Ordnance (Annapolis Text)
> U. S. Naval Institute, Annapolis, Maryland, 1937.

Elements of Ordnance
> By Thomas J. Hayes (West Point Text), Wiley, New York, 1938.

Hatcher's Notebook
> By Julian S. Hatcher, Stackpole, Harrisburg, Pennsylvania, 1947.

The Book of the Garand
> By Julian S. Hatcher, Sportsmans Press, Washington, D. C., 1948.

Textbook of Pistols and Revolvers
> By Julian S. Hatcher, Samworth, Georgetown, S. C., 1935.

Automatic Weapons of the World
> By Melvin M. Johnson, Jr., and Charles T. Haven, Morrow, New York, New York, 1945.

Textbook of Automatic Pistols
> By R. K. Wilson, Samworth, Georgetown, South Carolina, 1943.

Pistols and Revolvers
> By Walter H. B. Smith, NRA Book of Small Arms Volume I, Military Service Publishing Co., Harrisburg, Pennsylvania, 1946.

Rifles
> By Walter H. B. Smith, NRA Book of Small Arms Volume II, Military Service Publishing Co., Harrisburg, Pennsylvania, 1948.

The Book of the Rifle
> By T. F. Fremantle, Longmans, Green, London, 1901.

The Gun and Its Development
 By W. W. Greener, 9th Edition, Cassel, London, 1910.
In the Gunroom
 By Major Sir Gerald Burrard, Greenberg, New York, New York, 1951.
Notes on Sporting Rifles
 By Major Sir Gerald Burrard, Arnold, London, 1953.
The Modern Shotgun (in three volumes)
 By Major Sir Gerald Burrard, Jenkins, London, 1950.
Small Arms Design and Ballistics (Volumes I and II)
 By Colonel Townsend Whelen, Samworth, Georgetown, South Carolina, 1954.
A Basic Manual of Military Small Arms
 By W. H. B. Smith, Military Service Publishing Company, Harrisburg, Pennsylvania, 1943.
Notes on United States Ordnance
 By James E. Hicks, Hicks, Mount Vernon, New York, 1946.
Weapons of World War II
 By General G. M. Barnes, Van Nostrand, New York, 1947.
The Rifle in America
 By Philip B. Sharpe, Morrow, New York, 1938.
Rifles and Ammunition
 By H. Ommundsen and E. H. Robinson, Waverly, London, 1915.
The Book of the Springfield
 By E. C. Crossman, (Revised by Roy F. Dunlap), Samworth, Georgetown, South Carolina, 1951.
The Machine Gun
 By George M. Chinn, U. S. Government Printing Office, Washington, D. C., 1951.
Mannlicher Rifles and Pistols
 By W. H. B. Smith, Military Service Publishing Company, Harrisburg, Pennsylvania, 1947.
Mauser Rifles and Pistols
 By W. H. B. Smith, Military Service Publishing Company, Harrisburg, Pennsylvania, 1946.
Keith's Rifles for Large Game
 By Elmer Keith, Standard Publications, Huntington, West Virginia, 1946.

Miscellaneous

Photo-Micrography
 By R. M. Allen, D. Van Nostrand Co., Inc., New York, 1941.
The Story of Colt's Revolver
 By William B. Edwards, The Stackpole Company, Harrisburg, Pennsylvania, 1953.
Legal Medicine and Toxicology
 By Gonzales, Vaner and Halpern, Appleton Century, 1937.

Metallurgical Analysis by the Spectrograph
By D. M. Smith, British Non-Ferrous Metals Research Association, London, 1933.

The Use of the Microscope
By Belling, McGraw-Hill, New York, 1930.

Gunshot Injuries
By La Garde, Wm. Wood and Company, 1916.

The Chemistry of Powder and Explosives, Volumes I and II
By T. L. Davis, John Wiley & Sons.

The Universe of Light
Sir William Bragg, MacMillan, New York, 1934.

Mathematics for the Million
By Lancelot Hoyben, W. W. Norton & Company, Inc., 1937.

The Manufacture of Explosives, Volumes I and II
By Oscar Guttman, Whittaker & Company, London, 1895.

Fire and Explosive Risks
By D. Von Schwartz, Charles Griffin & Company, Ltd., London,

Encyclopedia of American Hand Arms
By George M. Chinn, Jr., and Bayless E. Harder, Standard Publishing Co., Huntington, W. Va., 1942.

The Microscope and Its Use
By F. J. Munoz and H. A. Charipper, Chemical Publishing Company, Brooklyn, New York, 1943.

Explosions, Their Anatomy and Destructiveness
By Clark Shove Robinson, McGraw-Hill, New York, 1944.

Textbook for Small Arms (Three different texts)
His Majesty's Stationery Office, London, 1904, 1909 and 1929

American Gun Makers
By L. D. Satterlee and Major Arcadi Blackman, The Stackpole Company, Harrisburg, Pennsylvania, 1953.

A History of Firearms
By Major H. B. C. Pollard, Houghton Mifflin Company, 1933.

Smith & Wesson Handguns
By Roy C. McHenry and Walter F. Rogers, Standard Publication, Inc., Huntington, W. Va., 1945.

Ammunition
By Melvin M. Johnson and Charles T. Haven.

The Blaster's Handbook
By Arthur LaMott, E. I. duPont de Nemours Company, Inc.,

Photographic Evidence
By Charles Scott, Vernon Law Book Company, Kansas City, Missouri, 1942.

Infra-Red for Everyone
By H. W. Greenwood, Chemical Publishing Company, London

Photography as a Scientific Implement (a collective work)
Blackie & Son, London and Glasgow, 1934.

APPENDIX 2

Organization and Operation of a Firearms Identification Laboratory

FOREWORD: We are quoting below in its entirety a bulletin of the New Jersey State Police in connection with their particular organization and procedure, including their own bookkeeping and filing systems.

In New Jersey, the State Bureau of Identification is operated by and under the control of the State Police. This State Bureau maintains the fingerprint records for the State of New Jersey. A State Law requires the taking of fingerprints of suspects in all indictable criminal offenses and provides a penalty for any police officer who fails to take prints. The Bureau also maintains a Chemical and Physical Laboratory, a Laundry and Dry Cleaning Marks Laboratory, a Photographic Laboratory, and a Firearms Identification Laboratory. Each section is headed by an officer who is directly responsible to the Supervisor of the State Bureau of Identification through the Executive Officer (Assistant Supervisor).

All reports and correspondence pass through the Executive Officer and are signed by the Supervisor.

There is considerable administrative work to be done in all such sections, and an Organization Chart will be of considerable help in keeping this work along the clearly defined lines necessary to avoid overlapping and confusion and unneccessary duplication of effort. For instance, each section head must maintain Civil Service records, time sheets, efficiency ratings, promotion recommendations, et cetera. These are sent to the Executive who consolidates them and submits those required to the Supervisor.

In the Firearms Identification Section, as in others, adequate records are absolutely essential. Applicable forms to facilitate the recording and compilation of pertinent data must be carefully worked out, and any unnecessary information kept out of the form.

Records, to be of any use, must be readily accessible, even in the face of sketchy or incomplete information from the people requesting the information.

It is a common occurrence for a policeman to come to the Firearms Laboratory to reclaim a revolver that has been processed. He will know nothing about the case and has forgotten the name of the victim, the make, caliber, or serial number of the gun, or even the nature of the crime. About all he is sure of is his own name and department.

Fortunately, in cases such as these, the department is usually a small one and has very few cases in the Bureau at the same time. By elimination, the gun can be found in less than a minute, if the records system is well thought out and operated.

Our Laboratory System was adapted, with only a few changes to fit local conditions, from that of the New York City Police, where the first Firearms Identification Officer from this organization studied and got his early training. We use the following forms:

Upon the arrival at the Laboratory of any and all evidence, the Expert or his assistant will fill out a receipt, listing on the receipt in detail the evidence submitted. This receipt is in triplicate. One copy remains in the laboratory, one copy goes to the Executive Officer, and the final copy is given to the officer submitting the evidence to the laboratory. At this same time, a laboratory Request for Information form is filled out or obtained. In theory, the submitting department or subdivision of the State Police should bring in a request for information on the proper forms concisely and exactly written. However, in practice, this seldom occurs. Two different forms are used; one for State Police requests and one for requests from outside agencies. It is important that the laboratory get down in specific language exactly what the submitting detectives wish to have done. Copies of the receipt and the two different request forms are reproduced herewith.

No. 12345   Date Dec 1, 1954

State Bureau of Identification
Division of State Police
Trenton, New Jersey

Description .32/C Hopkins & Allen Revolver

Case or Test John Doe, Murder   SBI # 5246

Source Lower Dunlap N.J. P.D. Det. I.Moe

Received at Laboratory   Returned to Det. B. Moe
N. J. State Bureau of Identification   Date Dec. 15 1954
Date Nov. 14, 1954   From: N. J. State Bureau of Identification

---

Receipt Card: This card carries a brief description of the firearm, including its caliber, make and model. It is signed by the person taking the gun from the Laboratory, and is filed by case number.

Second, immediately upon the evidence having arrived, a receipt card is filled out by the Expert or his assistant and the evidence properly secured. For instance, if it is already wrapped to protect it from damage and in a plain envelope or one filled out by the submitting department, it is left in this envelope in its original wrappings. If it is not, it is wrapped and placed in a plain envelope in the presence of the detective bringing in the evidence. He is asked to sign his name on the blank envelope. In some instances, in spite of everything that can be said, a policeman from some small town will bring in a bullet in his trousers pocket.

There are three distinct types of cases upon which information is desired in fairly large numbers. Special cases other than these must be treated in accordance with common sense. First, there are cases involving evidence in the form of bullets and/or discharged cartridge cases only. Second,

```
                    DESCRIPTION  OF  FIREARMS
......•32  S&W ..... .Caliber                    Twist 5  Right
Make Hopkins & Allen                       Serial No.    12345
Model  "Forehand"                          Capacity   5
                                           Barrel length  3¼"
Frame Hinge        Action  Double          Overall length      7¼"
Pitch 1 in 12" Land width .095             Groove width .095
                                                        45 Deg•
Ejection Simultaneous              Breech finish Filed 180 Deg.
Source  Lower Dunlap NJ P.D.                    Date Dec 1,1954

O
```

Gun Data Card: This card carries detailed description of firearm. The heavy circle at lower left is used to show location of ejector and extractor marks, where applicable. The card is filed by Caliber, Number of Grooves, and Direction of Twist.

there are cases involving suspect weapons without crime bullets or cartridge cases. Third, there are cases involving both a bullet and/or cartridge case and a firearm.

First, all cases of whatever origin in which bullets or cartridge cases are submitted receive immediately a number and a large manila file jacket stamped with that number. All evidence and paper work in connection with this case are kept in this file jacket.

Second, in a case where only a weapon is submitted, no number is

```
Case Doe,  John,            Ser. No. 12345

                            No.  5246
 Crime  Murder             Date Dec.1,1954

Source Lower Dunlap, N.J. Police Dept. Det.J.Moe

Cal. .32 S&W     Twist 5 Right
Desc. Hopkins & Allen Revolver
Type Hinge Frame, Hammer    Model "Forehand"
Action Double               Bbl. 3¼" Ribbed.
Capacity  5                 Finish Nickel, Worn.
Remarks Poor condition;does not range.Ratchet
damaged.
Test Load
2- .32 S&W Rem-UMC WP 88gr lead.
```

Nomenclature Card: This card gives details of crime, case, source, and description of gun and test ammunition. It is filed alphabetically, under name of Victim. Where there is no Victim, as in Concealed Weapons cases, the name of the Accused appears. Where there is neither victim or accused, the name of the department submitting it is used.

assigned, but the weapon is carefully receipted for with appropriate designations so that the weapon can be found and returned, even with a minimum of information as pointed out previously.

Third, where both components and a weapon are submitted, a combination of the above two procedures is followed, with all the paper work done in connection with both. A number is, of course, assigned.

In a case of the first type—that is, one in which components but no weapons are submitted—the procedure followed in the laboratory consists of an initial examination of all evidence. If this has not been marked by the investigating officer, the Firearms Identification Expert marks it or them with his own mark with a steel scriber. The bullets are then weighed and the caliber determined. The number of lands and grooves and the direction of twist are also determined. The pending cases of similar caliber, direction of twist, and number of grooves are compared. Since

```
Doe,John            Murder              5246

Dec 1,1954          Firearms

Lower Dunlap N.J. Police Dept. Det. J.Moe.
```

Cross Index File Cards: One buff and two white cards are used. The buff card is filed numerically by case number. One of the white cards is filed alphabetically; the other is filed by crime, chronologically.

there are no weapons, this is the only testing that can be done. A brief concise report is sent through channels to the submitting office or officer.

Two white and one buff card 3 x 5 size are typed bearing the name of the victim or, if no victim, the name of the accused as in the Concealed Weapons charge, the name of the crime, and the case number. If more than one victim or accused, a separate white card is made for each, with the alternate names on top.

One white card in each name is filed alphabetically; one white card is filed serially according to crime; the buff card is filed serially in one file, regardless of crime.

In a case of the second type in which a suspect weapon is received in the laboratory, test shots are fired from it. These test shots are then compared with all bullets of similar characteristics in the pending files.

If there is any reason to believe that the criminal may have been involved in specific crimes elsewhere, test bullets may be sent to cooperating departments throughout the country where the modus operandi is similar or where other evidence suggests the possibility of the weapon having been used in another crime. A file of requests for test bullets from other departments is maintained.

In a case of the third type—where both a weapon and ammunition components are submitted—both the above procedures are followed with, of course, a comparison of the crime bullet or cartridge case with test bullet or cartridge case from the suspect weapon done first.

When a positive match on a pending case is made, the laboratory report on the new case is written to include this, and a supplementary laboratory report on the pending case is also made. This is forwarded to the department concerned, the case is lined out on the list of pending cases, and the jacket taken from the pending file and put in the completed file. All evidence in the jacket—that is, discharged bullets and shells—remains there until the case goes to court. Thus continuity of evidence is unbroken from the time it is submitted until it is introduced in court.

When a case is closed out by going to trial and sufficient cases of this type are on hand, they are transferred to the Evidence Vault. When completed cases never come to trial, or for some other reason no disposition is received, and the files are crowded, the oldest ones are removed to the vault. They are similarly filed there and readily accessible. There is but one key to the vault available, and admission to the vault is made under strict supervision, so that there can be no suggestion of unauthorized entry or handling of evidence stored there.

A continual cross checking is done with every new suspect weapon and crime bullet or cartridge case against all other pending suspect weapons and evidence. This would be an impossible task if it were not for the classification possibilities in connection with the number of lands and grooves, caliber, and direction of twist. Even so, large departments handling hundreds of cases each year have accumulated under some classifications—for instance .32 caliber Smith & Wesson, 5 grooves right— an unwieldy number of unsolved cases, some many years old. Eventually, some common sense sorting out will be necessary.

Reports are made as promptly as is consistent with thorough processing. All useful information is passed on to those primarily concerned with particular investigations. In special cases, intelligence is necessary to inform the various departments of matches by means not only of regular reports, but also telephone calls or personal contact where of particular and immediate importance.

CHAPTER 11

GENERAL INVESTIGATION OF FIREARMS CRIMES

THE most important part of the investigation of most crimes is just after they have been discovered. Unfortunately, few police department experts and almost no independent firearms identification men are called in at this early stage. A notable exception is New York City where the Ballistics Squad is large enough to have men on duty 24 hours a day throughout the year; two members of this squad visit every firearms crime and gather the firearms evidence and data first hand.

Elsewhere the firearms laboratory personnel is not usually called upon to go out in the field to investigate crimes just committed, they should have however a rather detailed knowledge of the procedures followed by the general investigators and others, since these procedures have such a great bearing on the Firearms Identification work in the laboratory. The laboratory must continually stress the proper care and handling of evidence in communications to members of the police agencies they serve. Hardly a month passes in any big laboratory where careless handling of evidence has not ruined it or at least greatly reduced its value.

A deputy sheriff recently dug four crime bullets from the wooden walls of a shack in Southern New Jersey with a heavy pocket knife. In the process, he completely ruined any possible identification of the bullets as regards the individual weapon from which they were shot. The State Police investigator kept them for a few seconds and then gave them back to him suggesting they be used as sinkers for a fishing line. This was, of course, said ironically. The bullets were still evidence of a sort, but their value had been largely destroyed. Even if an identification had been possible, a smart defense attorney would have had it set aside on the basis of changed evidence. Even detectives who should know better frequently will damage a crime bullet by scrawling their initials in the wrong places. A detailed treatment of this matter is included at the end of this chapter.

273

Initial Crime Investigation

Let us begin at the scene of a crime. Any conscientious citizen who comes upon a firearms crime should first see if the victim is still alive. If he is, medical attention is paramount. If he is not, then the body should not be disturbed in any way. The second duty of a citizen is to report the crime immediately to the police in whose jurisdiction the crime occurred, or for that matter, to any police. Frequently in some sections of the country, the individual citizen will not understand the investigation procedures and the overlapping jurisdictions of the various police forces. The citizen reporting the crime should try to the limit of his ability to keep all evidence intact. However, as a general rule, this duty will fall on the first police officer to arrive.

If a single police officer arrives he will usually have an almost impossible task in a metropolitan district. He should maintain the scene of the crime in exactly the condition that he finds it; institute pursuit of suspected persons; inform his superiors of the crime; get medical aid for the victim if he is not dead; and finally round up all possible witnesses, keeping them separated so that they cannot communicate with each other until interrogated by the proper investigator when he arrives. Such a procedure, of course, is impossible; there is just too much to do. As a general rule, about all one man can manage is to see that nobody disturbs the corpse or the immediate surroundings. He will generally receive help in heavily populated areas in a relatively short time. Often today a patrol car with two police officers will arrive first at the scene of a crime; this simplifies matters, especially if the car has a two-way radio.

Formal investigations vary considerably in different sections and in different communities throughout the country. Actually, it may be easier for one man to investigate completely a crime in a rural section of the country than for a crew of a dozen men to do the same thing in a metropolitan area.

A hearsay account of a crime near the Mexican border comes to mind. A skeleton in rags was found in open country by a local rancher. A single sheriff was called who investigated all by himself. There was a skull fracture as by a blunt instrument and a hole through the vertebrae on a level with the heart. The remains of clothing were of a type worn by Mexicans. The amount of weathering could coincide with the time since the last harvesting season on truck farms not far away.

The sheriff with the rancher's amateur assistance marked off the area and searched it diligently. Even though about six months had passed, they found a billfold thrown into a bush, a tiny silver medal formerly worn about the victim's neck from a shrine not far across the border and, most important of all, a discharged .45 ACP shell. They dug beneath the spot where they believed the body to have originally lain before being disturbed by buzzards and coyotes and found a metal jacketed .45 ACP bullet quite close to the surface. No conviction was ever obtained; however, the shell and bullet matched those from two other crimes in the same area where Mexicans returning across the border with their season's wages had been murdered.

This sort of investigation is impossible in a large city. Evidence, if not obtained immediately is lost, sometimes within a few minutes. Investigations must be geared to actual crime conditions. Extra men and equipment are usually available where needed. Initially, a survey of the particular crime should be made by someone, preferably the detective in charge. A few basic factors should be determined before anyone does anything. The best possible way of doing this— a method that is frequently impractical—is to have the general investigator alone enter upon the scene of the crime, while all others are restrained outside, and carefully determine a few facts. Aimless wandering around should not be allowed.

The preliminary investigator will undoubtedly first want to know how the victim died. This will be determined legally at autopsy by the medical examiner or coroner; however, in most instances, it is reasonably obvious. At the same time the investigator will almost automatically decide what special steps must be taken in addition to routine sketching and photographing of the scene, searching for fingerprints and the like. Clues must be searched for intelligently and any firearms and ammunition components recovered with their position at the time of recovery carefully noted.

Wounds

Firearms cause several different varieties of wounds, only a few of which can be confused with those from any other weapons. The illustrations show several instances of three different groups of wounds. First, the average hand gun will create a relatively small entrance wound and perhaps no exit wound at all. There may be an exit wound, usually slightly larger. The edges of an exit wound are usually fairly obviously turned outwards if death followed reasonably soon. Sometimes the bullet will have tumbled and leave a characteristic 'sideways' wound at exit. As a general rule, the human body hit by the usual projectile from a criminal's weapon will not cause the bullet to deform drastically. The exit wound in most pistol victims will be only a little larger than the bullet diameter. It will not be easy to determine, however, from the size of the

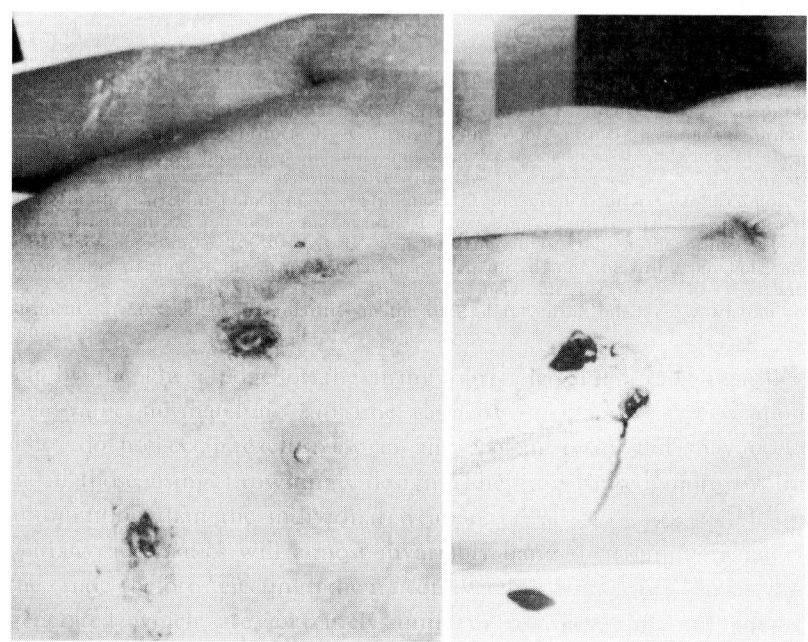

TYPICAL HAND GUN WOUNDS

Showing entrance (left) and exit (right) wounds caused by a .38 S. & W. Special cartridge from revolver in the hands of a suicide. Gun was held close to fully clothed body.

bullet hole in the victim what caliber was used. Where soft tissue lies below the area struck by the bullet, the actual perforation will be much smaller than that made by the same weapon where bone and sinew is immediately beneath the surface.

Except for some ultra high-powered pistols, the oft-seen large exit holes with a great deal of torn tissue are almost completely absent in hand gun crimes. The .357 Magnum, the .44 Special, and other weapons of this power are perfectly capable of creating these wounds. However, for some reason, criminals have not often used weapons of this type. The metal jacketed 9mm Lugers, 7.63mm Mausers and .45 ACPs do not usually expand in human tissue.

The second group of wounds are those caused by rifles. The low power weapons cause wounds a good deal like pistol wounds, particularly where solid, heavily jacketed bullets are fired or where ammunition of pistol power is used as in the case of the .32-20, the .38-40, and the .44-40. The heavier loaded cartridges of this type fired from a rifle barrel will frequently give a relatively large exit wound, since the ammunition is designed to upset readily.

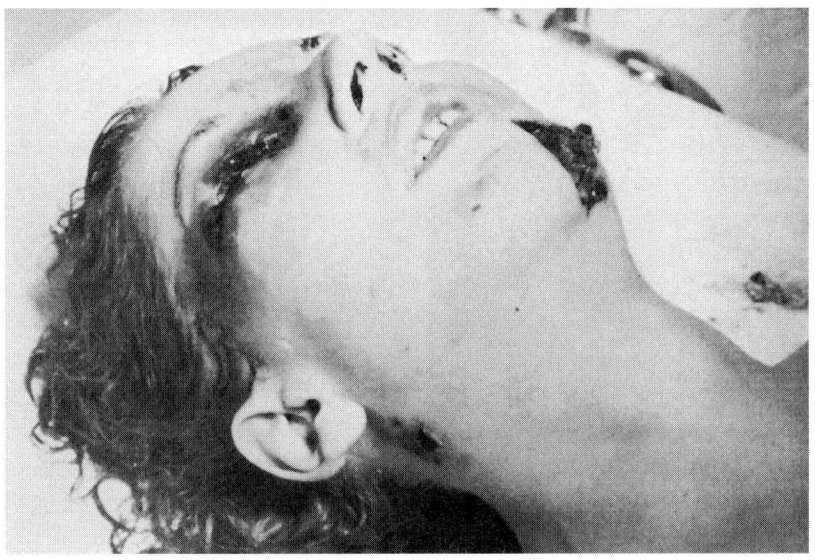

HOMICIDE POSTMORTEM EVIDENCE PHOTOGRAPH

In this homicide the victim was shot twice, once through the back of the head and once below the ear, this latter .45 caliber full-patch bullet coming out through the lower jaw.

Ordinarily, this particular bullet would make entrance and exit wounds of the same general size, but in this case the bullet struck and shattered the jawbone and made a much larger wound of exit.

High-powered rifles using sporting ammunition create characteristic wounds where the bullet upsets at all. Sometimes a solid bullet from a high power rifle will also give a unique wound. When a bone has been struck there may be several exit wounds, as pieces of bone may be put into motion by the bullet so as to form missiles which pierce the skin. These "secondary missiles" often enormously increase the damage done by a bullet. The dozens of pieces of bone driven in every direction like small bullets in themselves may be responsible for a large proportion of the injury. Occasionally, they may even cause a major injury to a person not struck

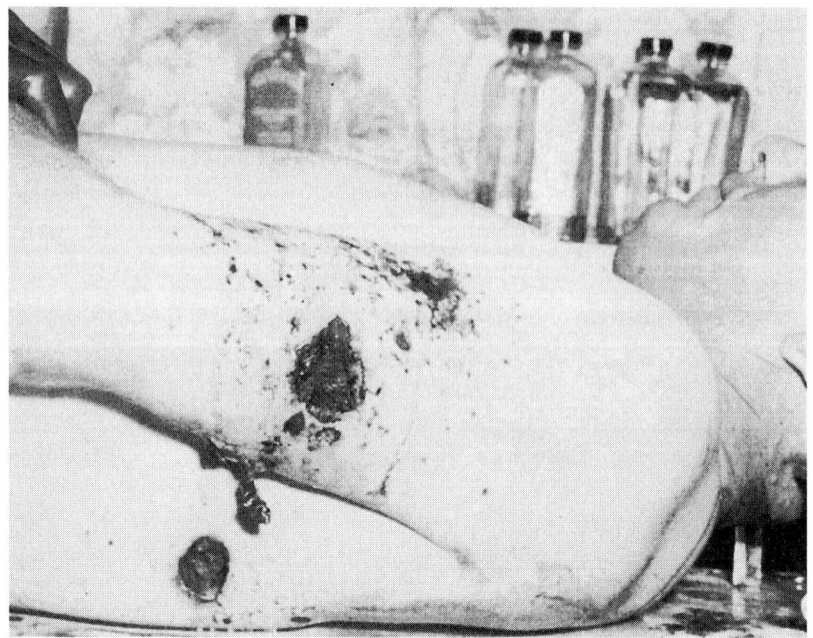

A 'SECONDARY MISSILE' TYPE OF WOUND

Here is a wound that would be hard to identify unless the facts as to the actual shooting were known. This wound was at first thought by the investigating officers to have been caused by a shotgun.

Actually, it was made by a .30 caliber Carbine bullet that struck this officer as he was getting out of an automobile. The bullet pierced the window glass of the car door, spattered and everted. Wound in upper left arm shows multiple small punctures made by small fragments of lead and glass. Large wound in the pattern is result of the bronze jacket of the bullet still retaining some of the lead; this large residue pierced the flesh of the posterior upper left arm and re-entered the left posterior side nearby and then continued under the skin to lodge at the spine. Death resulted from a second .30 caliber Carbine bullet through the upper chest that pierced the aorta. (From Alabama Department of Toxicology and Criminal Investigation).

by the original shot. A case occurred in the Army where a man was found in a tent apparently shot through the eye. He was taken to the hospital, and the wound washed out. No bullet could be found, though a large chunk of bone, that did not seem to be very well accounted for, was loose in the wound. After the wound was dressed, someone discovered another man lying dead in the same tent; it developed that the first man had not been shot at all, but instead, had been struck by a flying piece of bone dislodged by the bullet which struck the other man.

Third, there are the wounds caused by shotguns. These are the most characteristic of all wounds and are seldom confused with anything else, particularly if made at close range. The charge of

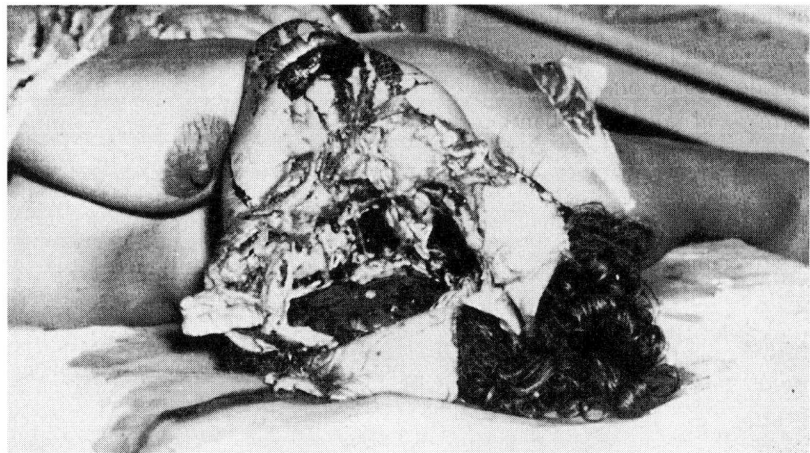

Effect of shotgun blast at close range—say within a foot or less. Victim was shot in back of head with a 12-gauge Bay State single barrel shotgun which had its barrel cut off to a length of 8¼ inches and stock cut off at grip. Shell was a buckshot load of oo.

The effect would have been about the same had an ordinary shotgun and small-shot cartridge been used at this same distance, as it is the gas from the powder charge which causes so disruptive a wound under the circumstances.

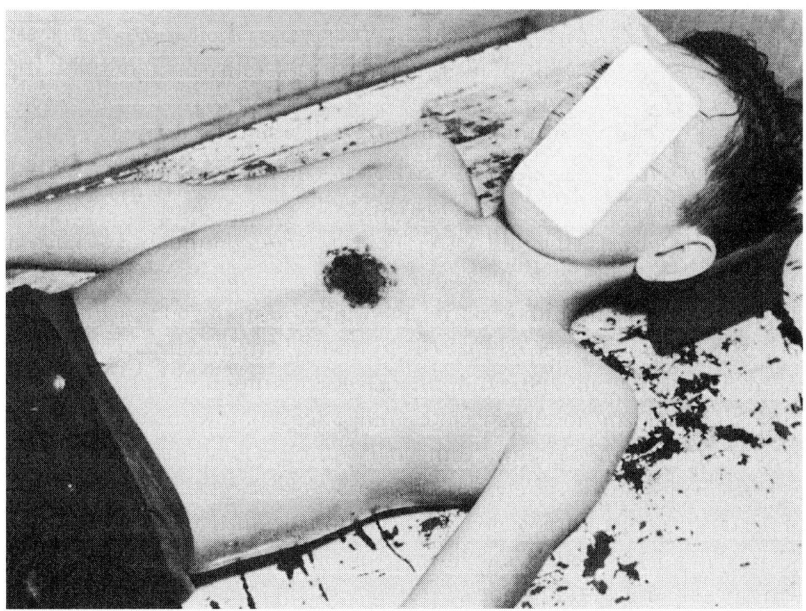

THE "RATHOLE" TYPE OF SHOTGUN WOUND

This boy was killed with a 12-gauge Stevens shotgun, full choke, 30" barrel. Range was about eight feet. Remington Express shell loaded with 1¼ oz. of No. 6 shot.

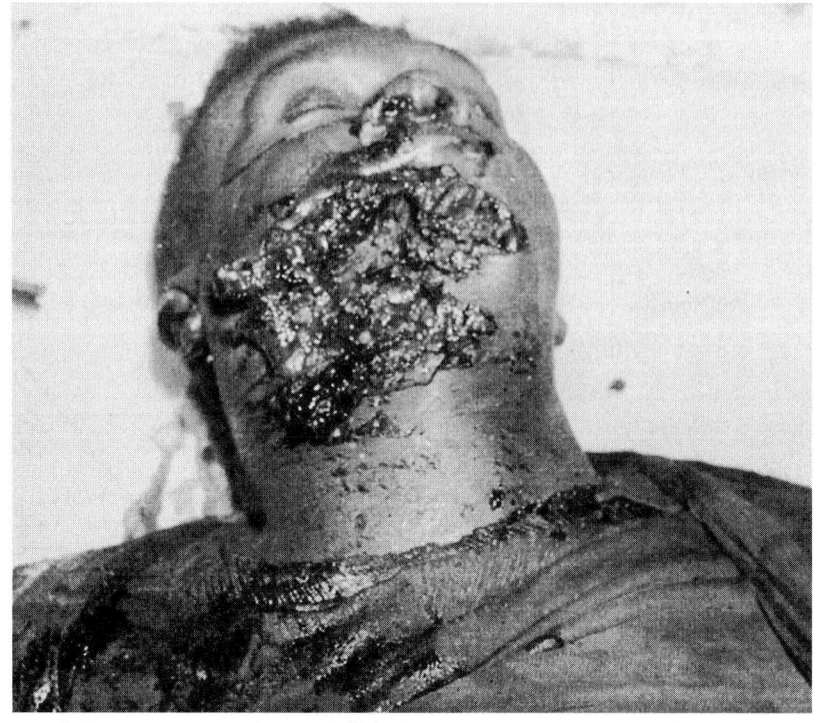

SMALL SHOT AT CLOSE RANGE

Shotgun wound in the face of a colored male, made with a 12 gauge shotgun and No. 4 shot, fired at a muzzle distance of 5 to 6 feet, the shot ranging from front to back and somewhat toward the right side of the body. (From State of Alabama, Department of Toxicology and Criminal Investigation).

shot accompanied by the wads and powder gases will tear into and through a human body creating the most instantly fatal type of wound. As was said before, a 12-bore shotgun loaded with No. 7½ shot can penetrate the skull of a lion if the shooter will but hold his fire until the range is about 10 feet. Our GIs in India found that one of their issue pump action shotguns loaded with ordinary trap loads was as effective on tiger at real close range as any of the British double rifles.

We will discuss in detail range determination from shotgun patterns later on, but in the case of human bodies used as targets, the entrance hole at a range of 2' or 3' or less will be a single perforation of large size. As the range increases, there will be a few outside pellets that will begin to spread away from the main charge and make separate smaller perforations around the main wound. The exit hole is even more characteristic. At close range everything is just blown out on to the surrounding area.

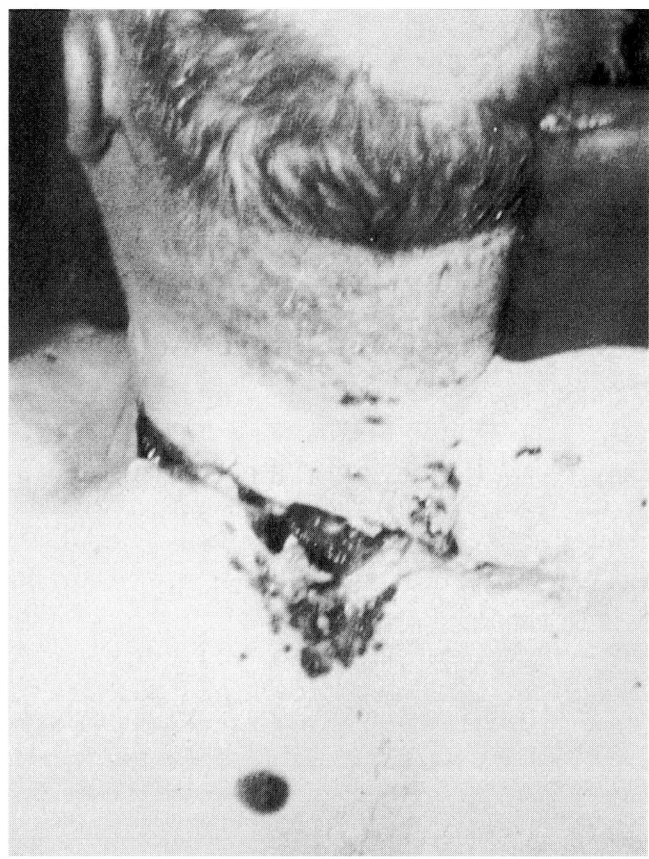

Here is an example of a shotgun wound at close range. Our information on this is not complete and we do not know many of the factors, but this wound was probably made at a distance of 10-12 feet or thereabouts, possibly a bit more. The angle of gun to body is not known, nor is bore and choke. Note the bruise below wound, caused by one of the powder wads striking; this often occurs at short ranges.

In all wounds it is important to determine at the post-mortem examination the exact path of a projectile in a body. This is necessary if the position of the shooter is to be established. Wire probes suitable for this work are shown in some of the accompanying photographs. Almost any bullet can change course even in a human body. If this is not understood, entirely erroneous conclusions based on the relative positions of entrance and exit wounds could be arrived at.

Sometimes, especially in the case of pistol bullets, the bullet may follow the contour of a rib or of the skull without penetrating the bone. General Hatcher knew a man shot in the chest at close range with a .38 Long Colt revolver who fell down and expected to die; all of his friends expected him to die too. He had a well defined wound of entrance in the left side of the chest, but he recovered promptly. The bullet had struck a rib and gone around to his back just under the skin without going into the pleural cavity at all. Again, it is rather common for a bullet striking the skull a glancing blow to travel around to the back of the head under the skin. There is a tradition in the family of one of the authors that an ancestor was so shot by a Frenchman at Trafalgar with a pistol. The bullet entered his forehead and came out at the very base of his skull. Just as he was about to be buried at sea he regained consciousness. The bullet is supposed to have traveled around between skin and skull; it could be true.

Suicide, Murder or Accident

Once it has been determined that death or injury is due to a firearm, the first question will be, "Was the wound self-inflicted, the result of a homicide, or just an accident?" Sometimes, of course, this question cannot be answered immediately or even at any future time. Perhaps most of the truly successful crimes are those in which a verdict of suicide or accidental death is reached and further investigation stopped.

Under old English Common Law, all violent or suspicious deaths were investigated by a magistrate known as the coroner. This office was common in this country until a few years ago, but is now frequently either abolished entirely or has far less authority than formerly. At the turn of the century, a coroner practically conducted his own investigation. The police were in a large measure under him in the initial stages of their inquiry. He would impanel a jury and actually more or less try the case in a preliminary way. He still so functions in some sections of the country, although the medical examiner, prosecutor's office and grand jury now usually perform his functions. There may, however, be both a medical examiner and a coroner in some jurisdictions. In general, the medical examiner is a doctor and will not only visit the scene of a crime to determine by preliminary investigation the cause of death, but also have the corpse transferred to an appropriate location for an autopsy. He performs this post-mortem examination. Any bullets or the like taken from the body are initialed by the medical examiner and turned over to the general investigator in whose custody they remain until delivered to the Firearms Laboratory.

In some localities, the medical examiner has several assistants, particularly where the territory involved is large. In the case of violent and unexplained death or where a registered physician will not sign a certificate of death from natural causes, the body of the victim is at the disposal of the medical examiner, the coroner, or the police throughout the United States.

The final decision as to suicide, murder, or accident is generally made by one man or a jury based, however, on several different opinions. Next to the actual inspection of the scene of the crime and

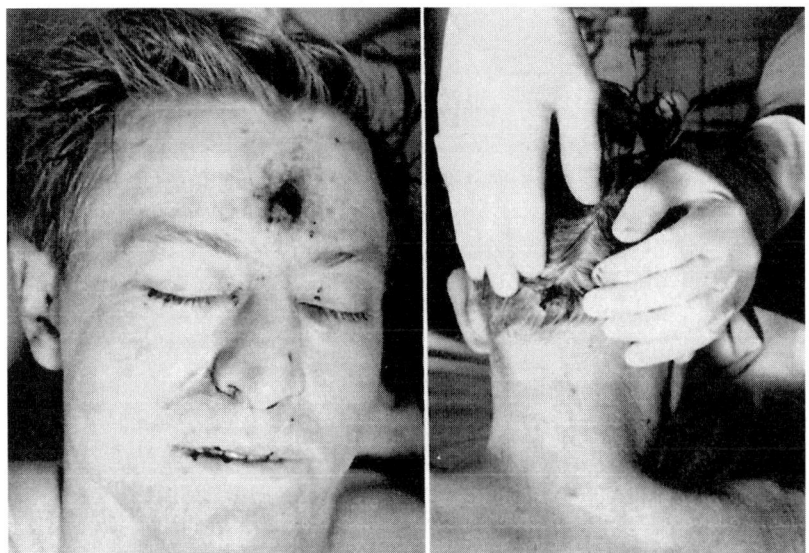

SUICIDE EVIDENCE

Postmortem photograph of a suicide who killed himself with a pistol firing the 9mm Luger cartridge, holding the pistol muzzle close to his temple. The "tattooing" marks, caused by burning powder grains, are very clear in this photograph.

The right-hand section shows this same wound at its point of exit, in this case the .38 caliber full-patched bullet made exit and entrance wounds of the same size.

victim, the post-mortem medical examination is most important. Generally, the investigating detective is present when it is performed.

Certain physical factors should be considered if suicide is suspected. The position of the wounds will ordinarily rule out many firearms deaths. More than two-thirds of the people who shoot themselves aim for the brain either through the roof of the mouth, at the temple, or through the forehead. Almost all the rest aim for the heart. A person will not normally kill himself in an uncomfortable or awkward position. The investigator should try reenacting the suicide himself with the weapon used or a duplicate of it. Make sure it is unloaded first. From the posting you will know the path of the bullet.

We have taken up in detail elsewhere range determination and the various marks left on the human body by firearms discharged close to the victim. If no marks are left on skin or clothing a suicide is unlikely. Usually, the muzzle is held in contact with the body when a wound is self-inflicted. Sometimes the barrel is pressed so tightly against the body that the powder burns are almost all inside the wound.

Obviously, the chance of suicide if no firearm is present is relatively remote since only the most unusual individuals will plan a suicide to look like a homicide. There are a few instances on record where this has been done; however, they are so very infrequent as to be almost negligible. Sometimes for religious or life insurance reasons, either a suicide or his family will try to have it appear an accident. Frequently a murderer will try to disguise his crime as a suicide.

The presence of a gun grasped tightly in the hand of a victim is not by any means positive evidence that he shot himself. Although rare, there are definite instances of a man's hand closing in death tightly around the gun used by a criminal to shoot him and then placed in his hand. Further, many suicides do not retain possession of the weapon after death; in some cases the firearm has been found several feet away.

It is not the purpose of this discussion to go in detail into the various possibilities of murder-suicide. Books can be written on this single subject alone. Most detectives, however, will know general principles and apply them with common sense. Every single crime is a little different from every other one; however, from a purely firearms standpoint, there are certain definite indications.

An accident is sometimes extremely difficult to differentiate from a suicide. In fact, the two fade into one another in some instances. In a recent New Jersey case a brilliant young professor shot himself demonstrating Russian roulette before a group of people. His chances were one in six since he loaded only one chamber of a standard Smith & Wesson revolver. However, an investigation among friends indicated that he had been playing this ghastly game at least once an evening for several days before he finally drew the loaded chamber. He had claimed that the one loaded chamber, because of the extra weight of the cartridge, would always end on the bottom. This is *not* the case. A similar weapon spun 120 times with one loaded chamber fired 23 times—according to probabilities it should have fired 20 times.

The Firearms Expert will frequently be asked whether or not a story of an accidental shooting, as told by witnesses or the person who did the shooting, is possible. All that can be done in such a case is to take the evidence weapon and try to reconstruct the accident. Two cases of this nature came up recently indicating the general extent of this type of investigation. In one, two brothers-in-law were shooting with an Ortgies .32 ACP automatic at an extemporized target in the New Jersey meadows. One shot and killed the other and told a story of the gun having gone off in his hand when he momentarily turned towards his brother-in-law. This gun was immediately submitted to the New Jersey State Firearms Identification Laboratory. The trigger pull on this particular automatic pistol had been reduced by someone, many years before, to below one pound. Further, during the course of 25 rounds fired with this pistol in the laboratory, it went off by accident without a finger being placed against the trigger in one instance. The findings of the laboratory were sufficient to confirm the general investigation. This was not in any way an intentional shooting, although instances of this type confirm more than ever the importance of never having a loaded gun pointed at anything you do not want to shoot.

A second instance involved an alleged accidental shooting of an elderly man with his own shotgun. Supposedly, it fell from his hands while he was out hunting. The sear mechanisms in this weapon were in perfect shape. The gun was subjected to extremely violent usage in an effort to get it to discharge in any way possible save by actual pulling of the triggers. The laboratory report in this instance said, "We have been unable to discharge this weapon by any means whatever other than actually pulling the triggers. It is thought unlikely that the victim shot himself by dropping his gun, unless some object outside the weapon pressed against the trigger surfaces." Since this shooting took place in the middle of an open country road, and the wounds were so placed as to rule out a part of the victim's body or clothing having caused the discharge, a full homicide investigation was instituted.

Investigation Procedures

If the crime is tentatively identified as a probable homicide, the body of the victim is disturbed as little as possible by the medical examiner or the attending physician in pronouncing the victim dead. Someone should then take photographs of the victim before he is disturbed. Sketches are made; minute searches of the entire area are undertaken as circumstances dictate. There are many instances where small quantities of foreign material from the scene of the crime have aided in its final solution. Tiny flakes of paint or even a few strands of hair have led to brilliant solutions of baffling crimes. Fingerprints, although not nearly so valuable as before the wave of detective fiction swept the country, are still important if they are found. Most detectives are copious note takers; some use the services of police stenographers.

However, to stick closely to the firearms part of this investigation, the wound itself will usually tell a good deal. For instance, will it be possible for the murderer to claim self-defense? Obviously a wound inflicted from the back will not allow such a plea. Direction as indicated by entrance and exit wounds, whereas not completely clear by external examination, certainly will be unmistakable at the autopsy.

It is often possible at the scene of the crime to find one or more places where a bullet or bullets have struck either after passing through the victim or after missing him. Sometimes shots are fired through windows or curtains before hitting the victim. For instance, where a bullet passes through a curtain and then into a wall, a straight line connecting these two points may determine the position of the gun at the time of discharge. Usually within certain limits, this will determine the position of the shooter at the time of the crime. Sometimes, particularly in rifle crimes at considerable range, this method will lead you to one and only one position from which the murderer could have fired the shot as, for instance, a window of an adjacent building or the top of a nearby hill.

Another bit of firearms evidence frequently encountered at the scene of the crime is the relative position of empty cartridge cases. These generally come from automatics which mechanically throw out evidence of their identity at each shot. The most important use of these expended cartridges is, of course, to determine the caliber and type of weapon and for later identification positively or negatively with suspect weapons. Their position, however, may be of importance in determining the spot from which the shots were fired. In some weapons and on some types of floor surfaces, sand, or the like, empty cases will be thrown in quite a tight group when fired in the same manner with a gun in the same relative position. However, in an average room where the walls and furniture interfere, and under the stress of crime conditions where the weapon may be held in almost any position, such evidence becomes of less and less value. Further, unfortunately during the firing of more than 50 pistols of this type

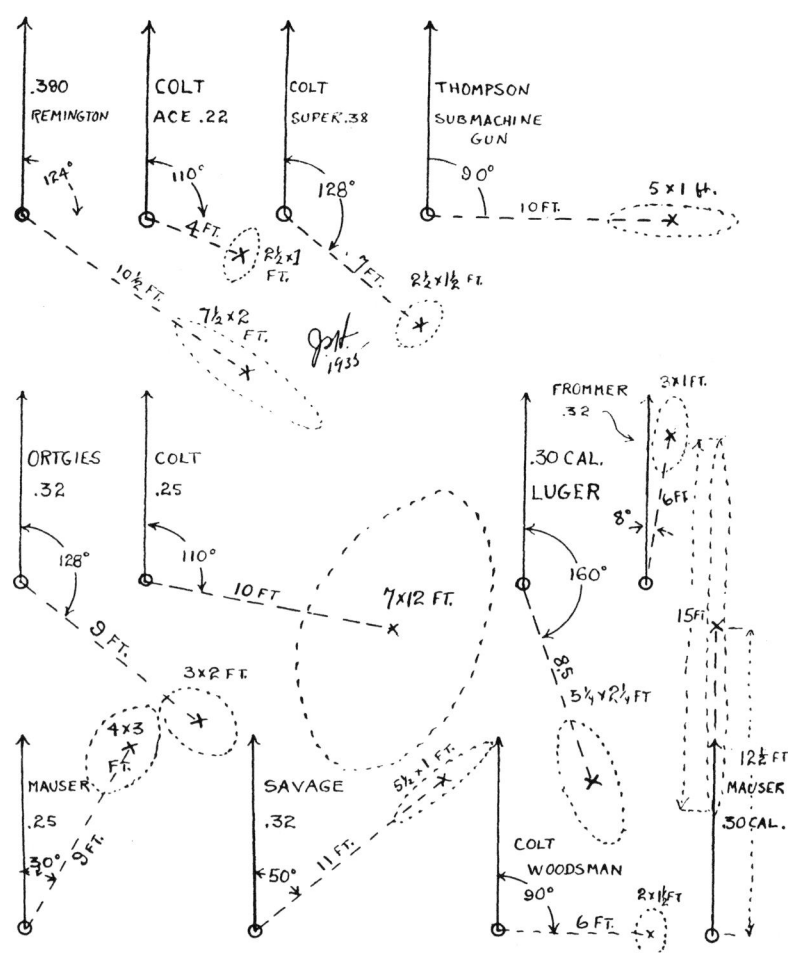

EJECTION OF EMPTY CASES BY AUTOMATIC PISTOLS

This diagram shows the results of tests made by General Julian S. Hatcher to determine the direction and distance of the ejection of the fired cases from various automatic pistols.

In each sketch, the circle indicates the position of the pistol, and the heavy line with the arrow at its end is the line of fire. Ten shots were fired with each arm and the empties were found distributed over the areas enclosed by the dotted lines, angles and distances also being given.

These tests were carried out in the open, there being nothing intervening to interrupt or change the flight of the empties.

recently, some of them were found to occasionally throw a cartridge case almost 180° from where most of the empties finally came to rest. The position of the empty cartridge cases may or may not be of importance. On the whole, investigators today are not giving it as much time as they did formerly.

Handling of Firearms Evidence

All weapons, bullets, cartridge cases, extra ammunition, and any components of shotgun shells should be carefully collected, labeled and packaged for delivery to the firearms laboratory. In an appendix to this chapter we have reproduced a description of the approved procedures in New Jersey. Three basic principles govern. All evidence must be carefully described as to position when found, identified, and later delivered to the laboratory. It must be preserved in its original state without damage. Finally, a continuity of sworn custodies must be proved from original taking of possession until trial.

We have already discussed in some detail the way that the bullet is marked when recovered by the medical examiner. Where a bullet is found at the scene of the crime, the finder should be asked to similarly initial the bullet. It is then taken by the general investigator and carried or sent by him to the laboratory where a receipt is given for it and, of course, appropriate entries made in the laboratory records. If and when the case comes to trial, there will be a complete chain of evidence linking the evidence bullet directly to the crime. It is important that this evidence throughout the time that it is in the hands of the Police Department be maintained with scrupulous care and without damage.

Unfortunately, lead bullets can be rattled around enough in boxes and even in envelopes, particularly where there are more than one, to quite seriously lessen the chances of a positive identification with

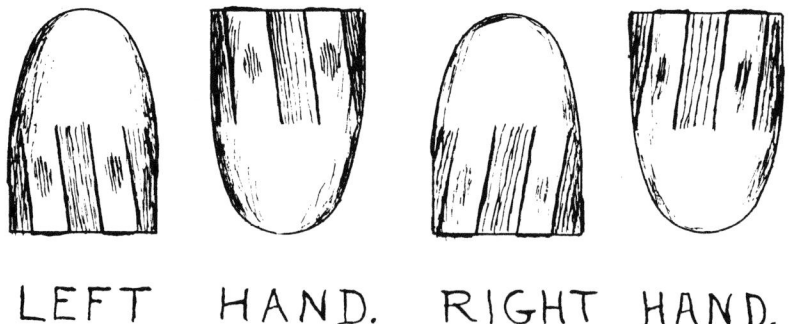

LEFT HAND. RIGHT HAND.

PITCH OF RIFLING AS SHOWN BY BULLET

This drawing shows how to tell from the rifing marks on the fired bullet whether the weapon from which it was fired had right hand or left hand twist. Note that on the gun with the right hand twist the rifling marks slope to the right as they go away from the observer; on the one from the left hand twist the marks slope to the left—this being true whether the bullet is viewed from the front or from the rear end.

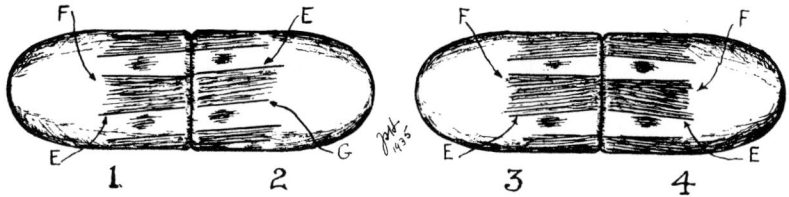

A ROUGH METHOD OF BULLET COMPARISON

A quick idea as to whether a given bullet might or might not have come from a certain type of gun may often be obtained by placing the bullet base to base with known samples, as shown in the sketch. 1 is the fatal bullet, and 2 is a sample fired from a .45 automatic pistol. While there is a difference in the width of the marks left by the lands on the two bullets, the conclusion is that the murder bullet could have come from either a .45 automatic or a .45 M 1917 Colt revolver, which has the same rifling dimensions as the automatic. (The two samples used as models for the sketch had actually been fired in two different .45 automatics. There is an unusually large difference in the groove width in these two guns). 3 is the same fatal bullet placed base to base with a bullet fired from a .45 S. & W. Model 1917 revolver. It will at once be seen that the two bullets came from different guns, as the direction of the twist is different. This rough method is of use chiefly for a comparison of such easily apparent things as the number of grooves, twist of rifling, etc.

E, E, E, E are the driven edges of the marks left by the lands.

F, F, F are the trailing edges of the marks left by the lands, showing "skid marks" caused by the bullet's plowing straight into the rifling for a short distance before it takes up the motion of rotation.

G is a trailing edge which does not show skid marks.

test bullets later on; the worse possible place to carry them is in one's pocket. Proper packaging of weapons and ammunition components is merely common sense, but a lot of people seem to be a bit short of this commodity.

Carefulness in marking and preserving from harm all evidence does not end once it arrives at the laboratory. In at least one major case, a fine laboratory has confused test bullets and crime bullets, to their extreme mortification. The laboratory, as well as all others having charge of firearms evidence, cannot be too scrupulously careful. The FBI procedure of never having anything on a table except the evidence from one case at a time is worthy of being followed everywhere. Even extremely competent men after years of practice have become on occasion not as careful as they could be in such particulars. This carefulness of handling of test bullets is of particular importance where a number of weapons are tested for positive identity with a single crime bullet or cartridge case. The actual comparisons may be relatively easy, whereas the mere clerical handling of the test specimens may be laborious. A fatality in ball cartridge maneuvers at Fort Dix in the early spring of 1953 was of this nature. The crime cartridge case was recovered. It was positively identified as being fired in one of a rather large number of M1 Garand rifles. The mere physical

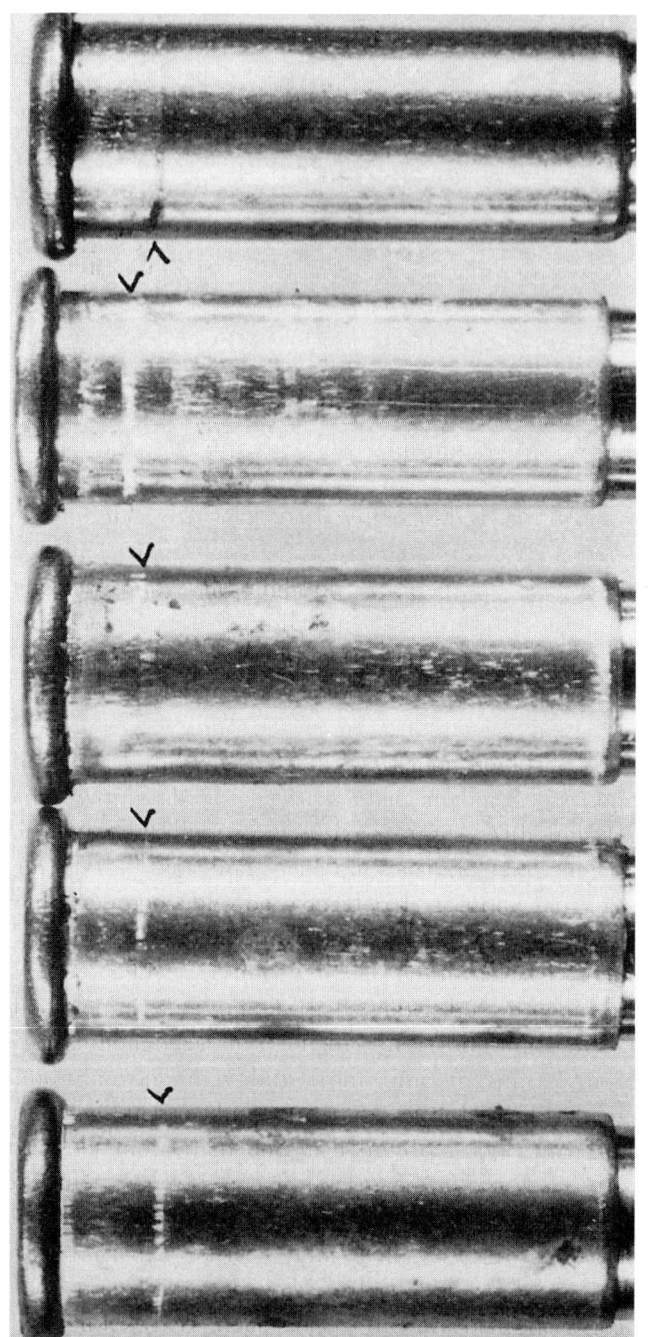

CARTRIDGE CASES FIRED IN REVOLVERS

This evidence was obtained by firing the above cases in five different .22 caliber revolvers.

These .22 cartridge cases were all fired in revolvers—no argument about that. The tip-off is that clearly-seen line etched across the case body, close to its rim. This marks where the extractor (for simultaneous ejection) fits in against the chambers of the cylinder. This feature in a revolver-fired .22 cartridge is unmistakable, as the line will extend halfway (or almost) around the body of the case and

(Continued at bottom of opposite page.)

handling of the rifles and the firing of test rounds from each with the marking and packaging of two cartridge cases from each weapon took more time than actual positive matching of crime and test rounds.

Preliminary Firearms Identification

Where trained ballistics men visit crime scenes they will provide themselves with a small kit of tools and be able to answer at once several types of questions. Where no ballistician is available, a detective will sometimes take along a jeweler's loupe and have a look for himself at the crime bullet or cartridge case or both, as the case may be. He cannot possibly do his investigation or himself any harm. He may get important information quickly where time is of the essence. If he has a little knowledge of firearms, he can soon learn to distinguish caliber and probable makes and models of guns. For instance, almost all Colt weapons have left-hand twist rifling. There are no other well known American makes of guns past or present with left-hand rifling, though some Forehand and Wadsworth and Hopkins and Allen weapons used this system a long time ago. Similarly, the general investigator will probably know from his examination of the wounds a great deal the medical examiner will later on tell him in more precise language.

In many instances, Police Departments, particularly those with small forces unable to provide a man to spend his entire time on

---

will *always* be positioned on that half directly opposite the firing pin imprint on the rim at 12 o'clock. In the cheap revolvers, or in those of much usage, this joint marking may be supplemented by an extractor swell of more or less depth.

Any comparison of evidence and test cases from a .22 revolver is likely to turn out to be considerable of a task. There will be from five to ten chambers to check-out, each of which may mark its fired cases in a different manner. Even the hammer-nose (firing pin) imprints may differ greatly, due to variations in the anvils of the different chambers and from the manner in which those anvils have been battered down and widened from faulty centering of the chambers as they were spun into firing position by the mechanism which, in some of the lower-priced revolvers, varies to a greater or less extent. As a gunsmith would express it "The timing is badly off."

Evidence of this type appearing on any fired .22 brass or copper cartridge case is likely to fade out; to be rubbed off from careless handling; discolored from oxidation; and so partially or wholly disappear after a brief interval as the metal tarnishes or oxidizes with but a bit of ageing. Hence the importance of taking photomicrographs promptly, once such signs are recognized, and then preserving such records while the firearm is being searched for or the evidence worked up. By the time the case reaches the Courts such positive material may not be visible on the evidence cartridge case.

Above is an example of the case markings of .22 caliber cartridges fired in the low-priced 'nail extraction' models of 'solid frame' revolvers. With such revolvers the main identification signs are likely to be on the head of the cases.

In these cheaply-made arms, the chambers are bored throughout the length of the cylinder, with no extractor fitted for simultaneous ejection, hence there are no extractor markings upon the sides of the fired cases. Extraction is by means of a nail or rod, one case at a time, through a port in the right bolster flange.

The inner face of the frame is roughly finished, with no recoil plate (firing pin bushing) backing up the firing chamber. Hence the area of the fired case base around the firing pin imprint is likely to be distinctively marked by the tool marks on this part. Furthermore, the timing on such revolvers soon becomes faulty, consequently the positioning of the cartridge-to-be-fired is irregular; this results in the anvils on each of the chambers becoming beaten-down to different degrees and individually marking each cartridge fired.

The hammer nose (firing pin) on this type of revolver is roughly shaped and finished, hence likely to leave a most distinctive imprint in all cases fired, as seen above.

The above case heads show an interesting sign which indicates rather clearly that these cartridges were fired in a revolver—and this indication can appear on any revolver-fired case, but more particularly on those fired in the cheaper, more roughly finished ones with uncertain headspace tolerances. This sign is those streaks running across the case head at right-angles to the firing pin imprint, it is caused by the cartridge heads dragging across the rough inner face of the frame as the cylinder is rotated.

Incidentally, the above three cartridges were fired in an entirely new revolver, shot but eight or ten times. Cases from these revolvers which have seen considerable usage are interesting studies indeed.

Firearms Identification, will employ part-time or call on regularly either an independent firearms consultant or someone with a detailed knowledge of firearms. This individual does not have to be a Firearms Expert since what he is doing is not in any way evidence but merely stating probabilities. It is perhaps easier for him if he is not. An Expert must always be conscious of his position and never say anything that he is not completely sure about. In fact, most Experts very rightly will not even state a probability. The independent investigator with a knowledge of firearms, however, can take a look at a .32 metallic jacketed ACP bullet with 6 grooves, left twist and tell the detective to look for a man with a pocket model Colt. All things considered, this is probably the best advice that can be given at this stage of the game; however, the Expert would have had to qualify his answer so much that many policemen, on reading it over, just become confused.

We are not speaking from hearsay. In a recent crime on which the authors collaborated, the detective receiving the laboratory report got so many names of German-made pistols that he could not pronounce that were remote possibilities that he threw it down in disgust, muttering; "All this says is that the bullets came out of a gun. Hell, we knew that." In the same case the other author had suggested strongly that the weapon was a German Sauer 7.65mm pocket automatic. The probability was at least six out of seven that this one single weapon had been used for the commission of the crime.

It is important, however, to realize that the quick judgments of the general investigator not specially schooled in firearms work and the amateur should be treated as tentative only and never counted upon for longer than necessary to get a real opinion. It is important, further, that the evidence not be hurt in the preliminary examination; for instance, a patrolman should never be allowed to measure a bullet with calipers. Further, this type of information is not so important as it once was. Firearms laboratories have increased in number throughout the country and speeded up their services. There is, however, an account of a case originally reported by General Hatcher that bears retelling.

About 8:00 a.m. one of the local police officers had been shot twice and instantly killed by a negro who made his escape. There were no immediate witnesses to the shooting, but the police at once set their net and in about two hours brought in the man that they were convinced was the guilty person. When arrested, the negro still had on his person the .45 S. & W. Model 1917 revolver with which the shooting had been done; its cylinder held four empty cases and two live cartridges.

But the accused stoutly insisted that he had last shot the revolver some four weeks previously, at which time he had fired all four shots at a mark in the woods, and he called on some three or four "cousins" to substantiate his statement. All these witnesses, with suave perjury, claimed to have been present on the occasion and to have seen all four shots fired at the mark in the woods. There was no shaking their testimony, either. Then the detectives thought of their firearms friend, and at noon, only some four hours after the crime was committed, they brought him into the case and asked him if it were possible to tell if the .45 caliber revolver in

The most readily available means of comparison on fired .22 cases will be their firing pin imprints. This illustration shows 24 cases which have been fired in 24 different rifles, a few of which rifles were of the same make and model.

Notice the great difference in these imprints. They vary in shape, size, depth and, to some extent, the striking area, all of which points are of utmost importance in matching. In addition, several of these case bases showed conditions not picked up by the camera; convexed centers, bulged areas or rims, caused by a loose action,

(Continued at bottom of opposite page.)

question had been fired recently. He gave hardly more than one glance at the weapon before stating that beyond question two of the chambers of the gun had been fired within the past six hours, whereas the other two had been fired at least some two or three weeks previously.

Now it did not take more than an elementary knowledge of firearms to perceive that fact, as the two cylinders that had just been fired showed this fact plainly by their fresh fouling and odor, and as the expert pointed out, they were "almost smoking"; whereas the other cylinders, which had been fired off at a mark in the woods some weeks previously, were coated with corrosion "set-up," and streaks of red rust had commenced to appear, as this was in the days of corrosive primers.

But the expert remembered a certain distinctive feature about the gun in question. He said to the detective, "I believe that I can go down into the cell room and pick out the man who did the killing without being told or shown who he is." This was gladly agreed to, and all hands proceeded to the cell room, where there were seven negroes in all locked up at that time. The expert said: "All of you face me and hold out your hands toward me so that I can see the palms." Then without a moment's hestitation, he pointed to one negro, and said, "There is the man who shot the policeman; he did it with his left hand!" The negro in question who was indeed the suspect in the case said "My Gawd, boss, how did you know I killed him?"

The detectives were much elated, and moreover, they were astounded at the ability of the expert to pick out the guilty man. Of course, after all it was childishly simple when explained. As most experienced shooters know, the .45 Smith & Wesson revolver of that day had a rather thin, narrow grip, with a straight nearly vertical part of the metal of the frame extending above the wooden

---

breech-face not fitting squarely, or by excessively sized slots or ejection ports being cut into the bolt face.

All of these different factors must be taken into consideration when evaluating the evidence given by a fired .22 case. In their aggregate, they present a great amount of information that is not likely to be duplicated by another .22 caliber firearm, certainly not by one in the locality in which the crime was committed.

However, the firing pin imprint should not be used as the sole or main basis of comparison unless a close check is made through the comparison microscope. A positive comparison made with the microscope is likely to be a sound one. There is considerable variation in the firing pins and breeching of all .22 firearms, especially of the low-priced .22 rifles, and there is no uniformity in size, shape and positioning of the striker profiles—even on rifles of the same make and model—as the illustration on page 297 will show.

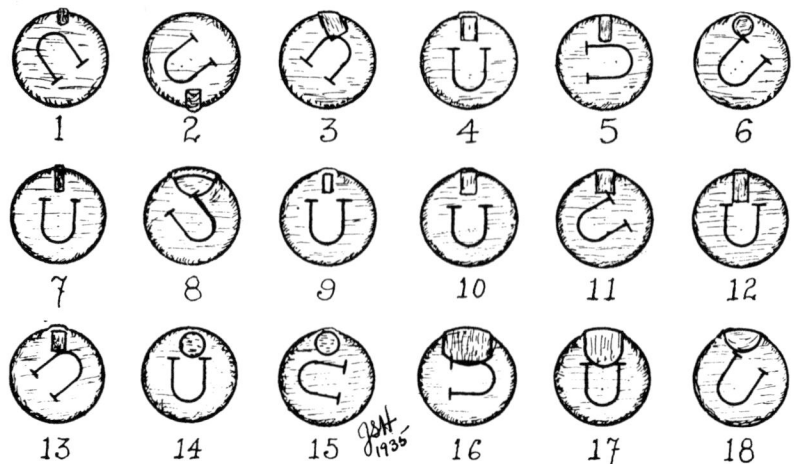

FIRING PIN IMPRINTS ON RIM FIRE CARTRIDGES

The point of the firing pin of a rifle firing rim fire cartridges has been shaped in a number of different forms over the past years, and it is sometimes possible to determine the make of a firearm by visual examination of an empty case fired in that arm.

The above sketch shows the difference in the firing pin imprints left on rim fire cartridges fired in different makes of .22 caliber arms.

1. .22 caliber Remington Autoloading Rifle, similar to model 24A, but a very early sample, given to the author by the late John Browning, the inventor of this gun.

2. Webley single shot target pistol. Note that the firing pin is on the bottom.

3. The Fiala Combination Rifle and Pistol.

4. Colt Woodsman Automatic Pistol.

5. Colt Ace Automatic Pistol.

6. The Reising Automatic Pistol.

7. The Winchester "52," .22 caliber rifle.

8. Iver Johnson Supershot Sealed Eight Revolver, early sample.

9. Same, later sample.

10. Colt Officer's Model Revolver.

11. Colt Banker's Special .22 Revolver.

12. Colt Camp Perry Model Pistol.

13. Harrington & Richardson U. S. R. A. Model Single Shot Pistol.

14. Smith & Wesson Single Shot Pistol, Perfected Model.

15. Smith & Wesson K-.22 Revolver.

16. Savage .22 caliber repeater, old model bought in 1916.

17. Winchester Single Shot .22 Musket.

18. Stevens Conlin Model Target Pistol.

handles at the back part of the grip near the top. This is rather hard on the hand of the shooter unless he is toughened to it and knows how to take care of the recoil. Almost invariably the straight part of the frame above the grip will bruise or split the "web" of one's shooting hand between the thumb and forefinger where that sharp vertical part of the grip bears. In this case, the expert simply looked at this point on the hands of every negro in the cell room, and then indicated as the murderer the only one whose hand showed this mark. And as it was the left hand that had been split between

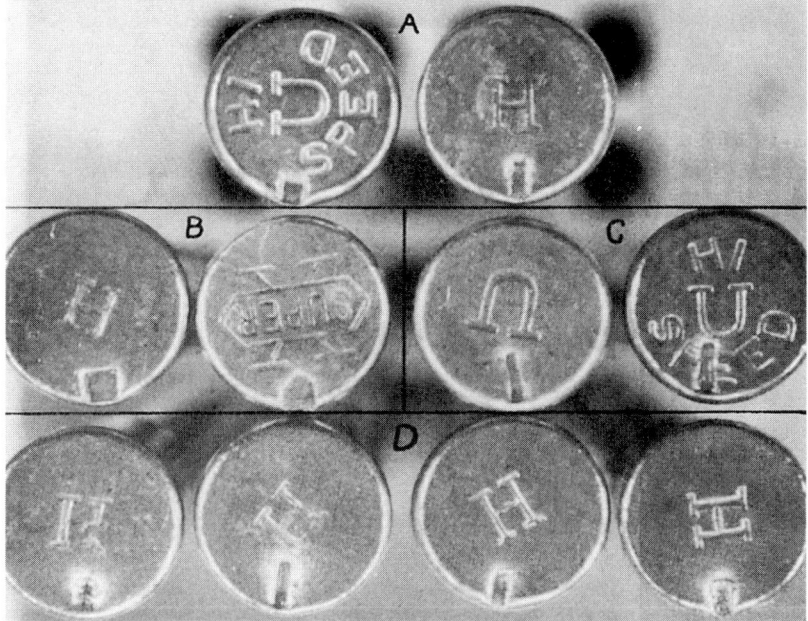

FIRING PIN IMPRINT VARIATION

Here we have four lots of cases fired in four different .22 caliber rifles. This illustration is intended to show the great differences in firing pin imprints in rifles of the same make and model.

A. This upper pair was fired in two Remington Model 33 rifles. Striker profiles are of same shape and approximate size but they imprint entirely differently as to size of depression and area struck.

B. Two cases fired in different Model 87T Stevens rifles. Big difference here in shapes of the imprints.

C. A pair of cases fired in two Winchester Model 67 rifles. Although resembling each other closely, the left imprint is slanted considerably more than the other, also narrower; differences readily observed by eye or under a magnifying glass.

D. Here, in the bottom row, are four cases fired in four different Remington Targetmaster rifles. Each imprint differs considerably from the others.

Although the quickest way to get on the track of a crime .22 firearm through a comparison of fired cases is by a visual examination of its primer imprint, it will be seen here that this may be only a rough and inconclusive method at its best. Further and closer comparisons are necessary. In a way, such positive comparisons can be compared to finger print evidence as far as accuracy is compared, but with this difference—you are not obliged to take into account hundreds or even thousands of comparisons—at the most there should be only a very limited number of firearms to check out over the area of the neighborhood involved. Probably, at the most, there will be only four or five similar rifles in the vicinity.

the thumb and forefinger when those two shots had been fired four hours previously, he also announced that the murderer was left handed.

"Elementary, my dear Watson, elementary!" And while it possibly was not evidence that could be used to advantage in court, nevertheless it broke down the prisoner's assurance, and was instrumental in securing a conviction.

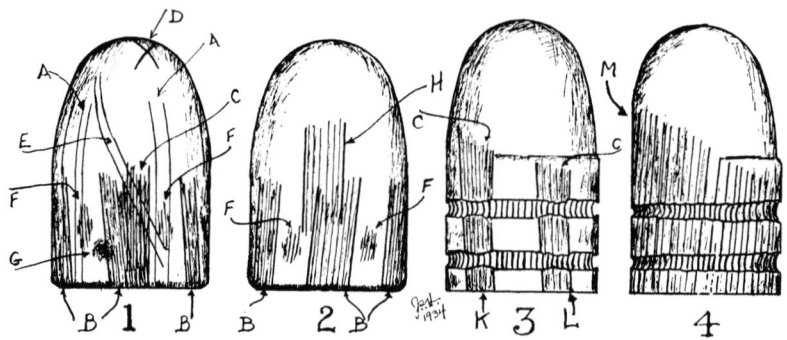

TYPICAL MARKINGS ON RECOVERED BULLETS

1. Bullet from a .45 caliber automatic pistol. A, A—Scratches produced during the operation of drawing the jacket during manufacture. They have no bearing on identification. B, B—Marks left on the bullet by the lands of the rifling. The width, depth, inclination, and number of these marks are class characteristics which may give the investigator a clue to the type or make of weapon used, or both. The surface of this mark is composed of parallel lines, scratches, and marks, and the pattern formed by them is an individual characteristic of the particular gun through which the bullet was fired, and forms the basis of the definite identification of the particular gun which fired the bullet. The lines forming this pattern are parallel to the DRIVEN EDGE of the land mark, which is the edge toward which the mark slopes when viewed from the base of the bullet. (In this case, the left hand edge.) C—"Skid Mark" caused by the failure of the bullet to take up the rotation as soon as it struck the rifling. It consists of lines parallel to the axis of the bullet on the TRAILING EDGE of the land mark, which is the side that the mark goes away from when viewed from the base of the bullet. D—Identification marks scratched on the bullet by police officials. E—Scratches made after the bullet left the gun, probably by contact with pieces of gravel or something of the kind. Note that these marks have an inclination to the axis of the bullet greater than the angle made by the rifling marks, which cannot be the case with any mark whatever which is made while the bullet is passing through the gun. F, F—Marks left by contact of the bullet with the bottoms of the grooves. They may assist in the individual identification of the bullet, but are neither as important nor as trustworthy as the marks left by the surface of the lands. G—A depression left in the surface of the bullet by a "stab mark" used in holding the bullet in the neck of the cartridge case.

2. A bullet from a .45 caliber S. & W. Model 1917 revolver. B, B, and F, F, same as already described under 1. H—An additional rifling mark, which has a different angle and extends higher on the ogive than the regular rifling marks. This is caused by the fact that the bullet struck the forcing cone, or origin of the rifling, AT AN ANGLE TO THE AXIS OF THE BORE. This imprinted the extra rifling marks on those parts of the bullet ogive that are furthest from the axis of the bore. As the bullet enters the bore, it is forced to right itself, and the regular rifling marks are then made. The investigator should become thoroughly familiar with this appearance on actual bullets, as these additional land impressions are often seen.

3. A .38 Special bullet with some of the rifling marks longer than others, showing that it entered the barrel parallel to the axis of the bore, but slightly to one side of the center. This is very common with revolvers, and always happens when the cylinder fails to line up perfectly. C, C—Same as described under 1. K—A land impression that is considerably longer than the average, showing that the bullet was closer to this side of the bore than to the other. L—A shorter land impression.

4. A test bullet fired from a murder revolver. The condition shown here is typical of the cheap, poorly constructed, or old and worn low-grade guns which so often figure in homicides. The cylinder lined up very badly, and much shaving of lead from one side of the bullet resulted. Moreover, the barrel is loose, and is so badly fouled and worn that there were absolutely no distinct land or groove marks discernible. Such a gun forms a difficult problem in identification. M—The side of the bullet where the shaving of lead took place on account of failure of the cylinder to line up properly.

What a Firearm Tells

Through long experience some detectives and all Firearms Investigators know many things in addition to make, caliber and the like, just through examining an arm. A weapon should be first tested externally for fingerprints, not forgetting the possibility of their being on detachable magazines from automatic pistols. The arm then can be examined.

After appropriate paper work, the weapon should then be placed by itself on a table and examined slowly, carefully, and thoughtfully. A general idea of the time since the weapon was fired can

Showing the bases of six .22 long rifle cartridge cases, of two different makes, fired in a Remington automatic rifle. Note how consistent and distinctive are these firing pin imprints. Observe also the dents and indentations put into the rims of these cartridges by the functioning of the extractor (at 3 o'clock) and ejector (at 8 o'clock) of the automatic action—which markings are not nearly as pronounced as the firing pin imprints but which may generally be picked out by the use of low power magnification in the comparison microscope.

The action functioning of almost any .22 automatic rifle is generally so violent as to mark up the bases and rims of its fired cases in a manner similar to this. At times, it is possible to arrive at a reasonable conclusion as to the exact make or model of such rifle; the positions of the extractor and ejector varying to a considerable degree in many models and thus affording a means of determination thereby. However, such conclusions should always be verified by a later comparison with the microscope.

In any such examination of case heads, care should be taken not to confuse these recently-made action-markings with somewhat similar batterings and scratches put there by the automatic machinery and subsequent handling during manufacture. Such markings impressed by the functioning of the rifle during firing will bear a fixed relation to the position of the firing pin imprint. There are also differences in the colors and shadings between the new and the old indentations—a bit of actual experience soon enables the investigator to determine between the two.

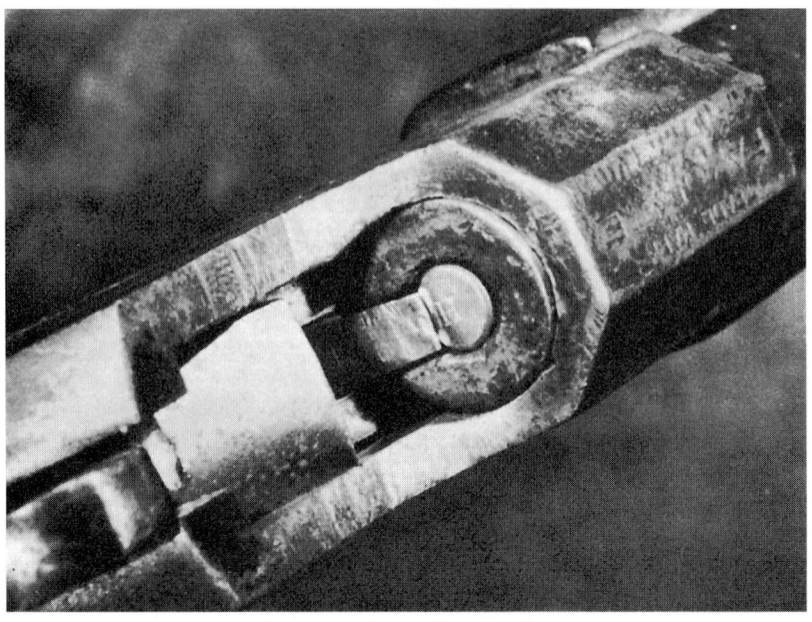

A HOME-REPAIRED RIFLE

Here is an example of a low-priced, single shot .22 caliber rifle that has been repaired by some "fit-it-myself" lad. The original extractor has been replaced by one that neither fits nor extracts, it closes down over the cartridge rim and a knife blade or awl is required to get out the fired case.

When this gun is fired, it results in a most distinctively marked case—showing quite the most mutilated case head this writer has encountered. Such home repairs often result in somewhat similar results and the investigator should familiarize himself with such "conversions" and know what to look for when similar cartridge cases are encountered. In under-lever, breech block actions of this type, another common repair is the use of a round, wire nail to replace a lost or broken firing pin, which also gives a highly distinctive case head imprint.

usually be determined but not, however, within any precise and close limits. For instance, a bore which shows dust, red rust, the accumulation of lint or tobacco fragments or the like, probably has not been fired for some time. A similar weapon which smells characteristically of powder was probably fired a relatively short time ago; however, this smell test is not really worth anything at all because few men are sufficiently trained both as smellers and as shooters to know the difference between one brand of powder and another. As we write this, we have before us six revolvers fired about five hours ago with six different types of ammunition. The amount and character of the odor remaining in the barrels of these six weapons is variable in the extreme. Bullseye as used by handloaders has a characteristic smell that many can recognize. On the other hand, the powder used in the Western .357 Magnum has an odor which disappears quite quickly. A good deal has been done with various chemical residue tests on fouling left in bores after shooting black powder; however, it is of little worth today because almost no one now uses black powder cartridges for crimes.

Marks of wear on weapons will frequently indicate the way they have been used in the past. For instance, holster wear will show up in characteristic places from different forms of holsters. A revolver practiced with in quick drawing will lose all blue from certain areas. Internal wear will be indicated in advanced stages by looseness. Under certain conditions it may be possible to identify from a holster the probable gun carried in it most recently.

Finally, no instrument ever used by man is so likely to indicate his true character as his gun. Probably one of the reasons that most crime weapons are such lousy examples of firearms is that the criminal himself is likely to be a lousy example of a man. This character stuff is, of course inadmissible as evidence, but you can bet your bottom dollar that if a crime is committed with a "zip" gun the criminal will be a young hoodlum. A "Saturday night" revolver will not have been used by a man of substance, integrity, and providence. If a fine weapon, clean, mechanically perfect and well cared for is used in crime—it seldom is—it is probably stolen. There are few exceptions save for new weapons recently purchased.

APPENDIX

Method of Handling Evidence for Firearms Laboratory

FOREWORD: A bulletin written by Lieutenant Jury on this subject and used in the New Jersey State Police School is reproduced below.

Care should be taken that no fingerprint evidence is destroyed. Never insert anything into the barrel in order to pick up the gun; the best method is to lift the gun with a pencil inserted through the trigger guard; another method is to grasp the edge of the trigger guard between thumb and forefinger. Care must be taken to avoid discharging the gun accidentally.

Unload, if absolutely necessary, but do not clean a firearm. If the gun is believed to be loaded, place in a suitable box, large and strong enough to protect the gun, and label both outside and inside of box with the word "Loaded." If the gun has to be unloaded for any reason, carefully note the contents of each chamber, if a revolver, or each barrel, if a double or multi-barrel arm, and correspondingly mark each shell or cartridge to identify it with a particular chamber or barrel. Care must be taken to avoid making any additional marks on the shells in unloading. Do not work the action, pull the trigger, or otherwise manipulate a firearm. Discharged shells remaining in the chamber will be scratched or marked by the breech face, or firing pin, and rendered worthless for comparison, and their value as evidence will be destroyed. Do not attempt to take the gun apart.

The gun should be tagged with a suitable identification tag, placed so as not to interfere with firing of the gun. If a serial number is present, this should be noted in the investigator's notebook together with other data related to case, including place, time, and date. Be precise in making these entries, as they will be needed by you in court.

Serial numbers alone are not always sufficient; some guns have numbers which may be duplicated on others of the same make and model. Others have patent numbers stamped on them which may be confused with serial numbers.

Cognizance should be taken of the value and condition of the gun, and no gun should be so marked as to deface it or detract from its value. Often, guns found in evidence have been stolen and have considerable value to their owners. If by careless or unnecessary marking this value is lessened, the officer responsible is liable to a damage suit by the owner when the gun is returned to him. Adequate markings can be made on a gun without defacing it. Close letter markings, made with a steel scriber, can be made in an inconspicuous place, which is also readily accessible. Automatic pistols can be marked inside the butt in the magazine well or opening, inside the slide, on bottom, with slide retracted, or on the barrel near the end of the slide with the slide retracted. The barrel should be marked in all guns with easily removable barrels, such as the Colt .45 automatic. Revolvers can be marked inside the frame where the crane or yoke fits, on swing out cylinder revolvers, or on the rear part of the standing breech near the bottom, on hinge frame revolvers, or in the same place on solid frame revolvers, with cylinder removed. The purpose of these marks is to enable you to make positive identification of the gun when on the witness stand in court.

Discharged bullets should be marked with a sharp steel scriber, preferably on the base, but if the base is so distorted that it is not possible to mark the base, it may be marked on the nose, in round nose bullets, or on the ogive near the point, in pointed bullets. The ogive is the curved portion between the cylindrical portion and the point. Never mark a bullet on the side, as this may obliterate existing striae left by the barrel.

If more than one bullet is found, the bullets should be numbered in the order found, with the investigators mark, i.e.: AB1, AB2, etc.

Empty cartridge cases (shells) should be marked inside the open end, if large enough, or on the side near the mouth, and also numbered in the order found. Care must be taken to avoid defacing or marking over marks made by the gun. In many .45 automatics, the empty shell is dented in on the mouth; distinctive marks often appear there. Keep away from any dented or scratched portion of the shell. Never mark on the head (closed end), or rim, or on the side near the rim.

Fired bullets and cases should be wrapped separately in soft paper or cotton, and handled as little as possible. They should be placed in an envelope, and the envelope (preferably of manila with a clasp) marked and contents listed. Bullets may be placed in small plastic boxes before putting in envelopes, but when this is done, the bullet or shell should be wrapped in paper to avoid contact of bullet with others or with the box. It is not necessary to encase the bullet in a mass of tape and seals.

Never carry evidence bullets or shells in the pocket or loose in any container.

Cartridge clips and magazines should be marked on the side, bearing in mind the precaution against damaging or defacing them.

Fired paper shot shells should be marked on the side in ink. Many are waxed or otherwise treated so that ink will not make a satisfactory mark; in this case, use a sharp scriber. Marks should, of course, not interfere with any on the shell. Do not open and close the breech on fired shells found in the gun, as this usually puts additional markings on the head of shell. In many guns, the firing pins fail to retract within the breech face, and these will destroy marks left by original firing pin blow on the primer. Sometimes the rims of the shells will get ahead of the extractors, and when the gun is closed, the rims are crumpled and original markings are obliterated. If the gun is opened, it should be unloaded, marking each shell on the side, to indicate from which barrel it was taken. Wrap each shell separately in soft paper, to protect it.

Shot pellets should be placed in a glass vial, or other stout container, filling the empty space with cotton to keep the pellets from moving about. The container should be labeled to identify contents and place, time, and date of finding. Shotgun wads should be placed in a cardboard box and the box properly labeled. A thorough search should be made of the premises for wads. These may be of paper, cardboard, felt, or other composition. They are often of vital importance. One or more wads may be found in a wound.

Where there are indications of bullet holes, shot patterns, or powder burns on clothing or other fabric, the material should be labeled or tagged and forwarded to the laboratory, as it is possible, by laboratory experiments and processes, to determine the approximate distance from which the shot was fired. In special cases, such as shotgun patterns on walls, trees or buildings the evidence should be protected from harm and application made to the Firearms Laboratory for personal investigation by an Expert.

Where a shot has been fired through glass, the glass should carefully be protected against further breakage and, if movable, should be brought to the laboratory since it is possible to determine from which side the shot was fired, and the sequence of shots if more than one was fired. Glass should be properly labeled, giving place, position from which removed, identifying which side was outside and which side was inside, right and left, and up and down. Time, date, and name of person who removed it should also be shown.

When it is necessary to remove a bullet which is embedded in a wall, tree, board or building, exercise great care so that the tool used does not come in contact with the bullet, or so close that pressure may distort the bullet. If in doubt, cut out the portion surrounding the bullet and send all to the Firearms Laboratory, securely packaged. Label and mark the container as described above in the section on Bullets.

Here are front and back views of four .22 long rifle cartridge cases fired in a Stevens automatic rifle.

This particular rifle is outstanding in the manner in which it marks up all cartridges fired in it. The owner of this rifle lives on the ocean front where minute particles of flying sand are a constant problem to cope with. Consequently, from grit lodged in chamber, cartridges dropped in the sand and then loaded in chamber, sand on cleaning patches, and the like, we have in this gun an evidence-producing chamber that may be hard to surpass.

There are innumerable points of comparison found on all cartridges fired in this rifle, as the above illustrations will show.

CHAPTER 12

LABORATORY INVESTIGATION AND PROCEDURES IN "NO-GUN" WORK

IN general, actual laboratory investigation of homicide and other shooting crimes involves work of two general types. First, there is all the miscellaneous investigation in which a comparison of crime bullet or crime cartridge case with test bullets or cartridge cases from a suspect weapon is not involved. This type of work is most easily called "no-gun" work. Second, there is the actual comparison of crime bullets or cartridge cases with test bullets or cartridge cases, or both, from suspected weapons. In this chapter we will consider in detail the "no-gun" work.

Evidence Examined

In cases of this type the laboratory will receive cartridge cases, unfired ammunition, bullets, and various components from shotgun shells. Less frequently clothing, wound photographs and in rare cases parts of the victim will be submitted for examinations and tests. Usually the "no-gun" laboratory work will be performed before the comparison work in a given crime, although exceptions to this rule are frequent. Often one or more bullets will be recovered at the scene of a homicide or from the body of a victim, although high velocity handgun projectiles, particularly metal jacketed bullets of small caliber, are sometimes lost due to passing through the body of the victim and not being stopped by any other object on the scene of the crime. Rifle bullets are even more likely to be lost in this manner. However, in shotgun shootings there is always a good deal of projectile evidence.

Cartridge cases of various types are also frequently found at the scene of a crime, particularly where automatic weapons of one type or another have been employed. Whereas revolvers are seldom emptied and reloaded at the scene of a crime, since few crimes involve as many as five or six shots, hunting field habits and instincts to have a loaded gun cause many men to reload a rifle or shotgun at the scene of a crime. Frequently, one or more shotgun shells or fired rifle cases are recovered quite early in a criminal investigation.

305

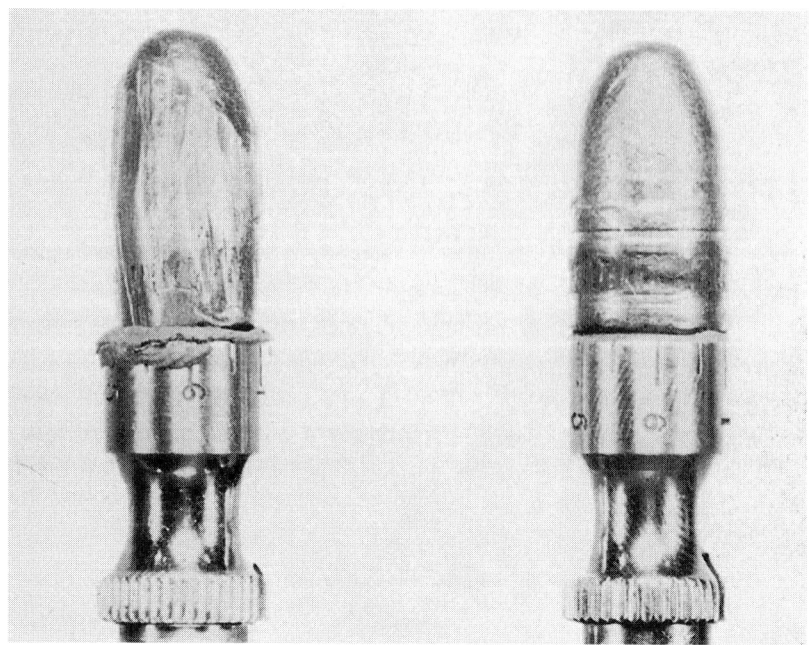

An evidence bullet mounted for microcomparison with an unfired bullet of same cartridge and caliber. The left bullet is one recovered in a bank holdup and it is going to be a difficult one to match, owing to bad skidding and disfiguration.

Laboratory examination of clothing, wounds, and in some few cases other surfaces is in this type of case usually in connection with range determination. This can be fairly accurately done with shotguns. Rifles and pistols leave indications of range only up to a maximum of about four feet.

Examining a Bullet

Treating crime bullets first—they are by all means most important—we will follow the usual procedure from the time of receipt in the laboratory. Visual examination, of course, is the first step. If the bullet has not already been cleaned with alcohol or other solvent to remove blood, mud, and grease, it should be soaked free of these. Actually, unconsciously the Firearms Expert will arrive immediately at certain conclusions. A metal jacketed automatic pistol bullet does not at all resemble a lead bullet fired from a revolver. Whereas there are endless varieties of handgun ammunition, in fully 99% of the actual crimes identification of the caliber and cartridge is practically immediate.

Each bullet or bullet fragment is weighed and the weights recorded. In theory, the diameter should be measured, but in practice, this is seldom actually done. Crime bullets are frequently rather badly smashed up. Diameters are difficult to measure physically. A precise measurement may easily damage the evidence itself; besides, in most instances it is just not necessary. After all, almost all handgun crimes are committed by at most five different calibers, although with an almost endless variety of different cartridges. In general, these are the .22 caliber, .25 caliber, .30 caliber or .32 caliber which in pistols are the same; the various .38s (.357 and 9mm weapons are approximately the same diameter) and finally the various .45s. There are other possibilities but save for the calibers from .40 to .43 called variously .38-40, .41, .44-40 and .44 Special these are extremely rare. Sometimes 6.5mm, .270, 7mm and 8mm rifles will be encountered.

Having once determined the caliber, the next step is to determine the actual cartridge. From a bullet only, this may or may not be possible to do with certainty. The weight will help. Save when a bullet is actually smashed to fragments as when a projectile hits concrete, usually very little weight is lost even though the crime bullet may be considerably deformed. Every laboratory should have some form of tabulation showing as many different bullet weights as possible, each with a description of the bullet and cartridge into which it was loaded. The table given in Appendix I at the end of this chapter suits our needs admirably. There are several thousand possible bullets; however, more than 99% of crimes are committed with far less than a hundred of them. Remember that variations in weight of as much as three grains are not unknown, even in the same lot of ammunition. General appearance of the bullet itself, as for instance shape, lubaloy coating, and the like, will go a long way towards eliminating many weight possibilities.

Family Characteristics

Once probable caliber and cartridge have been determined, the Expert will begin to examine the marks imprinted on the bullet during its passage down the barrel. In this "no-gun" work all we are interested in are the so-called family characteristics which are usually the same for all weapons of a particular make and model. These family characteristics in theory include the depth of the rifling, the width of the rifling, the pitch of the rifling, the direction of this pitch, the number of grooves, and the bore diameter. In general, depth and pitch of rifling can be determined only on crime bullets that are not badly battered. Even on these, it is extremely difficult to do. Direct

The comparison of bullets by the base-to-base method for family characteristics. It is stressed that this method is good only for family characteristics since the individual striae must be compared only when the bullets are pointing in the same direction. It will be seen, however, that only the middle comparison is worth continuing. The right and left comparisons are obviously incorrect. In one case, the bullets have six grooves each, but with opposite pitch. In the other case, one bullet has four grooves, and the other, six.

measurement is almost impossible. Various optical arrangements are resorted to such as a Filar eye-piece with a micrometer scale (as already mentioned under equipment). In general, however, such measurements are no longer common. They were found to be a tremendous lot of work and of no particular importance. Today, the Expert counts the number of grooves, records the direction of twist, and may note whether the lands in the barrel left wide, medium, or narrow grooves in the bullet. With this data, he will in theory consult a list that all laboratories are supposed to have, such as has been compiled in Appendix 2 at the end of this chapter, showing by calibers and by twist and number of grooves all firearms normally met with. The chances are, however, that after several years the Expert will not knowingly consult this list even mentally. Six grooves left usually means a Colt, although there are several other possibilities. Four wide grooves right will similarly automatically suggest to the Expert three or four other possibilities. He will usually mentally bridge this reference step and go immediately to his test bullets from various weapons, getting out the various possibilities for comparison. He will then discard some of them as negative matches upon visual examination, and check the rest under the comparison microscope.

Family characteristics examined under a comparison microscope may lead to a more precise answer than is warranted. Precise examination of family characteristics defeats its own end. The mounting of a test bullet from a known make and model under one-half of the comparison microscope and the crime bullet under the other will in general only confirm what most Experts intuitively guessed almost immediately. It is seldom possible to be exact as to make and model of weapon from a bullet only. Unfortunately, this tentative identification can never be more than a probability. All reports on crime bullets must always be made by the laboratory in such a way that the detective investigating the crime will not fail to send any and all weapons which appear firing approximately the same size bullets to the laboratory for actual comparison.

One of the writers prefers to put two bullets—the test and crime—base to base with a tiny bit of beeswax holding them together. The two are now brought into coincidence and looked at through a stereoscopic microscope. If the bullets came from the same make and model of weapon, the family characteristics will be coincident. This is not a particularly precise method; however, it is as precise as the maintenance of uniformity of family characteristics in the manufacture of any weapon made in reasonably large numbers. In other words, even with a bullet from each of two guns of the same make and model compared in this way variations can be considerable. In some instances, variations within the same model can easily be greater than between half a dozen different makes and models using one of the popular rifling systems. We want to emphasize that this base-to-base comparison is fine for family characteristics but cannot be extended to individual barrel characteristics since by this process we are reversing the order of striae from one bullet to another.

Frequently, a phenomenon leaving marks on a bullet can be used to identify the difference between the same ammunition fired in revolvers and in automatics. These are the so-called skid or strip marks sometimes found to a limited extent in singleshot and automatic pistols but almost always found in revolvers. Actually, in a revolver the bullet moves straight ahead for an appreciable distance before coming in contact with the rifling. Since it has already acquired velocity and inertia, it tends to move straight forward into

FIRED CARTRIDGE CASE EVIDENCE

When a cartridge is fired in a rifle, shotgun or handgun, the base of the case, particularly the primer itself, will be distinctively marked in more or less drastic fashion—the primer more so—the brass or copper case to a much lesser extent.

Such markings are caused: by the closing action of the breech block or the bolt; by the back thrust of the case base and the primer against the supporting portion of the action; by the battering or scraping action of portions of the mechanism, particularly the extractor and the ejector; from tool markings or foreign substances on the breech or bolt face; and in some instances from superficial markings from particularly violent automatic mechanisms, magazine arrangements or abnormal conditions of mechanism.

This illustration is intended to show certain case head and primer markings that may readily be apparent to the eye alone and which can be verified later by use of the comparison microscope.

It is regrettable that, in the above and in many of the accompanying illustrations in this book, the points and markings described are by no means as clearly shown or as readily observed as they are to the eye alone. Neither camera nor engraver have been able to bring out here the minute details, shadings, color differences or markings that are so readily and instantly seen by the eyes of a skilled observer.

The .45 Govt. cartridge, shown at the left, is one of the early center fire cartridges adopted about 1873, but one still in use by many sportsmen. Its pressure, in black powder, standard loads is 25,000 pounds or less. The cartridge shown was fired with a load developing considerably less than this. It will be noted that the primer has not flattened back to any extent and therefore shows only the markings left by the firing pin nose, plus a 'drag mark' caused by the firing pin remaining in its imprint under pressure of the heavy-non-retracting hammer. Also, underneath the primer (at 6 o'clock, in shooter language) there is a scored area where the breech block scraped across the edge of the brass base while seating the cartridge into the chamber. To the knowing investigator, this would indicate that this cartridge was fired in a rifle having a 'falling block' action; but the slope and peculiar indentation left by the firing pin also indicates that it was fired in a rifle having an outside hammer positioned on the side of the rifle, with a firing pin which came down at a steep angle and sloped towards the left. (This angle of fall and inward slope occurred on many rifles and shotguns of the 1870-1890 period). Actually this .45 Govt. cartridge was fired in an old, side-hammer, single shot Sharps rifle, made during the late 1870s.

The cartridge at the right is the .30 G 1906 (commonly known as the .30 Springfield). It is a rather modern one, originally brought out in 1903 and constantly modernized since then. It is a high velocity loading, working at pressures of from 45,000 to 55,000 pounds. The case shown fired a hunting load of around 50,000 pounds and at this pressure it forced the primer back against the bolt face and embossed it clearly; a slight ring remains around the edge of the firing pin imprint

(Continued at bottom of opposite page.)

the rifling for as much as ¼″ during which the trailing edge of the rifling will inscribe a straight line upon the bullet whereas the driving edge will impress its own helix. These skid marks are more easily explained by a photograph than in words. They are not a complete criterion, although they are certainly an indication. The .45 caliber revolvers of World War I are particularly likely to leave these skid marks because of the relatively long travel of the bullet before entering the rifling. The Model 1911A1 automatic pistols also will sometimes leave these marks, but usually nowhere nearly so pronounced.

Mismatching of Ammunition

One of the reasons for caution in precise identification of a crime weapon from a crime bullet only is the possibility that the criminal did not have proper cartridges for his gun. Many crimes have been committed in the past with weapons and ammunition that in theory did not match. A list of these mismatches is almost fantastic. The most common type of mismatch is where the caliber is the same but the cartridge is wrong. In general, almost all .32 revolvers will fire the .32 ACP cartridge, since this cartridge is not really rimless but semi-rimmed. One arrives at a peculiar situation in that the .32 Merwin & Hulbert revolver will not fire the .32 Smith & Wesson revolver cartridge but will fire the .32 automatic cartridge. This is because the original Merwin & Hulbert revolvers were made for special ammunition with a smaller rim than that of the .32 Smith & Wesson. The .38 Smith & Wesson cartridge will ordinarily not enter the .38 Special chamber; however, occasionally this ammunition will so chamber. Further, the .38 Long Colt chamber used

---

and at 55,000 pounds or more pressure this ring would be more pronounced and the outer edge of the primer would be more flattened out and might even spread over into the base of the case to a greater-or-less extent. The firing pin imprint is clear and sharp and comes in at an exact right-angle to the bolt face. This right-angle impact of the firing pin is a feature of practically all bolt action, automatic and other modern repeating rifles.

The firing pin on the .45 rifle shown on the left, and on all types of shoulder firearms up to the early 1890s was of larger diameter and left a noticeably larger (and seemingly shallower) imprint than with modern guns. The difference is readily observed in the illustration. However, many modern rifles are still made to fire these earlier cartridges, so this difference in imprint diameter cannot be taken as a fixed rule with old calibers any more than can one of the early caliber cartridges always be assumed to have been fired in an early model rifle.

The .30'06 cartridge clearly shows the type and size of impact from a modern firing pin—the point and profile of which are practically the same on all modern rifles, irrespective of their caliber.

It is the pressure of the cartridge, developed when the arm is fired, that, together with the primer hardness, determined the degree of primer flattening and which more-or-less accent its markings.

AN ODD BUT EFFECTIVE CARTRIDGE SUBSTITUTION

The Russians developed and use a most peculiar .30 caliber revolver known as the 7.62mm Nagant. It fires an odd looking cartridge, the bullet of which is loaded down within the case, as shown above at the left. The mouth of this cartridge protrudes from the end of the cylinder when it is loaded into the chamber. Upon being cocked, the cylinder slides forward and the mouth of the cartridge inserts itself into the rear end of the barrel; the idea being to prevent any escape of gas upon discharge.

A number of these Nagant revolvers have been brought into this country by returning service men, as war relics. Ammunition of this caliber is not manufactured here in the United States and the cartridge is pretty much unobtainable by our shooters.

In the above instance, the shooter substituted our standard .32 S. & W. Long revolver cartridge and with rather satisfactory results on the whole. Despite the American cases swelling and splitting upon being fired, this combination would be likely to have rather deadly effect.

The three shown to the right are the substituted .32 S. & W. cartridge and two of its cases after being fired in the Nagant revolver.

in the .38 Military revolver before and after the Spanish War will sometimes take almost every variety of .38 revolver ammunition including even the .357 Magnum cartridge, which will generally blow the old Colt to pieces. The .380 ACP, .38 ACP and .38 Super will chamber and fire in some .38 caliber revolvers.

Most automatic pistols can, in an emergency, be made to function with some revolver cartridge by the simple expedient of filing or grinding off the rims. Since there is no extractor groove, the automatic pistol may or may not reload properly. Cases of the use of this form of cartridge are far more frequent in Europe and the Near East where ammunition is relatively difficult to obtain and expensive. In America, in general, errors in ammunition are through ignorance rather than through difficulty of obtaining it.

Perhaps the most frequently used crime ammunition in the country after the .32 Smith & Wesson is the .45 ACP. This particular ammunition can be fired not only in the automatic pistol itself, but in sub-machine guns and several revolvers. The 1917 Colt and Smith & Wesson service revolvers take this ammunition; half-moon clips were provided by means of which the empty cartridge cases can be extracted easily. However, in general, these revolvers will function perfectly without these clips, although extraction is a minor problem. Some low serial number 1917 Colts will allow the .45 ACP case, which is a true rimless, to enter too far into the chamber so that the firing pin does not contact the primer at all. However, even these weapons can be rigged to function after a fashion without clips, particularly if the muzzle of the weapon is elevated before the cartridge is fired. Later Colts, and all Smith and Wessons, control headspace from the mouth of the case rather than with the clips. In the years since the end of World War II many British service revolvers made by Webley, Smith & Wesson, and Colt and originally chambered for the .455 Eley Webley revolver cartridge have been brought into this country and rechambered for the .45 Colt revolver cartridge, the .45 ACP with or without clips, and the .45 Auto Rim.

Further, with a bit of copper wire in the rim a .45 ACP round can be fired from a revolver chamber for the .45 Colt revolver cartridge. The .45 ACP will usually function in Webley and Colt automatics chambered for the .455 Webley Automatic ammunition.

One model of the Spanish Astra, the 900, was designed to function with all varieties of .38 or 9mm automatic ammunition from the .380 ACP through the 9mm long, the 9mm Luger, the 9mm Steyr to the .38 ACP and .38 Super. Many more substitutions in ammunition can probably be made. Never say then that a thing is not possible; some gun crank may make you out a liar.

In addition to the various misfit ammunition of correct bore size, quite a bit of ammunition will function in a larger bore apparently fairly satisfactorily. For instance, the .25-20 ammunition will fire in the .32-20 revolver; the .38-40 will function also in the .44-40 revolver and frequently in various .45 Colt revolvers; these undersized bullets, however, are not properly engraved in their passage through the bores. Further, no positive check of individual or even family characteristics is certain since the bullets may be partially tumbling, even in the bore. They will receive entirely miscellaneous contact with various parts of the barrel; however, an Expert familiar with the

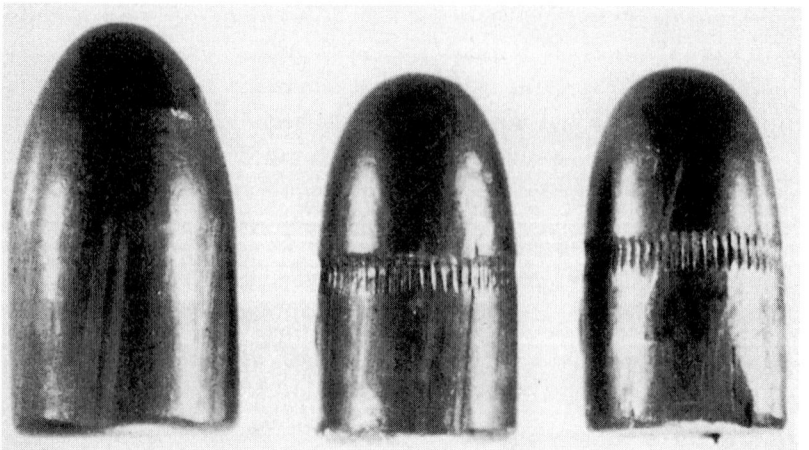

USE OF WRONG CALIBER CARTRIDGES

The left-hand and central bullets shown above are, respectively, 9mm and 7.65mm Luger bullets after being fired in the proper Luger pistols.

The right-hand one is a 7.65mm Luger bullet fired in a 9mm Luger pistol. Note that it has upset but little in skidding down the bore and that very few striae are available for identification.

appearance of these bullets will usually be reasonably certain of their identity and be able to make a good surmise at their origin.

A recent crime in New Jersey was committed with .30 caliber or 7.65mm Luger ammunition in a standard 9mm Luger pistol. The shell case is exactly the same diameter save for the neckdown portion. Velocity would be considerably reduced because of the windage between the bullet and the barrel. A weapon so loaded will not function automatically; however, the bullet was perfectly capable of inflicting a fatal wound. The markings upon it are characteristic and are shown in photographs herewith compared to three test bullets similarly mismatched. All four are differently engraved but have a rather characteristic appearance. It is important to realize that this undersized ammunition will reduce power and cause extreme inaccuracy, but since most crimes are committed at the very closest of ranges, this is of little importance to the criminal. A human being is not hard to kill.

This mismatching of boresize or undersize ammunition is not limited to pistols. The same kind of thing is possible in rifles. An amateur gun crank recently used a .22-250 Varminter in a .250-3000 rifle to scare a rival for his wife's affections. The undersized bullet was fired upwards through a window and recovered in the trim.

The most astonishing and rarest form of mismatching is that in which an oversize bullet is forced through a bore. A crime in New York City was committed with 9mm Luger ammunition in a 7.63 Mauser pistol. The breech pressure must have been very high indeed but the weapon stood it, not only at the scene of the crime but later in the laboratory where test bullets were obtained in this same way. A positive identification was made. The authors have not tried this personally and definitely do not recommend it.

Odd Foreign Combinations

Foreign revolvers present a nightmare of possibilities for mismatching. A service weapon made by the Israeli government is chambered for the 9mm Luger cartridge and uses half-moon clips in exactly the same manner as the Smith & Wesson and Colt Model 1917 revolvers. Until about 1910 most European armies were equipped entirely with revolvers which were superseded slowly with auto-

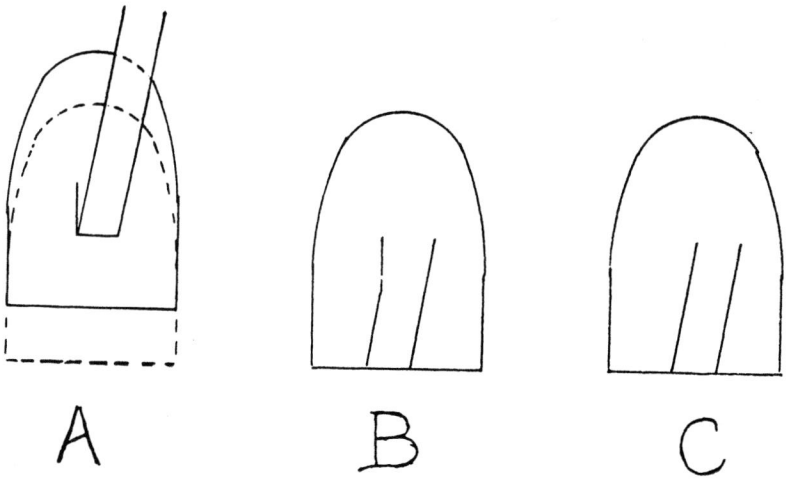

HOW "SKID MARKS" ARE CAUSED

Frequently, when examining a fired bullet, it will be observed that the marks left by the lands are wider at the top than at the bottom, as shown in B above. This is caused by the fact that when the powder charge explodes, the bullet jumps straight forward until it strikes the rifling, when it is suddenly forced to take up a motion of rotation. Very often, when it strikes the beginning of the rifling, it "skids" straight ahead for a short distance before the motion of rotation begins.

This is shown diagrammatically at A. The lands plow straight back into the bullet metal for a short distance before the motion of rotation begins and thus causes the straight mark shown at the top of the trailing edge.

"Skid marks" are likely to be much more pronounced in revolvers, where the bullet is already travelling forward very fast before it reaches the rifling, than in pistols where the bullet is practically seated in contact with the rifling as it lies in the chamber.

Very often, in a rifle or pistol, when the bullet is in firm contact with the lands before the motion starts, there will be no skid marks and the rifling marks will appear as in C.

Comparisons of four .45 ACP cartridge cases showing the difference in the class characteristics between revolvers and automatic pistols. A very clear impression is left, on the upper two cases fired in the revolver, of the recoil shield. Those fired in the automatic pistol show, in the lower two, a relatively smooth breech surface—although the automatic pistol marks its fired cases in other fashions, as is explained elsewhere.

matics. These old weapons are now coming to this country and being sold for about $15.00. Ammunition to fit them precisely is unavailable. However, one of the authors has half a dozen of these; five can be fired after a fashion with standard American ammunition.

The Austrian Rast-Gasser chambers the .32 ACP, although the bullet is undersize in the barrel and tumbles in flight. A gaudy Spanish piece of junk takes the .44 Russian. Two different Balkan revolvers fire the .38 Colt and the .38 Smith & Wesson respectively. An old French service revolver can be used with .41 Colt ammunition. Even the legendary Montenegrian Gasser would probably function with the old .45 Colt revolver cartridge, although with excessive pressure. A crime bullet from one of these combinations would be fairly obvious *if* the weapon were also submitted. The bullet by itself would present an almost insoluble problem for precise identification.

Examining a Cartridge Case

"No-gun" cartridge case work is neither so frequent nor quite so positive as that on bullets. More crime bullets are recovered at the scenes of crimes than cartridge cases. A bullet can be proved in most instances to have actually caused injury or death if it did cause it. A cartridge case on the other hand may, or may not, have at one time contained a particular crime bullet. The bullet, not the case does the damage.

However, a great many cartridge cases are brought into the laboratory. As already pointed out, these come usually from weapons other than revolvers. The automatic pistol cartridges found at scenes of crimes can on occasion give a good deal of information in connection with the arm from which they were fired. A cartridge case in addition to receiving an impression from both the firing pin and the face of the breech also receives certain other marks, particularly when fired in an automatic pistol. A number of cartridge cases fired in automatics are photographed and their particular characteristics pointed out in the caption. In large measure, these family

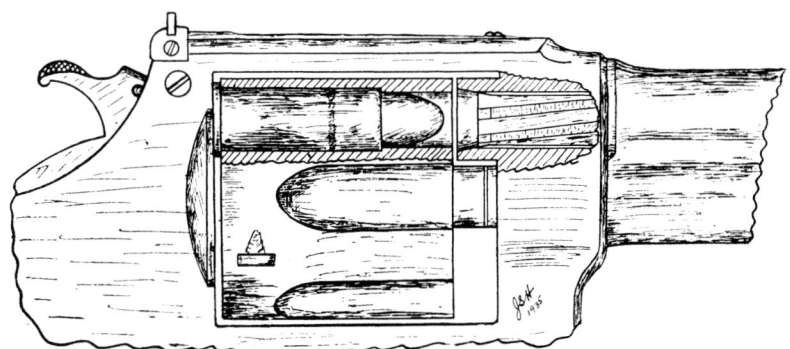

THE CHAMBERING OF A CARTRIDGE IN A REVOLVER

Note the manner in which a revolver cartridge chambers in the cylinder; observe that the bullet lies in an unrifled section of the cartridge chamber and must jump an appreciable gap (one-half inch or more) before it strikes the rifling in the barrel. Bullet first travels through front end of the chamber, then passes the slight gap between end of cylinder and barrel, then through "cone" at breech end of barrel and into lead before it slams into the rifling. This excessive jump, which does not occur in the automatic pistol, is what causes revolver bullets to "skid" for a brief interval upon engaging the rifling, as explained in the text.

This sketch illustrates the cartridge in a Smith & Wesson .38/44 Outdoorsmans Revolver. This rifling has a right-hand pitch.

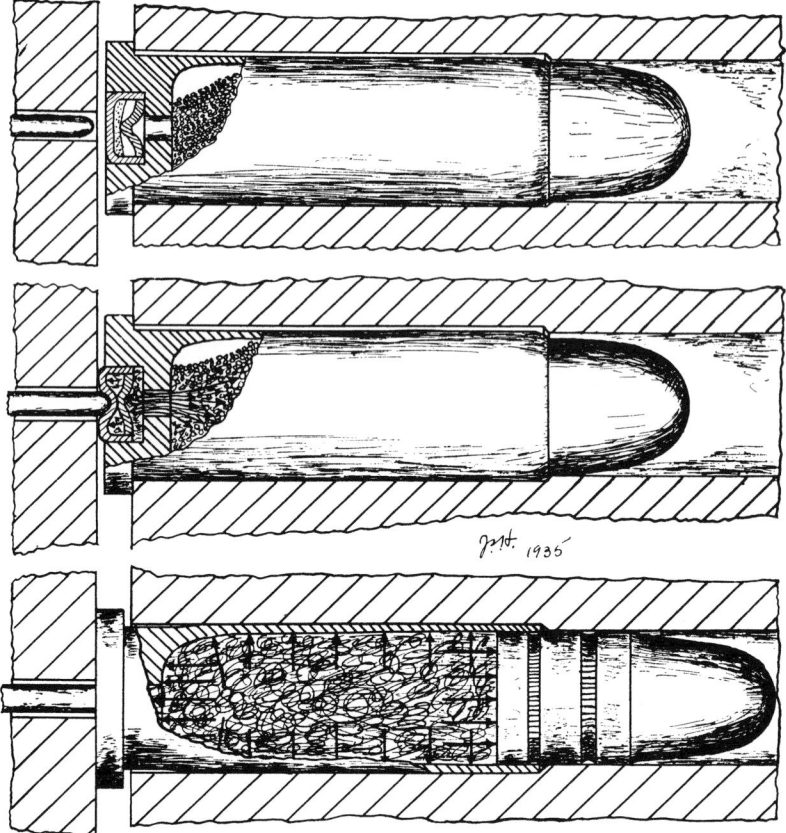

DIAGRAM OF WHAT HAPPENS WHEN A REVOLVER CARTRIDGE IS FIRED

Top. The cartridge is shown lying in the chamber of the revolver cylinder, ready to be fired. Naturally the cartridge must be somewhat smaller and shorter than the chamber, to allow it to be inserted at all, and as there is a variation in size of different chambers among the various makers, and even in arms of the same make, the cartridge may be quite a bit shorter and smaller than the chamber happens to be. This condition is shown in an exaggerated form in the sketch. The cartridge lies at the bottom of the chamber, and there is a vacant space both above the cartridge and behind it.

Middle. The firing pin strikes the primer. If the cartridge is not all the way forward, as shown, the blow of the firing pin will drive it as far forward and away from the breech face as it can go. The priming composition explodes, and the reaction of this brusque explosive drives the primer partly out of the primer cup and back against the breech face as shown.

Bottom. The flash of the primer has ignited the powder. The arrows indicate the powder pressure, which has driven the whole cartridge case back against the breech face thus smashing the primer against this surface, so that it takes the impression of any tool or file marks or other irregularities on this surface. The pressure is also driving the bullet out, and in doing so, it has straightened out the crimp in the mouth of the cartridge case. The impact of the gas on the base of the bullet, acting against the inertia of the bullet, "upsets" or expands the bullet so it is the full diameter of the neck of the chamber. The powder pressure, acting on the side walls of the case, which are thin and elastic, expands them until they bear tightly against the walls of the chamber and thus prevent the escape of gas to the rear.

characteristics consist of marks made by various parts of the mechanism during loading, extraction, and ejection. These are reasonably uniform between firearms of the same make and model since in most well-run factories at least the basic pattern of a weapon is standardized. This cannot be counted upon, however, in many makes of Spanish and a few makes of Belgian weapons. Further, even in a commercial .45 Colt, perhaps the best automatic in the world, there is enough variation from gun to gun, so that family cartridge case markings common in some samples are missing in others. After a

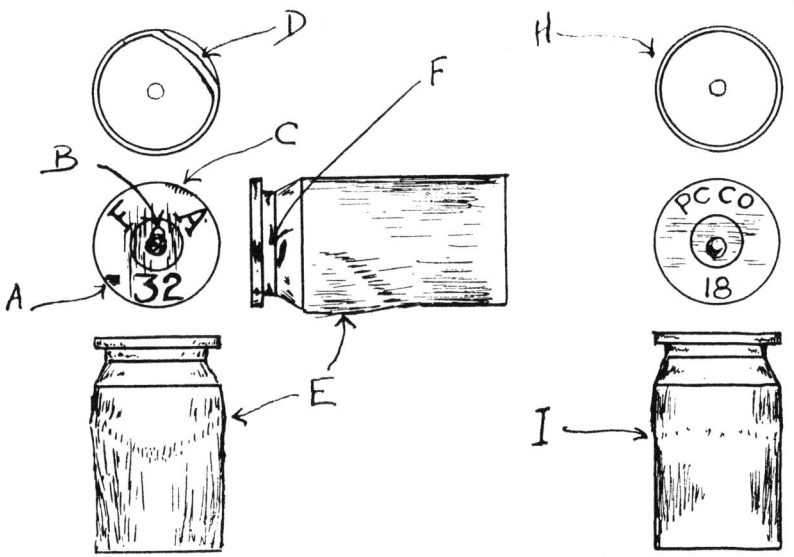

Characteristic appearance and marks found on cartridges fired in automatic pistols and in revolvers. Left, cartridge fired in the .45 caliber Government Model Automatic Pistol. Note the following points. "A," the ejector mark, usually visible to the unaided eye. "B," the secondary firing pin mark, or scrape, often found in automatics, especially the Super .38 and the .45. It is at the top of the regular primer impression. "C," the upper edge, opposite the ejector is often smashed and marked with striations from the tool marks on the breech. It is here that the case pivots as it is thrown out by the ejector. "D," the mouth of the case is usually flattened, from striking against the side of the slide as it is thrown out. "E," the cartridge is bulged on the bottom, as the chamber is tapered to the rear and cut away at the rear end on the bottom to facilitate feeding. "F," the extractor marks. They are usually most prominent on the forward or sloping part of the cannelure, and on the forward part of the rim. Sometimes there is also a mark, as shown, on the edge of the rim where the extractor has ridden over it from behind in taking hold. In addition, there will be vertical striations on the primer, and usually more faintly, on the head of the case.

Right, case from a .45 Model 1917 Revolver. Note, "H," the mouth of the case is perfectly round, unless it has been stepped on, or otherwise abused since firing. "I," the case is slightly bulged toward the rear, but the bulge is even all around, and not more pronounced on the bottom, as it is with the automatic. The firing pin dent is usually off-center, and most likely toward the bottom. The case head and particularly the primer will have horizontal striations or marks received from the filing marks which run across the breech face from side to side.

gun has become worn and perhaps damaged in cleaning or by abuse a different picture from the normal one of that make and model can be given.

The most frequently used family case markings are those left by the extractor and the ejector, which will be positioned at a constant relation to each other throughout a given make and model of well-made arms. The marks themselves can vary considerably from weapon to weapon; it is their relationship between each other and sometimes between other marks made by magazine lips and the like that is important.

Some designs and makes of automatic pistols have additional family marks. It is probable that there are more of these than any of us realize; we do not have enough to do with many makes and models to be conscious of their family characteristics. The authors are indebted to G. A. Gordon for his excellent and truly exhaustive monograph on the .45 ACP markings. We have reproduced this in its entirety as Appendix 1 following Chapter 13. If as much time had been spent with other weapons perhaps we would have similar knowledge. The family markings can be either impressed—that is a negative map of some area of the weapon—or generated on a surface of the cartridge case by moving past a series of high spots somewhere on the weapon. Some automatics, including the Model

PISTOL CARTRIDGE DRAG MARKS

These are two .455 British pistol cartridge cases after having been fired in a Webley & Scott automatic pistol. They are shown positioned just as they went into the chamber of this pistol.

The barrel and breech block of this .455 Webley are locked upon discharge and they recoil together for a short space while the mechanism is unlocking, during which period the breech block moves downward marking the fired primers as shown above. Note that the striae generated around the firing pin imprint are also in the vertical plane.

Firing pin drag marks. Usually found and duplicated only where the firing pin can not rotate. Notice that such drag marks, when made by the .45 Pistol as shown above, are the reverse of similar markings made by a top-break shotgun where the firing pin sticks in the primer. Top is evidence case, bottom half is test case.

1911 and 1911A1, have a portion of the rear of the cartridge case at the time of firing unsupported where the chamber wall has been cut away to make the feeding ramp. With some ammunition this will lead to a characteristic bulge at this point, which will of course be in a definite relationship to extractor and ejector marks. This bulge is of the impression type of mark but is not usually very clear; it is no good for individual identification.

Perhaps the best known and most interesting idiosyncracies of Browning's dropping breech design, such as used in the U. S. Model 1911, the F. N. 1935, the Radom, and the Tokarev, are the firing pin dragmarks and certain phenomena involving sheared primers. The most important thing to remember about these is that they do not occur very often in most specimens but may occur occasionally in any weapon of this type.

In Browning's high power design the rear of the barrel pivots

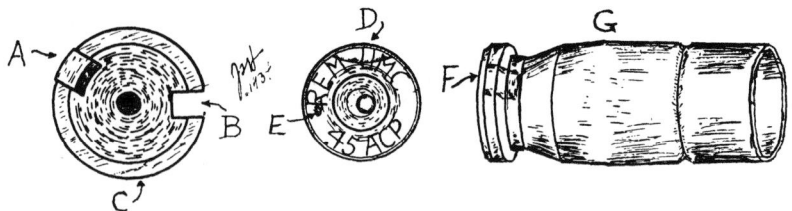

CHARACTERISTIC MARKS LEFT BY THOMPSON SUBMACHINE GUN

Left. Breech face of Thompson Submachine Gun. Note that this breech face is circular and is finished by machining, so that the surface is covered by tool marks in the form of concentric circles. A—the extractor, which snaps or springs over the edge of the cartridge rim when the breech slams shut, and in so doing, puts scratches on the rim. B—open space, or clearance cut, for the ejector, which is a rectangular piece fixed into the side of the breech recess. When the breech block flies open after the gun is fired, the empty case is drawn back by the extractor hook until the ejector is reached, when the impact of one side of the carrtidge head against the ejector causes the cartridge to pivot around the extractor and be thrown out of the gun. The impact of the cartridge against the ejector makes a bruise on the cartridge head, shown at E in the middle sketch. C—bottom rim of the breech block, which strikes the top edge of the cartridge lying in the magazine and drives it out of the magazine and into the gun chamber when the breech block slams home under spring action during the operation of the gun. The impact of this circular edge with the top of the cartridge head makes a circular mark on the cartridge, as shown at D in the middle sketch.

Middle. Head of cartridge which has been fired in Thompson Submachine gun, showing fine circular marks impressed on the primer and cartridge head by the circular tool marks left on breech face. Ejector bruise E and breech block mark D are also shown.

Right. Side view of cartridge fired in Thompson Submachine Gun, showing extractor marks F, which may be more or less visible, according to circumstances; and uniform circular swelling of case, G, which forms a brightened band, more or less visible, directly around the case, instead of around the bottom only, as in the .45 automatic pistol.

down to unlock the slide in recoil. If, for several reasons, the firing pin still protrudes slightly when this motion starts, the primer will be marked at 12:00 o'clock with a generated groove, varying from a mere touch of the top of the firing pin impression in the primer to a mark of considerable length. This mark has not, in the authors' experiences, ever been made by any pistol in any condition on as many as 50% of the cartridge cases fired with it.

The late Lieutenant Roche of the Massachusetts State Police based his tentative identification of a weapon used in a particularly atrocious murder of a police officer in part on dragmarks present on three of 12 crime cartridge cases. Eleven of the bullets struck the victim. The number of shots was of course most unusual; all bullets and cases were 9mm Luger or Parabellum. A sub-machine gun was at first suspected. However, all 9mm sub-machine guns are straight blowbacks and could not have these dragmarks, which are only possible in dropping-breech, recoil operated arms. Witnesses remembered only one series of shots. All bullets and cartridge cases were

fired from one weapon. The FN Browning 1935 is the only standard automatic chambered for this cartridge with a magazine capacity of 13 shots. It is interesting to note, however, that Lieutenant
Roche did not go completely out on a limb, even with this almost
certain identification. He called it a probability only, since the crime
just possibly could have been committed with a Radom with a
homemade lengthened magazine or a similar arm.

The sheared primer phenomenon is in some ways the opposite
of that of dragmarks. The combination of a soft primer cup, a
bit more powerful priming charge, a smaller flash-hole, or more
chamber pressure, will sometimes lead to the primer cup being
forced back against the firing pin and into the firing pin hole.
This metal will then be sheared off, either completely or partly,
when the rear of the barrel drops down. These are generated marks,

A crime and test .45 ACP cartridge case from a post-war Colt commercial 1911-A1.
The striae near the mouth of the cartridge case, caused by the striking of the
expended cartridge against the slide, is far more pronounced than is usually encountered. Further, the mark on the rim of the case, made by the extractor, is also
more noticeable than usual. A high school boy stole this weapon and used it in a
hold-up in which shooting was done.

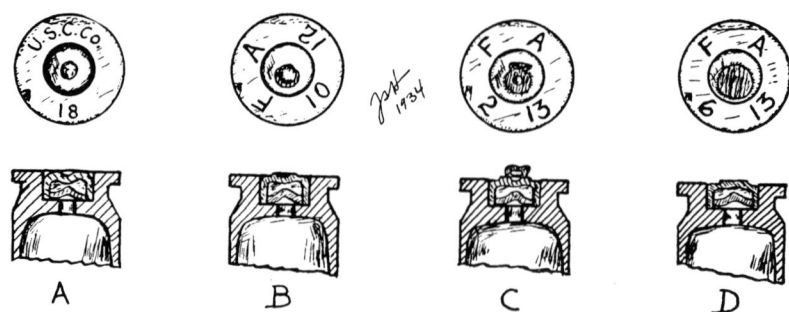

DIFFERENCES IN APPEARANCES DUE TO VARIATIONS IN PRIMER HARDNESS

These four .45 caliber automatic pistol carrtidges were all fired in the same pistol. A—Cartridge with brass primer. B—Cartridge with copper primer. Part of the primer has blown back into the firing pin hole, and then has been sheared off when the head of the cartridge was slid downward over the face of the breech by the downward swing of the barrel in unlocking. C—Still more of the primer surface has been forced back into the firing pin hole, and a little annular ring of metal which was sheared partly, but not completely off, may be seen adhering to the top of the primer. D—A considerable portion of the primer surface has been forced into the firing pin hole, and then sheared completely off, leaving a freshly cut metal surface exposed to view, with distinctive streaks left by irregularities in the edge of the firing pin hole which did the cutting.

left by the lower edge of the firing pin hole, and are of considerable value in matching a crime case with an individual weapon. They do not occur exclusively in any specific pistol but may occur in all those using Browning's design. They are caused more by the ammunition than the individual weapon.

Many automatics will almost invariably bang their empties against the slide before ejecting them completely. Sometimes striae are generated which are extremely uniform between different rounds fired in a single weapon. In addition, a Model 1911A1 has a tendency to flatten one side of the mouth of the case in the course of extraction and ejection from the weapon. This flattened mouth is not constant throughout the millions of different .45 automatics; however, when present, it is an indication of the round having been fired in this weapon.

Some automatic pistols have a rotating barrel in which the force of recoil has to revolve the barrel in relation to the slide as the slide itself moves to the rear. The best known weapons of this type are our Savages, which are no longer made but were for a period of several years very popular. Savage automatics are among the most troublesome today, since they can generally be purchased by criminals for a few dollars and are reasonably positive in their functioning. The Austria-Hungarian automatics of the Frommer,

The two cases photographed are, respectively: on the left, an 8mm Nambu case, and on the right, a 9mm Luger case. These are standard in every way and function perfectly in the weapons for which they were designed. However, in each instance an American .38 Special cartridge case has been used to make these cases. The original rim has been turned off on a lathe and a new extractor groove turned into the head of the solid case. The 8mm Nambu was forced into a sizing die and then cut off. The 9mm Luger was merely cut off. These two cases, as well as the photograph of them, were produced by George W. Spence, of Steele, Missouri, who has produced every single type of foreign ammunition reasonably simply and inexpensively from standard American cartridge cases. He will duplicate the shape and ballistics of any and all foreign ammunition and generally improve on the accuracy. He and other custom hand-loaders are most ingenious in reproducing even the long, straight, tapered German sporting rifle cartridges. It is emphasized that in very few instances will the original American manufactory's head stamp have been removed from the base of the case, which head stamp is, of course, meaningless after the ammunition has been changed.

Steyr, and Roth-Steyr types also have revolving barrels, although in those actions the slide and barrel are actually locked together for a short distance. In certain cases where the chamber is clean and oiled but contains a tiny trifle of abrasive matter, the pressure inside the barrel will seal the cartridge case to the breech so that the barrel will leave circular marks on the cartridge case itself. The Expert will need sharp eyes, however, to note this because none of the weapons which sometimes impart these unusual marks will be met with frequently. Unless one is actually looking for them, they resemble the accidental marks received in careless handling.

Cartridge cases fired in revolvers, even though seldom brought into the laboratory since they are not nearly so frequently recovered in "no-gun" crimes, can usually be identified. Revolvers, with the exception of the special types made to fire the .25 ACP, the 9mm Luger and the .45 ACP, seldom, save in the case of mismatched ammunition, use ammunition designed for automatics. Revolvers use rimmed cartridges. The .22 rimfire and .38 Specials for hand-converted .38

  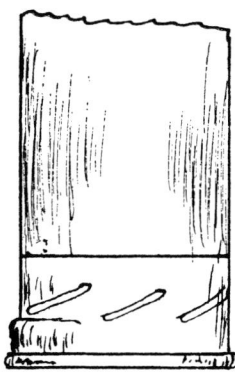

EXTRACTOR SWELLS ON FIRED CARTRIDGE CASES

Rim fire cartridge cases are very likely to show a clear indication of where the extractor segment or portion of the case lay when the cartridge was fired, as there may be an "extractor swell" showing on that section of the case. This swell may be very pronounced, especially in the case of a cheap, poorly fitted rifle or pistol or in one that has seen considerable use. The thin and rather soft copper or brass of which the case is made blows out and into such cavities or depressions, in almost any type of firearm.

In the above sketch, we show such extractor swells—which have been shown to an exaggerated degree for clearer understanding. The left illustration shows a .22 long rifle case after being fired in a H. & R. single shot pistol; the middle is the same cartridge when fired in a revolver—both showing those portions of the cases which lay against the extractors when they were fired. This same type of swelling may occasionally be seen on some thin-walled center fire cartridges which have been loaded to high pressures and where the extractor is badly worn or poorly fitted, as in some of the early single shot rifles.

The right sketch shows a shotgun shell with a bulge in the base next to the rim where the metal expanded out into the part of the chamber cut away for the extractor. This appearance can only be found with top-break shotguns, especially with some of the cheaper single shot models where the extractor fitting is poor. It will not be present where the shotgun is a repeating or automatic type.

Super target automatics are about the only fully rimmed cartridges in use in automatics today. Most revolvers have firing pins integral with the rest of the hammer or a semi-flexible nosepiece on the hammer which strikes a more or less inclined blow on the primer. These firing pins may impart generated striae. These are rather characteristic, particularly in the Single Action Colts. Many revolvers also have recoil shields or plates fitted into the breech face through which the hammer nose protrudes. These leave readily identifiable marks in many models. Most hammerless revolvers actually have concealed hammers, but then so do some automatics. Further, some revolvers including the new Colt .357 and some Smith and Wessons and Iver Johnsons have separate firing pins mounted in the breech frame.

Cartridge cases fired in standard model rifles are not usually confused with pistols save in .22 rimfire, .32-20, .38-40, and .44-40. We are reproducing a number of characteristic .22 rimfire firing pin impressions. From a cartridge case only of this caliber it may or may not be possible to say whether it was fired in a handgun or rifle. Because of the many crudely made pistols that started out as rifles most Experts will not make a positive statement. However, in the larger Winchester centerfire calibers the primer indentations are quite different in rifle cases from those fired in revolvers.

Usefulness of "No-Gun" Identifications

After examining bullets or cartridge cases or both, what have we learned? In general, no Expert will go out on a limb and tell a detective definitely what to look for. He will always hedge. He must do this; these things are not certainties. His official reports must always be accurate. He will protect himself and his department by qualifying his findings. Every Expert should realize the possibility of a nightmare weapon converted or fabricated by a crackpot which would make him look bad. For instance, a lot of .380 caliber automatics imported into this country immediately following World War I were rebarreled to convert them to .32 caliber ACP by using cutoff sections of Springfield rifle barrels. Further, innumerable singleshot pistols have been fashioned from rifle barrels and then chambered for various pistol cartridges. Rifles are in existence today that handle ammunition such as the .38 Special, the .357 Magnum and the .45 ACP. Finally, Spanish weapons of almost every type and description have been made with barrels and other parts actually produced in the kitchens of Spanish industrial workers.

Even if a weapon could be exactly described, it would do the general investigator a relatively small amount of good. He is looking for evidence of any type at all. If he is worth his salt, anything that he finds with even the remotest possibilities of being evidence, he will send to the appropriate experts for examination. This is particularly true in connection with firearms. It is a desirable condition because of the possibilities for error in preliminary "no-gun" identification and also to a lesser extent because any weapon belonging to a criminal, even though innocent of the particular crime immediately under investigation, may prove to be of use in the solution of some other crime from which there is bullet or cartridge case evidence in the pending files of the laboratory.

In spite of what some inexperienced prosecutors think, most "no-gun" examination reports are of little evidence value. One of the authors testified, much against his will, for a young prosecutor in a case involving a .45 lead bullet; this was the only firearms evidence. The defendant was known to have owned a .45 Single Action Colt which had disappeared. A box of 50 .45 Colt cartridges was found in the home of the suspect. The prosecutor wanted the firearms expert to testify that in his opinion the evidence bullet was fired by the suspect because he had in his possession some cartridges of the same caliber! The expert was able at a pre-trial conference to convince that such testimony would serve only to make the prosecution look ridiculous, and he was not called upon to testify. There was absolutely no way of positively tieing the crime bullet to the missing revolver other than to say that the bullet could have come from such a weapon.

Range Determination From Powder Marks

Burning, scorching, and the so-called powder tatooing of a victim or his clothing have long been recognized as useful in determining the distance between the muzzle of the firearm and the surface of the victim at the time of the shooting. In the old black powder days, these things occurred out to a considerable range. However, with the almost universal use today of smokeless powder, the evidences of this type perceptible to the naked eye are confined to much shorter distances.

However, new scientific methods of examination both of clothing and of skin by photographs under infra-red and ultra-violet light have increased considerably the distance at which the products of combustion of smokeless powders can be detected on skin and clothing. Further, certain polarizing microscopes can be used to examine clothing for nitrates directly. As referred to in connection with photography, it is now possible to use a contact method of photographic examination directly on the cloth.

As the results of quite a number of tests made in various police departments recently have shown, these tests are all extremely variable from one powder to another and from one barrel length to

A good view of the late Lieutenant George Roche, Officer in Charge of the Firearms Investigation Laboratory of the Massachusetts State Police. Lieutenant Roche has here an extremely accurate arrangement for determining the deposit of powder residue, and the like, from a specific firearm and type of ammunition. Such a rig is desirable not only in connection with pistols and revolvers, but also for pattern determination in shotguns. Inexact methods of measurement of distance from muzzle to target are deplorable but unfortunately fairly common.

another. Further, looseness of cylinders of revolvers, priming, the pressure of the loading, and several other factors are known to have some effect on this evidence. It is, therefore, more than ever necessary, in order to obtain accurate results, to use a weapon similar to the crime weapon with ammunition of the same type. Commercial ammunition may change considerably over a period of years without changing the label on the ammunition at all. The only sure way of running control tests is by matching in every way possible the actual condition at the time of the crime.

It is chemically possible to analyze powder residue, even in minute quantities. However, in general, this will probably not be found to be of any great use since most smokeless powders give almost identical residue. Some primers, however, particularly of the later non-corrosive type, will give quite different residue from other primers.

In general, burning is a phenomena which does not occur under any circumstances at all when the distance from the muzzle is greater than about 7 inches. Even this 7 inches distance is for a Service rifle. Few pistols will burn at a distance beyond about 2 inches, regardless of the loading, assuming, of course, that black powder is not being used.

Scorching, another phenomenon closely akin to burning, can be extended perhaps to twice the range of actual burning, which would limit it to a maximum of about 1 foot for all smokeless powder small arms.

The original black powder tattooing was the driving into the surface of a victim's body small particles of black powder. Actually, ornamental tatooing as originally indulged in by sailors a hundred years ago was frequently accomplished by introducing into the skin small particles of black powder. It was natural to call the same kind of thing by the same name when it occurred in connection with wounds. However, this phenomena rarely occurs with smokeless powder, although the depositing on the surface of the skin and on clothing of the products of combustion does occur in somewhat the same way. Occasionally these will actually be hard particles that are to some extent driven into the skin. The range of these in black powder used to be as great as 10 yards, particularly with a heavy overload in a muzzle loading weapon. However, at a distance of greater than about 4 feet, even the most sensitive of modern devices will not pick up smokeless residue. In general, powder residue is limited in most weapons and with most powders to considerably under three feet.

Several years ago, much importance was placed on the shape of deposits of powder around a wound. For instance, a revolver loaded with black powder ammunition would give a wound considerably offcenter with the powder tattooing concentration, since the revolver in recoil tended to have the muzzle rise after the bullet had left the barrel. It was also thought possible to determine the angle at

These are the fired cases from two very different types of shotguns—the old and the new. Their identification markings, in principle, follow the same course as on rifle cartridges.

The case on the left is an old U. M. C. (Union Metallic Cartridge Company) New Club, factory loaded, black powder cartridge, loaded prior to 1908—after which cartridges made in this factory were stamped with the 'Rem. U.M.C.' marking. This old cartridge was fired in a hammer, double shotgun, an early one with the plungers (firing pins) coming in from the side and at a downward angle, as is clearly shown by the imprint. This firing pin was the old, large diameter style, now obsolete. (This particular shell was fired some 60 years ago and its minor identification markings have all corroded away.)

The case on the right was fired in a modern, hammerless, double shotgun having the smaller diameter firing pins striking the primer on a parallel 'straight in' line.

Pressures in the shotgun are not nearly as high as in rifles and primers fired in a well-made, tight, modern shotgun do not flatten back anything like as is the case with rifle cartridges.

which a muzzle had been held to the surface of cloth or skin of the victim by the same method. A good deal of effort went into these two questions. However, in general, with smokeless powder such investigation will not usually prove very successful. There are too many unknown factors in connection with the movements of the victim and the person doing the shooting at the time of a crime for these patterns to be of any great importance, even if they could be interpreted correctly.

Shotguns

One of the reasons that Americans became free and maintained that freedom was because there was at the time of the Revolution a smoothbore or rifle in every home. The American flintlock rifle was always expensive, delicate, and reasonably rare; however, a strong sturdy flintlock gun capable of firing anything from a single ball down to a charge of tiny shot was a necessity. Throughout our history the integrity and solidarity of rural America has rested squarely upon these rather crude but tremendously powerful weapons

DOUBLE BARREL SIGNATURES

This is a view of the standing breech (or breechface) of a high-grade, ejector, double barrelled shotgun. This arm is in perfect condition and has been well cared for.

Observe how the automatic ejectors have worn smooth and thereby imparted individual characteristics to the area of the breech against which they sweep. Then observe the holes in this breech through which the two hammer noses (firing pins) strike; the one on the left (the right barrel) has been worn a trifle larger and oval shaped, due to that hammer nose not perfectly centering when it strikes.

Above each breechface is shown a primer fired by that barrel's mechanism. Observe what a great difference there is in the primer imprints made by these two firing mechanisms fitted to the same gun. Examination under the microscope would show many more identification markings but enough already exist to be noticeable by the eye of an experienced investigator.

and the men who owned them. One of the authors recalls in his youth a situation that developed in a small South Carolina town. Some outside labor organizers threatened to take over private property. The next morning a group of local farmers arrived in town with their shotguns. They were not for either side but private property was sacred; the organizers left hurriedly. It is remarkable today that those regions where the family shotgun stands in a corner of the living-room, or on hooks over the mantel in the style of our ancestors, are almost completely free of organized crime.

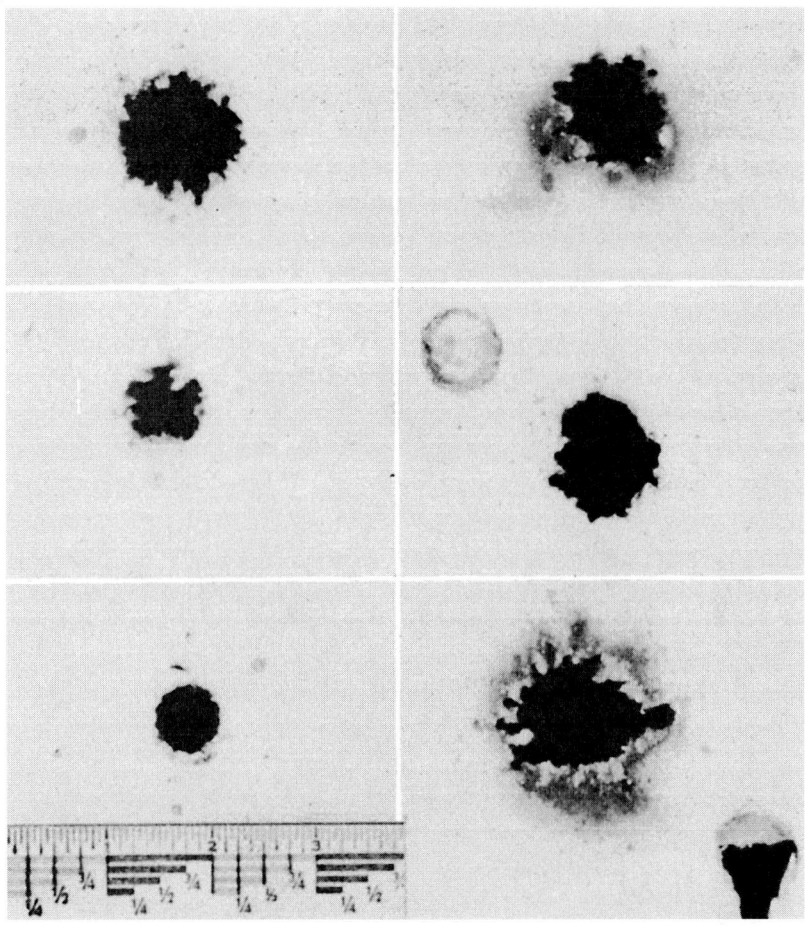

Various Shotguns and Loads and Their Patterns at 24 Inches.

1. A 12-bore full choke shotgun fired an ounce and a quarter of No. 7 shot.
2. A .410-bore full choke shotgun fired this with No. 7½ shot.
3. A 12-bore full choke shotgun fired this with No. 00 buckshot.
4. A 12-bore cylinder choke shotgun fired this with No. 4 shot.
5. A 12-bore shotgun caused this pattern with a Remington rifled slug.
6. This pattern was caused by 2 ounces of No. 4 shot from a 10-bore magnum shotgun.

General Comment: It will be seen from examination of these patterns that it is frequently not possible to determine accurately either bore or shot size from the pattern alone, although in most instances, the shot will be recovered both in the victim and from the surroundings.

Neither tyranny nor crime flourishes where citizens both own and can use weapons.

As pointed out already, although almost all shotguns are purchased today for sporting purposes in which they are designed to kill birds and small animals, they function at close range as perhaps

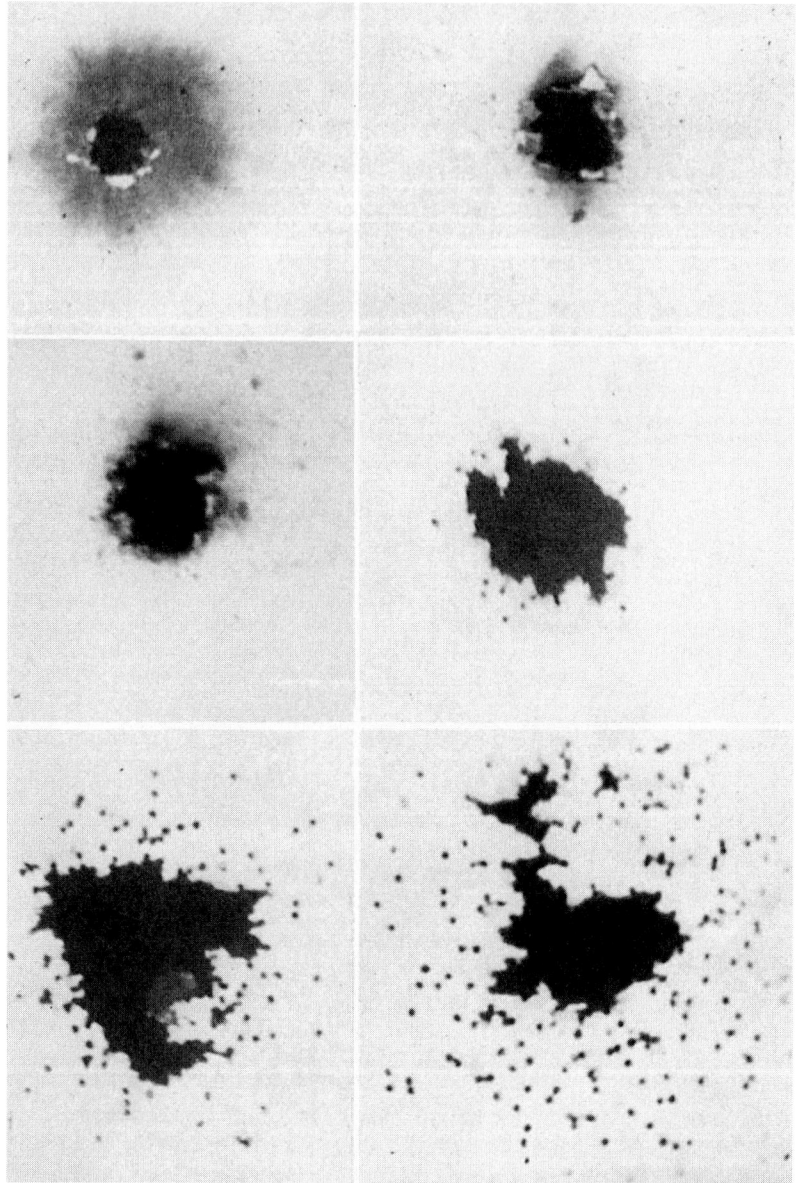

Six different patterns at different ranges, fired with a 20-bore pistol with standard commercial ammunition, ⅞ of an ounce of No. 8 shot. The barrel of the pistol is 10″ long and true cylinder throughout its length, which would about be the case where the barrel of a standard shotgun is so sawed off.

    1. Range 1 inch. Note the powder blackening visible even in standard light.

    2. Range 6 inches. Powder marks limited in black and white photography to unburned grains.

(Continued at bottom of opposite page.)

the most powerful weapon obtainable. All that is necessary to make the family shotgun into a veritable bazooka is to reduce the range sufficiently so that the charge of light shot, wadding, and gases tear into the body of the victim. Relatively few crimes are committed in this manner, and many of these are partially justified. The professional criminal is seldom involved. There are, unfortunately, a fairly large number of accidents with shotguns, but not nearly so many as with automobiles, farm machinery, or in the home.

Shotgun Projectile Identification

In shotgun crimes, all projectiles should be carefully retained, although the chance of learning something definite from them is limited. It was thought in Germany before World War II that the outside pellets in contact with the barrel of a shotgun would be sufficiently deformed to determine at least the diameter of bore. If this can be done, it can only be done at the expenditure of a tremendous amount of labor which is usually not warranted; the information is not particularly valuable. The shot itself will, of course, tell you the size loaded into the shell; the wads where recovered at the scene of the crime, or where the range is very short in the wound itself, will give an indication of the make of ammunition. Usually, little or nothing can be proved by pellet evidence. Knowing the manufacturer of the shot or shell proves nothing in itself. As a general rule, the Expert will not be frequently called upon to work on shotgun projectiles other than to identify the shot size and to estimate the range from the pattern made on the victim or his surroundings. This latter problem will be discussed presently.

Shotgun Shell Identification

Frequently, expended shotgun shells are found at the scene of a crime. The most obvious reason for this is that an automatic shotgun was used or that more than one shot was fired from a pump action shotgun or from some other form of repeating weapon.

---

3. Range 12 inches. The shot charge with this short barrel is beginning to disperse.
4. Range 24 inches. The powder blackening and particles of powder are no longer clearly visible in black and white photography. A few of the outer pellets are beginning to leave individual holes.
5. Range 48 inches. Outer pellets, because of complete lack of choke, are now considerably dispersed.
6. Range 10 feet. We now have a dispersion of considerable magnitude.

SINGLE BARREL OR DOUBLE? WHICH?

It is possible, in many instances, to determine from the empty shotgun case—*if* it was fired in a hinged frame shotgun—whether that gun was a single or a double barrel.

In a single barrelled shotgun the extractor or ejector (the same mechanism in either case) is positioned centrally, underneath the chamber, and is from ⅜" to ½" in width. This mechanism is fitted more or less loosely and it wears greatly from use. When the shell is fired the pressure from the explosion expands all portions of the case, and the brass adjoining the rim bulges out and fills the extractor segment as that extractor more-or-less gives away. Two such fired cases are shown above at the left, the first one being fired in a single barreled gun having an extractor ⅜ths of an inch in width and the next one from a single barreled gun having an extractor ½ inch in width.

In an 'ejector' double barrelled gun the ejectors are companion affairs, fitted separately to each barrel and working independently from each other, or in unison, according to whether one or both barrels are fired. On a 'non-ejector' double gun the extractor(s) is a combined affair, functioning on both barrels every time the breech is opened.

This principle of expansion of the brass works the same with either single or double gun (and some very expensive doubles will permit considerable expansion). However, on most double guns this extracting (and ejecting) mechanism operates against a much wider segment of the rim of the cartridge case, just about one-third of its circumference in fact; it bears against both the bottoms of the cases and up around their inner edges; it has about three times the bearing surface of the single barrel extractor-ejector. Two cases, fired in a double gun, are shown above at the right.

Occasionally, it is possible to tell from an examination of the shells from an old double gun which one had been fired in the right, and which in the left, barrel. These swellings in the brass of fired cases are caused by the extracting segment yielding under pressure of the expanding powder gases and from the back thrust of case. There is generally less side-play than down-play, thus the swelling in the brass often will be more pronounced at the bottom than at the side—which fact is shown in the two cases at the right, each of which is shown in the position in which it came from the gun; note how the swellings decrease on their inner edges. Also, on some old and extremely worn guns it may be possible to tell the order in which the barrels were fired—the shell which shows the greatest amount of swelling is from the barrel that was fired first.

Not all top-break shotguns will show these swellings and not all makes or lots of cases will swell, even in a gun which generally shows this condition. The investigator should be sure to try out test cases of the same make and loading and to fire a series of them before arriving at any decision.

However, this is not always the case since many sportsmen have acquired by careful training the almost habitual practice of immediately expelling an expended shell and reloading the weapon without thought. A criminal can hardly make a more grievous mistake than to leave an expended shotgun shell at the scene of a crime; however, because of the extreme nervousness undergone, particularly by the non-habitual, he will often do just that.

Unfortunately, for identification purposes, the pressure inside the chamber of a shotgun is far lower than that encountered even in handguns. This is a particular disadvantage in "no-gun" work since the family characteristics are even fainter than breechface and firing pin impressions from pistols. These markings, however, can be extremely illuminating in some cases. There are family characteristics common to all weapons of a certain make and model of pump gun. Similarly, automatics have family characteristics. Frequently, double and single-barreled weapons of the usual top break variety will be impressed by the extractors so as to show clearly the type of gun from which they were fired. Occasionally, a firing pin will stick in a gun of this type and leave a radial gouge in the base. A shell which killed a New Jersey State Trooper is illustrated with a marking of this type. The Identification Laboratory will not want to state definitely in any given instance the exact make and model firing a crime shell; however, no harm at all will be done to your prestige if you state a probability that turns out to be correct.

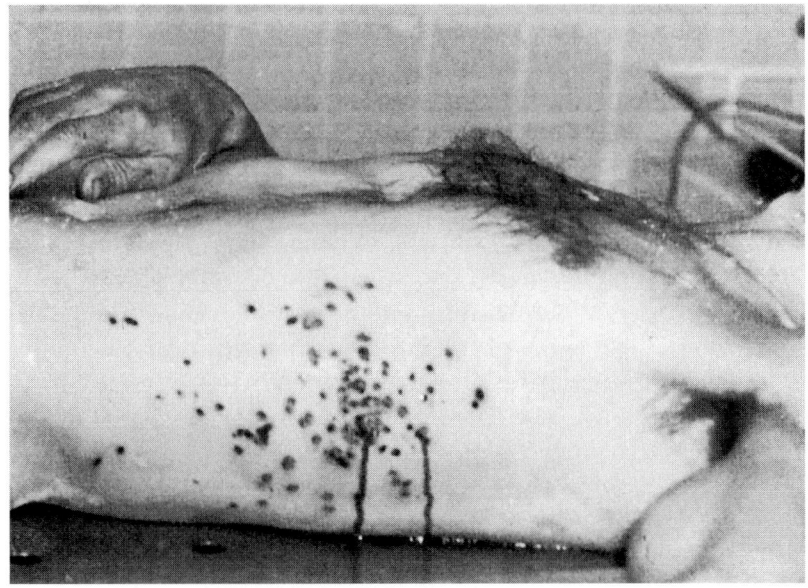

WOUND EFFECTED BY SMALL SHOT

This man was killed by a 12 gauge, full choke, shotgun, firing an ounce of No. 6 shot at a distance of 14-15 feet. Some of the charge grazed the underside of the left forearm and a portion of the load was lost between the arm and body. (From State of Alabama, Department of Toxicology and Criminal Investigation).

Range Determination From Shotgun Patterns

The authors have done rather exhaustive experimentating in connection with this subject. A representative group of patterns is reproduced with explanatory captions. We want to emphasize, however, that whenever possible, a new set of firings be made with the suspect weapon using, if possible, ammunition of the same make and lot as that used in the crime. Shotguns will vary considerably between themselves in their choking, barrel length, gauge, and the like. Further, certain types of ammunition will give different patterns. Because of the relatively small size of the average human target, ranges of beyond about 10 yards usually cannot be accurately determined since one may, or may not, have the center of the pattern. Where a victim stood near a flat surface that shows the entire shot charge, save for the pellets hitting his body, this naturally does not apply.

A number of conclusions will readily be drawn from the patterns indicated. For instance, gauge and shot size mean little at close range, whereas choke, although of extreme importance in killing a partridge at 30 yards or breaking a clay bird at 40 yards, never is of any great importance within the lethal range of the average shotgun.

A few words in connection with buckshot are important. Since there are only nine pellets in a 12-gauge charge of No. 00 buck, the pattern thrown by such a weapon so loaded is not good. It is only within about 35 yards that anyone is reasonably certain of hitting and killing a man-sized target. Although buckshot from a gun of this type may penetrate a skull at a range of 100 yards, it could not possibly be considered accurate at anything like this range. Patterns on human targets mean little beyond about 10 yards.

Occasionally, a crime will be committed with reasonably unusual shot from a shotgun. Actually, buckshot in many places is of this type, although because of the deer hunting laws in New Jersey, almost anyone may buy from time to time 12-bore 00 buck cartridges. Other loadings are rare, even here. In many sections, however, not more than one in a hundred sportsmen will ever buy any buckshot at all. Further, not many stores will stock it. Exhaustive detective work can lead to identification of buckshot purchasers of a particular variety. This will very greatly reduce the number of suspects. This, however, does not apply to many sections of the South where buckshot is the accepted deer load and may be carried for emergencies even by bird hunters.

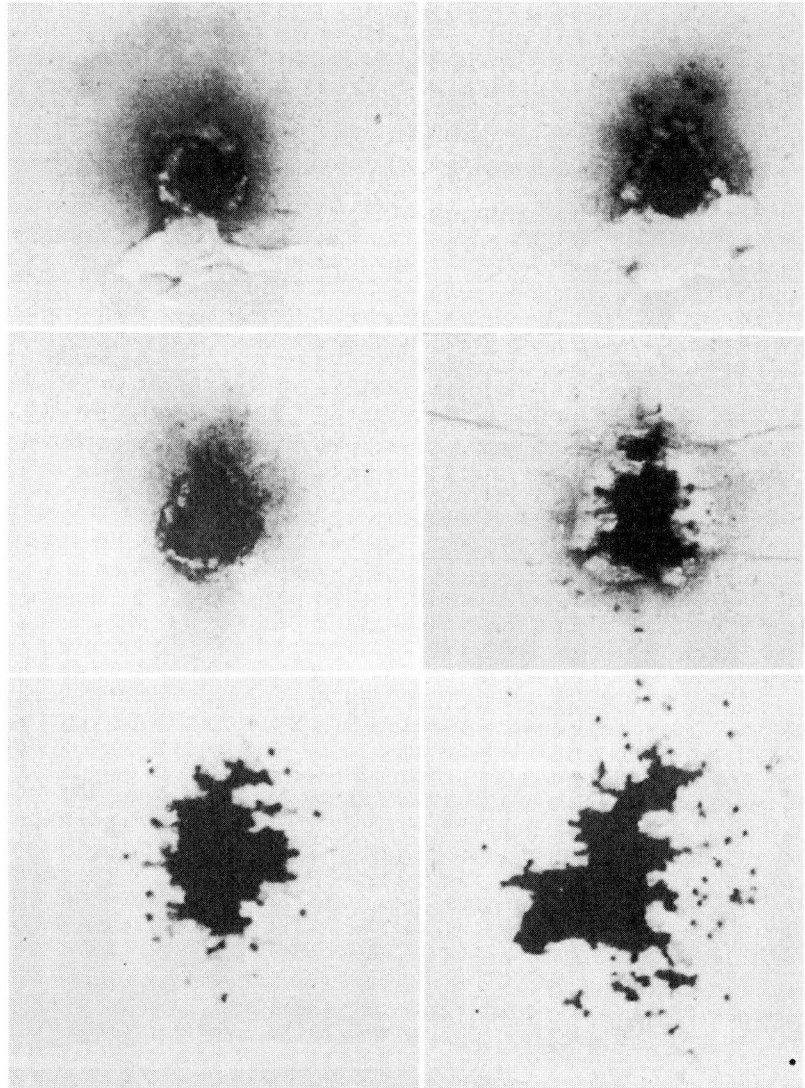

Six similar patterns with a 20-gauge shotgun with a 28" full choke barrel. Same loading as tests on page 334. The ranges are similarly: 1 inch, 6 inches, 12 inches, 24 inches, 48 inches, and 10 feet.

It will be seen that the full choke barrel allows far less dispersion at 10 feet. Although the first three ranges are remarkably similar, particularly in Views 2 and 3, the recoil of the longer barreled weapon has caused an inverted pear-shaped pattern. This weapon was intentionally held fairly loosely in order to endeavor to get this particular phenomena, which may be of importance in establishing the position of the weapon when fired. The trailing shot of the charge are discharged from the long barrel after this barrel has recoiled slightly upwards, as opposed to the main shot charge. This phenomena may or may not be present in a specific weapon, ammunition, and method of holding. It, of course, would not be present if the weapon were fastened down rigidly in certain bench rests used for obtaining these patterns.

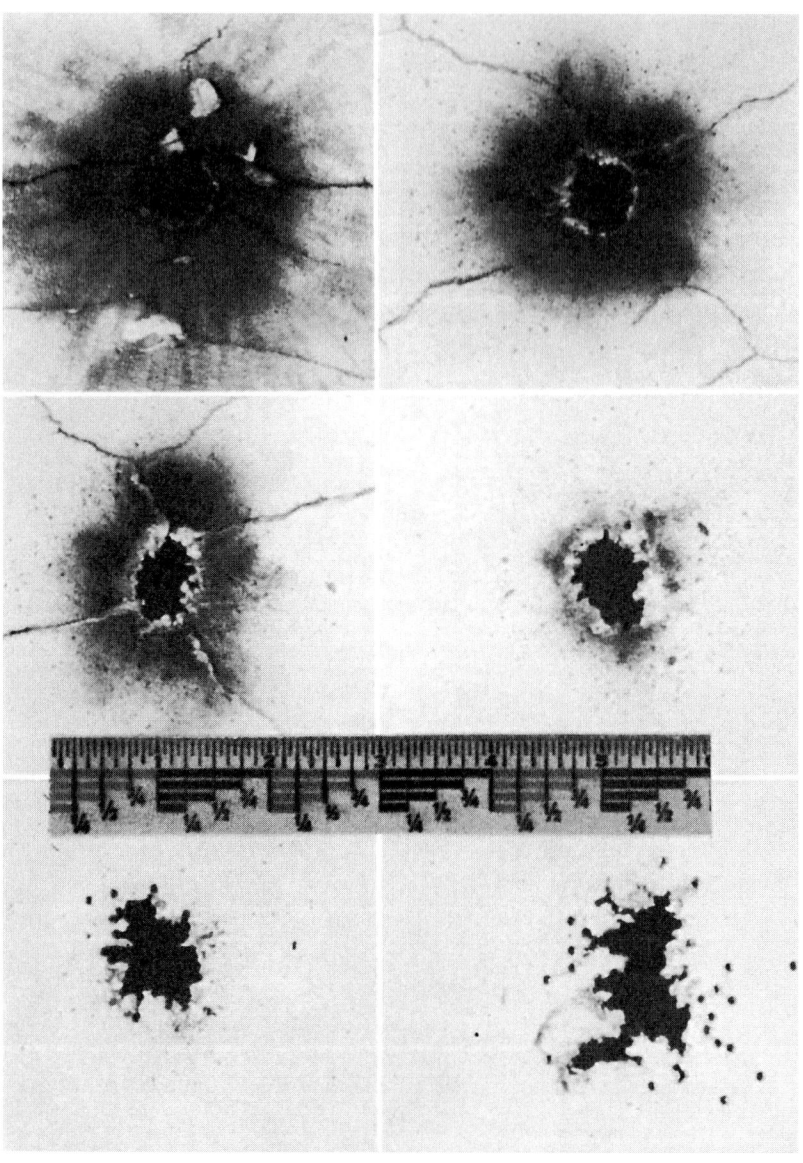

12-Bore Shotgun Patterns from a Full Choke Barrel at Various Distances with No. 7½ Shot.

These distances are respectively and in order: 1 inch, 6 inches, 12 inches, 24 inches, 48 inches, and 120 inches. It is, of course, extremely important in all range determination and other work on shotgun patterns, to be certain that you have either an identical shotgun and ammunition or, better still, the actual weapon itself whose use you are investigating and ammunition of the same make and lot number.

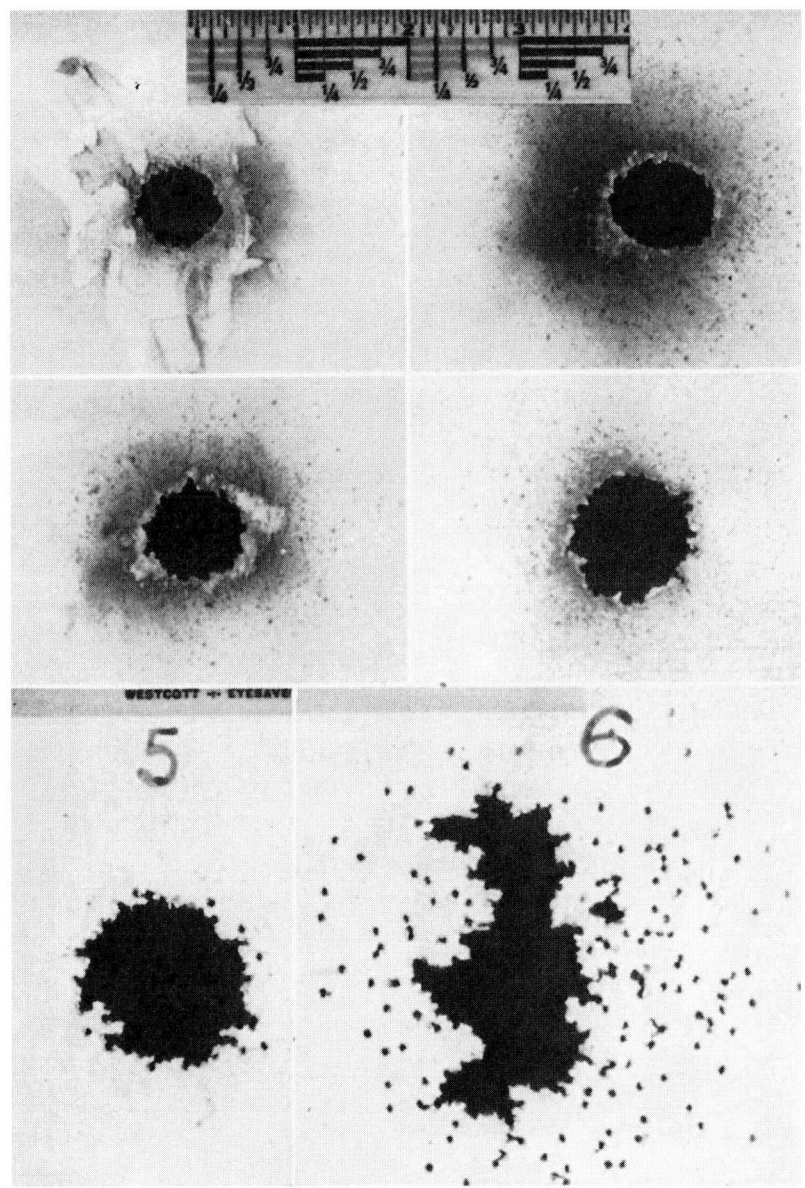

Pattern from a 12-Bore Shotgun with Cutts Compensator Attached and the Most Open Tube in Place, with 7½ Shot, at Various Ranges from the End of the Tube. This tube is called a spreader and was used for skeet shooting and the like. The distances are respectively and in order: 1 inch, 6 inches, 12 inches, 24 inches, 48 inches, and 120 inches.

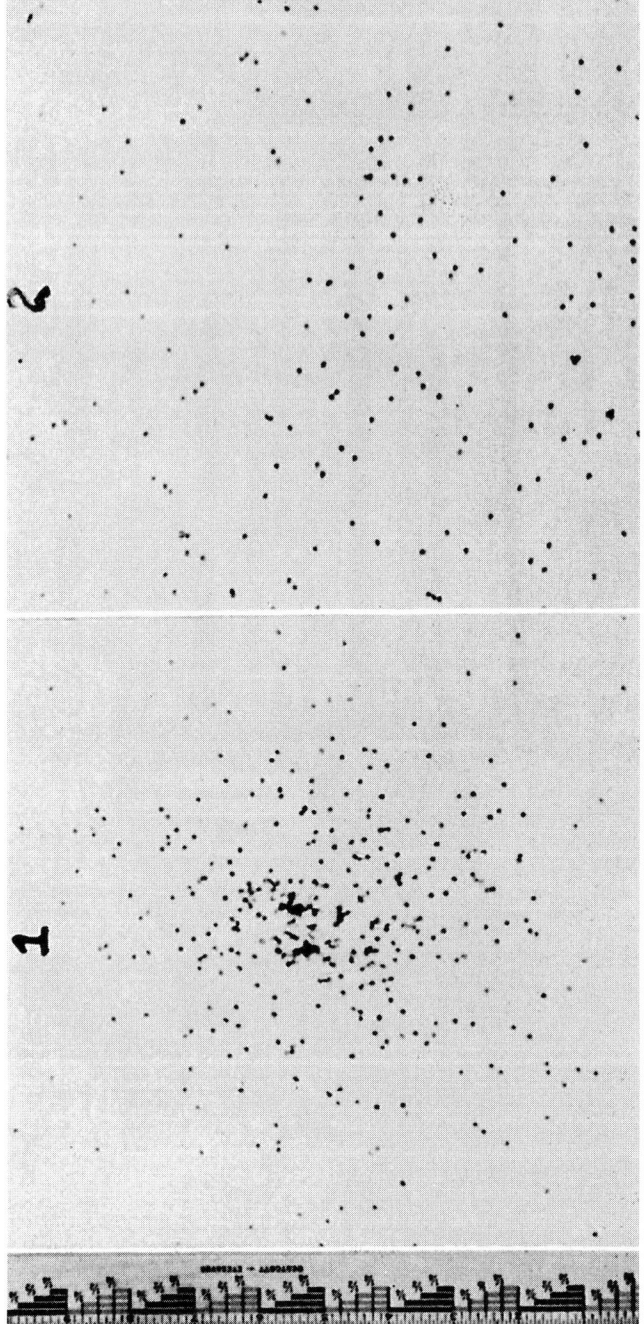

Patterns from Various Shotguns Using 1¼ Ounces of 7½ Shot at a Range of 45 Feet from Muzzle to Target.
1. A 12-bore full choke.
2. A 12-bore Cutts Compensator with spreader tube.
3. A 20-bore full choke.
4. A .410-bore full choke.

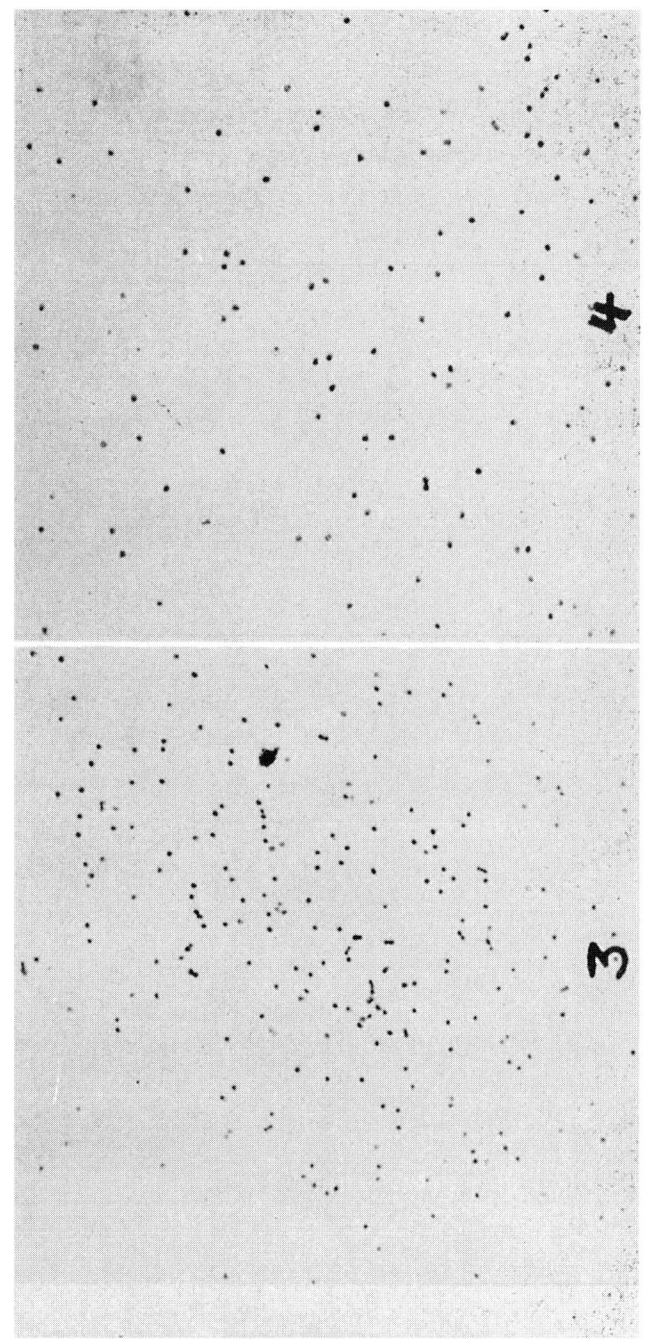

It will be seen from these patterns that any definite statement as to range beyond 45 feet is likely to be predicated on the amount of the pattern which you have. The squares shown are intentionally only 9″ x 9″ since this is believed to be about as large an area as will be usually exposed by most human bodies in the positions in which they are shot. However, where a wall or the like is hit, considerably more accurate range determination is possible.

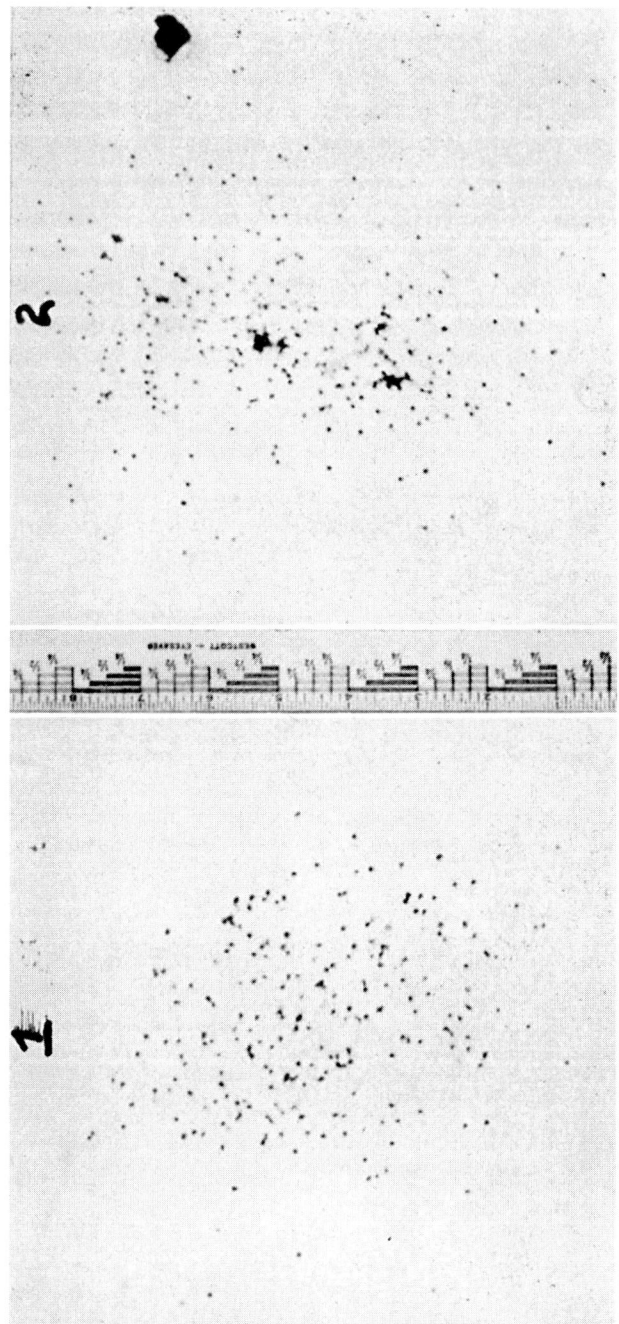

Patterns of Various Shotguns with 7½ Shot at a Range of 20 Feet from Target to Muzzle.

1. A .410-bore full choke.
2. A 20-bore full choke.
3. A 12-bore with Cutts Compensator and spreader tube.
4. A 12-bore full choke.

Obviously quite accurate determination of range can be done where the weapon and ammunition are known, when the distance is of this order and a reasonably large area is exposed to the charge.

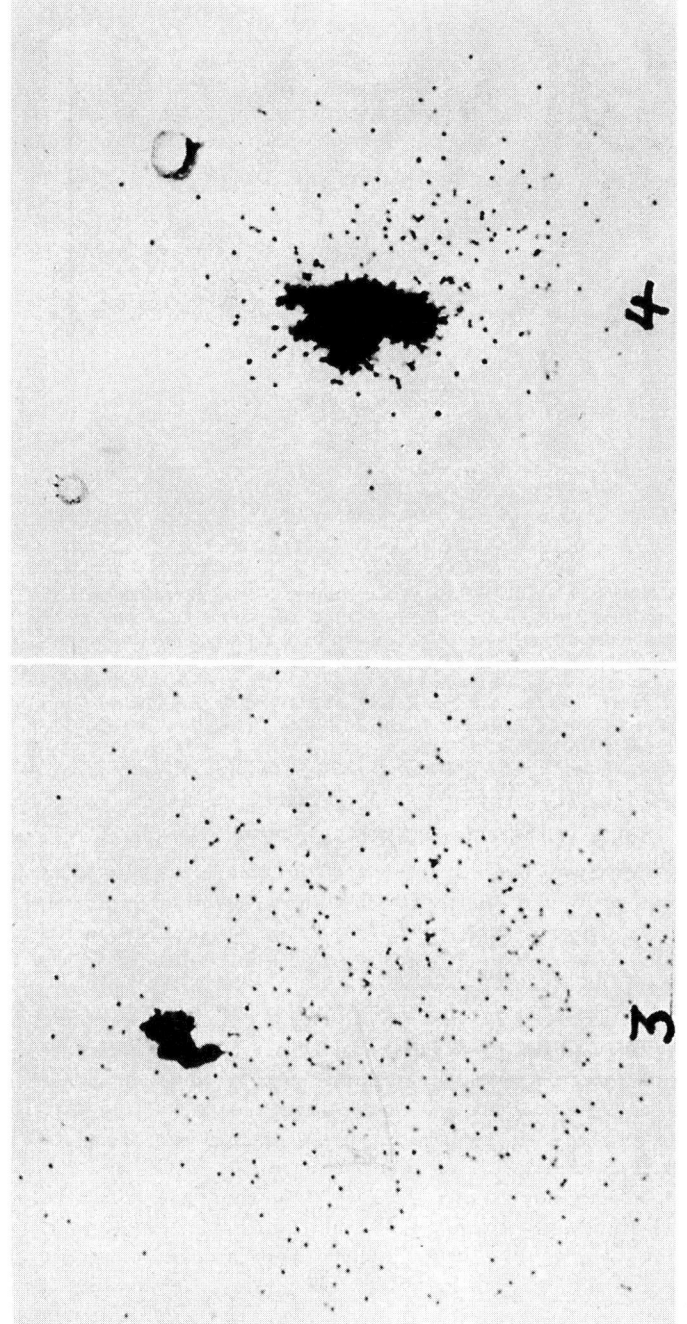

It is important, however, to make certain that the pattern from which you are working actually does contain the center of the charge. Where a wound on a human body is at the edge of the body or where more than one shot has been fired, the expert should be very reticent in making any positive range determination unless he has examined for himself not only the evidence, such as a pattern in a piece of clothing, but also the way the piece of clothing was exposed to the charge of shot at the time of the shooting. In most instances, a wound of this type cannot be satisfactorily evaluated unless the expert spends some time investigating the crime itself or in examining pictures and discussing with the detectives in charge the various possibilities. Remember that all things of this type are difficult because of the tendency of human beings faced with death to do extremely unusual things.

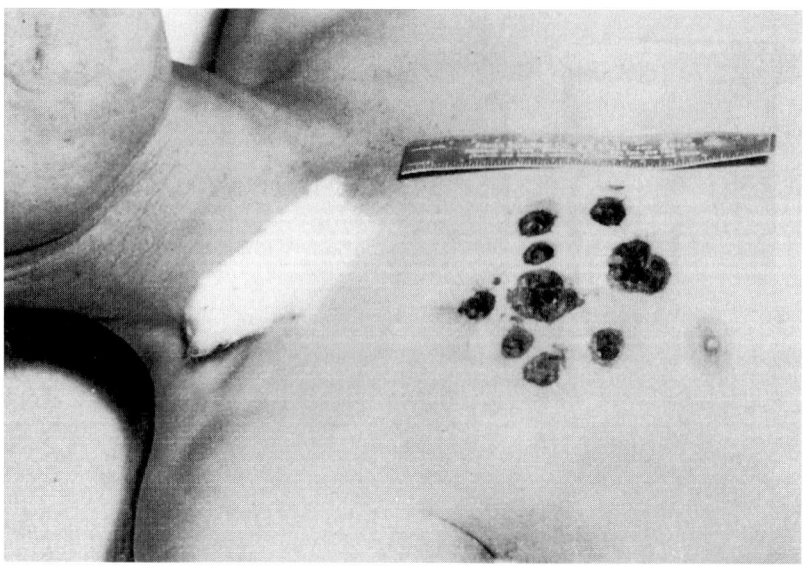

A BUCKSHOT WOUND

This fatal wound carried complete evidence as to the gun used, the cartridge, and the range at which it was fired. This pattern indicates a range of 12-15 feet. The buckshot (taken from body later) was No. o size; there were 12 pellets in all and the entire pattern entered the body; 12 pellets of No. o buckshot are loaded only in 12 gauge shells.

These findings were verified by the facts ascertained: The gun used was a double barrel 12 gauge shotgun with barrels cut down to 27 inches; Ammunition was Winchester Super Speed No. o buckshot loading. (From State of Alabama, Department of Toxicology and Criminal Investigation).

APPENDIX 1

Bullet Weights Arranged Progressively

The authors are extremely reticent to give tables of bullet weights since they lay themselves open immediately to all kinds of errors of omission. Many cartridge collectors will have bullets that we do not list. Some Firearms Experts will have encountered in actual crime, bullets that are not included. However, we feel that some form of listing is necessary and have therefore evolved the four tables which follow, endeavoring to show in these tables all bullets of American factory manufacture made within the past 20 years. Obviously, in some cases a complete list is impossible, even of American-made bullets, when one takes into consideration the various smaller commercial firms turning out these for handloaders. We have tried to use common sense in this matter and not clutter up the table with rarities and special bullets. Handloaders are, by the very nature of things, not criminals. We have further made no effort whatsoever to list weights of foreign bullets. For instance, it would probably be possible for the H. P. White Laboratory to make up a 9mm bullet-weight chart which would show more than 100 different specimens.

Our purpose has been to list the weights of all reasonably common bullets of American manufacture. Foreign ammunition is similar; the possibilities for variation are limitless.

TABLE I—LEAD BULLETS; PISTOL AND LOW POWER RIFLE

Approximate Weight in Grains	Caliber	Cartridge	Description and Remarks
15	.22 r.f.	Short	Disintegrating bullet; gallery use.
18-20	.22 r.f.	Bulleted Breech Caps	Round ball; rare in crime.
20-30	.22 r.f.	Shorts and C.B. Caps	Various types of disintegrating bullets; used in shooting galleries.
27-28	.22 r.f.	Short and Long	Hollow point; standard weight, all makes.
29-30	.22 r.f.	Short and Long	Regular solid point, sometimes loaded into C.B. caps also.
36-37	.22 r.f.	Long Rifle	Hollow point.
40	.22 r.f.	Long Rifle	Regular solid point.
45	.22 w.r.f.	Special Case	A rare cartridge that will not chamber in usual .22 rimfire weapons.
44-47	.22 r.f.	Special Automatic Case	Rare today; .22 automatics now take regular ammunition.
35	.22 r.f.	Extra Long	Special bullets for longs, now obsolete; today loaded same as shorts.

Notes on .22s:

1) The bullets are greased, waxed, or metal plated according to the maker's fancy; sometimes they are inside lubricated only.
2) Variations, particularly in weights of bullets loaded several years ago are common; standard loadings of the early 1930s frequently called for hollow points to weigh the same as solid points for the same cartridge.
3) During World War II Remington made .22 Long Rifle ammunition for the Government with full metal jackets.

65-68	.25 r.f.	Short and Long	Same lubricated bullet used for short and long; rare in crime today.
86	.25 c.f.	.25-20 (.25 WCF)	Seldom used in revolvers but a popular rifle cartridge in the old days.

Notes on .25s:

There are very few standard revolvers chambered for any .25 caliber cartridges; most automatics of this size regularly take jacketed bullets.

Approximate Weight in Grains	Caliber	Cartridge	Description and Remarks
80	.32 r.f.	Short	Lead or lubaloy coated; rare in crime today.
80-82	.32 centerfire	Short or Long Colt	These are similar to but not interchangeable with the S&W cartridges.
85-88	.32 centerfire	Smith & Wesson	An extremely common cartridge used in many cheap crime revolvers.
90	.32 r.f.	Long	Lead or lubaloy coated; rare in crime today.
95-98	.32 centerfire	Smith & Wesson Long	Rarer than above in crime today.
98-100	.32 centerfire	Colt New Police	Also called Police Positive; interchangeable with .32 S&W Long.
100	.32 centerfire	.32-20 (.32 W.C.F.)	Rifles and heavy revolvers.
115	.32 centerfire	.32-20 (.32 W.C.F.)	Originally intended for rifles but a powerful and accurate load in the old Colt Frontier.

Notes on .32:

These are difficult to report on with a bullet only, because of the various chamber shapes and the possibility of slightly mismatched ammunition. For instance, Smith & Wesson chambers will take Colt cartridges but not usually vice versa.

125-130	.38 c.f.	Short Colt	Rare in crime.
150	.38 c.f.	Long Colt	Obsolete and dangerous; chambers will usually take .357 **Magnum** cartridges which will usually blow up weapon.
130-200	.38 c.f.	Smith & Wesson	Also called .38 Colt Police Positive and New Police; 146 grain most common.
148	.38 r.f.	Short and Long	Very rare today. There may have been lighter bullets also.
146-200	.38 c.f.	Smith & Wesson Special	Extreme variety of target and special loadings; 158 grain most common. Usually just called ".38 Special."
158	.38 c.f.	.357 Magnum	Ultra-powerful; other bullet weights non-standard.
158	.38 c.f.	.38-44	A special and obsolete cartridge, longer than .38 Smith & Wesson Special.

Notes on .38:

Again mismatching is possible. The .38 Short Colt will chamber and usually fire in any of these weapons; however, the .38 Smith & Wesson will not usually chamber in any revolver designed for other cartridges. The bore diameters of weapons chambered for these cartridges are about .355" to .360".

Approximate Weight in Grains	Caliber	Cartridge	Description and Remarks
130	.41 r.f.	Short	Obsolete; usually for deringer type weapons; bore diameter about .390".
160-165	.41 c.f.	Short Colt	Obsolete; extremely rare; bore about .385" diameter.
180	".38" c.f.	.38-40 (.38 W.C.F.)	The .38 designation is a misnomer; bore diameter usually approximately .400".
195-200	.41 c.f.	Long Colt	The most common of this group, but still rare; bore about .385" diameter.

Notes on .41s:

These are seldom encountered in crime; no standard weapons have been chambered for these cartridges in many years.

115-120	.44 c.f.	Marble Game Getter	Spherical lead ball; rare in crime.
200	.44 r.f.	Henry	An obsolete cartridge developed in the early days of our Civil War.
200	.44 c.f.	.44-40 (.44 W.C.F.)	For rifles and revolvers; bottlenecked case.
246	.44 c.f.	Russian	Relatively obsolete.
205-220	.44 c.f.	American	Relatively obsolete.
246	.44 c.f.	Special	A fine powerful cartridge when handloaded.

Notes on .44s:

The American and Russian cartridges will chamber in .44 Special revolvers. Some .44-40 chambers will accept all three of these. Bore diameters about .427"-.430". There were also other .44 caliber cartridges for Webley and other "Bull Dog" revolvers with bullets of from 165 to 210 grains. Only the .44 Special revolver is still made; next to the .41s, this is the rarest class used in crimes.

Approximate Weight in Grains	Caliber	Cartridge	Description and Remarks
230	.45 c.f.	Auto Rim	Usually made for the 1917 Revolvers since lead bullets do not function well in the Model 1911 Automatic; the Auto Rim cartridge functions in these revolvers without clips.
250-255	.45 c.f.	Colt	A long, powerful but obsolete revolver cartridge still with us in large numbers particularly in some sections of the country. Two lengths of case may be encountered.
255-265	.455 c.f.	Webley	British Service revolver now common and cheap in this country.

Notes on .45s:

There are several lighter lead bullets sometimes used; however, the crimes committed with the Service Automatic usually involve metal jacketed bullets. There are also some other .45s such as the Smith & Wesson Scofield and the .450 Webley however, they are collector's items.

TABLE II—LEAD BULLETS: MEDIUM AND HIGH POWER RIFLES

Foreword:

No effort will be made to classify all the handloaded cast bullets with or without gas checks used in rifles ordinarily fired with jacketed bullets. All that remain, therefore, are the obsolete military and sporting arms of the .70s and .80s. These are now cheap; frequently a few old rounds accompany a gun.

Approximate Weight in Grains	Caliber	Cartridge	Description and Remarks
255	.38 c.f.	.38-55	Still fairly well thought of as a sporting arm.
300, 405, & 500	.45 c.f.	.45-70	These three weights were on occasion all used with the .45-70 Government case as well as the .45-60, .45-75, and .45-90 cases. The 405 grain bullet is most common.
350	.50 c.f.	.50-70	A fine cartridge for maximum stopping power on two-legged animals.
300	.41 r.f.	Swiss	For some reason the most common of all these old foreign military arms in this country; for years the only powerful rimfire cartridge made in America.
375	.43 c.f.	Spanish	Also called 11mm. Other noninterchangeable 11mms were the German Mauser and the Egyptian (Beaumont).

Notes on Table II:

In addition to the calibers mentioned, there are literally scores of sporting and imported military weapons using the same sort of ammunition. These are all rare or unknown in crime.

TABLE III—METAL JACKETED BULLETS; PISTOL AND LOW POWER RIFLE

Approximate Weight in Grains	Caliber	Cartridge	Description and Remarks
50	.25 c.f.	A.C.P.	Also known as 6.35mm; a common crime bullet quite uniform between different makes.
86	.25 c.f.	.25-20	Metal jackets are usually open in some way at point.
71-74	.32 c.f.	A.C.P.	Also known as the 7.65mm; very often used by criminals.
76	.32 c.f.	".35 Smith & Wesson Automatic"	This is a metal pointed bullet designed for the obsolete Smith & Wesson automatic pistol which will function with the standard .32 A.C.P. ammunition.
80	.32 c.f.	.32-20	Light bullet for high velocity; usually mushroom points.
85-86	7.63 c.f.	Mauser	The big Mauser pistol bullet; very powerful.
86	.32 c.f.	Smith & Wesson Short	Metal pointed ammunition for revolvers chambered for the .32 Smith & Wesson cartridge.
93	7.65 c.f.	Luger	The .30 caliber Luger used in commercial pistols and carbines; bottle-necked cases.
95	.38 c.f.	.380 A.C.P.	Bullet variations common; many military loadings.
95-100	.32 c.f.	Smith & Wesson Long	Same as above but for the .32 Smith & Wesson Long.
100	.32 c.f.	.32-20	Both solid and exposed points.
115	.32 c.f.	.32-20	Both solid and exposed points.
125	9mm c.f.	Luger	There are very many foreign 9mm bullets of this general type, some a good deal lighter.
110-200	.38 c.f.	All Revolvers	For many years loading companies have been putting up ammunition for these revolvers with metal pointed and full metal jacketed bullets for various specific purposes or to conform to treaties for military use. See your ammunition collection for possible identification. These bullets are not common in crime; some of the metal piercing varieties are not supposed to be sold save to police organizations.

Approximate Weight in Grains	Caliber	Cartridge	Description and Remarks
140	.44 c.f.	.44-40	High velocity, open point.
180	".38 w.c.f."	.38-40	Solid and open points.
200	.44 c.f.	.44-40	Solid and open points.
246	.44 c.f.	Special	Metal point; rare and obsolete.
173	.45 c.f.	A.C.P.	Metal penetrating high velocity bullet.
185	.45 c.f.	A.C.P.	Target wadcutting bullet.
200	.45 c.f.	A.C.P.	Old military and commercial bullet.
230	.45 c.f.	A.C.P.	New military and commercial bullet.

TABLE IV—METAL JACKETED BULLETS: MEDIUM AND HIGH POWER RIFLES

Foreword:

These are relatively rare in intentional crime. Further they are seldom recovered intact save at extreme ranges. The usual fragments will tell neither the original bullet weight nor the diameter. We are following a slightly different procedure in arranging this data than in the previous three tables.

Caliber	Cartridge	Usual Bullet Weight	Remarks
.22	Hornet	45 & 46	These are the only high velocity .225 for which ammunition is factory loaded. However, there are at least a dozen other Wildcat .22 cases. Handloaders use many different jacketed bullets varying in weight from 35 to 70 grains. Bore diameters are usually .223" to .226".
	Bee	46	
	Zipper	46 & 56	
	.220 Swift	46 & 48	
	Savage Hi-power	70	
	.222 Remington	50	
.25	.257 Remington-Roberts	87, 100 & 117	There is a good deal of handloading of these cartridges also. Bullets usually available weigh about as shown. Bores usually .257" in diameter to bottom of grooves.
	.250-300 Savage	87 & 100	
	.25-35	117	
	.25-36 Marlin	117	
	.25 Remington	117	
	.256 Newton	129	

Caliber	Cartridge	Usual Bullet Weight	Remarks
6.5mm	6.5 Jap	138	For the small bore Japanese rifles and machine guns.
	6.5 Rimmed	160	For military 6.5mm rifles particularly the Dutch and Rumanian Mannlicher.
	6.5 Rimless	139, 140, 150 & 160	For Mannlicher-Schoenauer and other Sporting Rifles. Fairly popular today.
.270	.270 Winchester	130 & 150	Very popular modern cartridge. Lots of non-standard bullets. Bore diameter usually .278".
7mm	Mauser	138 & 175	Still a popular hunting rifle. Special bullets usually lighter. Bore diameter usually .285".
	Mannlicher	138 & 175	
	.275 H & H Magnum	175	Rare cartridge in this country.
	.280 Ross	145	Also other bullet weights; rare.
.30	.30-40 Krag	150, 180 & 220	The old Krag is still common in some sections.
	.30-30	110 to 180	Commercially loaded in at least seven bullet weights before World War II.
	.30 Remington	110 to 180	Also formerly available in six bullet weights.
	.300 Savage	150, 180 & 200	Popular sporting caliber.
	.30-06	110 to 220	There are literally dozens of weights of jacketed bullets for this rifle.
	.303 British	174 and 215	Sometimes encountered in other weights.
	.303 Savage	180, 190 & 195	Really a .30 caliber.
	Cal .30 carbine	110	Government made.
	.300 H & H Magnum	180, 220 & 225	Powerful sporting caliber.
	7.62 Russian	145 and 150	Not common in this country.
	7.65 Mauser	216-219	For Belgian and other military rifles.
	.30 Newton	170, 180	Rare today.

Caliber	Cartridge	Usual Bullet Weight	Remarks
.32	.32 Winchester Special	110, 165, 170 & 180	A popular lever action cartridge.
	.32-40	165	Rarer than .30-30 or .32 Special.
	.32 Winchester Self-loading	165	For blowback automatics.
	.32 Remington automatic	110, 165, 170 & 180	Popular sporting caliber.
	7.7mm Japanese	about 175	Several variations.
8mm	Mauser & Mannlicher	154, 170, 200 & 236	Extreme variation possible.
	Lebel	170, 198	Sometimes solid bronze bullets.
.33	Winchester	200	Rare today.
.35	.348 Winchester	150, 200, 210 & 250	Powerful, for lever actions.
	.35 Winchester Self-loading	180	These are all fairly rare sporting calibers, save the .35 Remington which is quite popular.
	.35 Remington	150, 200	
	.351 Winchester Self-loading	177 & 180	
	.35 Winchester	250	
	.35 Newton	250	
Larger	.375 H & H Magnum	235, 270 & 300	Most powerful commercial American rifle.
	9mm Mauser	300, 280	Mauser & Mannlicher; rare.
	.38-55	255	
	.38-56	255	
	.38-72 Winchester	275	Unusual and obsolete.
	.40-65	260	Unusual and obsolete.
	.40-72	300 & 330	Unusual and obsolete.
	.40-82	260	Unusual and obsolete.
	.401 Winchester Self-loading	200 & 250	Unusual and obsolete.
	.405 Winchester	300	The "Teddy Roosevelt" rifle.
	.45-70	405	Usual smokeless loading bullet.
	.45-90	300	For the Model 1886 and others.

APPENDIX 2

Rifling Characteristics

FOREWORD: The authors are reluctant to handle this matter in a new way. However, much of the data published heretofore has been necessarily incomplete and of relatively little use. We refer particularly to many tables in which widths and depths of grooves, as well as pitch of rifling, is given in a great detail. These things are extremely difficult to measure in connection with most crime bullets. Such measurements are seldom taken by professionals whose laboratories handle a large volume of Firearms Investigation. We have tried, therefore, to revise and to give in usable form only the information that is of any considerable importance.

We call attention again, as in the text, to the extreme possibilities of variation in manufacture, the mismatching of ammunition, and finally special weapons. In general, more than 90% of all crime firearms will be rifled with four, five, or six grooves right-hand twist or six grooves left-hand twist. However, other systems are possible. Rarities, such as the Colt cap and ball revolvers, will have seven grooves either right or left-hand and gain twist. Recently a Frontier Colt was used in a crime and was found to have five grooves right-hand twist. It is probable that this barrel was never made at the Colt factory. No records exist of a modern Colt with right-hand twist or five grooves. However, it was indistinguishable from the standard production barrel, save for its rifling characteristics, both internally and externally. Never base your reputation on an identification of a weapon when you have the bullet only.

We are listing below all common rimfire weapons, both rifles and pistols, since there is a tendency in some areas not only to use a great many light rifles in crime, but also to cut them off and make crude but lethal pistols from rifles. We are then listing, in centerfire weapons, only the common calibers usually associated with automatics and with revolvers. We again caution the Expert that each and every common cartridge designed for one can be used in the other with little or no trouble.

RIMFIRE

.22 Cal—(Bore diameter .212″ to .217″; groove diameter .218″ to .224″)

Right-Hand Twist; 3 grooves	A system of rifling once popular for light, usually inexpensive, revolvers made by a number of companies under many names including American Standard Tool Company, Hopkins & Allen, C. S. Shattuck, T. J. Staffords of New Haven, and Iver Johnson for their Model 1900.
Right-Hand Twist; 4 grooves	A system used in rifles rather than handguns. Common in Savage-Stevens and Winchester weapons. Sometimes found in Remington rifles.

Right-Hand Twist; 5 grooves	A very popular system of rifling with Iver Johnson (makers of the old U. S. Revolver Company weapons). A dozen past and present Iver Johnson revolvers were so made. Also used by Harrington and Richardson for their Young America, Safety Hammer and Defender Models, and for "Liberty" revolvers whose actual maker is unknown to the writers. A popular system for rifles made by Remington, Marlin and some obsolete weapons by small arms companies now out of business.
Right-Hand Twist; 6 grooves	A system used today for handguns by Smith & Wesson, High Standard and Harrington & Richardson. Also used in obsolete revolvers by these firms and Hopkins & Allen, Marlin, Hartford, and Victor. Also used for current and obsolete rifles by Remington, Marlin, Mossberg, High Standard, Iver Johnson, Harrington & Richardson, and Winchester.
Right-Hand Twist; 7 grooves	Rare; Webley & Scott Saloon Pistol.
Right-Hand Twist; 8 grooves	Rare; Buchel "Tell" target pistol, also Pope-Stevens and Mossberg rifles.
Right-Hand Twist; 12 grooves	Rare; Hamilton firm produced polyagonal bores of this type.
Right-Hand Twist; 16 grooves	Marlin microgroove rifling.
Left-Hand Twist; 3 grooves	Rare; Allen & Wheelock Model 1858.
Left-Hand Twist; 5 grooves	Used in early revolvers such as American Arms Company, Defender "89," Hopkins & Allen "Prairie King," and Remington and Sharps four-barreled pistols.
Left-Hand Twist; 6 grooves	Used by Colt for almost all their .22 revolvers and automatics. Very few .22 handguns other than Colts have ever used this system.
Left-Hand Twist; 7 grooves	Rare; Colt single action "New 22."

.25 Cal—(Bore diameter .240"-.245"; groove diameter .250"-.256")

This caliber was used fairly extensively in inexpensive rifles until a few years ago. Now rare in crime. Very rarely met with in handguns save those made from rifles. Rifling systems varied a good deal even in the same model and make. Right-Hand Twist in 4, 5, or 6 grooves.

.32 Cal—(Bore diameter .300"-.306"; groove diameter .307"-.312")

A very old cartridge used in both rifles and handguns. Definitely obsolete and rare.

Right-Hand Twist; 4, 5 and 6 grooves — All used in Stevens, Ballard, and other rifles and in obsolete revolvers.

Left-Hand Twist; 5 grooves — C. Sharps four-barreled "deringer."

.41 Cal—(Bore diameter .390"-.396"; groove diameter .397"-.402")

At one time a popular cartridge for "deringer" type pistols. Both right and left-hand twists encountered, usually 5 grooves, in Remington and other deringers.

CENTERFIRE AUTOMATIC PISTOLS

.25 Cal ACP or 6.35mm Auto—(Bore diameter .242″-.248″; groove diameter .250″-.254″)

Right-Hand Twist; 4 grooves — Very popular in Europe; some of those using this system are Praga, Frommer, Lilliput, Dreyse, Schmeisser, Langenhahn, Kommer, Walther, Stoek, **Menta**, Zehna, Reform, Fehma, Haenel, and Jaeger Pocket automatics.

Right-Hand Twist; 5 grooves — Popular. Gloria, Princeps, **Vite**, Waldmann, Express, Astra, Helios, Regina, Stenda, Clement, and Vulcan foreign pocket automatics and the American Harrington & Richardson use this system.

Right-Hand Twist; 6 grooves — Most popular of all. These are some of them: Express, Ideal, Lilliput, Singer, Ruby, MAB, Clement, Royal, Marina, Sauer, Fiel, Waldmann, Premier, Victoria, Simson, Walther, Colonial, Libia, Ruby, FN, Stosel, Melior, Lighose, Fortuna, Helfricht, Jieffeco, Stern, Minerve, Heim, Phoenix, Seam, Vesta, Webley, Alkar, Titanic, Unique, Beretta, Kaba, Chylewski, OWA, Victor, Joha, F.N., Helfrich-Krausser, Mann, Atlas, Sauer, Monobloc, Simpson, Schmeisser, Pieper, Pickert, Steyr, Menz Menta, Ortgies, Continental, and Mauser. Some of the better known makers produced more than one model so rifled.

Right-Hand Twist; 7 grooves — Rare. Delv only one known to writers.

Right-Hand Twist; 8 grooves — Rare. Schmeisser Model 2 so rifled.

Left-Hand Twist; 6 grooves — No other left-hand twist known in this caliber. Colt by all means most common in America. But several European automatics are also rifled according to this system, among them, Omega, Unique, Mondial, Looking Glass, Bayard, F.N., Astra, Union, Jubala, Martz, Martian, Star, Sol, Allies, Pinkerton, Mondial, Marte, Danton and Marina.

.32 Cal ACP or 7.65mm Auto—(Bore diameter .302″-306″; groove diameter .308″-312″)

Right-Hand Twist; 4 grooves — Very popular in Europe. **Walther, Sauer,** Dreyse, **Rheinmetall,** Nordheim, **Mann,** Jager, Frommer, Langenhahn, Franz Stock, Praga, **Kassler,** Schwarzlose, **Melior,** and Stenda used this system.

Right-Hand Twist; 5 grooves — Relatively rare. **Melior** and Stenda only ones known to authors.

Right-Hand Twist; 6 grooves — Very popular indeed. Smith & Wesson. Also Walther, Sauer, **Singer,** Leonhardt, Stenda, Behalla, **Ruky,** Jieffeco, Titanic, **Express,** Alkar, Star, Warner, Clement, Stosel, Danton, Victoria, Premier, Martian, F.N., Astra, Walmann, Webley, Steyr, Pieper, Libia, Lilliput, Menta, **Ortgies,** and Mauser so rifled. Savage and **Harrington** & Richardson used the system over here. Also C Z Brno and Royal.

Right-Hand Twist; 7 grooves	Rare. Remington in their pocket model designed by Pederson.
Left-Hand Twist; 6 grooves	Colt most common but also used in European arms such as Star, Bayard, Vesta, Retolaza, Urrejola, Zulaika, Azanza, Destroyer, S. A. Alkartasuna, Fiel, Martian, Looking Glass, Astra Cebra, and Gabilondos. A. Errasti and Trust.

.380 Cal ACP or 9mm Browning—(Bore diameter .347″-.351″; groove diameter .356″-.362″)

NOTE: This caliber is not nearly so popular numerically as the .32 ACP or even the .25 ACP. However, practically all makers of the smaller automatics also produced their .32 models in small numbers chambered for the .380 ACP cartridge. They used in general the same rifling machines and systems of direction of twist and number of grooves.

9mm. Luger or Parabellum; also 38 ACP and 9mm. Steyr and Bergmann whose bullets are similar—Bore diameter .348″-.352″; groove diameter .355″-362″).

NOTE: Undoubtedly the most widely used cartridge in the world. However, there is only one rifling system at all common. Since this is a powerful military cartridge, not nearly so many different makes of automatics are encountered chambered for it.

Right-Hand Twist; 6 grooves	Luger, Big Mauser, P38, FN 1935, Radom, and most sub-machine guns.
Left-Hand Twist; 6 grooves	Colts of all models. Some French and many Spanish automatics.

7.63 Mauser and Tokarev—(Bore diameter .299″ to .303″; groove diameter .307″ to .312″)

NOTE: The Big Mauser and the Tokarev are fairly common in crime. Machine pistols sometimes use this ammunition.

Left-Hand Twist; 6 grooves	Spanish copies of Mauser usually, but not always, so rifled.
Right-Hand Twist; 4 grooves	Tokarevs and Russian sub-machine guns.
Right-Hand Twist; 6 grooves	Standard for Mauser.

7.65 Luger or Parabellum—(Bore diameter .300″-.302″; groove diameter .308″-.310″)

Right-Hand Twist; 4 grooves	Rare. Only one common weapon chambered for this cartridge—the .30 Luger.

.45 Cal ACP—(Bore diameter .442″ to .446″; groove diameter .450″ to .454″)

Right-Hand Twist; 4 grooves	Sometimes met with in sub-machine guns.
Right-Hand Twist; 6 grooves	Usual sub-machine gun rifling; old Savage pistol so made. Also the ever popular Smith & Wesson 1917 Revolver.
Left-Hand Twist; 6 grooves	All Colts and World War II Model 1911 A1 weapons made by other companies. Also Spanish imitations.

CENTERFIRE REVOLVERS

.32 Cal—(Bore diameter .301″-.308″; groove diameter .310″ to .314″)

NOTE: There are several cartridges such as .32 short and long Colt; .32 Smith & Wesson and finally the powerful .32-20. However, the bullets cannot always be identified so all weapons that we know of in this caliber are treated together.

Right-Hand Twist; 4 grooves	Rare. Foreign like La Unique revolver. Smith & Wesson.

Right-Hand Twist; 5 grooves	Very common. Smith & Wesson, Iver Johnson, Harrington & Richardson, Secret Service Special (made by Iver Johnson) and many others.
Right-Hand Twist; 6 grooves	Some foreign. Also Harrington & Richardson, Smith & Wesson and other older American weapons.
Left-Hand Twist; 3 grooves	Rare and obsolete.
Left-Hand Twist; 4 grooves	Rare and obsolete.
Left-Hand Twist; 5 grooves	Rare and obsolete.
Left-Hand Twist; 6 grooves	All Colts. Some foreign revolvers.

.38 Cal—(Bore diameter .345″-.348″; groove diameter .355″-.361″)

NOTE: There are several cartridges loaded with bullets of approximately the same diameter. Variations in weight and shape of bullets do not always allow identification of cartridges, so all weapons of this caliber are treated together. They are in part .38 short and long Colt, .38 Smith & Wesson in two lengths, the British .380 (same as .38 Smith & Wesson), the Colt Police Positive .38, and the ever popular .38 Special as well as the .357 Magnum.

Right-Hand Twist; 4 grooves	Rare. Usually obsolete arms.
Right-Hand Twist; 5 grooves	Usual for Smith & Wesson and Iver Johnson revolvers.
Right-Hand Twist; 6 grooves	Usual for Harrington & Richardson. Sometimes Smith & Wesson.
Right-Hand Twist; 7 grooves	Rare. Usually British.
Left-Hand Twist; 6 grooves	Colt and some foreign revolvers.

.41 Cal—(Bore diameter .393″-.396″; groove diameter .400″-.404″)

NOTE: Both the .38-.40 and the .41 Colt are obsolete. Rare caliber in crime but bores and often bullets the same. Colts are 6 left.

.44 Cal—(Bore diameter .415″ to .422″; groove diameter .420″ to .430″)

NOTE: This caliber includes the .44-.40, the .44 Russian, the .44 American and the .44 Special, which latter when properly handloaded is the most powerful handgun cartridge in the world.

Right-Hand Twist; 5 grooves	Some foreign revolvers.
Right-Hand Twist; 6 grooves	Smith & Wesson.
Right-Hand Twist; 7 grooves	Very rare. Some British revolvers used the system.
Left-Hand Twist; 6 grooves	Colts.

.45 Cal—(Bore diameter .440″-.451″; groove diameter .450″ to .459″)

NOTE: This caliber includes the obsolete but powerful .45 Colt, the .45 Auto Rim and a whole flock of old .45 weapons from Great Britain. There is no essential difference between .45 and .455 when referring to these weapons.

Right-Hand Twist; 5 grooves	Some foreign revolvers.
Right-Hand Twist; 6 grooves	Smith & Wesson.
Right-Hand Twist; 7 grooves	Very rare. Several British revolvers so rifled.
Left-Hand Twist; 6 grooves	Colts and imitations.

CHAPTER 13

MATCHING OF CRIME AND TEST BULLETS AND CARTRIDGE CASES

T HE real payoff in Firearms Identification comes when the Expert says in court, "The crime bullet was fired by that weapon." Sometimes a cartridge case instead of a bullet is positively matched to a weapon. Less dramatic, since they are usually not presented in court, are the negative matches. Innocent men are frequently saved from trial and possible execution by them. All these involve comparisons of crime and test bullets and/or cartridge cases. These comparisons are the very heart of the science.

General Bullet Comparison Procedure

Bullets are slightly more important than cartridge cases, since bullets actually do the physical injury. A crime bullet or bullets will arrive in the laboratory from the body of a victim or the scene of a crime. If a suspect weapon is also brought into the laboratory at the same time, test bullets from it will be compared to the crime bullet. If, after this comparison, negative results are obtained or if no suspect weapon is submitted, the crime bullet is compared to all other pending case crime bullets of similar size and family characteristics.

In most firearms suicides, and many crimes in which a weapon is abandoned at the scene, the first match is positive. However, 362 bullet comparisons were done with negative results in the Diskin-Wielandics murders. Cooperating departments throughout the East sent test bullets fired in all 9mm weapons picked up after the date of the crime.

As has been described, test bullets will be fired from a suspect weapon. These bullets are recovered in a near perfect condition. As a general rule today a small number of test bullets will be fired, four being about the maximum under normal circumstances. However, when working with a badly corroded bore or trying to duplicate a special condition, several times this number may be fired. G. A. Gordon, perhaps the best man in his particular field, fired 48 shots in a single suspect weapon before getting the type of sheared primer he wanted.

Once near-perfect test bullets from a suspect weapon of approximately the same type as a crime bullet are available, we generally mount two of them under the two sides of the comparison micro-

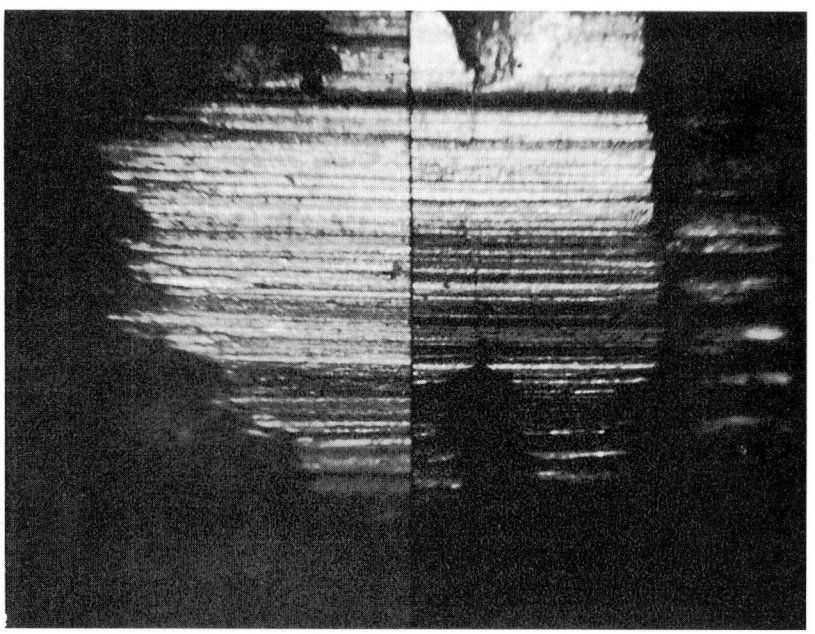

A GOOD POSITIVE COMPARISON

This is a good match. The print was too dark for good reproduction but very good for study and comparison.

While there are many perfect matching striae here, there are also many that, to the untrained eye, do not match. However, if the bullets were moved slightly along to right or left, many of the lines that do not match in this photo would match perfectly in another position, while some that do match now would not in another position.

This apparent discrepancy is due to a minute difference in the focus of the two bullets and would cause no concern or misjudgment in the minds of trained investigators—but it could lead to doubt in the minds of some jurors. This is one reason why they no longer use photomicrographs in court.

scope. We now adjust them in their holders and focus the stages. We are looking at two bullets from the same barrel; we just finished firing them both. We slowly revolve the bullet on the left until there is a particular unusual mark and leave this uppermost. We then revolve the bullet on the right in the same manner until coincidence is obtained.

A more normal microcomparison of a crime and test bullet, cal. .32 ACP. It will be seen that even though this is an interesting comparison, it is not positive, as shown in the photograph. It would be necessary to increase the power and examine in detail the entire circumference of both bullets in order to decide whether these two constituted a positive match, although the family characteristics are obviously the same.

This preliminary examination should be done at relatively low magnification, not more than 12 diameters at most. After practice, bullets can be set in their holders sufficiently precisely to stay in focus as they are revolved through a full 360°. Experts vary as to where they start their examinations of striae. We start with the dividing line of the comparison eyepiece close to the bullet base.

After finding a spot of identity the bullet (or both of them) is twisted in its holder until the etched numbers on the holders indicating groove numbers are in coincidence. Now all grooves in both bullets should match, the raised portion of the bullets will also match if there are sufficient striae on these raised portions. By examining two, three or four test bullets under the comparison microscope the Expert prepares himself for the degree of similarity in individual characteristics in that particular weapon. When he has compared these, he chooses the best, mounts it on the right and puts the crime bullet on the left.

Of course, this comparison will not be made unless the family characteristics match roughly. It is obviously unnecessary to mount a crime 5-groove bullet and a test 6-groove bullet upon a comparison microscope to substantiate a negative match. If both have six grooves but different directions of twist, microscopic comparison is dispensed with. Even when the number of grooves and direction of twist agree, many negative matches are apparent to an experienced man almost immediately. For instance, the grooves in the two bullets being examined may vary considerably in width or pitch. Even in the case of a positive match an experienced man will usually suspect an identity within a couple of minutes, though an hour or more may elapse before he is certain.

A bullet under a comparison microscope does not look like much to a layman. However, it takes on a distinct character to the trained eye. The Expert is able to ignore marks made by contact of the bullet with the victim, walls, dirt and the like; he sees only the striae coming from contact with the barrel. It is impossible to describe this adequately. One must use a comparison microscope for a good many hours before a description would mean anything. However, in general, the experienced investigator sees not even, individual striae, but rather fairly complete pictures. It is truly remarkable how smashed a crime bullet can be in some cases and still be easily and positively identified.

In digression, it should be pointed out that the grooves in the bullet are formed by the lands in the barrel. A great deal that has been written of rifling processes and their influence on bullet identification is only partially correct. Most active investigators today look first at the impression of the lands of the barrel in the bullet— that is, the grooves in the bullet itself. Various marks down inside the grooves of a barrel often leave no marks whatever on the bullet for the simple reason that the bullet does not touch them. In fact, in many weapons, particularly those fired with metal jacketed bullets, there are extremely few striae on the projectile made by the bottom of the grooves. In the course of an argument between the authors and an independent expert, we fired half a dozen bullets from each of six different 9mm automatic pistols taken at random from our collections and found that in four of these six barrels, using the imported Spanish ammunition which was all somewhat undersized, we got no identifiable stria from the grooves in the barrel at all. Some of the bullets would lack all marks on the raised portions that fitted into the barrel grooves. These bullets, as well as

Photomicrograph of two 9mm Luger bullets, fired in Luger pistols to illustrate differences in striations of lands and grooves. The one on the left shows that this bullet did not completely fill the grooves. Bullet on right shows more complete bottoming.

many others, just do not bottom in the grooves of some barrels. In every instance, however, the bullets were well engraved by the lands of the barrel and were therefore completely identifiable.

Soft lead bullets, on the other hand, are sometimes engraved almost equally by lands and grooves. Usually lead bullets are a press fit in bores—that is from .001" to .004" larger in diameter than the bottom of the grooves. Generally the impressions of the driving edges of the lands of the barrel are clearer and more distinctive than the trailing edges.

Once the crime bullet is in place under the microscope the Expert

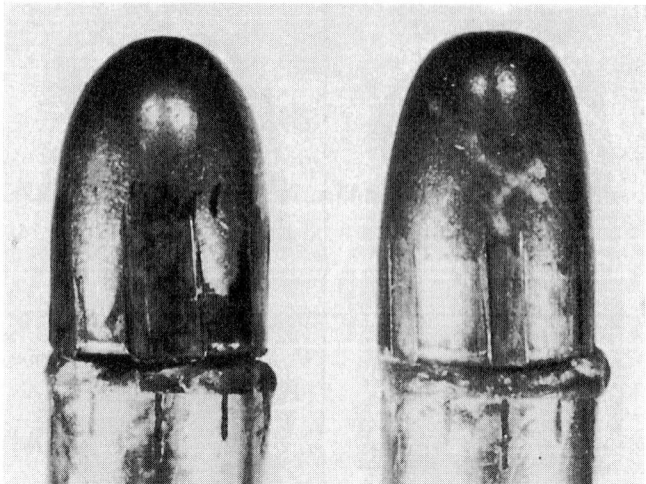

Photomicrograph of two .32 ACP bullets, the left one fired from a Walther 7.65mm and the right one fired from a Colt .32 Automatic.

Shown to illustrate the differences in the bottoming of bullets in grooves of barrels. Note that this failure to so bottom is not limited to one caliber only, or to one type of rifling. It is more frequent in metal jacketed than in lead bullets but occurs with both types.

This also points up the fact that the marks left on the lands by reaming and lapping operations, or by wear, rust, dirt or careless cleaning are of more value than those in the grooves of the barrel, contrary to popular opinion.

Another point not emphasized in these illustrations is the real value of marks in the edges of the bullet lands, and in the corner down at the bottom of the groove where the driven edge of the bullet land begins.

will usually examine it all around, paying little attention to the test bullet also in place. He will be unconsciously comparing the patterns of the striae. He will then usually pick the most distinctive portion of the crime bullet and leave it uppermost. The test bullet is then revolved slowly to see if a similar portion appears. If the bullets came from the same barrel, in at least 90% of the cases these will appear quite quickly. The experienced man will have the patterns of both the crime and test bullets pretty much in his mind by this time. However, this first match is important. He will check it very carefully, looking at the general areas of the two bullets in two or three different adjustments and perhaps changing the angle and intensity of the illumination. If he is satisfied that these two areas indicate identity, he will grasp one of the bullets and twist its mounting so that the number coincides exactly with the number etched on the other mounting fixture. In other words the two will have similar grooves numbered in the same way. The focus will now be rechecked by rotating the axis of the bullet whose position was changed in its holder.

The Expert will now compare each groove in turn of the two bullets, turning them in the same way and, of course, getting the same groove number on both mountings. In the vast majority of positive matches, the patterns of striae in each groove all-around the bullets will check. The Expert may be convinced; however, a man's life may rest in balance. He will generally rest his eyes a few minutes, perhaps while he writes his notes. He will then return to the microscope and compare every groove in the bullets again. If there are sufficient marks on the raised portions of the bullets, made by the grooves of the barrel, he will compare these too.

In laboratories having more than one man, the Director or Commanding Officer will generally check every positive comparison. Even when he does the preliminary work himself he will usually have at least one assistant check the whole comparison carefully. There is always a chance that leaves, sickness, or two cases being tried in different cities on the same day might make this double or triple knowledge of each positive match of great convenience and considerable importance. By the time two or three men have gone around the two bullets, there will be no doubt in their minds.

It is important to stress this 360° comparison. It is probably possi-

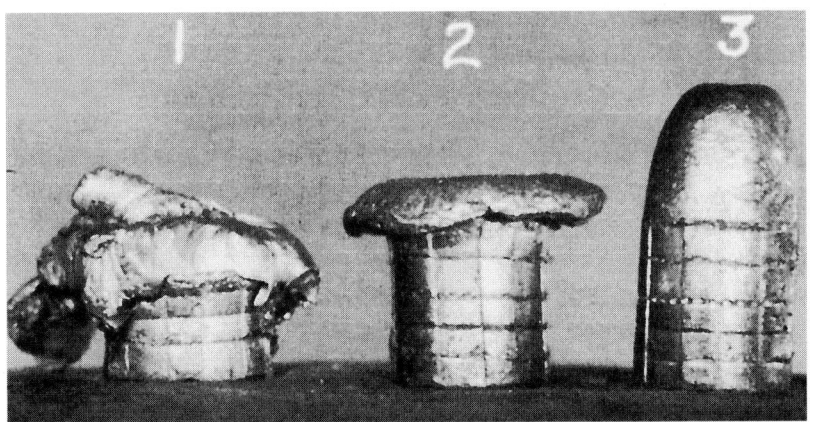

MUSHROOMING OF HOLLOW POINT BULLETS

Showing the unpredictability of mushrooming of hollow point bullets. The three bullets above are identical .22 long rifle, hollow point, high velocity, Lubaloy coated bullets.

No. 1 was the fatal bullet, fired into the head of the victim. Nos. 2 and 3 came from the same box of ammunition and were fired from the fatal rifle into water five feet deep. No. 2 mushroomed very well, No. 3 showed very little distortion.

Do not attempt to duplicate the mushrooming of any bullet.

This is a photomicrograph, taken with a long-extension bellows fitted with a short focus lens. The actual photograph was about three times as large as our illustration and showed all markings and other detail equally clear as this reduction.

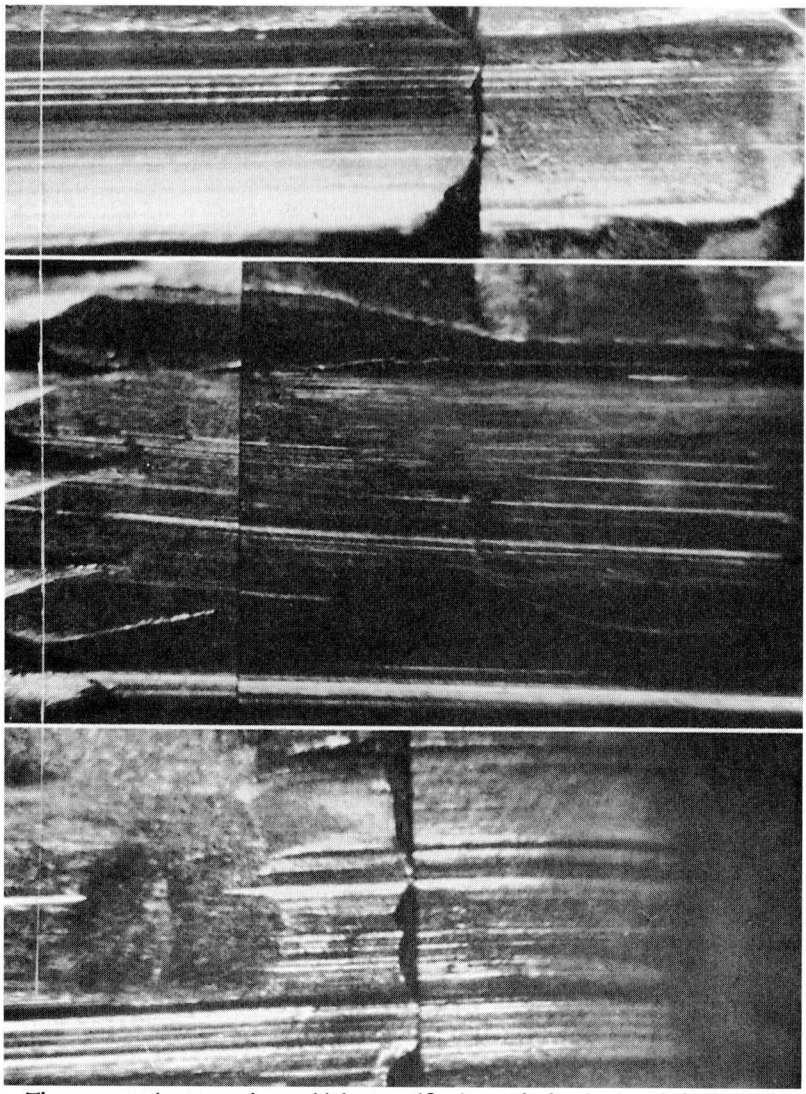

Three comparisons, made at high magnification, of the land striations only, of barrel signatures. While to the uninitiated these might appear to be complete comparisons of both lands and grooves this is not the case—they show only the impressions of a single land. This, and the comparison illustration of groove impressions, shown on opposite page, give an idea of the danger of basing one's judgment on photographs made from too high a degree of magnification.

ble to produce artificially a similar mark on two bullets from different barrels by nicking the muzzles correspondingly. An even more apparent similarity in bullets from different barrels might be caused by a small defect in manufacturing tools leaving marks which might go undetected for several barrels. Smaller striae are con-

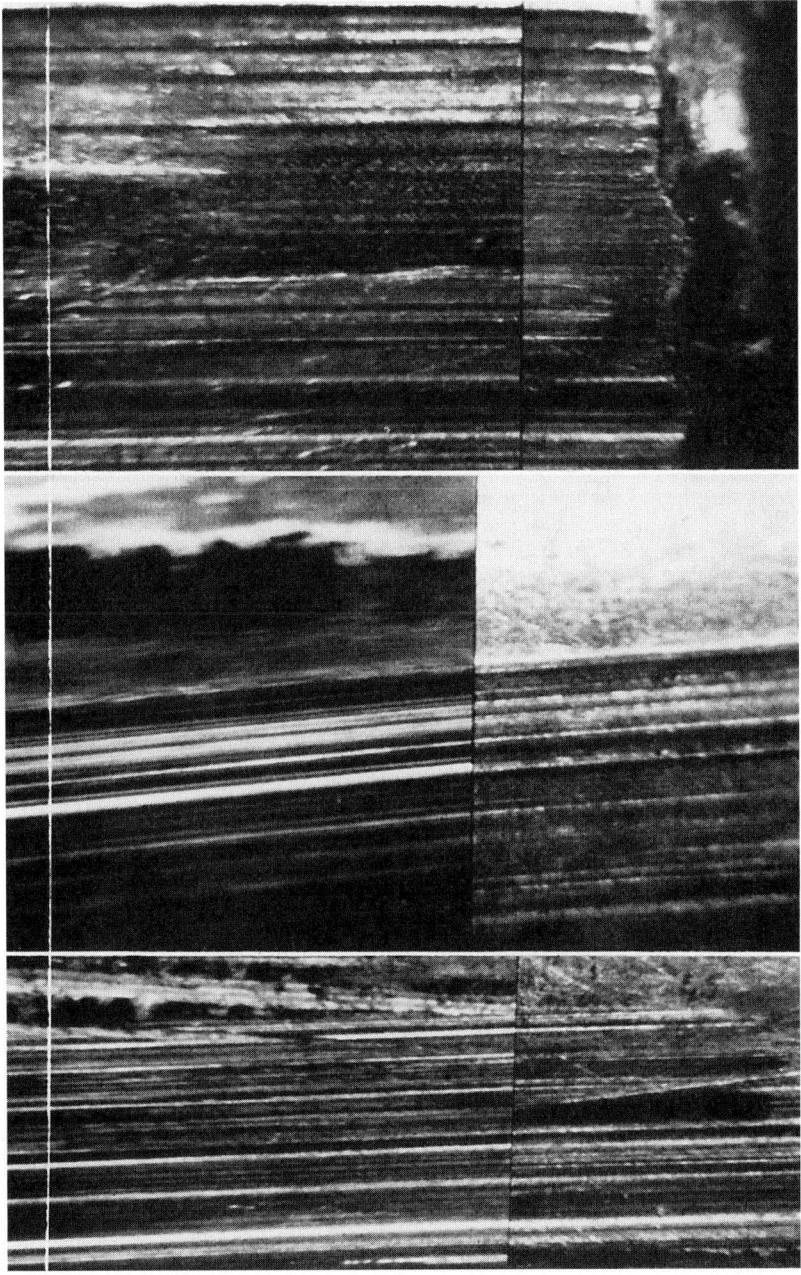

Three comparisons, made at rather high magnification, of the groove portions only of barrel signatures on bullets. The bottom comparison is unusually strong for groove striations. In all three cases the test bullet is at left—evidence at right. These are not full comparisons of the bullets, only a portion of one groove of each bullet is shown in each example.

sidered more important by many Experts than the larger more obvious ones.

When all desired people have examined each groove and each raised portion in turn, the man doing the preliminary work should make a practice of indicating coincidence in his laboratory book with a short, perhaps cipher, note as to that found in each groove, giving their numbers. The bullets should receive a tiny nick at No. 1 Groove so that at some future time they can be placed easily in the proper position without delay. This detailed note making is not universal; however, it can not do any harm. It is a better record than most pictures.

A matter of considerable importance in comparison microscope work is the actual power used. The individual Experts vary considerably in what they have available and what they actually use. One of the authors favors doing most of his personal work at about 20 diameters magnification, though he has his assistant start with a lower power. The other author prefers about 8X to start, with a check if necessary at about 16X. Save where the Expert has years of experience and great dexterity, a magnification of 12 diameters should be sufficient in the first comparison. Some Experts make one examination under higher power, some will actually make two. In general, however, the highest magnification used should not be more than 30 diameters. Above this amount, the instrument has such a tendency to magnify so greatly that apparent similarities are perhaps possible in places upon bullets from different barrels, particularly where the same type of weapons in the same condition are involved.

A rather astonishing photographic match under 300X was introduced in a case in Jerusalem. There was only the one picture, but to the layman it seemed convincing. The British Expert for the defense asked the court to count the grooves in the crime bullet and the revolver of defendant. One had three; the other four. The case was dismissed in spite of the beautiful photograph.

The authors doubt the possibility of getting such a comparison photograph from bullets from different barrels at below 30X; however, one apparently similar area is not what proves an identity to the competent operator today. It is the succession of similar patterns of striae all around the two bullets that really determines an identity.

But to return to actual bullet comparisons. If no readily identifiable area of similarity between test and crime bullet can be picked up in examining them both in a more or less haphazard way, every groove in one bullet must be compared to every groove in the other. Most Experts endeavor to do this in three places, the forepart of the striae, the middle, and finally just in front of the base. If there are any recognizable striae on the raised portions of the bullets—there usually will be some on even metal jacketed bullets— the same procedure is followed in connection with these. If this is all completed, and it will take at least an hour by a skilled, careful operator to do the operation properly, one can be completely certain that these two bullets did not come from the same barrel, although it is almost certain that they came from the same make and model of weapon. If the two had a difference in family characteristics this would have shown up in the first few minutes of examination.

In general, the bullets with no apparent strikingly individual character will be found to have come from relatively well-cared-for, carefully made weapons of the same make and model. This does not mean, however, that every brand new barrel will have anything like the same individual characteristics that other barrels made at approximately the same time by the same machines under the same conditions. As a matter of fact, some brand new first class commercial pistols have very individual characteristics.

Usually weapons in reasonably good condition will, under normal circumstances, produce positively matching test bullets reasonably easily if the arm in question fired the crime bullet. Rifling processes make no difference. For instance, two .45 automatic pistol barrels used to be machined from one forging; this is still the case with certain .22 automatic pistols. The barrels are completed internally including final reaming and rifling and then cut apart. The old .45 pistol barrels had their breeches back to back so that identity in patterns of striae from them was not expected. As pointed out previously, each mark would be inversely positioned in the second barrel; however, many other pistol barrels are merely cut apart after a long section of stock has been bored, reamed, and rifled. In this case, bullets from different completed pieces may easily be passing the same direction down different sections of the same original bore. We have recently done some experimenting in connection with two such barrels that were originally positioned adjacent to each other and finished together with their chambers placed in the same relative position. Actually, distinguishing between bullets from these two is not nearly as difficult a task as it sounds. So much of the individual characteristics of every barrel comes not from the operations proper performed on it, but rather from haphazard scratches; for instance, from chips in the reaming of the original bore, in the chambering of the barrel, or in the finishing and countersinking of the muzzle. Few of these adjacent barrels tested are any more similar in characteristics they leave on bullets than barrels made at intervals of several years.

It was thought at one time that two barrels with rifling pressed into them by the same mandrel on a type of centerless grinding press used to produce the M3 machinegun barrels during the war would be identical. In this case, it was thought that because of the extreme pressure, the relatively soft steel would bear a mark of every irregularity of the rifling mandrel and none of its own individual original irregularities since these would be pressed out completely. Whether this was so in fact inside the bores before the toolsteel, stellite, or carbide mandrels were removed is not known. However, once the mandrels were taken out, the barrels were no longer in any way similar. Bullets from these weapons are easy to identify.

The newer broached rifling also has not produced any more similarity in individual barrel characteristics than the old hook and scrape type rifling cutters. Perhaps again this is in large measure the result of most individual characteristics coming from the original lands. The writers are positive that even though the finish on the lands is still good after the rifling has been broached, it is not so good as it was before the broaching took place. Even though the steel removed by each individual broach is of small thickness, it would seem that it can still score the top of the lands slightly. Another explanation is that the broach actually rides on the top of the lands. In artillery broaches, this riding action on top of the lands is taken advantage of in order to be able to push-broach rather than pull-broach.

Although we are not positive yet what effect the buttoning of the bore of a barrel will have before buttoning of the rifling, we believe in this case also that there will be no great trouble. No matter how smooth a reamed barrel may be and how well it is buttoned—that is, expanded and smoothed by the passing of an oversize carbide button through the bore—you will not ever be able to remove all traces of irregularities from the bore. Even if one could remove these irregularities before rifling, they would

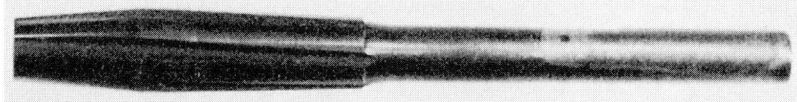

Showing a 'button' for the rifling of .22 caliber barrels. This button is made of tungsten, titanium and tantalum carbide. The actual constituents of this particular grade used for swaging bores is a secret. However, it is a sintered substance, extremely hard, but still capable of standing considerable tension and even shock. It is produced from a suitable blank by grinding with a diamond wheel on a universal cutter grinder and then mounted in a shank by silver soldering.

probably be reintroduced during the rifling operation, or at least later on in the manufacturing procedure. See Appendix No. 3 of Chapter 5 for details of the Buttoning Processes.

Bullet Comparison Difficulties

While most positive and negative comparisons are completed with no special troubles, difficulties do occur sometimes. There are cases in which neither positive nor negative identification is possible. Such cases do not amount to as much as 1% of the matches that an Expert will be called upon to do. They generally involve either crime bullets extremely damaged or a crime weapon in poor shape. The damaged crime bullets most usually encountered in which identity is impossible are damaged during recovery as, for instance, by a surgeon and his forceps when the victim is still alive, or by a police

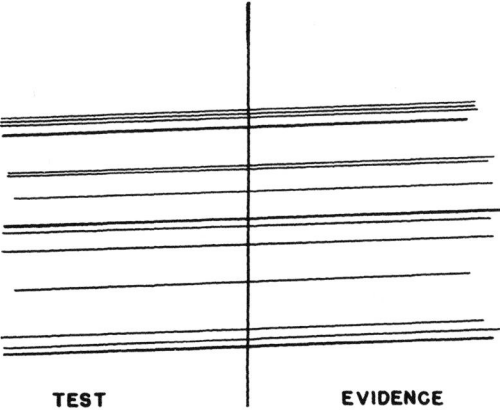

TEST | **EVIDENCE**

The above, and the five following drawings, were made by G. A. Gordon to use in small training classes for field agents. They illustrate the difficulties encountered in firearms identification and show why photomicrographs are often not convincing; are out-of-focus in portions; are poorly lighted; and are often impossible to make. In fact, they can comprise a good case against the use of photographs at all. They also demonstrate why it is a dangerous field for a person without experience.

A good photomicrograph made as perfectly as this example could be a splendid identification for court purposes, because the striations are of varying width and depth and spacing as well as evenly colored. The alignment can be discerned readily by a layman.

officer not knowing his job, like the deputy sheriff who dug the bullets out of boards with his knife.

Where a gun in extremely poor bore condition is used in a crime, the first shot may drive out accumulated rust to such an extent that succeeding shots will not resemble it sufficiently closely for positive identity. A .44 Smith & Wesson Russian revolver was used in a New Jersey crime in 1952. The first bullet fired from this weapon in about 50 years was fatal. The next five all fired in the laboratory each drove out a little more rust so that absolutely positive identity

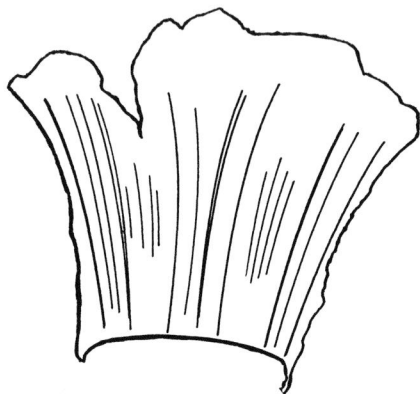

Illustrating a bullet in which the forward portion of its jacket has stretched, so that the striations diverge from the base of the bullet; often the forward portion of the jacket will have curved back with the markings inside the curl.

between the crime bullet and any one of the test bullets or even between the first three test bullets was not possible. There were indications and similarities, but the New Jersey Laboratory would not certify an identity, although in an off-the-record opinion to the investigator in charge, they stated the conditions that they met with in detail.

Another difficulty often encountered is getting satisfactory test bullets from a weapon hurt by exposure or deliberate damage. Because of the wide-spread knowledge of the possible matching of a crime bullet with a test bullet from a suspect weapon, the crime weapons are frequently disposed of. Fortunately, the disposition of an article the size of even a small pistol is difficult. Perhaps about 50% of them are ultimately recovered. This is particularly true in cases of confessions brought out by continual questioning of one type or another.

It is important to realize that a confession obtained in this manner

is easily repudiated later and has very little weight in court. However, the criminal confessing should be asked a great deal, once he
has broken down in an endeavor to get positive evidence. The most
positive of all this evidence is the crime weapon. If he tells where
he threw or buried the gun and then leads police officers to it, his
confession will be difficult to repudiate. A competent detective will
make every effort to recover a weapon, in a case of this type, in the

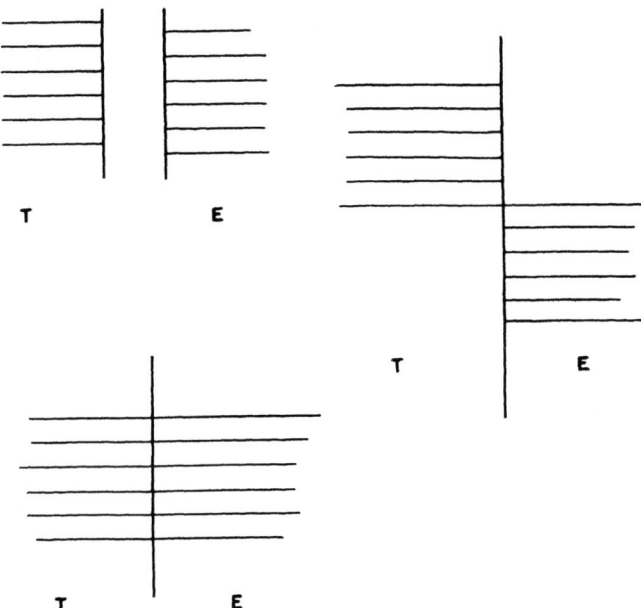

An example of how difficult it can be to align fine, even striations of approximately
the same width and depth without a good initial point or index. They can be
aligned incorrectly.

presence of several people able to testify later. A confession alone
can be easily repudiated; a confession including the site of the disposition of a crime weapon, with concrete evidence of the recovery
of that weapon at that site, and finally conclusive proof that the
crime bullet was fired from that weapon, will make repudiation
extremely difficult.

Weapons recovered in this way will frequently be damaged, both
by exposure and by the criminal before disposition. At any event,
considerable ingenuity may be required to get a test bullet from
the gun. Where elements of exposure and corrosion only are involved, soaking and scrubbing over a period of days or weeks will
probably be effective. We have gone into this in some detail in

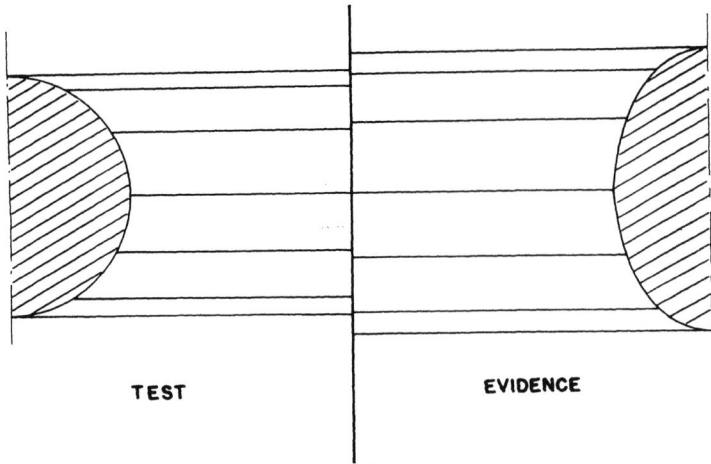

TEST   **EVIDENCE**

Illustrating the comparison of a bullet smashed out-of-round when compared with a round test bullet.

Appendix No. 1 of this chapter. Physical damage presents an even more difficult and varied problem.

In general, bullet identification is dependent upon the barrel only. If the barrel is, or can be restored to operating conditions, all is well. Flattened barrels can be made approximately round again fairly easily. Bent barrels can usually be fired as they are, although straightening would present no problem. Mutilated muzzles can be counterbored or cut off and refinished. Positive identification is sometimes easy when the damage is not excessive.

A remarkable identification was made in a .45 crime in which the pistol had been cut up and smashed and only a relatively few

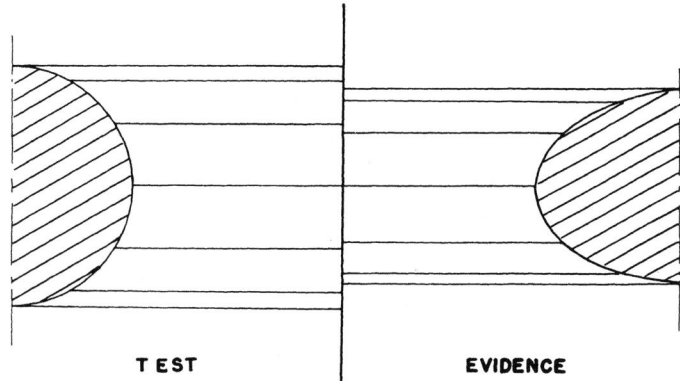

T EST   **EVIDENCE**

Effect of comparison of the out-of-round bullet shown previously when rotated 90 degrees and presenting its "thin" side for comparison.

parts of it recovered, including only about 90% of the barrel. One of the reasons for the large collections of firearms maintained by the leading laboratories is in order to be able to make such tests. They assembled the remainder of the barrel, properly straightened and muzzle refinished, into one of their own weapons and fired test bullets. There were more than sufficient individual characteristics all around both the crime and test bullets to establish identity.

Occasionally, perhaps only once in two or three thousand times, two test bullets fired from the same barrel in apparently reasonable condition with the same ammunition will not show positive identity. This must be clearly understood. After all, no two bullets fired from the same barrel are ever identically marked. Occasionally a variation will be of such magnitude that positive identity cannot be shown. In general, this happens with metal jacketed bullets. The authors have in their possession two test bullets fired consecutively

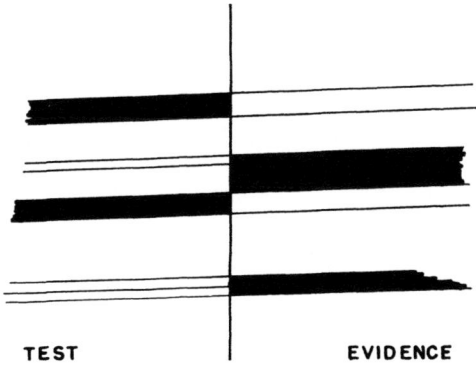

TEST   **EVIDENCE**

Example of an uneven discoloration of bullets—and many photographs suffer badly from this defect. Although the match shown above is good, such a photograph would not be very convincing in court.

in the laboratory from a British Sten sub-machine gun that definitely do not show enough similarity for a positive identification, although each of them can be identified with the remaining three of the original group of five bullets.

We want to stress, however, that the converse just is not so. No two bullets from different barrels ever are similarly marked. We cannot say this after examining all the bullets in the world, any more than the Fingerprint Expert can say from actual experience that no two sets of fingerprints are ever identical. However, bullets are even more variable in their characteristics than fingerprints.

This whole question is intimately connected with the theory of

A fatal .22 caliber lead bullet, removed from the skull of a suicide. Although badly smashed up, the mushrooming of the forward portion gave good protection to the rear half and sufficient striae remained in good condition to enable an accurate check and comparison to be made with a test bullet—as shown in the following illustration.

probabilities, and because the Expert may possibly be subjected to cross-examination designed to expose his unfamiliarity with this subject, he will do well to get out his advanced mathematics and brush up once more on what he learned in school on that subject.

The theory of probabilities is founded on the fact that if there is an equal chance of having a thing happen in any one of several different ways, then the chance that in any particular case it will happen in a certain designated one of those ways is unity divided by the number of ways it can happen. Further, if there are several different possibilities involved, the chance of their all happening in the designated way is the product of the chances of each separate happening, all multiplied together. Thus dice have six sides; any side can equally well be uppermost on any throw. Therefore, the chances of getting say a six on any one roll of a single die is 1 to 6, and the chance of getting three sixes if three dice are rolled at once is $1/6 \times 1/6 \times 1/6$, or 1 in 216. If a coin is tossed, the chances are 1 in 2 that it will come heads. Therefore, if three dice are rolled and a coin is tossed at the same time, the chances of getting three

A perfect match between the suicide and a test bullet. While this is obviously a good match, it would have been better practice to have brought the two bullets into line, then rotate them until they matched. This would have brought centers of each bullet to the highest point and permitted even and identical lighting.

Sometimes grooves will appear to be of different widths solely because they are not on the same axis. Taking this illustration as an example, to an experienced eye the left bullet is as much above center as the right bullet is below center, thus one error compensates for the other. If however, one were on center and the other above or below center, the groove in the one on center would appear wider than the other.

Note the "foreshortened" effect of the lower part of the left bullet which is in shadow of the rest of the bullet. The other bullet is in better position, as shown by visible striae, but is not in the best possible one.

sixes and "heads" at the same time are 1/216 x 1/2, or 1 in 432.

The same laws apply equally well to the occurrence of combinations of markings on bullets. If, when a certain mark appears on a bullet, it has a certain exact location on the bullet 1 in 400 times, and if only 1 in 600 arms gives this kind of mark, then the chances of finding another bullet with the same mark in the same location are 1 in every 600 x 400 arms tested, or 1 in 240,000. Now suppose that this mark is of a certain depth, and that only 1 in 10 of the marks of that kind are of that exact depth. Then the chances of finding a mark of that kind and depth in that location on the bullet are 1 in every 2,400,000 arms tested. But suppose that there are 120 such marks around the entire circumference of the test bullet and evidence bullet. The chances of finding another gun that will make every mark exactly like those on the bullet in question will be 1 in 120 x 2,400,000 or 1 in 288,000,000 arms tested. Surely, these are sufficiently great enough odds as they stand at this impressive figure. But they may logically be carried still further by supposing that the arm is of a certain caliber, with a certain kind of rifling, and that only one man in 1500 has such a gun. Then our chances of finding a man whose gun will duplicate the bullet in question becomes 1 in 1500 x 288,000,000, or 1 in 432,000,000,000 men examined.

Of course the details given in this discussion are all purely speculative, for no data are available as to the exact probability of the existence of any particular make at any definite location on a bullet; but the discussion is intended to show how the mathematics of the theory of probabilities work in a simple case. It is precisely these principles that cover the practice of basing a conviction on circumstantial evidence. If the number of the circumstances surrounding a crime all point to a single suspect, the chances that anyone else can have done it becomes increasingly remote as the number of incriminating circumstances increases. It is the same way with the occurrence of coincident marks on bullets. It may be quite common for two or more prominent individual marks on bullets from two entirely different guns to match exactly, but the chance that there will be a correspondence of a great many of the individual characteristic marks on two bullets that came from different guns is so remote as to amount to a practical impossibility.

A Firearms Identification Laboratory should not go to the limit of their positive opinion, in their reports and testimony. If there is any doubt at all as to a match, it should be reported as negative. While cases of this type are extremely rare, it is important for young men entering this field to realize that they must continue this tradition. Their entire professional reputation, integrity, and their own self-respect could not possibly be other than completely ruined if they ever certified something of which they were not completely positive. If there is the faintest doubt, do not ever take a chance.

Unfortunately, years ago there were so-called experts that would take chances. Even worse, a few of them would appear in court for either side, willing to testify not according to facts, but according to which side was paying their daily fees and rather exorbitant expenses. The true Expert will be warned of these men. Unfortunately, they can be extremely clever. One instance might be of interest. Perhaps the most capable Expert in the world at the time was inveigled into making a statement that he could match six bullets fired from six different revolvers of the same make and model. The opposing expert was neither scientific nor respectable; however, he was clever. He fired his six weapons—all .38 double-action Colt Army Model 1892s—with cartridges loaded with undersize lead bullets and equal parts of black powder and No. 35 emery. Before firing the pistols, he had corroded the bores with a salt solution, left in overnight. The six bullets to be identified had an appearance that was, to say the least, extraordinary. Further, the weapons themselves were in terrible condition. The real Expert, although he did accomplish the almost superhuman task of positively matching the six bullets to the six guns, missed by a wide margin the time limit set. His side lost the case.

We should mention also that identity will be far harder to establish positively where a revolver has a loose cylinder, with successive shots being fired with different alignments. The writers have done some experiments with a gun similar to those used in the above case, deliberately forcing different amounts of misalignment. The results can be astonishing.

New Developments Affecting Bullet Comparison

In addition to Buttoning of bores already referred to and discussed in detail in Appendix No. 3 of Chapter 5, there are several other recent innovations that may cause considerable changes in our science. Bullets are being made of plastics and other substances. Whereas certain .38 Specials fired at high velocity for target work developed by one of the authors are nicely engraved by both lands and grooves, some .22s available commercially for shooting gallery use are extremely poor. These bullets are made of pressed iron dust and a relatively weak plastic; they disintegrate into tiny fragments upon striking the steel backstops, with no chance whatever of a ricochet. Unfortunately, these bullets, even when caught carefully in loose cotton, show almost no striae at all. One can scarcely tell how many grooves the bore had, let alone what individual characteristics. The lack of power, however, of the round coupled with its tendency to fly apart will probably prevent its use in crime.

A far more important potential trouble is the new microgroove barrel being produced by Marlin, and discussed in Appendix No. 4 of Chapter 5. We have three groups of bullets shot from three new barrels. They are the toughest identification job that we have ever tackled. The bores were beautifully smooth; all striae on each of the three groups are remarkably even and uniform. There are so many grooves that negative comparisons take about two to three days. Further, positive comparisons seem to show up far less clearly.

Fired Cartridge Case Comparisons

In general, matches both positive and negative are far easier and more quickly made with a crime case and a test case from a suspect weapon than with bullets. Although family characteristics are not so apparent in cartridge cases as in bullets, individual characteristics are even more obvious. Frequently, identity can be determined by examination of two fired cases with a pocket magnifier. One of the authors has more than 30 .32 automatic pistols. He can tell which of these pistols fired a given .32 case, picked up at random, with a jeweler's loupe only. Such a facility, which is reasonably easily acquired, should not, however, be extended to formal positive identification. The Expert must be intensely conscious of his position. A man's life may depend on his findings.

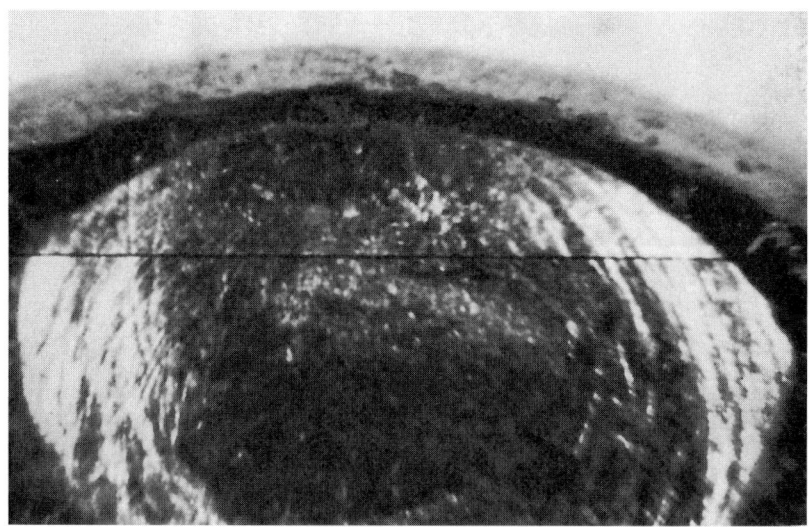

Firing Pin Impressions on a .22 rim fire cartridge. Test shell is at the bottom, evidence shell at top.

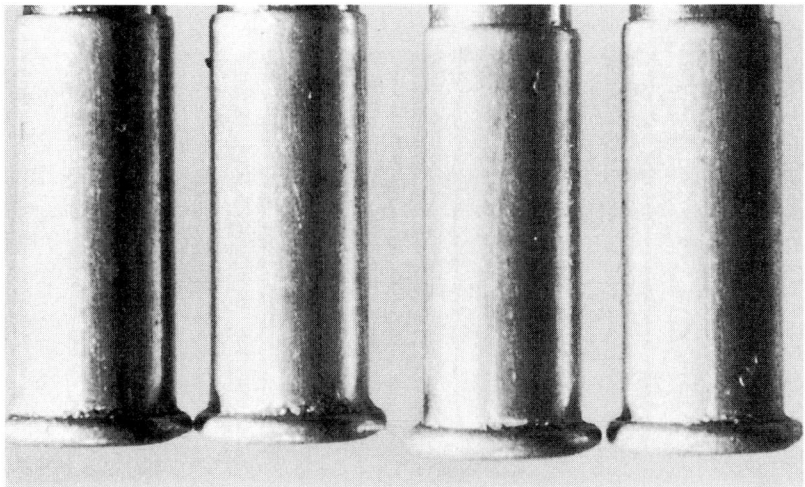

These four .22 long rifle cases were fired in a chamber, the counterbore of which had been battered down and slightly burred at 3 o'clock—as a rifleman would express it. (Probably done with a knife blade or a nail in jacking out stuck cases.) This rifle has its anvil at 6 o'clock (bottom of the chamber), there the firing pin strikes; where much snapping is engaged in this peens down the counterbore at this point.

This slight burring puts a distinctive embossing on the junction of the case body and rim at 3 o'clock; but one sharper point in the burr scores a scratch the entire length of every case fired as it is extracted, as seen above.

Such a scratch, positioned accurately with relation to the firing pin imprint, will invariably give a positive check with the evidence case; in the four cases shown above it is certain, as the distinctive burring at beginning of the scratch is conclusive when viewed under the comparison microscope.

There are two distinct methods of cartridge case identification used about equally throughout the world. In one the crime case and a test case are placed side by side under a stereoscopic microscope of low power. The base markings, including particularly those on the primers, are compared. The cases are manipulated by hand into a coincident relationship—the ejector mark on the base of most cartridge cases from an automatic pistol is often placed at 6:00 o'clock then the cases are rotated together until the striae left by the breech face are horizontal. The head-stamped letters on the case itself have, of course, no meaning in this relationship since they do not in any way position the round in the chamber unless the Expert makes a special effort to obtain a similarity of test to crime case head stamps, which is not usually important.

Both positive and negative matches are quickly determined, particularly in pistols. A positive match should be confirmed by a second examination. The usual laboratory personnel should check the comparison. Appropriate notes should be taken.

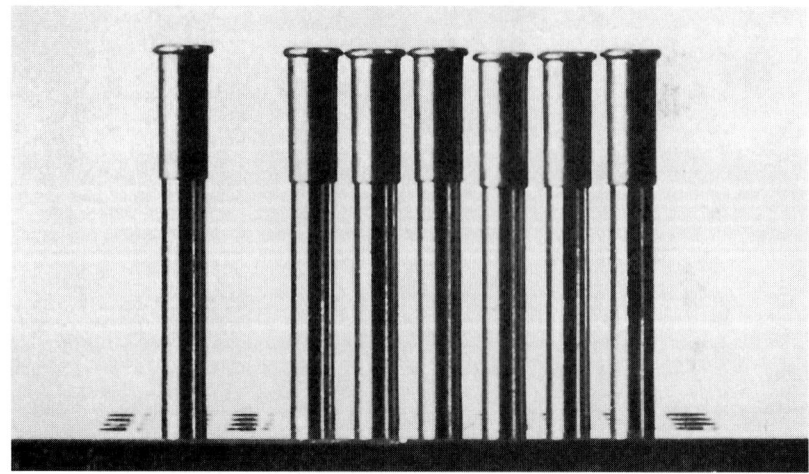

EXAMINATION OF FIRED .22 CARTRIDGE CASES

This gadget is a homemade, steel affair, designed for the purpose of examining, comparing and photographing fired .22 cartridge cases. It can be made of wood, fitted with pins made from round nails of proper size to fit the inside of the cases.

What we are looking for on the fired cases are the various markings imparted upon or into the sides of the cases by minute faults in the chamber of the gun; the abrasions, scratches, indentations, pittings, bulges and nicks left by the tools in the final stages of manufacture, or put there by neglect or harsh treatment on the part of the owner of the arm. Such markings invariably will be found if the investigator knows what to look for. Furthermore, they are likely to be more readily visible to the eye alone, if viewed properly, than when searched for under a magnifying glass or microscope. The eye will also pick up shadings, abrasions and impressions that neither microscope nor camera may register.

In use, the evidence case is positioned at one end of the row, with the comparison cases alongside. Pick out the most prominent identifying mark on the evidence case and position that mark to the front; then observe the location of its firing pin imprint and orient all the comparison cases accordingly. Then carry the block into a suitable light and closely compare all markings by going completely around the cases. Twist and tilt the block in all directions. It may be necessary to throw either light or shadows across the cases, and from different angles, in order to pick up everything; the brightest light available may not always be the most suitable for these purposes.

Rim fire cases invariably will show one or more matching points, in addition to the firing pin imprints. The important thing is to catch all foreign markings that maintain an exact relation to the location of the firing pin imprint—when such are found it is well to go further with the examination, using a comparison microscope if one is available.

The student need not wait until a crime investigation comes along, in order to get the necessary experience on .22 caliber cases. Take every advantage and opportunity to pick up, examine and compare the fired cases thrown away by folks hunting or plinking around with their .22s—particularly the cases from the rifles of boys and casual hunters. Avoid, at first, the empties thrown out on the target ranges by skilled marksmen, as these are likely to come from expensive and well-cared-for rifles, where both action and chamber are kept clean and well greased; under these conditions the fired case is not likely to shown many, if any, markings to the inexperienced eye. It is the neglected, rusty, much-used or worn-out rifles which invariably impart their individual characteristics very plainly upon the fired cases.

The much heavier, center fire cartridge cases are also susceptible to such examination, but not to the extent that the softer and thinner rimfire cases are. It takes a rather serious blemish in the chamber of a center fire arm to impart its individual "sign" onto the thicker and stouter center fire case. At times, the imprint into the fired primer may be the only matchable area and this invariably calls for the use of a comparison microscope.

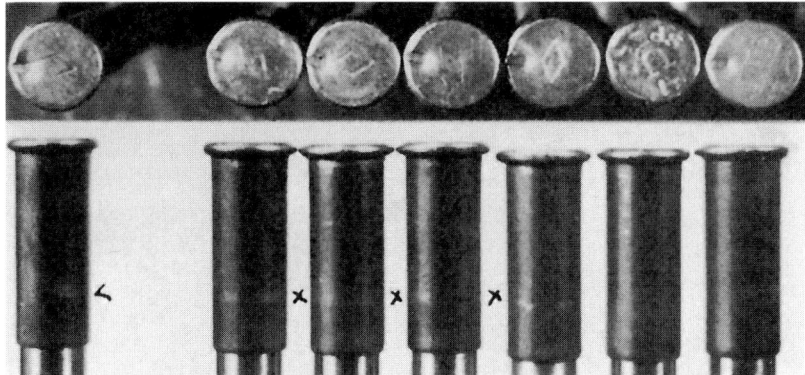

Here is an evidence .22 long rifle case positioned (at the left) and ready for comparison with six other cases fired in the same rifle. Note from the upper illustration that all seven cases are oriented properly with relation to their firing pin imprints.

Observe that on the first four comparison cases, and also on the evidence case, there appears a light spot some quarter-way up from the case mouths. This is the sort of indication we are looking for in order to obtain a positive comparison with our evidence case. This spot was caused by some slight fault in the chamber of the rifle, possibly a hard place in the steel that did not cut out smoothly when the finishing reamer was in operation. This one spot, in conjunction with its fixed position with relation to the firing pin imprint, is sufficient to justify the investigator going to work in earnest upon the suspect.

This is not the kind of examination that is likely to be carried out in any of the well equipped laboratories, these folks preferring to use their microscopes whenever possible. However, not every person owns, or has access to, a comparison microscope and many investigators will be obliged to make out with what is at hand. Conclusions arrived at by the method given here should always be verified by a later laboratory check, if possible.

Bright areas, similar to these shown above, are also often found upon the heavier, much stouter center fire cases and are a condition to be looked for and taken into consideration, although these heavier cases do not show anything like the chambering defects that the thinner and more ductile rim fire cases will present.

Getting back to the above six .22 comparison cases: Note that the last two cases (the pair on the right) do not show any bright spot, although fired in the same rifle. This is because they are of two brands of cartridges different from the other five, possibly they are of stouter metal, or of a different temper—at any rate they did not take-off this chambering fault as did the five cases to the left. Which illustrates the necessity of the investigator always using comparison cartridges of the same brand, and preferably of the same ammunition lot, as is the evidence case; a most important point and one always to be kept in mind.

The second method of cartridge identification is to mount both crime and test cartridges in the comparison microscope. This has the advantage of gaining a continuity of markings from one to the other or placing impressed markings side by side for close examination. However, cartridge base marks are not generated so continuity is of less importance than with bullets. Further, some men feel the lack, under the comparison microscope, of seeing the whole of both case bases and the helpful three dimensional field. Perhaps the very best procedure is that followed by Chief Ballistician William Del Torre in Philadelphia, and others; who uses the steroscopic micro-

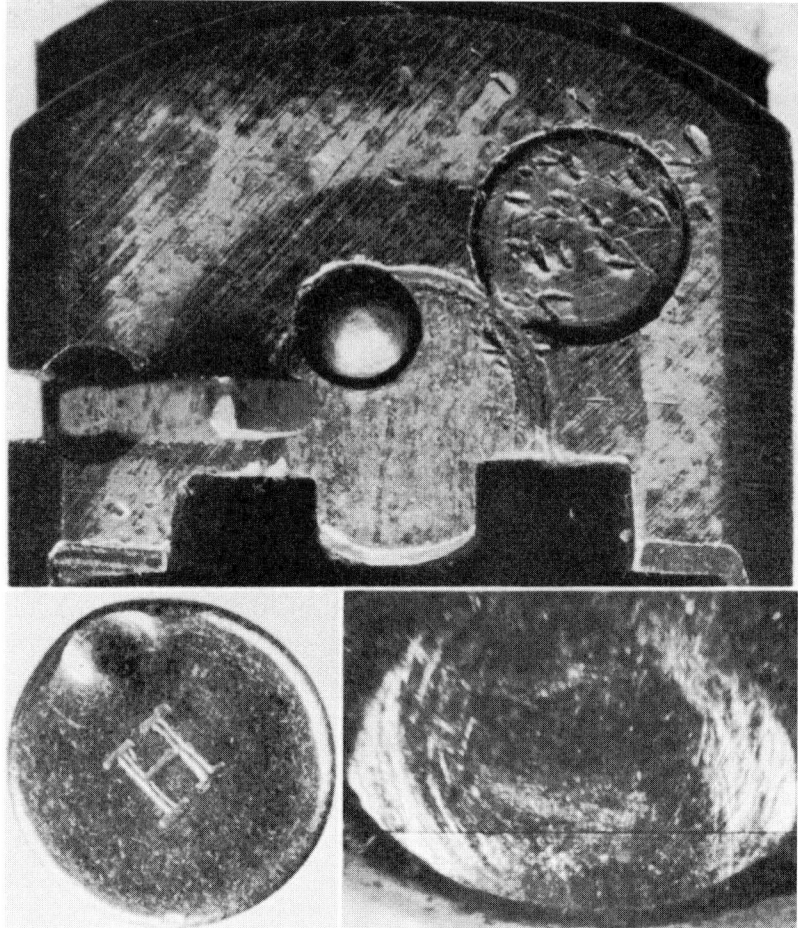

This composite illustration gives an example of breech-face and firing pin markings of a .22 automatic pistol taken from a criminal and matched up with an empty fired cartridge case found at the scene of a holdup. Upper is an enlargement of the breech-face of the pistol, showing clearly the battered irregularities on that part which resulted from the vigorous but misguided attempts of the crook in trying to take it apart.

Lower left is a very low enlargement of the crime cartridge base—and at the right is a micro-comparison photograph of the firing pin indentation markings of both crime and test cases. Under higher magnification than possible to show here, and with use of the reflected shadows, the individualized extractor and base markings were clearly apparent.

scope for preliminary examination and then checks all positive matches with the comparison microscope.

The more one sees of the breech faces of different weapons and their firing pins, as well as the way that these surfaces are finished, the more one becomes convinced that similarity of cartridge case markings is rare and identity never encountered at all. For some reason, almost every revolver and automatic will impart some striking mark or marks. The presence of these on two different cartridge cases is a strong indication of identity; however, in the absence of other similar points, it should never be taken as positive.

Several photographs are shown of cartridge cases from the same weapon. In each instance, literally dozens of identities can be picked up on the actual cases. These are just as important as one relatively prominent marking. In fact, some of these prominent markings are family characteristics rather than individual to the one single specimen. The recoil shield on a .45 Model 1917 Smith & Wesson revolver is a well known source of such family characteristics. Similar marks on cartridge cases will be made by many other makes and models. Some revolvers leave circular marks on the bases of cartridge cases fired in them, caused by irregularities in the breech faces. As the cartridge is revolved in the cylinder, something will abrade the cartridge case base. The mark itself will show a radial

Just below the middle of each of the above four .22 cases, all fired in the same rifle, will be seen an irregular-shaped lighter spot of variable outline. Spots such as these are a means of identification, and whenever they are present on an evidence case the investigator should make further and closer checks of the rifle suspected.

Spots of this nature are not any too distinct in most instances, appearing generally as a lighter, burnished area. When taken into consideration as to their fixed position with relation to the location of the firing pin imprint they are a valuable point to check and compare. They are caused by a rusted spot formed on the chamber wall, or by a fragment of unconsumed primer composition becoming lodged and caked on the bottom of the chamber. Such a rust spot will remain in a fixed position; if it's a bit of caked fouling it may move about a bit and ultimately will drop away or be cleaned out.

The outlines of such spots will not always be consistent, due to slight variations in the wall thickness and temper of different cases. But if a string of shots be fired, a pattern of considerable uniformity will be obtained.

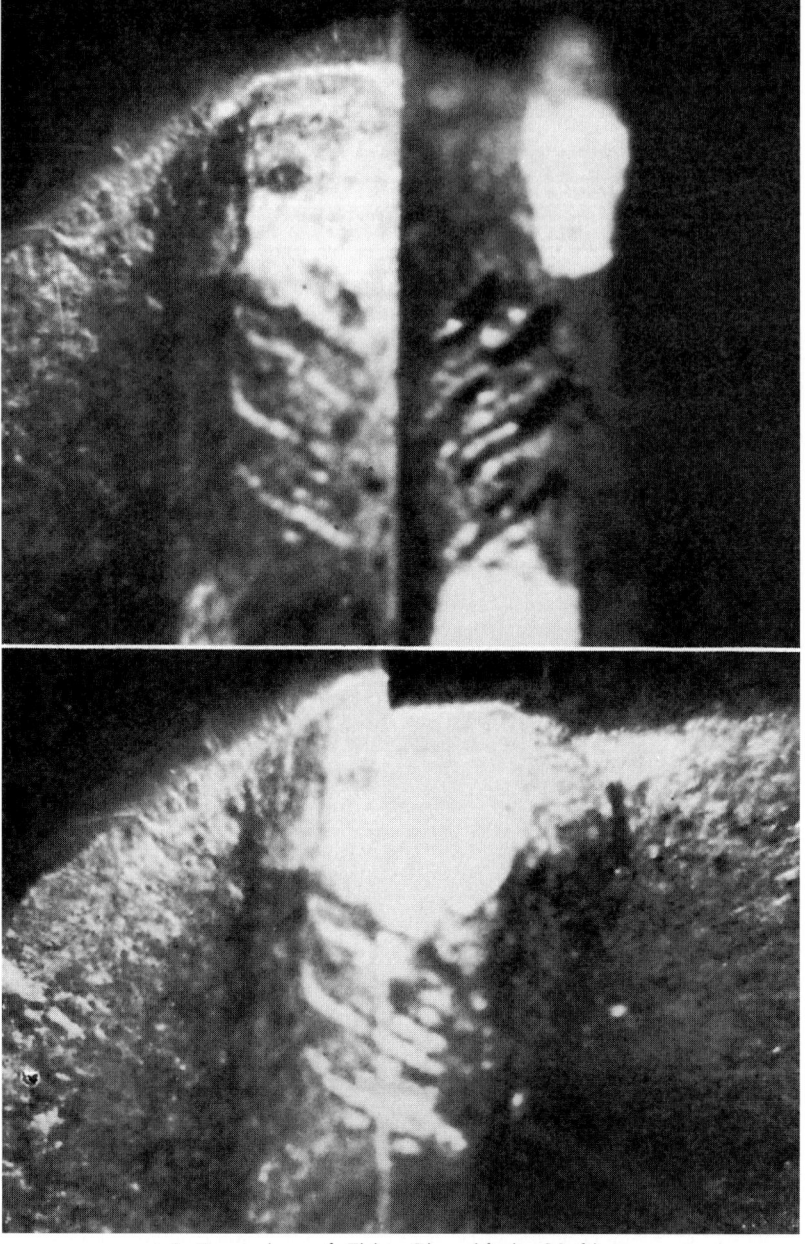

A Comparison of Firing Pin with its Markings.

In this investigation, involving a .22 pistol, there was a series of markings on the firing pin; this firing pin was removed and photographed beside the evidence cartridge case as shown above in the top section—therefore the markings are going in the opposite direction in this picture. In the lower picture, these firing pin markings on the test and evidence cartridge cases are compared; test case to the left, evidence case to the right. (Courtesy of Mr. Robert E. Millhoff.)

distance coincident with the distance from the center of the revolution of the cylinder in the weapon. These may or may not be permanent marks good for individual weapon identification. A slightly bulged hammer nose hole may occur in two different revolvers leaving similar marks on cases fired in both.

The Expert should remember that firing pins can revolve in some weapons. A breech and primer impression appear quite different, though fired in the same weapon, if the firing pin has been turned 90°. A few firing pins and hammer noses will be quite distinctive in individual weapons; however, circular marks left by tools in screw machines should be viewed with caution, since these form tools tend to produce the same pattern on many different firing pins.

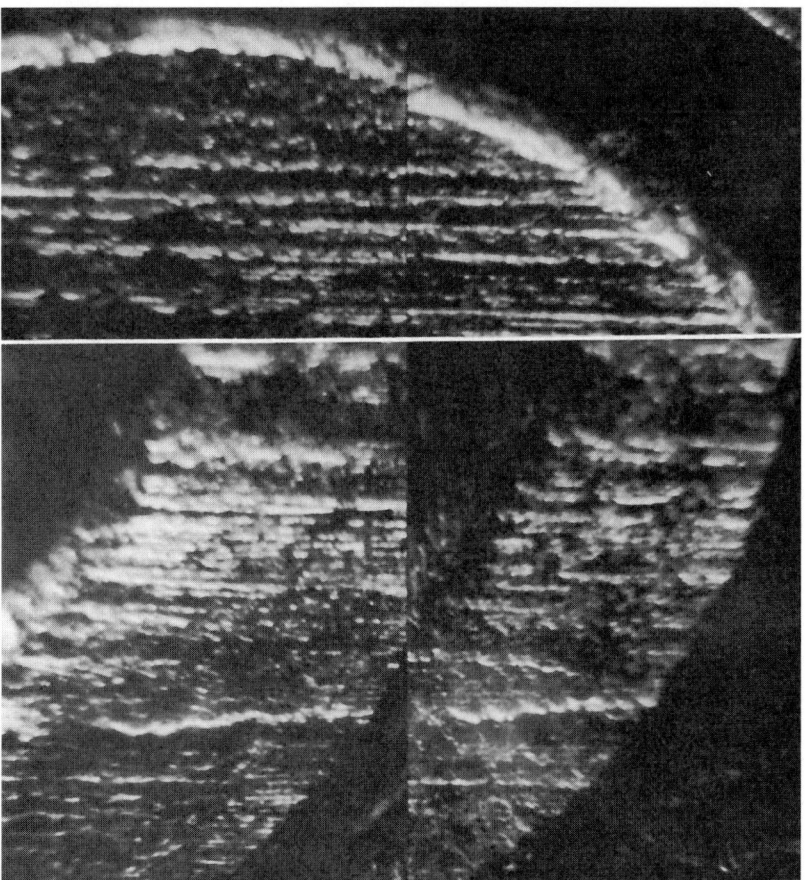

Two examples of breech-mark comparisons on primers—often spoken of as "breechblock signatures," although this same term can also be applied to the breech-face markings on the base of the brass cartridge case. Test primers at left—evidence primers at right.

In this particular investigation the brass cartridge cases took the impression of the badly corroded chamber of the rifle, leaving a distinctive pattern on each fired case. (From G. A. Gordon.)

Cartridge case identification does not have to be confined to base comparisons. In theory, identity can be positively or even negatively established by marks on other parts of the case. As pointed out previously, extractor and ejector marks are frequently encountered in automatic pistol cartridges; however, these alone should not be used for positive identity since at best they can only prove that the individual shell case was chambered, extracted, and ejected by this particular weapon. This same cartridge may later have been fired in another automatic pistol or even in a revolver.

There are, however, definite marks elsewhere than on the bases of the generated type that may lead to positive identification of cases. The best known is the pierced primer type, but that is actually a base marking. With careful work an Expert can match the slipping-over-rim marks by an extractor in an automatic pistol. Further, a generated mark on the side of a fired case can be made only when the weapon is functioning after actually firing. In a busy laboratory the secondary markings are usually confirmatory only. The authors have, however, attached to this chapter as Appendix No. 1 G. A. Gordon's superb monograph on the Government Service Automatic.

Shotgun Matches

In shotgun crimes, one has, at best, only a fired shell for positive comparison. As already pointed out, due to the low pressure and con-

Many shotguns of the top-break, hinged-frame type, both double and singles, develop a fault which can lead to positive identification, as shown above. The firing pin, in one or both barrels, often sticks securely in the fired primer and must be shoved back by the prying action of the shell-base developed when the gun is opened. This cuts a groove of more or less degree in the brass base of the shell, running downward towards the rim. Some actions will streak the brass badly, each firing pin showing its distinct markings as it drags across the primer and case base.

sequent irregularities in markings due to this and other factors, shotgun shells are not always marked in the same way even in the
same gun.

In all shotgun case matching, every effort should be made to have
identity between the test ammunition and the ammunition used in
the crime; this is even more important here than with pistols and
rifles. A clean, slightly oily chamber will aid breechface markings
in test ammunition; the case will slide back in the chamber to the
extreme limit of the headspace in the gun, therefore, the base and
primer will take the impression of the breechface and firing pin
irregularities more easily. However, if the crime shell base is not
well marked there is not much that can be done. Variations in hardness of brass bases or primers will cause extreme variations in breechface markings.

We should state frankly that many Experts will refuse to identify
positively a crime shotgun shell from a suspect gun far more often
than in a pistol or rifle crime. Unless the markings are clear and
sharp in the crime shell there just is not the certainty that one
always feels in the latter comparisons. These words are meant only
as a caution. If you are sure you should not hesitate to say so, but
be sure before you express an opinion.

Permanency of Markings

A few words should be said of the permanence of individual characteristics within a single weapon. Some tests with a .50 caliber
machine gun, firing a single long burst, indicate that the limit of
positive identity between successive bullets from this weapon was
in the neighborhood of a dozen rounds. For instance, the first bullet
could be positively identified with the tenth and the tenth with the
twentieth. However, there were only vague similarities in the first
and one hundredth rounds fired. These experiments were in the case
of an unfortunate accident involving a branch of the Armed Forces.
Evidence bullets were in this manner traced back to the gun responsible.

On the other extreme, a .22 target pistol will undoubtedly fire
several thousand rounds without any drastic change in its individual
characteristics unless, of course, it is neglected. Between these two
weapons, there is a rather open field. The writers conducted a number of experiments with several different guns to determine what
is possible. In general, identity can be made by minor characteristics

An example of really first-rate use of the comparison microscope in cartridge case markings. It is unusual for a photograph of such a comparison to show this extreme similarity.

a good deal longer than it was formerly believed possible. Most revolvers and low-powered automatic pistols will change their individual barrel characteristics with firing only relatively slowly. The limit of positive identification of a .45 sub-machine-gun barrel was, in our test, from 300 to 600 rounds fired in bursts. There is reason to believe that .38 Special target weapons with first class ammunition can be positively identified with bullets fired at least 10,000 rounds before. Although some bronze bullets used during World War II in pistols are said to have changed barrels within 50 rounds, most weapons are still positively identifiable after years of normal use. Neglect introduced an entirely different factor into the picture, although even here reasonable neglect seems to do no harm at all and positive identification can be made after some forms of intentional damage.

Another factor which sometimes must be considered is in connection with bullet and cartridge case identification after elapsed time. In general, cartridge cases, primers, and jacketed bullets will preserve their individual characteristics so long as they are kept clean and dry—perhaps they would be identifiable after a couple of de-

These three .22 long rifle cartridges were fired in an old Stevens "Lord Model" single shot target pistol, and they showed four distinctive individual markings—only two of which are visible above.

This pistol has a "tip-up" action with the conventional inset extractor, and the fired cases all show the usual extractor bulge. However, in the cases fired in this Stevens, the extractor bulge is definitely a one sided swelling, deep on one side and fading out entirely on the other side, where it was held down by a sharp burr which had been thrown up at the right forward corner of the extractor. Evidently the extractor had suffered an accidental blow at some unknown time in the past. The projection thus formed has clearly imprinted its form into the fired cases.

It is seldom that so clear and individual a marking is to be found on fired cases. If these cases had been found at the scene of a crime, the owner of a gun producing similar marks would have a lot of explaining to do.

cades. Lead, however, that has been plated with another metal, seems to deteriorate fairly rapidly. Even relatively pure lead will start to lose the clearness of its striations after a period of years. Many bullets recovered from Civil War battlefields show a whitening from oxidization that may have removed a few grains in weight from the exterior of the bullet. This is not entirely consistent. One of the authors partially identified a .45 caliber hollow-base cylindrical bullet from Kennesaw Mountain as from a small-bore Enfield rifle rather than the more publicized Whitworth and Kerr types. The individual striae are gone, yet the five land barrel impressions remain. As a general rule, an unplated bullet can remain in a laboratory for several years at least without danger of losing individual characteristics. We know of a bullet that remained in a human body for a period from 1903 until 1938. The bullet was enclosed in a covering secreted around it in the body of the victim. He lived happily and healthily for this entire period and then began to get pains in his chest. An X-ray revealed the cause and a relatively minor operation removed the bullet. As astonishing as it may seem, this bullet was completely identifiable, since the covering of natural secretions had prevented any deterioration of the surface of the bullet whatever.

Most Experts will be asked fairly frequently, by policemen and others, why every firearms maker is not required to keep a file showing a test bullet from every barrel that he makes. First, this is utterly and completely ridiculous since it would add tremendously to the cost of firearms. A certain department of the Government because of lack of funds has had to close down one of the largest fingerprint collections in the country. Second, individual characteristics of new barrels and as shown on lead test bullets vary quite quickly. Finally, there is no practical method of classifying bullets in accordance with their individual characteristics such as that evolved for fingerprints. There would be no alternative save checking every filed bullet of the same general characteristics in the country with a given crime bullet and hoping that the individual crime bullet was not fired from the gun in poor condition, a gun that had been fired a lot, or a weapon that had been doctored in some way. Such a check, for instance, of a .38 Special 6 groove left-hand twist would require a large force of men to work at the Colt Company for a period of several centuries. A positive check can be sometimes made in a matter of a few minutes. A negative check takes an Expert at least an hour. Further, the human eyes cannot stand working under a comparison microscope for more than an average of three hours a day.

APPENDIX 1

G. A. Gordon on U. S. Model 1911 and 1911 A1 .45 ACP Cartridge Cases Identification

FOREWORD: Mr. Gordon has a wide and varied experience in connection with Forensic Ballistics involving military arms, and of .45 Service Automatic evidence in particular. The authors are gratefully accepting his kind offer to reproduce below, slightly changed, his extremely valuable monograph on this particular subject.

The United States Service Automatic has been made in larger numbers than any other handgun ever produced in the world. Further, it is available to military and civilian personnel, not only in the United States but in foreign countries through aid to friendly powers. Finally, in connection with cartridge case identification, these weapons almost always leave concrete evidence of themselves at the scene of any crime committed with them. Very few criminals are in a mental state sufficiently calm to search for and recover expended cartridge cases.

Whereas the identifying and matching of bullets is important, this work is largely routine and does not differ greatly from other Firearms Identification. The cartridges, however, are more rewarding than those fired by any other weapons, in part at least, because of the familiarity of Experts with the various markings on these cases. These family and individual characteristics are discussed in detail below. In general, the brass cartridge cases, normal in commercial ammunition and in peace-time Government ammunition, are more desirable for identification purposes, since they take better impressions than the plated steel cases used for some of the war-time military ammunition.

Certain nomenclature should be understood in the specific discussions to follow. We have referred to the position of a cartridge in the chamber according to usual usage; 12 o'clock is at the top, immediately under the sights. Therefore, 6 o'clock will be at the bottom, next to the magazine. We will refer to the positions of all marks on the cartridge cases in accordance with this original position in the chamber of the weapon of the cartridge case when it was fired.

We want to make clear the use of the term "generated" as opposed to the term "impressed." A negative map of a surface, such as that obtained when a soft metal is forced against a harder one, is called an "impression." However, when one surface slides against another, the softer will be marked by a series of high points on the harder. We call such a method of marking "generating"; the generated marks are known as "striae" or "striations."

We will list below specifically the various possible marks on a .45 ACP cartridge case fired in one of these weapons.

1—**A Bulge in the Rear Portion of the Cartridge Case:** In order to make a feeding ramp of the proper inclination to accomplish feeding without trouble, metal must be removed from the bottom of the chamber of this pistol. The removal of this metal leaves the cartridge case unsupported at 6 o'clock. A slight bulge will frequently take place with a brass case at this point. It is useful mainly to determine that the round was fired in this weapon. It rarely, if ever, gives any individual identification details.

2—**Out of Round Cartridge Case Mouth:** In ejecting a round, this weapon frequently bends inward the mouth at the case by contact with a portion of the chamber or slide. This bent condition is quite characteristic of the pistol in question and will seldom be confused with any other form of bending that an expended cartridge case may undergo before it is recovered. However, it indicates only the use of an automatic weapon. It generally is at 2-3 o'clock.

3—**Striae on the Side of Cartridge Case:** Sometimes specific weapons of these types will leave very characteristic marks about one-third of the way back from the mouth of the case to the head. These are caused by the case striking against the edge of the slide, where a roughness generates striae on almost every brass case fired in the weapon. Although not by any means a majority of these weapons give these clear-cut markings, when they are present, they are useful not only for family identification, but also for specific individual identification. These marks occur at 1-2 o'clock.

4—**Extractor Marks:** There are four separate possibilities in connection with extractor marks, one or more of which will generally be found in every cartridge case fired in these weapons. When a round is

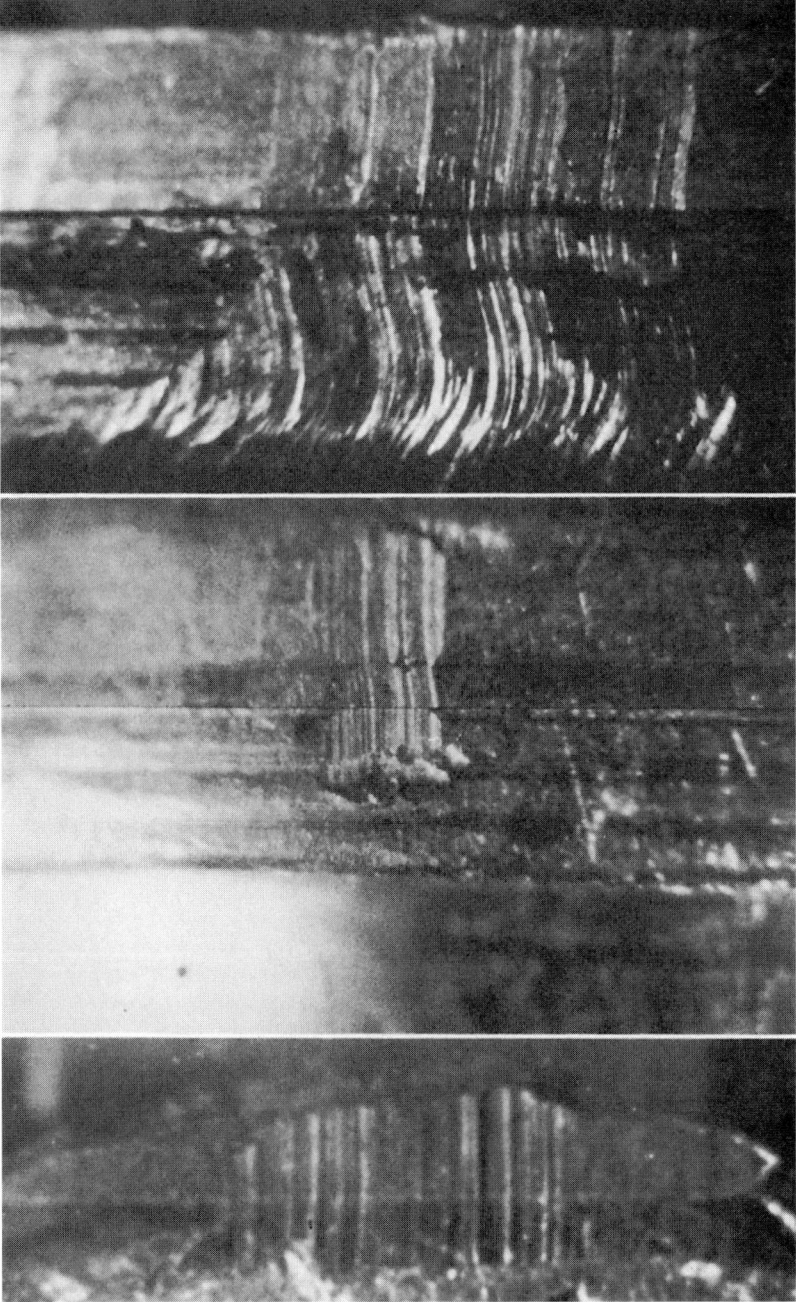

Three examples of comparisons of extractor markings in the .45 Automatic Pistol. In each comparison the evidence cartridge case is the top section—bottom section is the test case.

chambered, the extractor will slip over the rim of a cartridge case, either before or after the round has reached the chamber. Many cartridge cases will have generated striae on the rim of the case caused by the edge of the extractor. Under high magnification (30 X), striae of this type are of considerable importance in individual weapon identification.

After sliding over the rim, the extractor may mark the cylindrical portion of the groove and also the sloping portion of the front shoulder of the grooves. Both these marks vary considerably and are generally not of use for individual identification. Further, at the time of extraction, the vertical back wall of the groove of the cartridge case may be impressed by the relatively flat surface of the underside of the extractor hook. These markings are rarely sufficient for positive identification. Extractor marks occur at 2-3:30 o'clock.

5—**Pivot Marks:** When the ejector hits the cartridge case, still supported at one side by the extractor, the case will be pivoted about a point on its rim at approximately 2 o'clock. In some weapons, there will actually be striae along a flattened portion of the rim at this point. Where the striae exist, this also may be of use in a secondary positive identification.

6—**Slide Closing Mark:** When a round is pushed from the magazine by the closing breech and forced along the ramp into the chamber,

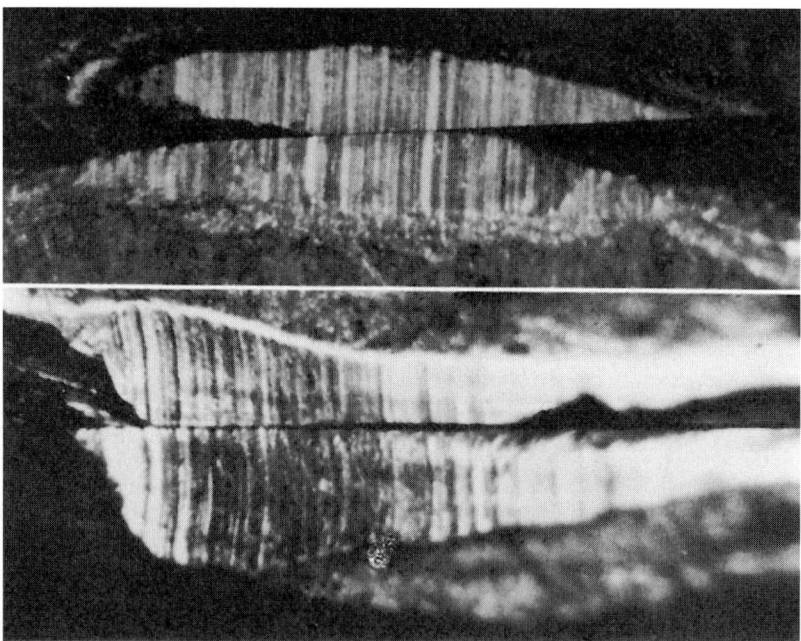

Showing two examples of comparisons of pivot marks on the edge of a .45 ACP case, made by the ejector as the extractor grips the fired case and slams it back against the ejector. Test cases are at top, evidence cases at bottom in both examples.

there may be an impressed mark on the base of the cartridge case extending from approximately 10:30 to 1:30 in a straight line. It is pointed out, however, that this particular mark is of use only in identifying an automatic weapon, and is usually not clear enough to be distinguished from other breechface marks; it may not be entirely coincident with them in two consecutive cartridge cases fired in the same weapon.

7—**Breechface Marks on Cartridge Case:** At the time of discharge, the rear of the cartridge case will be pressed back against the breech block or slide for a small interval of time before the weapon starts to recoil. In general, the pressure will have dropped sufficiently to release the cartridge case from such close contact with the breech block before it begins to move down; these are impressed marks, not to any appreciable extent generated. These marks, which are best on the brass cartridge case, depend upon a number of factors, and may or may not be clearly visible. The conditions causing this variability in clearness of markings are, in addition to the hardness of the brass, in part, pressure, condition of chamber, and lubrication. They are rarely visible on steel cases. They are, of course, of extreme importance in connection with individual identification since the breech faces of these weapons are, in part, hand-finished and give unique markings for every one of the millions of weapons produced. The ejector mark on the base of the cartridge case is a part of this general breechface markings and is useful to orient two cases for comparison. The ejector mark is at 7:00-8:00 o'clock according to our established nomenclature.

8—**Marks on Primers:** Since the priming cup is the softest metal in the cartridge, it will usually take on the clearest and most individual markings of any made on a discharged case. These vary considerably from round to round, even in the *same* box of ammunition. We will discuss each of three types briefly. The most frequently encountered marks in primers conform to a pattern, which may be called regular, in which the firing pin impression is clearly visible; the primer cup has been pressed tightly against the breech face throughout almost its entire area. The primer will have taken on the normal impression of any primer fired in the usual way in most any weapon. The firing pin impression is regular and clear; no excessive flattening of the primer cup or extruding of its metal will be noticed. An examination of this whole area will be of extreme importance for individual identification, and may be for steel cases almost the only available area suitable for individual identification.

A second general class of fired primers will show about the same as the first or regular class, save that the firing pin will not have been drawn back into its housing by the time that the cartridge case starts to move down the breech face in recoil. There will be remaining at 12 o'clock a mark called a "drag mark," where the nose of the firing pin dragged against the top of the primer cup imme-

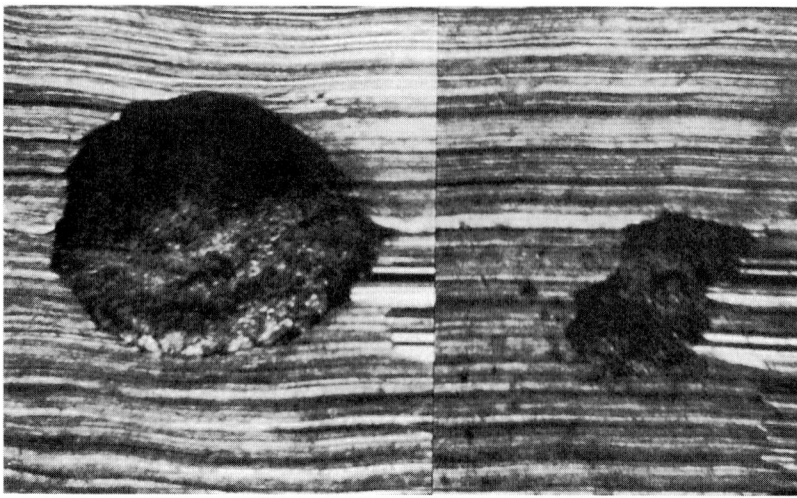

This shows what is called a sheared primer and is found most frequently with the U. S. Pistol Cal. .45 M1911 and M1911A1. The torn out portion varies in size from a thin crescent, through a narrow ring, a large circle with a small center, and sometimes is a solid circle including the whole top of the primer. Sometimes this sheared portion remains attached by a small lip and looks like the top of a round tin can which has just been opened and the lid turned back a bit. Usually the top has a hole in the center where it was originally struck by the firing pin. The striations on these sheared primers are pronounced and almost invariably provide an excellent means of identification.

In several instances the same pistol will leave expended cartridge cases which can be separated into two or more groups which have few if any features on their primers in common. Two of these groups are obvious, the impression type marks and the sheared type.

In two instances within a short time, I encountered evidence cartridge cases which included both impression and sheared primer marks. In order to obtain a test cartridge case with a sheared primer, one sufficiently clear and with enough striations to identify the evidence case, I had to fire 46 rounds in one instance and 49 in the other. But the resulting test cases were worth the effort. The photomicrograph shown above is of one such comparison. (G. A. Gordon.)

diately above the firing pin impression in the primer. This drag mark can vary from a mere touch, to a considerable furrow. It is important to remember, however, that these drag marks do not ever seem to be consistent from round to round in a given pistol, and may occur infrequently in almost any specimen of this weapon. Where the furrow is considerable, the generated striae here may be of extreme importance for individual identification purposes.

The final type of primer markings is called a "sheared primer" since a portion of the primer cup is actually sheared off by the edge of the firing pin hole, or perhaps recoil plate, during the movement of the cartridge case downwards in recoil. These sheared primers can vary from a small crescent, at top or bottom, to the removal of almost the entire central portion of the primer cup. An unusually soft and thin primer cup may be pressed back into the firing pin hole and be sheared off clean. Sometimes the portion of the primer cup removed will remain attached at the top (12 o'clock) to the primer cup in very much the same way as the top of a tin can is sometimes folded back after being opened with an old-fashioned can opener. These sheared primers are extremely important for individual identification. However, many rounds must be fired in some weapons before the condition can be duplicated.

It is particularly important to remember that these primer marks of the three different categories can occur in any weapon at any time, and never occur in any weapon all the time. Where a crime cartridge case is of one type or another, test rounds should be fired in a suspect weapon until a similar class of primer cup markings is obtained.

There are three other conditions of variability in connection with these weapons. First, it has been impossible to lay down any hard and fast rules in connection with ejection from even currently produced .45 Service Automatics. The direction, trajectory, and range of cartridge cases thrown out by a specific weapon is variable in itself and will be extremely variable from weapon to weapon.

Second, the markings on primers covered with waterproof lacquer can be deceiving. For instance, the lacquer may so cushion the primer that the lacquer itself is marked and then falls off, leaving the primer practically unmarked. This lacquer is not an aid to identification.

Third, thick oil, or even thin oil in cold weather, will so cushion the contact between cartridge case and primer with the face of the breech, as to leave very few marks on the former. This is particularly encountered in the first shot from a well cared-for weapon. Firing the weapon, even once, tends to remove this oil or grease from the breech surface. Therefore, when only one shot is fired in the commission of a crime, it may be difficult to positively match crime and test cases.

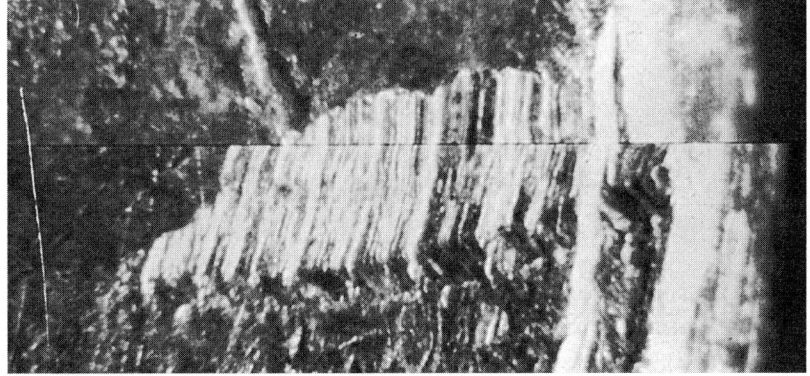

Comparison of a portion of the breceh signatures found on a cartridge case—the ejector markings of test and evidence fired cases from a .45 Automatic Pistol.

CHAPTER 14

AUTOMATIC WEAPONS AND RIFLE IDENTIFICATION

BECAUSE of their greater bulk and therefore smaller potentialities for concealment, shoulder weapons are not nearly so frequently used in crime as pistols and revolvers. However, any department will have a fairly large percentage of both full automatic weapons and rifles submitted to them for examination, of a type designed to be fired from the shoulder. Perhaps half the rifles will be picked up in connection with unintentional shootings or instances involving infringements of game laws. However, because of their supposed superior power, certain criminals habitually choose automatic weapons for execution purposes and the like.

Sub-machine Guns

Perhaps the most frequently met with and certainly the most highly publicized automatic weapon used by the habitual criminal is the sub-machine gun, or machine pistol. This weapon is almost always equipped with some form of stock, although it may be either detachable or folding. Frequently, however, the sub-machine gun is fired cradled in the arms rather than placed against the shoulder. At reasonably close range, criminals in particular seem to feel that it can be employed more or less in the same manner as a garden hose.

As pointed out previously, sub-machine guns really function with pistol ammunition, sometimes stepped up to higher velocities and greater breech pressures than the average pistol will stand. These crimes captured the imagination of the country in the late twenties and early thirties to the point where some criminals were actually putting extra handles and dummy magazines on shotguns to intimidate bank personnel and others. Fortunately, organized crime of this type is no longer so wide-spread; however, a person high up in a national law enforcement agency pointed out recently the possibility of a revival of the sub-machine gun in the hands of criminals. Unfortunately, there are now many more of them in

402

complete working order than law-abiding citizens and most police departments realize. After all, not by any means all of the returning souvenirs were actually made inoperative. Further, some measures taken to render the guns inoperative were not really effective. A barrel would be welded closed in one arm; in another of the same make and model the firing pin might be ruined. These two arms can be reassembled to give one souvenir and one workable sub-machine gun. A gunsmith or even an amateur handy with tools can frequently repair or replace a defective part from a sub-machine gun without too much trouble.

These sub-machine gun crimes are generally reasonably easily detected. If any witnesses were present when the shooting was done, they will remember the bursts of fire. The victim and the area surrounding him will generally show the effects of the dispersed barrage. Finally, unless the firing was done from the inside of an automobile, a lot of cases will be strewn around the position occupied by the gunner at the time of the crime.

Sub-machine gun crime in this country is usually limited to three cartridges; namely, the .45 ACP used in our own sub-machine guns, the 9mm Luger or Parabellum used throughout the world in sub-machine guns including even our own with special barrels, and the relatively rare 7.63mm Mauser or Tokarev used in certain German and Scandinavian arms and a number of different Russian automatic weapons. Whenever a number of bullets or cartridge cases are recovered in these calibers, a sub-machine gun should be suspected. We show photographs of the more likely weapons along with captions stating the family characteristics found on both bullets and cartridge cases. Do not forget that as pointed out earlier almost any standard self-loading pistol can be converted to a full automatic. Occasionally someone will convert a rifle to full automatic; a model 1907 Winchester .351 was so changed and used in crime recently; it had been re-chambered for the 9mm Luger cartridge and functioned very well. Sub-machine gun identity was indicated as probable by the lab report. Later when the weapon was found, the Expert in charge thought he had been a trifle lucky since he had not tried to specify which one.

A bit more detail is perhaps worthwhile in connection with our own .45 caliber sub-machine guns. These are the M1 or Thompson, the rare M2, the Reising, and the M3 or "grease gun." In addition to these, there are at least two more that were tried out experimentally but never adopted. All these weapons have right-hand twist rifling. Further, the cases themselves generally receive far more distinctive markings than are usual with pistol use. The original Thompson guns, because of the long travel of the heavy breechblock, give a characteristic impression where the block hits the cartridge before forcing it into the barrel.

Further, the more recent Thompsons as well as the M3 have fixed firing pins that have a tendency to fire the cartridge an instant before it is fully chambered. The pressure pushes the primer from its pocket, which is, within a tiny interval of time later, smashed back into the pocket by the closing bolt; primers so acted upon have a distinctive appearance. These fixed firing pins in the M1 show characteristic tool marks; however, the breech face of most M3s is sandblasted so that the appearance of the primer is amorphous with few definite marks.

Sub-machine guns, because of their value, are not considered expendable in the underworld. Further, they are a little harder to dispose of satisfactorily than a small pistol. Occasionally a criminal will use one for some time, taking the precaution to change the barrel between different crimes. The barrel change, of course, does not in any way change the breechblock impressions on fired cartridge cases.

Sometimes full automatic fire can be delivered from a weapon smaller than the usual sub-machine gun. An autoloading (usually called just an automatic) pistol sometimes is capable of this form of fire. About the only commercial weapon of this type ever sold to any great extent was the big Mauser Model 1932, which had a lever to allow full automatic or semi-automatic fire. This weapon also had provision for attaching the wooden holster to the stock of the pistol, making a very satisfactory shoulder stock. In addition to this usual arrangement of shoulder stock, this weapon was sometimes specially equipped with a stock designed to remain in place whenever the weapon was operated, but detachable so that it could be carried compactly, for instance, in a reasonably small dispatch case.

Several other pistols have been made in limited quantities for military use, with a lever for selecting full automatic or semi-automatic fire. Among these is the copy of the big Mauser described above, made in Spain by Astra, and the French Browning similar in design to our own .45 automatic, but capable of full automatic fire with or without a holster stock. Quite a number of other weapons have been at one time factory produced experimentally to deliver full automatic fire. In fact before the turn of the century, John M. Browning experimented with pistols of this type.

Further, it is quite possible to convert most automatic pistols, to fire full automatic, although it is, of course, illegal unless registered. The authors feel most strongly that such a change is not advantageous. A few moments experimenting with such a weapon held in one hand will convince anyone that accurate fire is impossible. If the first shot from a weapon of this type does not hit, none of the rest will either, even at a range of 20 feet. The magazine will be

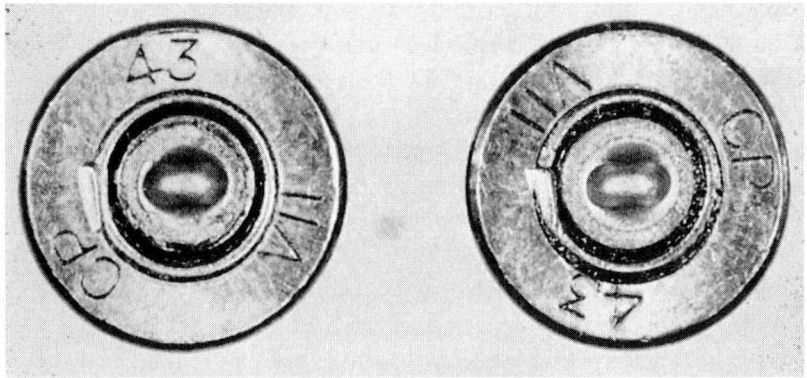

Family characteristics of the Bren gun. Note the oval firing pin impression and the clearly stamped impression of the ejector in the rim of the primer pocket.

expended at a tremendously high cyclic rate of fire, with the weapon quickly pointing straight up in the air.

Perhaps no one in the world has had enough experience with sub-machine guns and full automatic pistols to give in any way a complete list of characteristics. However, the authors have investigated, with the collaboration of the West Point Museum, all likely weapons of this type. Some few have distinctive family characteristics such as extractor and ejector marks and primer impressions; most do not. Since all full automatic weapons are, for all practical purposes, illegal, there is no way of knowing what types may exist in the form of war souvenirs, which can on occasion get into the hands of a temporary or habitual criminal. Many dozen different makes and models were used in World War II.

Sometimes there is an inconsistency of names. For instance, the Sten gun issued in the British army is officially known as a machine carbine, yet it fires the 9mm Luger cartridge. However, two other cartridges midway in power between those usually fired by sub-machine guns and those for rifles and full machine guns have come into use to a limited extent since World War II. Our .30 carbine M2 can fire full automatic; although rare and not a success from a military standpoint these are widely sought after by some criminals. They use the M1 carbine cartridge. The 8mm Short is a German product firing a .32 caliber bullet weighing 120 grains at about 2400 feet per second. It is used in the German so-called machine pistols already described.

Machine Guns

There is, of course, another class of full automatic weapons; namely, those firing rifle ammunition and, in general, being referred to as light machine guns. Also, it is always possible, although extremely unlikely, that someone may decide to employ a heavy water-cooled machine gun for some specific crime. However, the weight, bulk, and difficulty of obtaining and transporting such a weapon for criminal purposes seems to eliminate them from consideration.

Whereas the sub-machine gun uses, in general, only three different cartridges, almost every type of ammunition used in any service rifle anywhere in the world has on occasion had one or more light machine guns adapted or designed to fire it. Even some rifle cartridges that have not been used for military purposes have occasionally been used in home-made automatics. For instance, the first automatic weapon of the great David Marshall Williams was a conversion of the Remington Model 8, although this eventually handled the .300 Savage cartridge.

Practically, however, in this country the Firearms Investigator will most likely be concerned—if he is concerned at all with light machine guns—with one of the following cartridges: The .30-06, the 8mm Mauser, the .303 British, and one of the two Japanese cartridges—the 6.5mm or the 7.7mm. We believe that the order of likelihood is roughly given by the arrangement of listing.

The .30-06 cartridge is far more likely to have come from the Browning Automatic Rifle (BAR) than any other weapon. Literally millions of Americans have learned to shoot this very fine gas-operated rifle. It was produced commercially for Police Departments by Colt and called the Monitor. However, the .30-06 was used a short time ago in a shooting incident in New Jersey in a Lewis gun designed for and issued in the American Navy. Those of our readers who are old enough will remember the *Panay*. The Japs in the course of their undeclared war with China attacked one of our naval vessels.

A most unusual crime cartridge, successfully identified in advance by Frank Jury as having been fired in a sub-machine gun with an obstructed chamber. Actually, several cartridges, all with this characteristic blown-out appearance, were found at the scene of the crime. The weapon that fired these cartridges, when found, had the forepart of the cartridge case stuck in the front of the chamber. The fixed firing pin was setting off the cartridges, 9mm Luger, before they were chambered.

Newsreels were shown throughout the country of the incident. Our only anti-aircraft armament on this little river gun-boat was a battery of .30-06 Lewis guns. Conceivably, since a number were used during World War II, the Johnson light machine gun also may be in the hands, or may get into the hands, of criminal elements. Quite a number of other machine rifles have been chambered for the .30-06 at one time or another more or less experimentally. To a limited extent we and our Allies have chambered foreign designs for this cartridge. However, the chances of such a weapon being used in crime seem remote. Photographs with corresponding explanations are given in connection with the Browning Automatic Rifle, the Lewis gun, and the Johnson.

Quite a number of light machine guns have been designed around the 8mm Mauser cartridge for both the German and other services. Actually, this cartridge is the one most frequently met with throughout the world, save where nations get their arms for free from Uncle Sam. Even the English use it universally in their armored divisions for their Besa machine guns; however, these weapons are not designed for use away from their vehicle. The most likely weapon to be found in America in the hands of a criminal firing the 8mm Mauser would probably be the German MG-34 or the MG-42, or variations of these two. They are light, fairly portable, have a high cyclic rate of fire and came to this country in fairly large numbers.

The .303 British conceivably could have been fired from several different automatic weapons; however, the Bren and the Lewis are by far the most likely. Actually, neither of these has ever come to this country to any considerable extent. We show similar photographs of the Bren gun only; the Lewis is indistinguishable from the .30-06 variety externally.

Perhaps the best weapon the Japanese Empire produced was their Nambu light machine gun. There were actually several different variations of two distinct models. The one illustrated first is by all odds the best, although the earlier model is almost as frequently met with in collections in this country. Each of these has been test fired and individual characteristics of the weapons available to the authors indicated.

The authors want to emphasize that machine gun crime is rare in this country. There have been few instances of full automatic guns of rifle power being used by organized criminals since the final apprehending and death of Baby Face Nelson and some of his contemporaries who used BARs in some of their criminal efforts. However, hardly a Firearms Identification Laboratory in the whole country will not have on exhibition one or more machine guns that have been captured in raids and the like. Many criminals have a tendency to accumulate an armory of weapons, most of which they do not use. They seem to be unable to resist the purchase of a machine gun, even though they probably never fire it.

Occasionally a headline will be read in some paper about a subversive organization whose cache of arms was raided, revealing supposedly large quantities of extremely lethal weapons and the like. The authors recall some pictures of some organization with a name like "The Christian Front" showing several obsolete rifles and pistols, together with perhaps ten or twelve usable modern weapons, a relatively small quantity of ammunition, and some smokeless powder. In any state in the Union, many individual gun cranks could produce ten times the quantity of weapons, ammunition and powder. We cannot actually remember whether any machine guns were in this particular capture, although they frequently are. At the present time, the Government of the United States is not in great danger of being overthrown by force of arms. The criminal elements are far more dangerous with their revolvers than they ever would be with a machine gun. It is only the unusual crime in the future which might be committed with one.

Rifles

The authors hesitate to enter into the subject of high-powered rifles because of the extreme multiplicity of possibilities. Hundreds and hundreds of different models could conceivably be used for crime. Very few of these have family characteristics which can be identified at the present time, first, because the characteristics are not present, and second, because few people have given any great time to the subject because of its small importance in practical Fire-arms Investigation.

The various commercial sporting weapons available in America today have been described in fair detail already. Common sense will generally indicate the possibilities when a fired cartridge case is available as evidence. The same will be true with a crime bullet;

Two comparisons of .44-40 bullets showing similar and dissimilar family character-istics. One of these comparisons (upper) shows both bullets having been fired out of a Winchester Model 1892 rifle. In the other comparison (lower), one of the weapons that had left-hand twist rifling was a single-action Army Colt (showing skid marks), while the other was the Model 1892 Winchester.

however, crime bullets are not usually recovered, save in a smashed condition.

"No-gun" work will, of course, depend on a cartridge case or a bullet. Not many experts will be willing to hazard their reputation on an identity of a commercial rifle by its cartridge case and/or bullet. Further, even though the amount of rifle crime is small, the possibilities are almost unlimited because of the handloading already referred to, which is far more common in sporting rifles than in any form of pistols, save perhaps the .38 Special.

It is sometimes thought necessary to catch full-velocity rifle bullets. This is entirely possible where they are full metal-jacketed. It can be accomplished with three oil drums welded one on top of the other, with the bottoms taken out of the two upper ones, if $2\frac{1}{2}$ of the drums are full of water. Oil may be preferable. The only laboratory that the authors know of with facilities of this type, however, is that in Philadelphia where there is unlimited head room in the top of the City Hall. This same thing was accomplished by Colonel Goddard in Japan with a long bullet trap filled with cotton and waste. However, the bullets do have a tendency to change direction in traveling through this material, and may easily penetrate the sides of the box unless it is lined with steel or made entirely of steel. The personnel of the laboratory over there found that perhaps one bullet in five fired from the Garand would be spoiled by hitting against the sides. However, this is, of course, of no importance since extra rounds can always be fired. Even though there is no scientific evidence to support it, a defense attorney may object to comparisons made between a full velocity crime bullet and a reduced velocity test bullet. If this is thought likely, the Expert will recover full velocity test bullets where possible.

Test cartridge cases can be quite a problem in rifles, particularly those chambered for military ammunition. Some hard brass in some chambers will give so very few marks of identification as to be almost worthless for all purposes. For instance, the authors were asked to destroy for a local police department an accumulation of .30-06 ammunition. There were several hundred rounds contained in boxes, paper bags, and the like, mixed in with .303 British, 8mm Mauser, and a few rounds of other stuff like 7mm Spanish Mauser, and 6.5mm Dutch Mannlicher. We separated it and put the different sizes each together in one box, and then destroyed the ammunition as we feel it should always be destroyed; namely, one by one in the

THE EFFECTS OF PRESSURE UPON FIRED PRIMERS

The Astra 400 is a foreign .38 caliber automatic pistol which will shoot almost any kind of .38 automatic cartridge—such as the 9mm Luger, .380, .38 A. C. P. and the likes. Here are fired cases from two different cartridges, shown to illustrate the effect upon the primers of high and low pressures in the same gun.

To the left is a fired case from a .38 A. C. P. cartridge, which develops about as high a chamber pressure as any pistol cartridge known. Note how the firing pin imprint has been flattened back and protrudes beyond the adjoining surfaces of this primer.

The case to the right is from a foreign 9mm Luger cartridge of rather moderate loading and its imprint is normal and not unusual. From a visual examination of these two cases, seen together, the average investigator would be likely to decide that they came from different guns. It would take a close examination of the ring-shaped markings generated around the firing pin imprints by the bolt face, and this with the microscope, to arrive at an honest conclusion in this instance. And that, considering the different makes of ammunition and hardness of the primer cups, might prove a very difficult task to undertake.

chamber of a weapon for which it was intended.

We took 64 odd assorted rounds of .30-06 ammunition and loaded it haphazardly into eight clips for the M1, and then fired it at 100 yards prone position with sling, using the same hold for every round. Our tremendous respect for this rifle was even greater after looking at the target, which was of the 100 yards small bore type. We had only one 5 and 17 more 6s and 7s. The rest of the shots were in the black. The ammunition which we had used included everything from Frankford Arsenal 1914 through various National match loadings, including commercial ammunition, up to Denver 1945. There were different powders, different primers, different bullet weights, and radically different amounts of both muzzle blast and recoil; however, the arm functioned perfectly with the accuracy already indicated.

We took these 64 cartridge cases and examined them all under a stereoscopic microscope. We were aghast at the difference in the markings. Some of this brass was so hard that the shell case itself was marked almost not at all. In some cases, even the primer cups were sufficiently hard to take little in the way of markings. In other cases, however, both the individual and the family characteristics of the weapon were plainly shown. The Garand does, of course, have fairly easily determined family characteristics.

As already pointed out, good test case markings can frequently be obtained by polishing and oiling a chamber. There is another method, however, that can be used, and has been used by the authors in order to get extremely clearly marked cases for photographs. That is to anneal the head of a .30-06 case, and then hand-load it to close to its normal pressure. This soft case could conceivably let go. It is certainly not safe to fire from the shoulder. However, no harm can be done if you put the action in a vise and fire it by remote control. Actually, we merely put on a pair of shatter-proof glasses and reach out forward to pull the trigger.

Extremely clear markings, however, on a test cartridge case cannot be compared, in general, to faint markings on a crime case. It is far better to get similar markings from firing the crime weapon with similar ammunition, including cartridge cases.

The authors have endeavored to find family characteristics in various other military rifles with a singular lack of success. The Mauser type weapon, and practically all other bolt action rifles, leave little or no identification, save that, in general, the bolt heads will have been finished with some form of end mill. However, this is of little or no good since fully 99% of all the rifle ammunition fired in the world has been fired in weapons whose bolt faces have been so finished.

The Johnson Rifle and some few others do have distinctive family characteristics. We have shown two or three photographs. However, as pointed out previously, family characteristics are of use only in no-gun work and are not of any great importance even there.

CHAPTER 15

MISCELLANEOUS INFORMATION

FROM time to time, the Expert will be called upon for opinions in connection with firearms and shooting which are not, strictly speaking, within the usual boundaries of Firearms Investigation. However, all defense attorneys, most jurors and many judges will expect a Firearms Expert to know about most anything to do with guns. General experience and common sense are extremely important in questions of this type. We give below a few remarks that may be of value in connection with subjects sometimes not clearly understood by the average person.

Maximum Range

Perhaps one of the most frequently encountered matters of this type is in connection with extreme range. Several different general ideas are sometimes confused under this heading. For instance, any gun held in one hand and fired by the average criminal will be unlikely to hit a man-sized target at a distance of greater than 25 yards. However, almost any arm—even a .22 pistol—will reach out and inflict a dangerous or fatal wound, by chance, a lot further than most people realize. Extreme range—meaning the greatest distance that a bullet will carry in the air—even for a .22 will be nearly a mile. However, the elevation for even the most powerful handguns necessary in order to carry this distance will be such that the target will be completely obscured by the gun itself when held in the normal firing position. Accuracy at such ranges is gone completely. Even a good target shot cannot hit a man-sized target with a handgun with certainty beyond 100 yards.

Rifles are much more accurate than pistols, in part because of their shoulder, cheek and both-hands support and their longer sighting radius. A .22 rimfire rifle is far more accurate to 200 yards than a pistol firing the same or any other cartridge. However, a .22 rimfire rifle at 1000 yards is little better than a pistol of the same caliber. The range is so great that the path of flight of the tiny bullet at low velocity resembles a rainbow. Initial preciseness of aim would mean little.

The mid-range trajectory of the .22 long rifle bullet over 1000 yards will be in excess of 400 feet. On the other hand, with heavier bullets and higher velocities rifle precision can be considerably extended. Before the days of machine guns, riflemen frequently fired successfully at ranges of a mile. In the Boer war, a number of sharpshooters from the British Rifle Brigade kept a battery of guns silenced by long range rifle fire alone. The distance between the sharpshooters and the guns amounted to 2900 yards. The mid-range trajectory of these shots was about 400 yards. Sights of the conventional type would be no good at anything like this range; however, the particular rifles used by the English Army at this time had a peep sight elevated on the receiver and a long range forward sight which swung down to a distance of 3″ below the barrel. Such fire obviously would be inaccurate. A rifleman could not possibly hit a single man very often. However, it was of use against this battery of guns extending over a front of almost 100 yards and of considerable depth.

Today, the extreme range usually fired at in combat with military rifles is about 600 yards. Our recent enemies, the Japanese, felt that they were safe at ranges of 700 or 800 yards, even to the extent of washing clothing in the open. Our GIs seemed to agree with them in most instances, although in at least one case a Camp Perry rifleman inflicted several casualties within a few seconds.

Machine guns, on the other hand, have been used for firing at long range targets in both World Wars. Our original .30 caliber 172-grain boattail bullet was adopted in order to give the .30-06 cartridge ranges similar to that of the 8mm Mauser heavy bullet cartridge used in the German machine guns. In some firing at Daytona Beach in the 1920s, it was shown that most machine guns firing reasonably heavy bullets of approximately .30 caliber have ranges of around 5400 yards at the extreme elevation of about 30°. Any rifle firing the same cartridge with the same length barrel would also carry this distance.

Most people would estimate maximum range at far under what it actually is, even in spite of the well-known warning on .22 caliber long rifle rimfire ammunition boxes that it is dangerous to a distance of one mile.

Penetration

A second subject that is frequently encountered is penetration. Anyone having the price of Stoeger's catalog can easily look up penetration in ⅞″ pine boards spaced 1″ apart. In general, this figure is almost useless since one is never firing in earnest at anything even roughly approximating such a target. If the boards are put together, the penetration is very greatly reduced. Actually, a bullet that will go through 15 ⅞″ boards when spaced as indicated above will not penetrate more than five if they are clamped together. The reason for this is, of course, that the spacing allows the bullet to clear itself of debris from the one board before entering the next all the way along. In a reasonably solid medium, the target material tends to pack in front of the bullet and greatly increase the resistance.

Metal jacketed bullets naturally penetrate far further in most mediums than do lead bullets, since the metal jacketed bullet does not so easily deform. There are, however, all kinds of exceptions to any penetration rules. If the matter is really important, tests should be conducted approximating as nearly as possible the range, the type of ammunition, and the type and form of the target.

Even in lead revolver bullets there are variations that are occasionally important in crime. A lead bullet from a revolver is likely to be unstable at the muzzle and for a reasonably short distance beyond. The penetration of this bullet will be greater when it hits a target point-on than if it hits the same target at the same range when it is temporarily in maximum yaw to one side or the other. For instance, it has been said that penetration for a revolver was greater at 20 yards than at 5 feet. If someone fired only a couple of shots, they might easily reach this experimental result.

Another unusual circumstance in connection with penetration, particularly in a medium like dirt or sand, is that a high velocity bullet will go through more at a range of 500 yards than it will at a range of 25 feet. The reason for this probably is that at 500 yards, the velocity is so far decreased that the bullet bores through, whereas at 25 feet the bullet tends to explode or expand. At the shorter range a .45 ACP bullet sometimes penetrates further than a .30-06 because the latter breaks up or spreads its base and tips sideways.

Men have worn armor and put up projectile shields for a long time. Occasionally, both the criminal and protective agencies have made use of devices of this type. Bullet-proof vests have been sold for a long time in this country and have finally been brought to near perfection for use by the Air Force in World War II and the ground forces in Korea. The most recent garments of this type are made of nylon and certain metal foils. Where weight is of less importance, as in suits made for waist gunners in the later B17s, the armor was made by fitting metal plates into pockets of special aprons so that they overlapped. This body armor used today is far less bulky than the types used years ago; however, it still encounters considerable sales resistance where the soldier wearing it has to move around. Further, modern body armor is not intended to stop high velocity bullets. It obviously cannot. However, a single .45 automatic pistol bullet will not penetrate that issued in Korea at a range of 10 yards. Needless to say, most fragments from mortar shells and grenades are likely to be no more powerful than pistol bullets.

At one time or another, banks and other similar institutions have put up screens of bullet-proof glass. Unfortunately, these screens of glass were proof against one bullet but not against two or three more hitting in the same general area.

Another well-known armor is that installed in various delivery trucks operated by private protection agencies. This is definitely proof against most pistol and sub-machine gun ammunition. Some of it will withstand several bullets from the .357 Magnum. However, nothing whatever that would be in the least practical for commercial armored cars or banks will withstand even a single ultra high-velocity rifle bullet. The authors recently took a ½″ piece of homogeneous armor plate that was proof against a single .50 caliber machine gun bullet and penetrated it easily with a .220 Swift. Velocity is far more important for actual penetration in steel than bullet weight or even bullet composition if the said velocity is high enough. At a velocity of about 3500′ a standard bullet will penetrate almost as well as one with a tungsten carbide center. At about the same velocity, bullet weight also ceases to be so important although, of course, at 2800′, the 715-grain bullet of the .50 machine gun will penetrate a good deal more steel than 152-grain .30 bullet at the same velocity.

A word as to penetration holes in armor plate. A mere measurement of the diameter of a hole will not give even a rough approximation of the diameter of the bullet making the hole. For instance, the 45-grain .220 Swift bullets referred to above made round holes with raised edges at both entrance and exit, of approximately .36 caliber. This same plate was penetrated with the fifth 500-grain .470 Rigby bullet traveling at about 2500 feet per second. However, in this case, the first four had created cracks and the fifth merely knocked out a portion of the steel leaving a rather jagged hole.

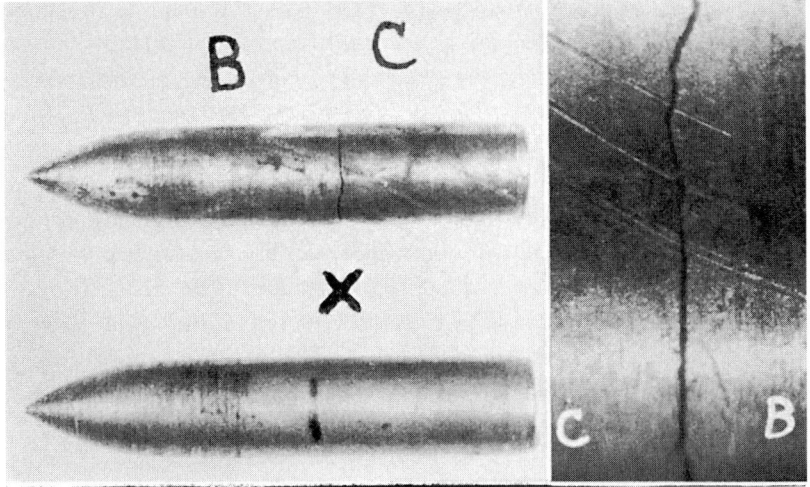

AN ODD RIFLE PROJECTILE IDENTIFICATION

In this case, an armor-piercing 8mm rifle bullet ricocheted off a brick building; on striking the building the jacket was stripped off the bullet and the core broke in two pieces. One fragment killed the driver of a passing vehicle, entering his head; the other fragment killed his companion who was riding in the front seat beside him, entering his heart. The left section of the above illustration shows a comparison of the two fragments with a normal core from a known 8mm rifle bullet. The right section shows a matching of the two fragments. (Courtesy of Mr. Robert E. Millhof.)

The second ball round from a rifle chambered for the .50 machine gun cartridge went through. A special 715-grain, cored .50 at 2900 feet per second penetrated the first time.

Rifle penetration in the human body is, of course, dependent upon velocity, bullet type, shape, weight, and the particular body. However, in general, a full metal jacketed rifle bullet will not stop in the human body, even though it traverses the body from end to end. There is an unsubstantiated story of seven casualties inflicted by an American Sniper with single bullet. The bullet is supposed to have passed completely through the seventh body. This is logical, even though it may have been wishful thinking on the sniper's part.

Firearms Noise

Another subject frequently discussed in criminal cases is the report made by various types of weapons. People will go into court and swear on occasion that a weapon fired was of a certain type and even make and model. Unless a great many other factors are known, such evidence may be sincere but it is utterly ridiculous.

In order to understand the entire subject of noise in connection with the discharge of firearms, let us analyze noise itself. The actual explosion of the propellant charge forms only one part of the noise of firearms. This in general is the muzzle blast. The mass of extremely hot gases smash into relatively stable air at the muzzle of the gun creating vibrations of greater or less magnitude. This sound is not, however, the only part of firearms noise in many weapons. When a projectile travels at a greater speed than sound, a pressure or shock wave will build up ahead of the nose of the bullet and will give a "crack" well-known to big bore target shooters.

This shock wave noise is most unpredictable. The first few shots on a still morning at a firing point next to a line of telephone poles will create a separate crack at each pole as the bullet passes down beside them. Later on the air will be so disturbed that these sounds become less apparent.

There is also, of course, the sound of a bullet striking, which will depend upon the bullet, the velocity, and most of all the target struck. Finally, in automatics there will be the noise of the working of the action. These all blend together to give a single sound with which someone may be familiar; however, this one total noise will be completely different at different points. For instance, rifle fire sounds entirely different on the firing line and in the target pits. The components of the noise of discharge vary greatly in these different positions.

One of the authors received a most impressive demonstration of this phenomenon while in a small pit to operate a target of opportunity in ball cartridge maneuvers several years ago. As the firing line approaches, the various components of noise will be changing dramatically. At 600 yards, for instance, the first thing that a man in such a pit will hear is the bullet striking his target or passing close to it, almost immediately followed by the crack of the shock wave in front of the bullet and followed some interval later by the noise of discharge. As the line approaches, and other targets are fired on further down the range, the cracks from bullet impact behind may blend with the sound of discharge in front. As the line draws even with the pit the sound will become that heard by the firer.

An interesting phenomenon was that some .22 rimfire ammunition for target shooting in the 1930s was loaded to almost exactly the speed of sound which, of course, varies with temperature and to some extent with barometric pressure. This ammuntion, even though remarkably accurate in its day, would have slight velocity variations from round to round. These variations were sufficient on some days so that about half the rounds had a shock wave in front and the other half did not. Peculiarly enough, this did not affect the accuracy. On a cooler day, sound would travel more slowly so that almost all this ammunition would have a shock wave. On a warmer day, the opposite would be true. The sound of a round below the speed of sound was noticeably different from that of one above the speed of sound.

It is extremely difficult to tell direction by sound of discharge of a firearm. Further, the distance that this will carry audibly is even more variable. At one time, Devout and Bernadotte, both marshals of France under Napoleon, were but seven miles apart when the former fought Auerstädt. This was a big battle with 100,000 men and more than 100 guns firing. Yet no one in Bernadotte's Corps heard it. At the other extreme, the first battle of Ypres was audible in Yorkshire, 175 miles away.

The noise of a single discharge of a pistol can sometimes carry two miles or more in a still dry night. However, one of the authors while deer hunting in the Pennsylvania timber country had occasion to actually measure the next morning the distance apart between two halves of a hunting party when they had both fired shots to attract the attention of the other. They were less than 500 yards apart yet the conditions of a foot of new fallen snow, thick growth and a heavy fog muffled even the bellow of a .300 Weatherby Magnum so as to be inaudible to the other half.

In the trial of Eugene Monahan, for the murders of John Diskin and Sebastian Weilandics in a tavern in Elizabeth, New Jersey, the defense attempted to get an opinion from the Firearms Identification Expert on the sound of shots. Seven shots were fired from a 9mm Luger, in a tavern. No one seems to have heard anything although it was a busy Saturday morning about 11:00 o'clock. The tavern was next door to a food Super Market, and a traffic policeman was on duty not over 50 yards away.

In the cellar of the home of the defendants, 27 empty .32 short rimfire shells, two 9mm Luger shells and one .32 Smith & Wesson shell were found. There were numerous bullet holes in the pillars, a screen door, and other pieces of wood. Detectives questioned neighbors and could find no one who admitted hearing any shots at any time. The Expert (one of the authors) was asked, "How far from the house could these gun shots be heard?" The State's attorney very properly objected to the question on the grounds that there were too many factors involved; the Expert had never been in the cellar. The court sustained the objection. The next question was, "How far away could *you* have heard the shots?" Again objected to. This time the court over-ruled and instructed the Expert to answer if he could. The witness said, "I do not know. It simply is not possible to determine such a thing without first knowing all the factors. The shooting could have been done when all the neighbors were out."

As referred to at the beginning of the chapter, witnesses will sometimes state that they recognized a particular weapon by sound only. Some men can recognize certain general categories of weapons. If a man familiar with firearms is blindfolded he will usually be able to recognize by sound different arms. However, even this is not certain. A man stationed 50 yards behind a firing point can tell the difference between a shotgun, a high-velocity rifle, a heavy pistol, and a light pistol. He cannot, however, be much more precise than this. Even an Expert might not know one rifle from another or the difference between a .30-30 rifle and a heavy pistol such as a .44 Special. Take the same individual a quarter of a mile away in the woods, fire just one gun from an unknown position, and he will be guessing pure and simple. Little credence, therefore, should be put in what anyone says about a shot or even the number of shots. These things coming upon him suddenly are generally extremely inaccurately recorded in his memory. One of the authors asked one deer hunter last fall how many shots another hunter less than 100 yards away had fired. The answer was "five." Actually only two shots were fired. A Philadelphia surgeon once said he had just shot a 200-pound buck with a single round from a modern lever action repeater. The deer was not a buck. The entire tubular magazine of cartridges had been discharged in a wild fusilade without any damage whatever being done to the doe which ambled away, somewhat frightened.

Silencers

These are far more common in mystery stories than in Firearms crime. Their operation is dependent on the principles of sound already discussed. Many different forms of silencers have been tried over the years. None of them works completely satisfactorily. In general, the only component of the noise of discharge that you can reduce by silencing is the muzzle blast. If the bullet has a velocity greater than sound, you will, of course, have the crack from the shock wave undiminished and far more pronounced, since the muzzle blast will be partially eliminated.

The silencers of the Maxim type had a cylinder attached to the muzzle of the gun with different baffles set at intervals in this cylinder. The bullet itself passed through holes in these baffles; however, the angles of the baffles were usually varied so that the powder gases were dissipated and muffled in very much the same way that an automobile engine exhaust is muffled. These silencers reduced the muzzle blast considerably where the volume, pressure and velocity of the muzzle gases was not too great. They will even to some extent reduce a part of the noise of a .30-06 or 8mm Mauser rifle. When installed on a .22 rifle firing low velocity .22 short rimfire ammunition, the only appreciable sound will be a "putt" and the mechanism working if it is an automatic. You will now hear distinctly the sound of the bullet hitting the target. The sound that remains might not be identified as a shot by the layman. These same silencers were installed on various low-powered automatic pistols by criminals of a couple of decades ago. They were extremely inconvenient to carry around and were not often used in crime.

At the beginning of World War II, however, the OSS and certain other special forces were extremely interested in the best possible silencing of firearms. Several different arms were more or less silenced. Some of these arms have been found in the hands of criminals and are a source of potential danger. Both the Thompson and M3 submachine guns, the Model 1911A1 automatic pistols, and the M1 carbine were silenced, as well as certain pistols firing .22 rimfire ammunition.

These were all silenced along the same general lines. The forward half of the barrel, or extended barrel in the case of the .45 automatic, was perforated with four rows of holes set at 90° from each other. Stainless steel or galvanized wire mesh was then wrapped around these holes. A chamber was provided in front of the muzzle proper that contained around 150 stainless steel gauze disks through which the bullet passed. These disks took up in large measure the muzzle blast gases. A silenced .22 rimfire pistol was made by High Standard; it was their regular "military" model with the usual barrel changes mentioned above. With .22 shorts the noise of firing is caused principally by the operation of the slide. It does not sound very much like a gun going off. The M3 sub-machine gun silenced—the silenced barrel is interchangeable with a regular barrel—makes a considerable noise because of the operation of the breech, but very little noise of

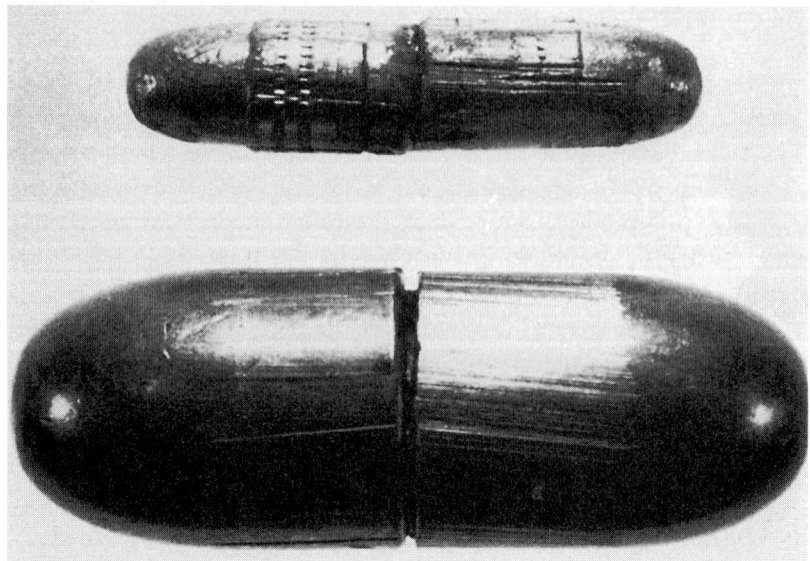

Showing the difference between bullets fired through silenced weapons and conventional weapons. The .22 bullet with deep grooves in it (top, right) was fired from a High Standard silenced automatic pistol. One of the washers was dragging on the bullet during its passage down the silencer. However, this condition is very frequently encountered to a greater or lesser extent in weapons which have been assembled by inexpert persons. The .45 caliber bullets are, respectively, from the Service pistol (bottom, left) and the M3 silenced sub-machine gun (bottom, right). The large number of holes in the silenced barrel cause an extreme number of striae on the circumference of a bullet. This appearance is so characteristic that it can usually be picked out without great difficulty.

actual discharge. According to the OSS manual this weapon silenced is audible in general at a distance of only 50% of the audible range of the unsilenced weapon.

The British Sten gun was also silenced by using in part rubber disks. One of the authors saw one of these fired several times. The silencing was about equal to ours, but the weapon heated very badly indeed.

Ricochets

Another subject frequently discussed and sometimes not well understood even by Experts is ricochets. In general, any flight beyond the initial contact of a bullet and some other substance of sufficient solidarity to cause a bullet to change direction is called a ricochet. In the old days of smoothbore cannon and round cannon balls, ricochet fire was about as accurate as direct fire and had a good many advantages for certain types of actions. However, with the advent of elongated projectiles, rifling, and high-velocity bullets, ricochets are completely unpredictable; shootings resulting from them are usually accidental.

On occasion, almost any bullet can be made to ricochet. In general, however, it is a phenomenon of low velocity rather than high velocity. The way to avoid ricochets is to shoot only at targets and surroundings that will be struck by the bullet at near 90°. In other words, do not shoot at a snake swimming in the water at considerable range. One is almost sure to get a ricochet with bullets striking the water at a relatively acute angle. Stone, frozen ground, and on occasion almost any other surface will lead to ricochets. The late Capt. E. C. Crossman once said that a .44-40 would ricochet off a custard pie.

A bullet at high enough velocity and of sufficiently weak jacket metal will generally break-up upon hitting almost any surface. For instance, you cannot get a .220 Swift bullet to ricochet from water at any reasonable range. However, if you fire at sufficiently long range so that the bullet has slowed down, even a .220 Swift soft-nosed projectile would probably bounce off.

It is not at all uncommon, in observing coast artillery fire at towed targets on water, to see a half-ton 12-inch seacoast projectile skipping over the water like a flat stone thrown by a boy. They may bounce two or three times.

In general, a bullet that has ricochetted will bear marks of this when it finally comes to rest. A bullet that was extremely important to one of the writers did bear marks of this type in the form of sand inclusions from a ricochet from a piece of sandy building stone. Water ricochets are less easily identified although in some cases, they are said to have been made oval by the first contact.

The sound frequently associated with a ricochetting bullet is the whine that can be of greater or less intensity, depending on a number of factors. A bullet will probably not whine unless it is reasonably deformed on the first contact. For instance, a metal jacketed bullet may have the case burst open. The continued spin of the irregular metal jacket would then create a very easily identified sound; however, if the metal jacket is not broken, this sound will probably not be recognized.

Ricochets from water, ice, or like substances may actually carry as far, or it is claimed by some people, a little further than direct fire; however, the usual ricochet off dirt or similar substances considerably reduces the range. A loudly whining ricochet quickly expends its energy in spinning against air resistance though it is dangerous at close range.

Spent Bullets and Shot

An Expert will frequently be asked about spent bullets. Unfortunately, the questioner may not have in his own brain a very clear idea of what he means by spent bullets. In the machine gun tests at Daytona Beach a machine gun was fired vertically into the air; the bullets when they came down did not have sufficient velocity to bury themselves in sand. Presumably, no bullet from any handgun or shoulder weapon whatever when fired vertically would inflict a fatal wound upon returning to earth. The same would be true for spent bullets from airplane machine guns to a greater or lesser extent, depending upon altitude and original direction of fire. However, as pointed out previously in this chapter, bullets coming down from an elevation of 400 yards in the descending part of their trajectory were still capable of inflicting fatalities.

Most hunters are familiar with the phenomena of shooting at a squirrel or similar target in a tree directly overhead and having the shot patter around them. One of the writers, while in a hunting field received in his posterior parts half a dozen bird shot, probably about No. 9, at a measured range of 85 yards. These momentarily stung but did no harm whatsoever and did not penetrate the fairly heavy pair of brush pants that he was wearing at the time. On the other hand, one should not conclude that shot from a shotgun is incapable of long range damage since there is a phenomenon, better known in the old days of soft shot than today, in which for some reason several pellets will ball together and carry a very long way. One of the English shotgun authorities of the last century describes a case of this balling phenomena having inflicted an extremely dangerous wound at a range of almost 400 yards. Frankly, this range seems incredible; however, balled shot certainly will reach out a lot further than the single pellets.

Wound Phenomena

Sometimes bullets will not make round entrance holes. The reasons for this can be varied. The first and most important is that improper ammunition was used so that the projectile tumbled in flight. Undersized bullets from the wrong ammuntion will tumble. Lead bullets fired at too high a velocity in quick twist barrels will also tumble since they strip in the bores. For instance, certain foreign rifles in first-class condition have a twist so quick that cast bullets cannot be made to operate satisfactorily in them at any reasonable

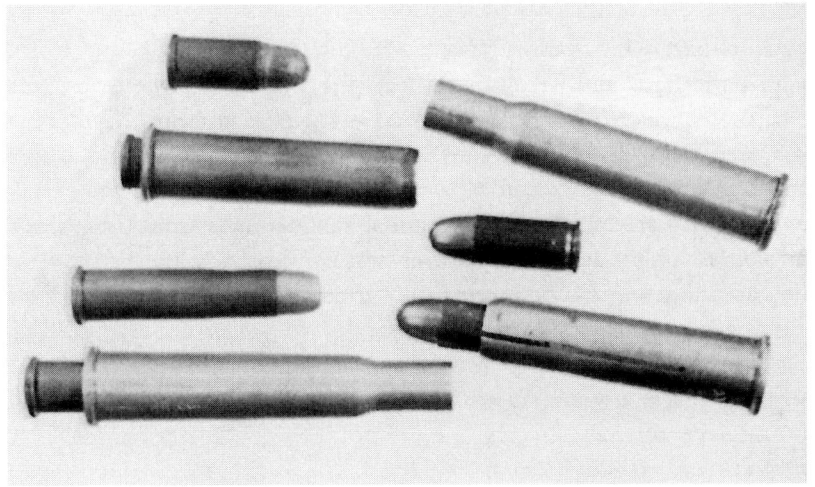

SHOOTING PISTOL CARTRIDGES IN RIFLES

These are auxiliary, or supplementary, rifle chambers, intended to permit the shooting of less-expensive pistol cartridges in high-power rifles, for short range practice or small game shooting. Any pistol cartridges so used must have bullets of the same diameter as the rifle they are intended to be used in.

Bullets fired through such chambers will take the impressions of the actual lands and grooves of the rifle used. Today, such adapters are not used to any great extent, but they are still available.

Formerly, similar devices designed to permit the use of .22 caliber or even more powerful rifle cartridges in shotguns were fairly common, but are not often seen any more. They were known as sub-caliber tubes; rifled shells; or cartridge adapters. They came in two styles, fitting into both bore and chamber, and ranging in length from about three to 16 inches. At this time—1956—the only ones available on the American market are imported, and are usually intended to permit the use of .22 short cartridges in large bore shotguns, though some for the shooting of 410 gauge shells in 12 gauge guns can be had.

velocity. Some weapons firing conical bullets are unrifled, particularly the very cheap .45 caliber ACP singleshot pistol furnished to the European Underground by us during World War II. These, of course, tumble in flight and cause irregular wounds.

Another reason for imperfect holes is during the initial stabilization period of flight close to the muzzle, depending on a number of factors, bullets will yaw. If a number of targets are placed several feet apart and shot through, the bullet may leave alternately evidence of the base first on one side of the point and then on the other. If the rifle and bullet are properly matched, this yawing will cease and a stable flight ensue within a few yards. However, at extreme range, for one reason or another instability may again occur.

An Expert should be on the lookout for the possibility that some clever individual has loaded a bullet already fired in a rifle or pistol into a shotgun shell or the like to create extreme confusion.

It is possible in theory to fire a test bullet from someone else's weapon and then load it into a shotgun in order to commit first a given crime, and second to place the blame elsewhere. Such a projectile, however, certainly would not have normal penetration nor accuracy. Except with extreme luck, the entrance wound would be most unusual.

The path of a bullet in the body can be determined definitely, as already pointed out. If the position of the body at the time of the shooting can be determined also, the detectives will know the position of the axis of the bore of the weapon. The same thing can be determined by two holes from the same bullet some distance apart if the materials penetrated were not so strong as to change the bullet's direction.

However, nothing can be determined by the axis of bullets at the time of their recovery in wood or the like as the scene of a crime or in the victim. In a recent New Jersey case the defense was trying to prove that the pistol which fired fatal shots was in a specific position. By questioning the Firearms Expert, the defense attorney tried to elicit the testimony that the position of a bullet lodged in a wall 8′ above the floor would indicate the exact position of the gun when fired. The maximum range possible in the room was 32½′. The witness refused to accept the defense claim that the position of the axis of the bullet was conclusive proof that the gun was 3′ from the floor. The Expert explained that the axis of the bullet could be in almost any position relative to the original axis of the bore of the crime weapon at the instant it struck the wall, after penetrating the back, shoulder blade, and chest of the victim. Even if a bullet strikes nothing save a final wall, yaw at instant of impact and other factors might bring it to rest in most any position.

Use of Magnets for Retrieving Firearms from Underwater

Occasionally it becomes necessary to search underwater for firearms which have been thrown there in an effort to avoid detection or to destroy such evidence. This operation calls for the use of a suitable magnet.

It is well to have about three sizes of magnet available, say 75, 125 and 200 pounds pull. These weigh about a pound for the first weight of pull and three pounds for the last and cost respectively

$6.50 to $16.50. They can be purchased from the Miami Magnet Company, 3240 N.W. 27th Avenue, Miami, Florida.

Work from a rowboat and drag the magnet slowly along the bottom over the suspected area. The magnet will attach itself to any iron or steel, including water mains and old automobiles. A swimmer or diver can go down to bring the gun up if the magnet does not hold-on or pull it up.

The 200 pound pull is a nice size to keep around the station house for loaning out to duck hunters who have dropped overboard a good grade shotgun.

GUN FAILURE AND THE FIREARMS EXPERT

Since the end of World War II, the firearms makers have received an unprecedented number of blown-up guns, along with implied or actual claims for damage either to the gun, or to the gun and shooter. Fortunately, gun failure does not frequently lead to serious personal injury to the shooter; however, there are occasional exceptions. Further, spectators are occasionally injured. Civil suits may become more frequent in the future. It is well for any expert to understand the basic problems involved.

Gun failures may be divided roughly into three classes: handguns, rifles, and shotguns. The causes of these failures may also be broken down into three separate categories: excessive pressure, obstructions in bores, and mechanical weakness in the weapon itself.

HANDGUNS

The American manufacturers of revolvers and automatic pistols are receiving a steady stream of blown-up weapons. Sometimes the gun is in such a condition as to almost defy explanation. However, not a single weapon of this type leaves any American factory without having been fired several times with proof loads developing far higher than normal pressure. In the opinion of the authors, no weapon has gone out of any of these factories within the past quarter century, at the very least, that could have been blown up by normal loads.

It is interesting to note that one almost never sees a blown-up revolver of the so-called "cheap" variety—the top-break and solid-frame H. & R., Iver Johnson, U. S. and such. This is not because these guns are made of exceptionally good steel, but rather because these two calibers are rarely reloaded; factory ammunition is used

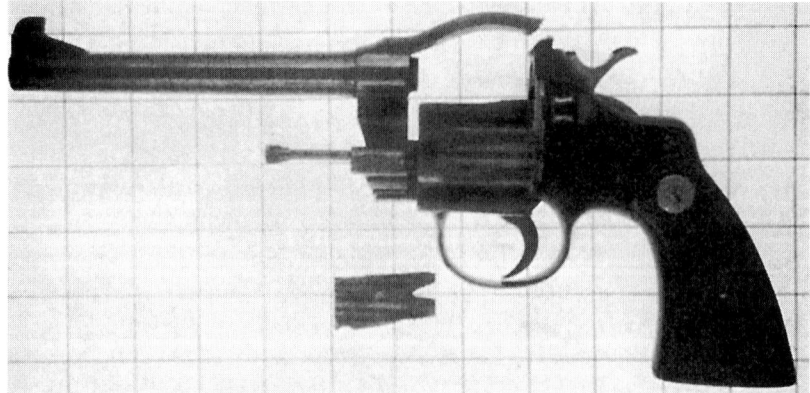

An Officer's Model Colt revolver blown apart by an excessive charge of smokeless powder. The party who reloaded this ammunition blew up three guns in a period of ten days (his own guns) and finally gave up the handloading of his ammunition. Bursts in a revolver such as this one invariably are the result of either an excessive or a double charge of powder.

almost exclusively and these low-priced handguns are amply strong to handle the cartridges for which they are intended.

The hand loaders usually load their ammunition for target or special-purpose shooting, almost invariably buy high-priced, first quality guns—and frequently blow them up with an excessive charge. The best handgun made will be wrecked by a case full of Bullseye powder behind a crimped-in bullet. Bullseye powder has been successful for over 50 years because it ignites easily and burns uniformly in reduced charges, for one reason. Yet these same properties will make it appear to detonate if the charge is so large that it is compressed or fills all air-space in the case.

Obstructions in revolver and automatic pistol barrels not always lead to blown-up weapons. A bullet can become lodged in a barrel without causing disaster. A second bullet either pushes out its predecessor, or stops behind it, sometimes bulging the barrel. Very few handguns are damaged at all by obstructions when using normal loads. Handguns are blown up by excessive pressure, usually from hand loads.

RIFLES

The authors feel that it is highly unlikely that any American-made rifle, or any foreign commercial rifle, will be blown up with proper ammunition unless, for instance, there are obstructions in bores. Even then, bulges are far more frequent than total ruptures, except where the obstruction is right at the breech. Neither normally causes personal injuries to the shooter, save in unusual cases. It is sometimes possible to fire a normal round of ammunition in an American-made bolt-action rifle with the barrel closed positively at the muzzle, and have the weapon hold the pressure without rupture.

Improper ammunition, however, is an entirely different story. While at the Marlin factory recently, we were shown one of their strongest rifles which had been blown up completely in some manner. This rifle was chambered for a .22 rimfire cartridge and was of the type that would stand a pressure of 80,000 pounds per square inch. The excellent research facilities of the Marlin Company were strained to the utmost to try to duplicate this blown-up weapon with a new weapon of the same type. Full charges of the various types of detonating compounds were tried. Pressures sufficient to practically melt the little .22 caliber rimfire cartridge cases were tried without in any way damaging the weapon. No claim was made by the owner of the damaged gun; he just wanted the ruined parts replaced at his expense. We have received no information as to what was in the cartridge that he fired at the time of the accident; however, it must have been something approaching fulminate of mercury in intensity.

SHOTGUNS

Most blown-up handguns and rifles are caused by improper hand loads. Shotgun ammunition, on the other hand, is not nearly so frequently hand loaded. Blown-up shotguns are generally the result of obstructions in the barrels and weaknesses in the weapon itself. Pressures developed by standard smokeless-powder shotgun ammunition are far below those in rifles and pistols. However, the strength of shotguns is also far below the other two classes of weapon.

A Reising submachine gun, with muzzle blown apart by a defective cartridge, the bullet of which lodged in the bore and was followed up by three more shots fired rapidly.

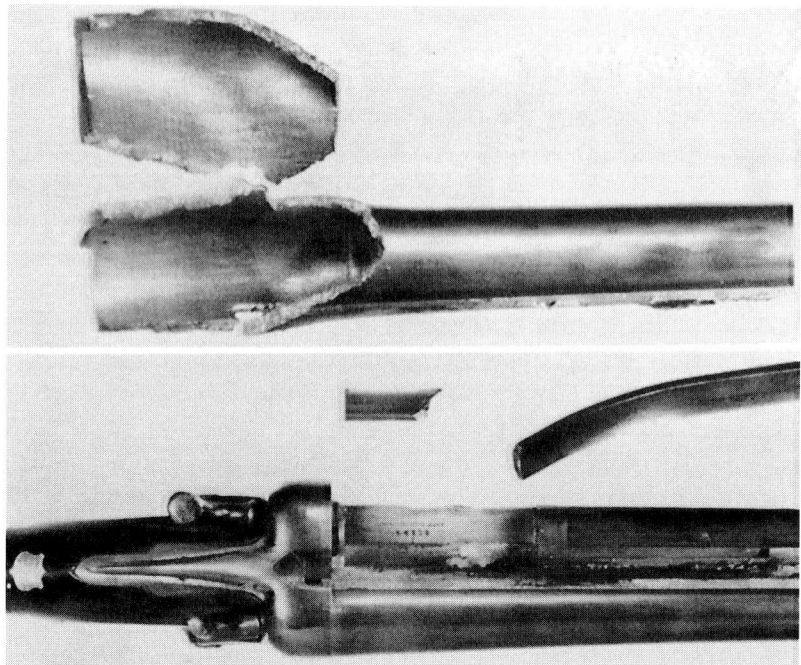

Showing a shotgun made about 75 years ago and fired with modern, high power ammunition. This kind of burst can occur with any of these old shotguns having Damascus, twist or laminated barrels. This particular burst was due entirely to failure of the old metal in the barrels of the gun.

This was a rather high grade shotgun in its day, but the fact that it cost around $150 in a decade when the dollar was worth one hundred cents does not justify its use with present-day high-pressure ammunition. Any old shotgun, such as this one, is safe today only when used with black powder cartridges—and some are not safe even then, as time, the elements, and long usage may have taken too great a toll since the day the gun was made.

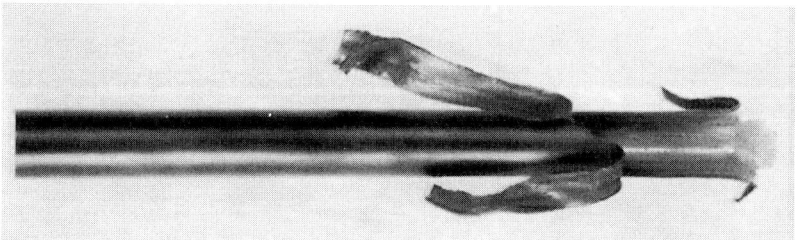

A rifled slug lodged just back of the muzzle of the barrel shown above and the gun was then fired with a load of No. 5 shot.

This illustrates the sort of burst caused in a shotgun barrel by a substantial obstruction such as wadded cleaning rags, a heavy scouring brush lodged in back of the choke, or a long column of wet mud extending back from the muzzle. Any weighty obstruction may do the same.

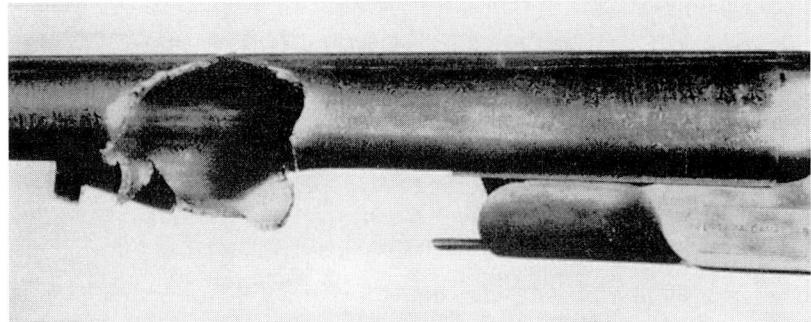

This illustrates a very frequent type of burst barrel in 12-gauge shotguns, which is caused by dropping a 20-gauge shell into the chamber mistakenly for the proper 12-gauge round. The 20-gauge shell drops forward into the bore, seating itself far enough ahead to permit the proper 12-gauge round to be seated and the gun action closed without hindrance.

Upon the gun not firing, the user looks in the chamber, sees nothing and assumes that he forgot to reload that barrel; he then slips the correct 12-gauge shell into the chamber without glancing through the barrel.

In this type of burst the origin of the blow-up invariably occurs about 4½″ forward of the breech, where the 20-gauge shell is positioned by its rim contacting the walls of the forcing cone and preventing it from falling down the barrel.

In the particular gun shown above, the magnitude of the burst was increased because of the old Damascus shotgun barrel, this gun being at least 50 years old.

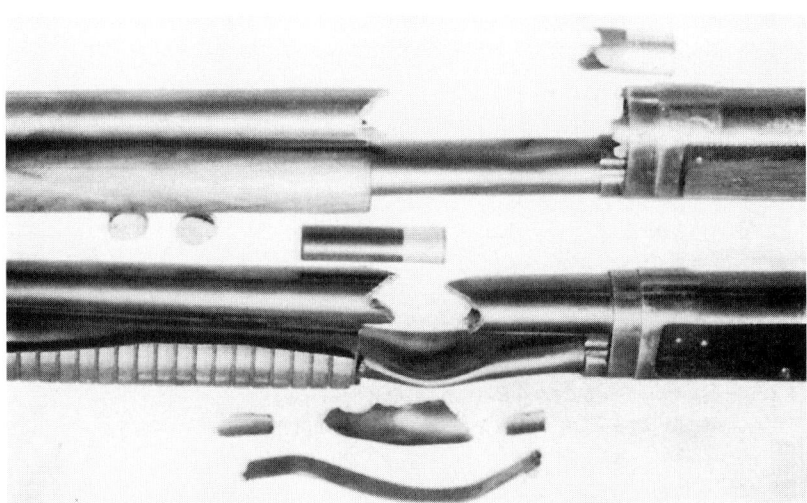

Here are shown two 12-gauge repeating shotguns with similar types of blow-ups. The upper gun was submitted by a user, it exhibited all the usual signs of a blow-up caused by a 20-gauge shell having lodged in the forcing cone of the barrel.

The lower gun was deliberately blown up in the laboratory by loading a 20-gauge shell ahead of a 12-gauge shell; the characteristics of the burst are almost identical to those of the gun in the upper section.

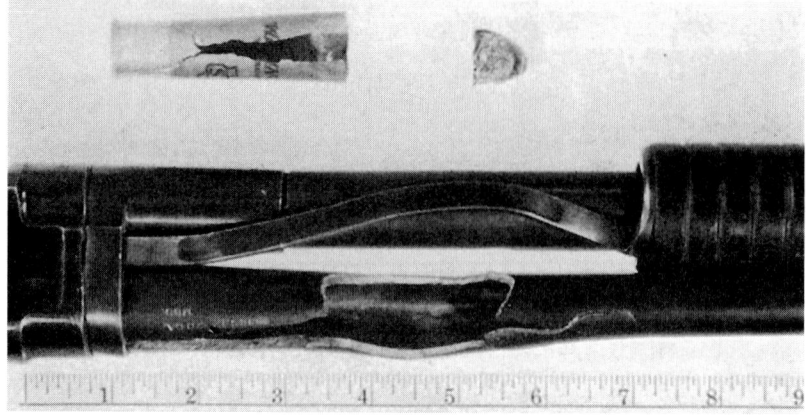

Showing a 20-gauge pumpgun barrel which was blown up by a 28-gauge shell being ahead of the chambered 20-gauge round. This condition is similar to that shown in the previous picture for 12-gauge guns, where a shell slightly smaller than the correct one for the gun drops down into the barrel and contacts the forcing cone yet permits the second correct shell to be chambered and fired.

Very many different things can obstruct shotgun bores. Mud, snow, cleaning brushes, and the like, are common. However, by all odds the most common is the introduction into a 12-bore weapon of a 20-bore cartridge, which will fall forward through the chamber and catch on the forcing cone in front of the chamber. This is certain to blow up the gun if a 12-bore shell is then fired in it. A 28-bore shell will usually do a similar job in a 20-bore gun.

Another very frequent cause of shotgun failure is in connection with Damascus barrels and modern nitro powder. Damascus barrels are no longer produced. They were beautiful and were, to most extent, safe during the era of black powder cartridges. However, it is pointed out that the stress-strain diagram for this metal always

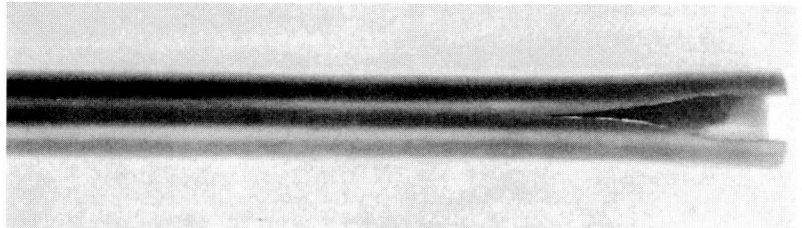

Here is shown the most common type of burst shotgun barrel. This particular shotgun had six inches of snow in its muzzle and was then fired with a rifled slug. A load of shot would have done the same.

Dirt, mud, muzzle plugs, snow, ice and water all may give a similar burst or split. Often a thin layer of ice formed inside the barrel, caused by dipping the muzzle into water in freezing weather, will cause this same split muzzle.

This illustration was taken of an old 20-gauge shotgun having Damascus barrels, and the burst is typical of failures in these early breech-loading guns which are now well below present standards.

In this particular instance, the barrel wall was paper thin, which, together with the low yielding strength of the Damascus steel, resulted in the eventual failure of the barrel. It is usual for these old Damascus barrels to perform satisfactorily for indeterminate periods of time, and then fail suddenly and unexpectedly and with no preliminary indications or warning. None is safe with today's modern, higher pressured ammunition.

had some of the characteristics of iron. Many of these Damascus, or laminated, steel barrels were actually welded strips of iron and steel placed alternately. Iron is not perfectly elastic; it will not return to its original condition after a heavy stress is relieved.

The net result of this is that many Damascus shotgun barrels burst, even in black powder days, after firing perhaps several thousand rounds satisfactorily. They were unpredictable and could not really be proof tested in the way that all-steel barrels can be.

The introduction of smokeless powder shotgun shells has considerably increased the danger of gun failure in this type of material.

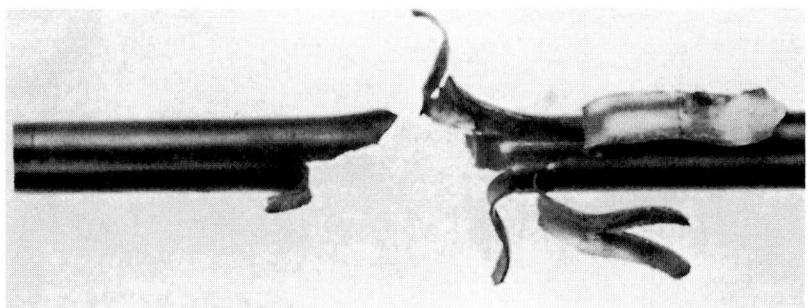

Above barrel had a rifled slug lodged at the muzzle and the gun then fired with a load of small shot.

This is the type of burst that can be caused by such substantial or well-lodged obstructions being in the muzzle of the gun.

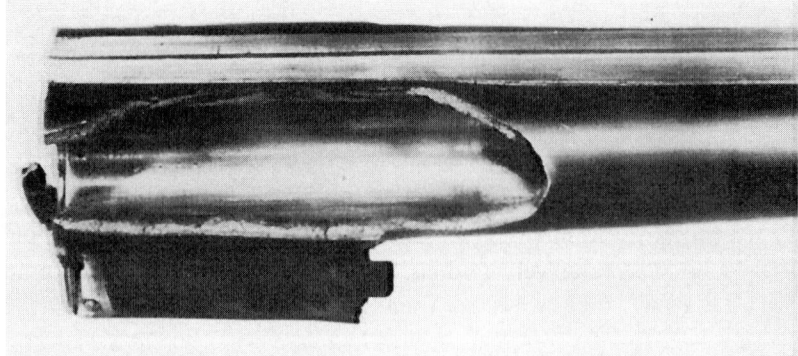

Here is shown the type of damage done to the chamber section of a shotgun by the action of an extreme pressure load—one entirely excessive. Subsequent to this burst, the barrel metal was examined and found to be of correct composition and heat treatment.

This failure is typical of those caused by extreme pressure loads, as opposed to bursts caused by obstructions down the barrel.

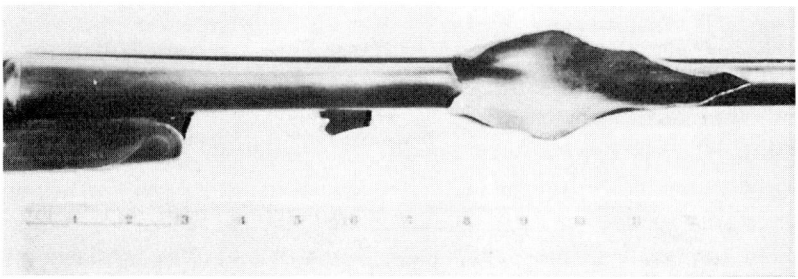

This shotgun barrel was burst open by an obstruction lodging down in the breech section of the barrel. Obstructions in this section of the barrel can be caused by any foreign substance having enough weight or frictional resistance to raise the bore pressure to the blow-up level.

However, obstructions in this portion of the bore generally come in the form of wadded wiping rags, cleaning patches, broken-off cleaning rod or stick wedged in the barrel, or a stuck metal scouring brush or appliance.

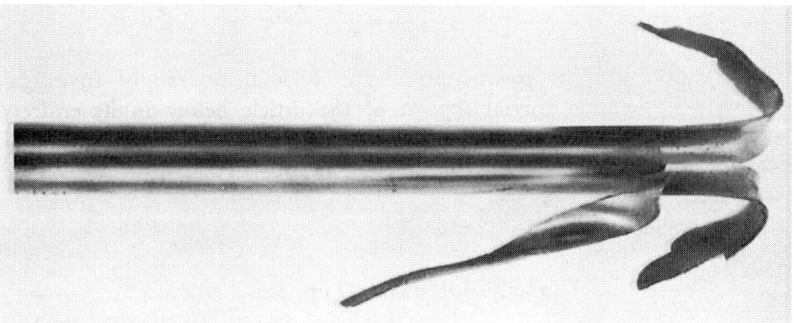

This barrel had six inches of wet sand packed in its muzzle and the gun then fired with a rifled slug. A load of small shot would have caused about the same sort of burst.

Mud wasps often build their nests in the muzzles of guns stood away in garrets, outbuildings or on porches, and can be the cause of a burst of this nature and extent.

Whereas there are in existence a few high quality shotguns with Damascus barrels that were actually nitro-proved and are therefore reasonably safe with smokeless powder loads, it is far better to look with disfavor at all such weapons.

Paraffin Test For Nitrates

In some ways, this is one of the saddest episodes in police history. Whereas in this country these tests were never completely accepted as final, and were never used as the entire conclusive evidence in homicide trials, they were so used in some neighboring countries to the south. Men were convicted and executed because of a positive paraffin test.

This test consisted of pouring hot paraffin over the hand or hands of a suspect and then chemically examining this paraffin for nitrates. If a man fires a revolver or automatic pistol the products of powder combustion will be deposited on the hand firing the weapon. These deposits can then be picked up by chemical analysis of the paraffin. This was the way that the test was originally supposed to work. Apparently no one ever bothered to do any control tests on it in this Latin American country.

The Federal Bureau of Investigation and other departments, however, have done control tests here. They have found the most alarming inconsistencies. For instance, a woman who could not remember ever having had a firearm in her hand in her life gave positive tests for both hands. Further, men who fired guns all day were sometimes able to remove most of the evidence of this firing by an ordinary washing of their hands with the usual soap provided in the wash rooms.

APPENDIX 1

FOREWORD: With permission of the Federal Bureau of Investigation, we are giving a partial reprint of the article below in its entirety which originally appeared in the *FBI Law Enforcement Bulletin* for October, 1935. The authors wish to point out particularly the early date of this article which is still completely authoritative on this subject, which has been in other countries a source of rather serious miscarriage of justice.

The Diphenylamine Test For Gunpowder

In view of the current widespread use of the diphenylamine test for gunpowder, this article has been prepared in order to set out for the information of law enforcement agencies the experience of the Federal Bureau of Investigation Laboratory with respect to such tests. In recent years, considerable publicity has been given to chemical tests used by law enforcement officials in homicide cases for the purpose of determining whether a deceased person was the victim of a murder or a suicide. The usual method has been the diphenylamine test. In this test the hands of the deceased person are coated with melted paraffin which upon cooling is peeled from the hands, the theory being that the paraffin will adhere to any particles of gunpowder which might have been blown back on the hand from the weapon, if it was in fact fired by the deceased. The surface of the paraffin which touched the hand would then be treated with a solution containing diphenylamine and sulphuric acid. If gunpowder was present on the wax, its presence would be shown by the appearance of specks of a distinct blue color, the normal reaction for nitrates.

The theory of the test is that the nitrate contained in gunpowder forms oxides of nitrogen at the time of the explosion which become embedded in the pores of the skin, particularly on parts of the hand exposed to the escape of gases from the gun. It also unites with the salts of the perspiration to form salts which readily react with the reagents used.

At the outset it should be understood that when this test is used and a positive reaction takes place, that is, specks of a blue color are found in the paraffin, the only conclusion that can be drawn is that some nitrate was present on the hand. In the case of a deceased person, whose hand gives a positive reaction to the diphenylamine test, it should be remembered that the deceased may have had on his hand one of a large number of substances with a nitrate content other than gunpowder. Substances which give the positive reaction include:

> Ammonium nitrate
> Ammonium molybdate
> Mercuric nitrate
> Potassium nitrate
> Potassium nitrite
> Silver nitrate
> Sodium nitrate
> Sodium nitrite
> Uranium nitrate
> Urine
> Nitric acid

From this statement, one might expect to obtain positive reactions from the hands of miners who handle explosives, farmers or home gardners using fertilizers, employees of nitrate, fertilizer, and fireworks plants, and in some instances chemists or pharmacologists. These facts should be kept in mind in order that a false conclusion that the deceased fired the weapon may be avoided. Such a false conclusion would, of course, indicate a suicide in a case which might actually be one of murder.

On the other hand, it is well to know the limitations of the diphenylamine test which are due to the fact that in a great many instances one may fire a revolver or pistol without leaving any trace of gunpowder on the hand which may be detected by this test. The automatic pistol, of course, is almost completely enclosed and practically all powder particles are forced away from the hand of the person firing the weapon. The revolver has considerable space between the chamber and the barrel which permits a greater quantity of gas and other products of the explosion to be freed near the hand of the person firing the gun. With the revolver, therefore, one would expect to find larger deposits about the thumb and forefinger. In spite of this possible escape of gas and powder particles, numerous tests have indicated that a person may be found to give a negative reaction to the diphenylamine test even though he has fired as many as 40 rounds of ammunition. Consequently the mere failure to find a trace of gunpowder by the diphenylamine test on the hand of a deceased person does not prove that he did not fire the weapon causing his death.

It should also be borne in mind that the finding of actual traces of gunpowder on the hands of a deceased person does not conclusively determine that he fired the weapon in question. It is possible that the gunpowder particles may have been blown on the hand directly from the barrel of the gun being fired at the victim. An attempt to shield the body by raising the hand would in some instances result in the implanting of powder particles on the hand.

Among the many laboratory tests made of the diphenylamine method, the one described below is illustrative of the usual results. Tests were performed on both the firing and idle hands of 17 men following a course of five shots with .38 caliber revolvers using smokeless powder. Six of the men, all firing with the right hand, showed negative results when the test was applied to both right and left hands. One man who fired with his left hand showed negative results on both hands. Three of the men who fired with the right hand showed negative results on a test of the right hands, but positive test on the left hands which were not used for firing. One man who showed a negative result on the right hand, when firing with that hand gave a doubtful result on the left hand. Two men firing with the right hands showed positive results on the right hand test and negative results on the left hand test. Four men showed positive results on both hands when firing the right hands only.

A number of experiments performed in the FBI Laboratory were directed toward the effect of tobacco residue on the hands of smokers. As a result of these tests, it was found that persons smoking in any considerable amount gave a positive reaction to the diphenylamine test when applied to the fingers and palms. This result would be expected, of course, in view of the presence of nitrates in the products of the combustion of tobacco.

In spite of the obvious impossibility of drawing positive conclusions from the diphenylamine test, when properly conducted and with a full understanding of its merits and limitations, it is of some value in criminal investigations.

The following is a procedure for conducting the test:

Clean paraffin is melted and painted over the hand, particularly the palm and about the thumb and index finger. This is allowed to cool and is then removed by carefully rolling the edges and peeling the paraffin, endeavoring to keep it in one piece. After the wax has been removed from the hand, it is placed in a white porcelain dish or china saucer. As a precaution, the use of enameled ware or other metal ware is to be avoided because of the action of strong acids on the metal. To test for the presence of nitrates, add a portion of a solution composed of the following:

> Diphenylamine 1 gram
> Sulphuric acid C.P. (free from nitrates) 100 c.c.

If nitrates are present, a blue color appears on the paraffin.

In order to be sure that the test is being correctly performed, a control test is conducted. The melted wax may be poured over a small area, such as the bottom of a pan or dish, and allowed to cool. The test solution is then applied to the upper surface. No color change insures the examiner that the wax itself does not contain nitrates. In performing this test, caution should be taken to avoid any contact between the body or clothing of the person performing the test and the sulphuric acid, inasmuch as such contact would result in severe burns.

No reported cases involving any attempt to introduce evidence concerning the diphenylamine test have come to the attention of the Bureau.

APPENDIX 2

Toolmark Identification

FOREWORD: The Firearms Laboratory in almost every Police Department will be occasionally called on to do work in connection with matching evidence from various crimes with suspect tools. Other cases may involve the matching of toolmarks on two different samples when the tool itself is not available. Some departments consider Firearms Identification as a special division of the general science of Toolmark Identification. Toolmarks are present on almost every material object in the country. Almost everything is made with tools which, to a greater or lesser extent, leave marks.

Scientific identification of toolmarks is now accepted without question in courts. The Supreme Court of the State of Washington ruled on 28 April 1930 in an opinion quoted below:

"Courts are no longer skeptical that by the aid of scientific appliances, the identity of a person may be established by fingerprints. There is no difference in principle in the utilization of the photomicrograph to determine that the tool that made an impression is the same instrument that made another impression. The edge of one blade differs as greatly as the lines of one human hand differ from the lines of another. This is a progressive age. The scientific means afforded should be used to apprehend the criminal. (Case of Washington vs. Clark.)"

Tool mark identification. Positive and unusually good comparison from the blade of a brush hook.

Tool marks are basically of two types. First, there are negative impressions such as those on molded objects, surfaces hit by blunt tools, and impressions made by one surface on another. Fingerprints, footprints, and breech face impressions of cartridge case bases are of this type. Second, there are generated impressions or striae made when one surface moves past a series of points on another. Such impressions are made by cutting of a softer material with a tool or the passing of a bullet down the bore of a firearm. Sometimes a combination of these two types of marks remains, as in a glancing tool blow which generates striae but also leaves a final negative impression of the tool face. In firearms, a case of this type is a primer indented by the hammer nose of a Colt Single Action Revolver.

The identification of suspect tool with marks left at the scene of a crime is the most frequently met with problem of this type. For instance, any stroke of an axe, knife, hatchet, chisel, scissors, or the like, will leave striae in the material cut matching the edge of the tool at the time the stroke was made.

The edge of even a new or freshly sharpened tool when viewed under a microscope will show ridges and hollows from the processes of manufacture and sharpening. A worn edge is even more irregular. Each ridge

TOOLMARK IDENTIFICATION

A most accurate and positive comparison of an axe blade, made by microphotography of evidence and test cuttings on cedar fenceposts.

and irregularity leads to a corresponding negative impression on matter worked upon.

Unfortunately, the material which may be cut and the instruments which may be used to do the cutting are so diverse and variable that a great deal of initiative is needed. Sometimes it is possible to make a comparison directly between the object cut and the instrument doing the cutting, although in this case you are using a positive against a negative with consequent possible difficulties. In general, a negative to negative comparison is preferable. A chisel mark in paint or a photograph of it can be microscopically compared to a test mark made in lead or plastelina with a suspect tool. Much common sense is needed in this work. Nobody gets enough of it of any one type to establish definite procedures.

Nails cut with a cold chisel in a burglary-murder were received a short while ago in the New Jersey State Police Laboratory along with a suspect tool. A positive identification was made with similar cuttings of small lead cylinders. Test nails could have been cut and used instead, but there would have been danger of changing the evidence chisel.

In another somewhat similar case, a wood chisel was used to cut the hasp out of a lock. The tool blade was so badly dulled that no positive identification could be made as to pattern of striae. However, a small portion of the original tool edge found at the scene of the crime fitted perfectly into place in the suspect tool.

Unfortunately, general characteristics of striae made in the material by an edge may not be uniform in distance apart. For instance, if the axis of an instrument is not at right angles to the material cut, the striae left will be closer together than the points which caused them on the instrument itself. Further, if there is a change in the relative position of the axis between the time that one portion of a blade passes through an object and another portion passes through the same object, the striae remaining will not be constant in their distance apart.

Signatures of tools used months or years before on objects when manufactured may be of importance. Finished lumber carries the characteristics of its planer for a long time. A piece of telephone cable recovered at a junk dealer's was identified as stolen by the die marks on the inside of the lead sheath, made when it was originally manufactured years before. These matched exactly the marks inside the two cut ends still in place.

Other things besides tool marks can be identified. For instance, teeth will leave characteristics in certain substances bitten. Progress has been made in the appearance of hair subjected to one or more blows with various types of instruments; however, these are far beyond the scope of the present study.

It is quite possible to use photography to advantage in comparing marks, particularly when the material marked cannot be moved to the laboratory. For instance, a door of a vault may have been impressed by a chisel. To take the vault door to the laboratory is obviously impractical. A photograph will frequently serve the purpose, although some form of cast of the mark may be superior. A cast has the added advantage of being directly comparable to a tool.

We wish to stress again the common sense approach to this special work. It will not happen very often. When it does happen, the Expert will be pretty much on his own since no two comparisons are ever exactly alike. A clever man will not only think of a way to make his comparison, but also present it concretely and dramatically in court.

APPENDIX 3

Restoration of Badly Rusted Firearms

FOREWORD: We are reproducing exactly a memorandum written some time ago for the New Jersey State Police Firearms Laboratory.

Frequently a murder weapon is thrown into a stream, lake, or the ocean, or is buried in woods, fields, or swamp, and not recovered for many months. When recovered, it is usually an almost unrecognizable mass of rusted iron and rotted wood, deeply encrusted with mud, rust, and slime.

It would seem an almost impossible task to clean such a gun sufficiently to fire it, let alone to identify bullets and shells fired in it. However, it frequently is possible to restore such weapons to working condition, fire test shots, and make positive identifications of test bullets and shells with the evidence bullets and shells.

First, the gun is carefully examined to determine whether or not it is loaded; then an effort is made to unload it. Usually this is not possible because every movable part is frozen fast in a thick encrustation of rust, dirt, and mud. As much of this encrustation is scraped off as can be done with scrapers and wire brushes. The gun is then soaked in a solvent until the rust and corrosion have been softened or dissolved sufficiently to permit removal. This soaking may take anywhere from two days to several weeks. The .32 Smith & Wesson revolver, illustrated from the Jessie Fitchett murder, was soaked for 18 days, and cleaned up remarkably well.

The rust is scraped and scrubbed off until the gun can be opened and unloaded, disassembled, and the internal parts cleaned and, if necessary, replaced in order to restore the gun to operatable condition.

The bore is wiped as clean as possible by using cloth patches, removing all the loosened rust and fouling. The use of abrasives and wire brushes for this first cleaning should be avoided; some additional damage might be done to the bore beyond that already done by the rust.

Surprisingly enough, most barrels are in far better condition than the rest of the gun would indicate. This is at least partly due to the use of non-corrosive priming in present day ammunition. The inside of a barrel is, because of its position, better protected than any other part of the weapon.

The solvent used should be one that is economical as well as effective, and while reasonably fast in action, not so harsh as to remove or alter the surface of the metal. For example, a 10% solution of hydrochloric acid will quickly dissolve rust, but will also nearly as quickly remove most of the irregularities in the bore that put the barrel signature on the bullet.

We have conducted many experiments along these lines, using various commercial solvents, strippers, and acid and alkaline baths with varying degrees of success. While we have not arbitrarily ruled out the caustics, acids, and hot bath strippers, we do most of our restoration with a solution of Kotol and Magnus-Metaffin. Kotol is an oil base preservative for paints and metals, but is quite expensive: $9.00 per gallon. Magnus-Metaffin is a commercial penetrating oil, used in garages. It is relatively inexpensive and is purchased by our Automobile Maintenance Shops in 55 gallon drums. We obtain it from them as needed, a gallon at a time, and make our solution one part Kotol to four parts Magnus-Metaffin, which works very well. This Kotol has many other uses and requires some watching as people have a habit of "borrowing" it.

Test shots through a weapon rehabilitated after extreme exposure are far less uniform than those from an undamaged weapon. Test firing is done in accordance with no standard procedure. Each weapon will react slightly differently. For instance, frequently the first shot removes some of the new rust that has accumulated since the original crime bullet passed through the barrel so that it does not match the crime bullet. However, if successive shots are fired, some of the later shots may give a positive identification. There is no definite rule for this. It is far better to fire several shots, even up to as many as 20, and compare them all to the

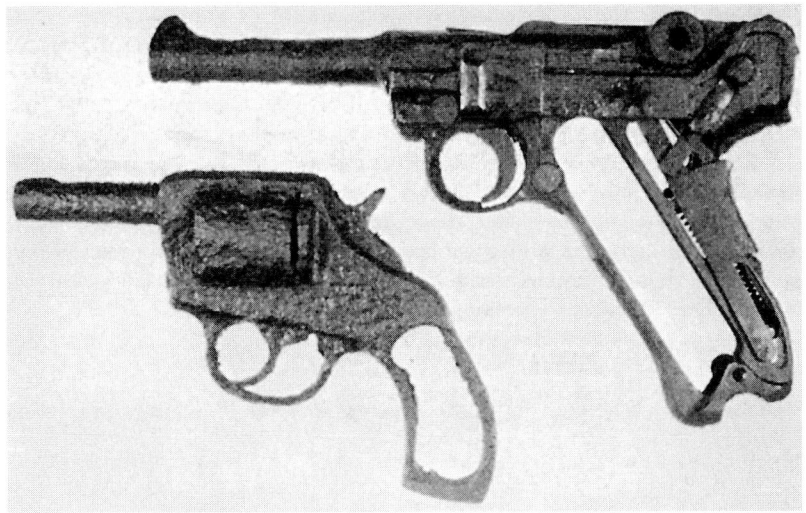

RESTORATION OF BADLY RUSTED HANDGUNS

These two handguns had been thrown into a river after being used in the commission of a crime. They were recovered some months later, through confessions of the criminals.

Both guns were partially cleaned up with a wire buffing wheel before being put into the rust-removing solution. The Luger pistol had been soaking for 10 days at time picture was made; it should come out in fair shape for evidence purposes. The .32 H & R revolver had been soaking for 54 days when photographed; results in its case do not look promising.

crime bullet than to be satisfied with a negative match after one or two test bullets.

A good deal of experience is necessary to know exactly how much cleaning to do of the barrel before beginning to fire test bullets. However, in general, it is far better to do too little than too much. A bristle brush is about the limit. Abrasives of all types should be avoided because of their possible effect in court later on. Evidence should not be changed in any way unless absolutely necessary.

There is no hard and fast rule in connection with the changing of evidence, so long as the result of the change cannot be considered as unfair to the defendant. For instance, if a weapon is recovered in such bad condition that it cannot be fired, almost all courts would allow the transferring of the barrel from that weapon to another similar weapon, and the firing of test bullets in the original barrel. This is frequently quite easy in some automatic pistols. It is more difficult in many revolvers. The transferring of a barrel will not in any way change the marks made on the bullet, although, of course, entirely new marks would be made on a cartridge case since a different breech would be in use.

Sometimes a slightly oversized lead bullet can be pushed through a barrel and give striae sufficiently like those on a crime bullet to positively prove an identity. However, this basic arrangement, whereas it will work in the case of soft lead bullets sometimes, is not by any means beyond question. It should not be used whenever it can be avoided. There is no established precedent that will apply in all cases in regard to validity of a positive comparison between a crime bullet and a test bullet pushed through a barrel by hand slowly.

In conclusion, it might be pointed out that perseverance and gentleness are the qualities most needed in the restoration to shooting condition of firearms after long exposure. It is astonishing what can be done in some instances, although in others, restoration will be completely impossible. For instance, a weapon after a few weeks in salt water is very likely to be gone beyond reclaiming, particularly if it has lain on the beach where it was sometimes wet and sometimes dry.

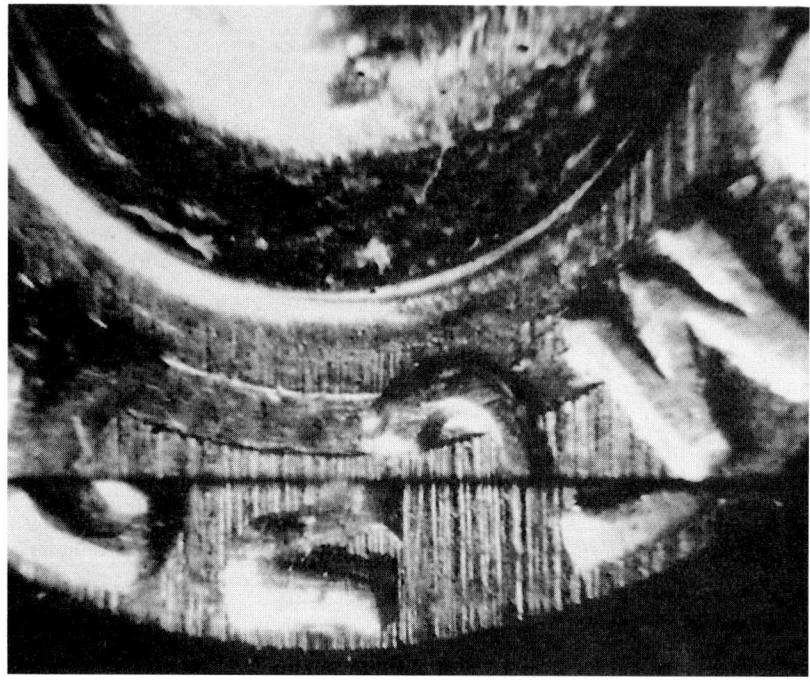

An example of really first-rate use of the comparison microscope in cartridge case markings. It is unusual for a photograph of such a comparison to show such extreme similarity as appears above.

CHAPTER 16

PRESENTATION OF LABORATORY FINDINGS AND COURT TESTIMONY

THE Firearms Expert must not only do his work meticulously, accurately and efficiently; he must also report his findings in the same manner. The public is the judge of so many things in America. Your particular public is officers in your own department, the cooperating departments, and finally the judges and jurors before whom you give expert evidence. Unfortunately, very few of these people are in a position to really evaluate your work. They judge you and it through your reports and by your appearance upon the stand. These things are vitally important.

Laboratory Reports

In general, every Firearms Laboratory will have special report forms which must be used. These forms will include in some instances blanks for information not vitally necessary; however, until you can get the forms changed, fill in all the required information meticulously. One well known department requires a mass of rifling data. Their Ballisticians fill this in scrupulously though they are personally sure it is unnecessary.

If a new form of report is to be set up, simplicity is just as important as formality. There should be plenty of space for descriptions of evidence and the like but specific spaces labeled as to what should go into them should be avoided. The source, date, and method of delivery of evidence as well as all other pertinent information should be shown. Details will vary with different laboratories.

The heart of any report is the actual ballistic answer. This should be kept as short and concise as possible consistent with clarity. For instance, if a crime bullet and suspect weapon are submitted, one of the three answers listed below will probably suffice.

Positive: "A comparison was made between the evidence bullet marked ABC and test bullets fired through the suspect weapon (identify fully) with positive results; the evidence bullet was fired in the suspect weapon."

Negative: The same as above, concluding "with negative results, the evidence bullet was not fired in the suspect weapon."

Indeterminate: The same, excluding the words "the evidence bullet was not fired in the suspect weapon," but concluding with a statement: "There were not sufficient similarities of personal and accidental characteristics on the evidence and test bullets to establish identity either positively or negatively." This is rarely necessary.

Cases Where Appearance in Court is Not Necessary

In at least 75% of all cases, the final report will close the case as far as the Firearms Identification Laboratory is concerned. There are several reasons for this. Perhaps the most important is that most investigators of firearms suicides send the weapon, fatal bullet, and cartridge case to the laboratory as a precaution, even though this may be a mere formality. If the match is positive and no other unusual circumstances are disclosed, there is not anyone to prosecute. Murders followed by suicide of the murderer leave the same sort of situation.

There is another form of case also that frequently is settled without an appearance of the Expert in court. In many political subdivisions of the country, the prosecutors or district attorneys have a great many cases; the courts are overcrowded. In certain cases a compromise may be reached with the defendant accepting less than a maximum sentence in order to save the overworked courts' time and the taxpayers' money. For instance, one criminal may have shot another under circumstances that make it difficult or impossible to prove conclusively a first degree murder charge. With full approval of the judge and without any underhandedness or chicanery, the prosecutor and defense attorney may get together in the presence of the judge. The laboratory report may be given to the defense attorney to read. If it is conclusive, the defense attorney may ask if a second degree plea of guilty, which will automatically save the defendant's life and the expenses of a long trial and the time of the court, will be accepted. If this is satisfactory to the defendant himself, the Firearms Expert will, of course, not be called upon to testify, since there will be no formal trial.

Similar settlements are often made for third degree murder and even manslaughter. A trial is always a gamble, both for the defendant and for the prosecution. They will frequently each sacrifice something of their maximum opportunities in order to be certain not to receive the maximum in unfavorable results. In many non-capital cases, such as holdup and robbery, atrocious assault and battery with intent to kill, carrying concealed weapons, and the like, the prosecutor will accept a plea of guilty to the indictment. The case does not go to trial; the prisoner is brought before the judge on a day set aside for sentencing. Usually a dozen or more cases in a calendar are disposed of in this manner in one day.

The honesty, integrity, and complete efficiency of modern Firearms Investigation, particularly when conducted by laboratories with many years of experience, is such that in more than half the cases, the defense attorney accepts the report of the Firearms Laboratory without contesting it. If the case comes to trial, the Expert will present his testimony. However, his qualification will not be contested nor his evidence questioned. This frequently happens when both the Expert and the defense attorney are old hands. The defense will admit that the suspect weapon fired the crime bullet but claim that the defendant did not have the weapon in his possession at the time of the crime. They may also claim that though the defendant actually fired the shot, there was no intent to kill. A ricochet from the sidewalk or a building can be claimed. In a recent Eastern case a clever defense attorney introduced medical testimony to this effect and considerably reduced the penalty paid by the defendant.

Court Appearance For Defense

In certain circumstances, the report of the laboratory may be subpoenaed by the defense. They may actually call the Expert to testify in their behalf. Cases of this type are not common, but they do occur. The prosecution cannot legally withhold evidence even though it is unfavorable to their case. However, they have a rather wide discretion as to what is pertinent and what is irrelevant. For instance, if a pistol captured on the defendant two hours after a crime did not fire the crime bullet, he is not automatically innocent. He may have changed guns. It may be possible to convict him on other evidence.

In a case of this type where the Expert reports a negative or inde-terminate match, the defense attorney will probably be able to find out the content of the report before the trial. If he cannot do so by apply-ing to the prosecution—these reports are not secret—he can by court order. If the evidence of the Expert will help the defendant, the Expert will be subpoenaed and will testify as to his findings. Further, the prosecution cannot very well question the qualifications of their reg-ular Firearms Investigation Personnel.

Court Appearance for the Prosecution

By far the most important cases for the Expert are those in which he appears for the prosecution. These are almost always intensely inter-esting. Each case, even after years of testimony, is a little differ-ent. Usually the firearms evidence will be developed out of the medical testimony although where the Expert brings the crime bullets and/or cartridge cases into court, this may require two appearances on the witness stand. First, the coroner or medical examiner will be called upon to testify as to cause of death and to identify the bullet or bullets that he took from the body at the time of autopsy. If the bullet or bullets traveled through the body and was found at the scene of the crime, the person actually finding this bullet will be asked to testify. One of these two individuals should have marked the bullet or bullets with his initials so that continuity of evidence can be shown. Second, the police officer receiving the bullet either from the coroner or person finding it at the scene of the crime will testify as to the identity of the bullet or bullets which are being introduced as evidence. He will state that he delivered the evidence bullet to the Firearms Laboratory and surrendered it against a receipt which he will have in his possession. Third, the Expert representing the Firearms Laboratory will then be called. In general, he will be either the head of his laboratory or the man who has made preliminary examinations of the evidence. In any case, since he is appearing in court to give evidence as an Expert, his qualifications for this special consideration by the court must be established. This qualification procedure can take from five minutes to more than five hours de-pending upon the reputation of the Expert and the experience of the defense attorney.

The prosecutor will establish the Expert before the court. The first questions asked will bring out to the judge and jury the qualifications of the Expert. Naturally, these questions will be formulated to show the Expert's experience, training, and general familiarity with the work that he is doing. They will, of course, put the matter honestly but in the best light possible. The defense can then question the witness. In general, this is the toughest possible time for the Expert, especially if he is reasonably new and inexperienced in court appearances. Even a man with years of it behind him must be on his toes. Although the defense attorney may know the Expert is a good man and even like him personally, he will do all in his power to put the Expert in the very poorest light in the eyes of the judge and jury. The attorney must in conscientiousness do his very best for the defendant. If it is at all possible to create a doubt in the mind of the court as to the Expert's qualifications, he will do so. This is perhaps the best way to cast as much doubt as possible on the final damning testimony that the Expert will make later.

The questions asked during this qualification ordeal are impossible to anticipate. They range from just plain stupid interrogation based on the defense attorney's one day deer hunting last fall to diabolically clever traps. A keen mind may have stayed up most of the night absorbing the most technical details of some scientific apparatus like the Filar eye-piece or the theory of breaking up light of various wave lengths in a spectroscope. He may introduce this subject haltingly and innocently into the examination and pounce hard on the first words that the Expert says that are open to misinterpretation.

Unfortunately, some Experts in the past have tried to outwit defense attorneys and win arguments in court. Even the best of the Experts have had their troubles, at least a part of the time, with this procedure and in no circumstances can it be recommended. It is impossible to foresee the actual questions, even in a specific case; however, certain general rules should be kept in mind by every Expert who testifies. First and most important is to keep one's temper and be as calm as possible. The defense attorney will try in every way at his disposal to get you flustered and cause you to lose your temper. Remember, he is very good at this sort of thing; otherwise, he would not be a trial lawyer. Words and their misinterpretation are his specialty. You are an expert in your field too, but do not try to meet him on his own ground.

The Chief Firearms Expert with more than 20 years court experience on the largest police force in the world points out that he never replies to any question until he has allowed a certain space of time to elapse for the prosecution to object to the question if they see fit to do so and for him to think carefully. He points out further that he replies with a very minimum of words and introduces no new ideas whatsoever. One of the serious mistakes made in the past has been for the Firearms Investigator to volunteer information. Every word that the Expert says increases the opportunity of the defense attorney for malicious misinterpretation.

The Expert, particularly a man new to the job, should never under any circumstances allow himself to get out of his depth in generalities having to do with general education or to endeavor to bluff in connection with his training, his general arms knowledge, or his reading. Do not be afraid to state, "I do not know." For instance, in the past, some men have been questioned at length about production procedures in connection with the manufacture of firearms. No one should be goaded into guessing or hastily remembering what was done in a factory which he may have seen briefly years before. After all, you are being examined for your ability to compare crime bullets with test bullets under a comparison microscope, not in connection with your ability to organize a factory and manufacture arms.

Unfortunately, you cannot go to the other extreme with replies such as, "I don't know," and "I don't remember," since if you adopt this as a standard procedure, the defense attorney may get you to answer the firearms equivalent of, "I don't remember whether I had breakfast this morning or not." However, of the two, it is far preferable to risk the jury thinking that you are slightly soft in the brain, than to being caught in a misstatement of fact or in an opinion contrary to known facts.

It is important to remember, however, that some extremely well thought of publications such as the Encyclopedia Britannica and some of the early text books on Firearms Identification have voiced opinions not now generally accepted. In the Fourteenth Edition of the Encyclopedia Britannica published in 1938 we find on Page 561 of Volume 12, "After examination of a bullet, they (the Experts) are able to state the make, caliber, type, approximate date of manufacture and approximate serial number of the gun from which it came." This is pure gibberish, as is pointed out at length elsewhere.

Of particular importance is the caution not to get caught in any lengthy explanations of any type whatever, since they are so easily turned to your disadvantage by a clever attorney. In the past, Firearms Experts have tried to explain in court things like vernier scales and the way that measurements are made with them as well as the actual mathematics in connection with their construction. A vernier is familiar to us all; we can all use it. Yet how may of us can clearly explain the principles upon which it operates universally in court with a prosecuting attorney continually trying to change the meaning of our words? Emphasize the fact that you know Firearms Identification but are not either a professor of science nor a debater.

Perhaps some instances of contested qualifications in court recently may be of interest. One of the authors had a slight operation on a day when he was scheduled to appear in court. He sent his assistant, a capable young Trooper already an Expert in the very best traditions of the science. However, the defense thought they saw an opportunity to discredit a new man without much court experience.

"I have here three cartridges, Trooper. If I were to fire them in a gun could you tell me the muzzle velocity of the bullets?"

"No, sir."

"Not even if I let you examine these cartridges?"

"No, sir."

"Your Honor, I object to the qualifications of this witness."

"Please tell the court why you can not give these muzzle velocities, Trooper," said the judge.

"Yes, Your Honor. Unless I examined the cartridges before they were fired I would have no way of knowing the amount and nature of the powder charge or even if there was any powder inside the cases at all. If I was allowed to disassemble the unfired rounds an approximate answer could probably be given by weighing and analyzing the powder. However, if the Defense Attorney fired the rounds there would not be anything left to analyze. The only sure method of determining velocity is by use of a chronograph. The rounds in his hand there may or may not be factory produced. Even if they are, there are considerable variations between different brands and even lots of the same brand. I do not remember the published velocities for all the more than 200 cartridges loaded commercially even today; there were many more varieties 15 years ago."

The court accepted the witness with very little further questioning. This was rather a simple problem well handled. An example of a more difficult one badly handled comes to mind. A witness was asked if he was acquainted with a list of 10 books; eight of these titles and authors were well known texts, but two were entirely fictitious. The Expert said he was acquainted with all 10.

The defense attorney then read a passage purportedly from one of the well known authorities and asked the Expert for his agreement or disagreement, with particular emphasis on one sentence cunningly changed from the way it had orginally appeared. The Expert again fell into the trap. The Expert was eventually qualified but the confidence of the jury in his knowledge was shaken.

Actual Court Testimony

After qualification, the Expert will be asked to state his findings in connection with the evidence. Frequently he will physically bring this evidence into the court and present it. The prosecutor will bring out the opinion of the Expert by direct questions. There seems to be a difference in procedure between different Experts in the matter of court testimony in connection with their findings. The more numerous group prefers to use the word "opinion," whereas some men treat the same findings as facts. It is merely a psychological argument between different schools of thought. The same can be found among Fingerprint Experts. One school of thought has an opinion that no two sets of fingerprints are alike because during the entire history of fingerprinting, no identical fingerprints have ever been found. No identical barrel characteristics have ever been found either. There remains, in theory at least, the slim possibility of finding at some future time identical fingerprints from different people and identical bullet striae from different weapons. Perhaps a statement as follows is about as good as can be made. It would follow questions of the prosecuting attorney.

> Prosecutor: Have you examined the evidence bullets and the evidence firearm?
>
> Expert: I have.
>
> Prosecutor: What were your findings?
>
> Expert: I have evaluated the similarities of striae between the evidence bullets and test bullets from the evidence firearm and reached the conclusion that the crime was committed with the evidence firearm.
>
> Prosecutor: Please tell the jury how you performed your examination.

The last is not necessary, though in general it will be of value in many cases. The Expert will appear in the best light when discussing his actual laboratory procedures. A tendency to lecture the jurors should be curbed, however; it can lead to undesirable results. It is not necessary today, although it was years ago.

Pretty much the same thing can be said for the various properties used in the presentation of these lectures in the past. For instance, models were used sometimes as already referred to made of wood to simulate bullets with their individual striations. These were cut apart to simulate what might be seen under the comparison microscope. Photographs of bullets, cartridge cases, and even casts of the barrels of suspect weapons were formerly introduced as evidence. As discussed elsewhere, these were expensive, time-consuming, subject to misinterpretation, and finally unnecessary.

Comparison microscopes have been brought into the courtroom in the past with varying results. The essence of the matter is that the average juror, being completely at sea in matters scientific, has no notion whatever of microscopes in general or the pictures that he sees when he looks through them. Further, these people are human. They may be distrustful of something they do not understand and be unfavorably impressed by even the best comparisons all set up for them. Even where there is an extra eye piece and pointer for explanation by the Expert, there may be subconscious distrust of the unfamiliar. As referred to before, many lines match in a good comparison; however, many more do not. The layman has no way of evaluating the evidence since undoubtedly he has never seen another bullet under the magnification of even a hand glass. He undoubtedly takes away with him at least a reasonable doubt.

On the other hand, human psychology will react to another stimulus which, even though it is favorable to the prosecution, may not be favorable to justice. In one instance during the past two years, a defense attorney insisted that a comparison microscope be brought into court so that the jury could decide for themselves. The Expert explained the instrument, after which the judge inspected the comparison. Apparently, he was satisfied. Every member of the jury followed suit. Probably not one of them really knew what they had been looking at. Human nature being what it is, no one of the jurors wanted to be the first to confess his lack of understanding. The defense attorney's strategy had backfired and the defendant received a second degree murder conviction when the prosecution might have been willing to settle for less.

On the other hand, there is a definite precedent for the defense objecting to the prosecution bringing a comparison microscope into the court on the grounds that jurors are not qualified to examine evidence, but merely see the relatively small area the Expert puts before them. All these things lead to the same conclusion: The reason for Experts is to interpret evidence for the court. They should do this and not merely prepare evidence in certain ways for evaluation by unskilled laymen.

Today the prosecutor will question the Expert only sufficiently to bring out "the suspect weapon fired the crime bullet." Nothing else is usually important. Save in New York City, most Firearms Experts know nothing of the crime or its scene. Only in special circumstances will an Expert in most departments do much work outside his laboratory and range. He does little or no general investigation.

The defense, however, will cross examine. Again the Expert must be on his toes, although now questions must be relevant to the particular case. The defense will first try to shake the specific testimony of the Expert by endeavoring to get him to qualify it in some way or lead him to some inconsistency. If a report is positive, the crime bullet came from one and only one weapon in the whole world. No qualification is possible. If qualification is possible today the Expert will seldom be called at all.

The defense, if unsuccessful in shaking the basic testimony, may ask all manner of miscellaneous questions in part to establish some facts favorable to the defendant but mainly to confuse and confound the jury. The defense will take full advantage of any possible inconsistencies in any statements by the Expert to discredit his basic statements.

Among the irrelevant questions that may be asked an Expert are interpretation of wound details more specifically medical matters, sound of firearms and the like. An Expert should know as much about a case as possible but should limit his testimony to Firearms Identification. He should be particularly careful not to attempt to evaluate positions of victims at the time of the shooting. People do incomprehensible things when they know they are going to be shot, including trying to find protection under small pieces of furniture.

Expert Against Expert

Some of the most interesting firearms cases were those in which Experts testified for each side. There were, 25 years ago in America, men of small knowledge but great courtroom presence. These charlatans would hire themselves out to either side with equal ease and testify as technical experts without any real knowledge or even facilities for examination. Men cheerfully gave testimony in firearms cases which either convicted an innocent man or turned murderers back on society; they had little knowledge and no instruments other than a hand magnifier, possibly a micrometer and a scale. Should one of these men appear in court today on the side of the defense, a real Firearms Expert could generally coach the prosecuting attorney in a cross examination procedure that would undoubtedly either prevent his qualification or at least discredit his testimony. These self-styled experts are never in the position of a real Expert. It is necessary for them in their testimony to impress. Allow them to speak sufficiently and they will undoubtedly hang themselves with their own statements. For instance, in the past, one of these men said that the barrels of revolvers complete with the rifling grooves, threads, and everything else were cast in a mould. Another charlatan confounded himself by pointing out that shotguns were rifled; he was referring to all shotguns, not the special British Paradox bores.

The charlatan, of course, may not so obligingly make himself ridiculous. However, the prosecution should endeavor to find out his equipment and his familiarity with it, together with where he received his training as well as his previous experience in court.

There is, of course, the chance that a real Expert will appear for the defense, although such a procedure is at present unusual. Actually, there are many men in the country with reputations and many years of experience ready and willing to appear for the defense. These are in the main retired Firearms Identification police officers. There are a few independent Firearms Consultants with both the equipment and knowledge to do extremely competent work. But to what would they testify? If a positive identification is made by one competent Expert, every other qualified man would reach the same conclusion. The science is at present in a state so that positive identication cannot be contested. No self-respecting full-time professional will make a positive identification unless he is absolutely sure. If there is any doubt at all, he will protect his own reputation as well as his piece of mind by saying frankly that the markings are not sufficient for a conclusive opinion.

Sometimes a defense attorney will endeavor to retain an independent Expert of known reputation to refute the testimony of the prosecution's Expert. One of these men would probably telephone the Firearms Laboratory making the initial examination. If there is not any

A cut-away view of an Iver Johnson Hammerless Revolver, opened to show the operation of the lifter, hammer and firing pin, and prepared in this manner to disprove in court the defendant's claim that the gun went off accidentally when dropped. Dissections such as this are occasionally necessary for exhibition in the courts.

question in the minds of the laboratory personnel as to a positive identification, the independent Expert will refuse the job. In general, one real Expert can refute the findings of another expert only by fraud; however, Experts are human. A case is still pending in a neighboring state in which a retired Expert and the present holder of his old position are opposed to each other. However, the contested point is not a positive identification, but the importance of certain firearms evidence which did not come to light in the first trial. The defense is claiming suppression of evidence. The prosecution claims the evidence was and still is irrelevant.

Defense attorneys, particularly young men with little knowledge of firearms, might on occasion profit very greatly by the advice of independent Experts, even though these men cannot refute the findings of the prosecution Expert. Firearms, particularly when used by individuals in the highly nervous state commonly accompanying the commission of a major crime, are complicated in the extreme. Each is different. Even the defendant may not remember clearly what happened. Few lawyers know weapons well enough to bring out points in connection with them favorable to the defense. The independent Expert may be able to testify as to some of these. He will certainly be able to suggest questions to the defense attorneys for witnesses in relation to firearms.

Independent Experts in Great Britain

In a case involving Firearms Identification in England, the defense is allowed to have their own Expert examine the evidence, including firing a reasonable number of test bullets through a suspect weapon and then comparing these bullets with the crime bullet under a comparison microscope, but in the presence of a police officer. The impartiality of the law in England is such that if four unexpended cartridges are recovered in a suspect revolver, two of them will probably be saved for the use of the defense Expert. The only lack of impartiality is necessarily the fact that the prosecution will know exactly what steps the defense has taken in its investigation, since the policeman present will, of course, report these. The prosecution Expert can conduct his examination of evidence and tests in private.

In this country, if a defense attorney wishes, he can petition for an examination of the firearms evidence by an independent Expert and receive a court order allowing practically the same procedure as followed in England, although only after a good deal more trouble. Further, the defense Expert may not be allowed to fire the suspect weapon but only to examine particular test bullets already fired by the Firearms Investigation Laboratory by the prosecution Expert. This procedure, although theoretically possible, is seldom used. It is not often necessary.

The Independent Expert

Although sometimes in the past the Experts testifying have been charlatans, they have a definite function in any system of justice in the same way that defense attorneys have. Lack of trained opposition could conceivably lead a lazy Expert to ease up in his exhaustive searches of evidence. The two great deterrents to independent Experts developing, however, are first, the difficulty of getting a start; and second, the lack of business. Police laboratories usually make no charge whatever for their services. Neither does the FBI. The defense attorney would have to pay the independent Expert out of his own none too large fees. He is far more likely to ask a friend to find out about the Firearms Laboratory Report, or even apply to the court for a copy of it. If the findings are unfavorable to his client, he will try to create doubt in the minds of the jury and concentrate on other points in his defense.

In spite of all difficulties a few independent Experts do exist, although on a part-time basis. Some of them are good men. Sometimes, these individuals make up in cleverness what they lack in training and equipment. Eminent authorities have in the past suffered most discomforting setbacks at their hands.

One of the foremost early authorities in forensic ballistics was at one time testifying in a case against an elderly unlicensed doctor, who was apparently an authority on many different subjects as well as having occasionally run afoul of the law personally. This individual, without a comparison microscope or much else in the way of really first-rate equipment, somehow maneuvered the true Expert into making a statement about identifying bullets fired from each of six revolvers of the same make and model after a short but specific number of hours. This story was told in Chapter 12.

Another coup of a similar nature made by the same self-styled expert has not yet been explained. A positive identification by a police laboratory was made in connection with a relatively new American made revolver presented as evidence and a fatal bullet recovered from the body of the victim. Somehow, the wily doctor was able to present affidavits from a jobber and a dealer that the particular weapon, whose serial number was recorded and obviously on the evidence weapon which had not been tampered with, had not been sold at the time of the crime. These records showed that the revolver was, in fact, in the jobber's possession more than 200 miles away. These affidavits together with the records were then examined and found to be completely in order. The defendant was acquitted.

Subsequent reexamination of the weapon and the crime bullet confirmed the original identity beyond a question of doubt in two other cooperating laboratories. Apparently, there had been a fortuitous mistake or an alteration of records in the bookkeeping of the jobber and dealer. There was even the possibility that the charlatan had in some way changed serial numbers on a weapon delivered to the jobber and then had the spurious arm purchased by a confederate. At any rate, the modern Firearms Laboratory will want to check with the factory, jobber, and dealer on any traceable arm as a matter of routine, if for no other reason than to prevent a similar occurrence in the future.

Another unfortunate mistake was made some years ago by a college professor who used to sometimes testify in ballistic cases, usually for the prosecution. Somehow, photographs were taken for comparison purposes of two bullets. An independent and far less worthy man testifying for the defense cleverly picked up the fact that both bullets were the test bullets, although one was purported to be the crime bullet in the presentation of evidence in court. There was no effort whatever at fraud. It was an honest mistake. The professor had returned the crime bullet clearly and properly marked, which had been introduced as evidence in the court along with both test bullets. The picture looked at carefully, however, revealed the comparison to be between the two test bullets and not one of the test bullets and the crime bullet. In this case, there was a positive identity between all three bullets. It was checked by cooperating laboratories; however, the photographic error did the damage. The man was acquitted.

Preparation for Court

Regardless of which side an Expert is to appear on, there are certain basic measures that he should take in preparation. He should go over in his mind carefully the entire work that he has done in the case. He should take with him to court both any evidence that may have been entrusted to him and the entire details of his examination.

Some laboratories favor one procedure, some another, in regard to the records that go to court. All records in the case file jacket and the original laboratory book may be brought along by the Expert. Although he will almost never need refer to them if he is properly prepared, they give confidence. Other laboratories prefer a single special 3″ x 5″ blank card on which the Expert to appear in court carefully lists all pertinent data. A man of great ability and many years experience not only does this himself, before every court appearance but checks the comparisons in detail a day or two before appearing.

Certain other things are important also. Be sure you can disassemble the evidence arm quickly and expertly, and then get it back together expeditiously. Some men like to take a small tool kit into court, although they almost never use it. At any rate, take a jeweler's loupe or a pocket magnifier.

Most prosecutors and defense attorneys review with their witnesses the questions and answers before court appearance. This is as it should be; only by so doing can either side do an efficient job of presenting their case. However, a very usual question from a defense attorney will be, "Have you rehearsed your testimony with the prosecution before your appearance in court?"

You can probably answer this, "No." However, the defense attorney will come back with, "But you did go over your testimony with the prosecutor, did you not?"

We suggest if possible that you say about as follows: "I have communicated my findings to both the investigators and the prosecution in writing and verbally."

And remember, you are a Firearms Expert. You are not a debater nor a college professor.

.22 BREECHBLOCK SIGNATURES

The various .22 caliber firearms are becoming increasingly used in crime and the soft lead .22 bullets are often so badly mashed or disfigured that they are useless for identification purposes. In many instances the fired .22 cartridge case will give more evidence than the bullet.

Shown above are the bases of three .22 long rifle cases fired in a Colt 'Woodsman' automatic pistol. The firing pin imprints on these are clear and sharp, under the microscope they give a positive identification. This is supplemented by another positive check, also a very clear match under the microscope and one that can be instantly picked up by the eye—that nick in the rim of these cases at 2 o'clock.

The location, extent, angle and depth of nicks and batterings such as these all contribute to their matching identification. And they generally are there on the soft-metal .22 cases—if you know what to look for.

CASE HISTORIES

CASE NUMBER 1

Contributed By: Colonel Calvin Goddard, USA, Director, Far East Criminal Investigation Laboratory at Tokyo

Nature of Case: In a murder investigation, a laboratory staffed by only partly trained technicians made an erroneous identification in connection with cartridge cases fired from a U. S. carbine, cal. .30M1. This erroneous identification was fortunately rectified by Colonel Goddard who later made a positive identification in connection with another weapon which led to the apprehending and conviction of parties responsible.

Principals: Major General Castaneda and an aide vs. three Philippino assassins

Crime: Murder

Time: Early morning, 15 November 1949

Circumstances: Major General Castaneda, Chief of Staff of the Philippine Army, was riding with an aide, a Philippine Army colonel, on the outskirts of Manila in a political subdivision known as Quezon City. The two horsemen rode into an ambush and were fired at several times. The aide was killed. General Castaneda put spurs to his horse and rode through untouched.

461

Evidence: Several fired U. S. carbine cal. .30M1 cartridge cases. Eventually, an M1 carbine. Still later, a second M1 carbine.

Preliminary Investigation: The Prosecutor of Quezon City and his office, as well as the Philippine National Police, investigated this crime with the aid of the usual sources of information used where political unrest makes matters of this type not uncommon. Four Philippine natives were arrested and charged with the crime. A carbine was found in the possession of one of these.

A comparison was made by the Firearms Expert of the Philippine Constabulary of cartridge cases fired in the suspect weapon and all cartridge cases recovered at the scene of the crime. A positive identification was made by the technician in connection with every crime cartridge, which was in itself unusual from the position in which these rounds were found and the general information as to the number of shots fired. One would suspect at least two different weapons fired from different points. The prosecutor was not satisfied with this identification, even after he had secured a conviction of the four men accused of the crime. He was looking around for help. Either he or someone in his organization had an idea.

On 24 January 1950, Major General Anderson, Chief of JUSMAG (Joint U. S. Military Advisory Group) radioed GHQ, Far East Command, Tokyo, that the Philippine Government had requested that Colonel Goddard be sent to Manila on temporary duty to deliver a series of lectures to the Philippine National Bureau of Investigation. He added that his headquarters would like also to call on Colonel Goddard "to provide essential advice to Philippine ground forces and Philippine Constabulary on technical investigative matters."

Colonel Goddard, soon after his arrival in Manila, was visited by the Quezon City Prosecutor, who brought a large number of enlarged photographs of the heads of crime and test cartridges supposed to have been fired from the same M1 carbine. All these photographs did show the family resemblance of the model of weapon in which they were fired; however, Colonel Goddard disagreed completely with the findings of the Philippine technician. The suspect carbine did not fire the crime cartridge cases.

The case was reopened. A promising lead was uncovered. An actual eye witness to the crime was finally persuaded to answer questions. Part of these answers revealed that one of three assassins had thrown an M1 carbine into a stream. The stream was dragged and the carbine found.

Laboratory Procedure: Colonel Goddard was asked through channels to examine this second suspect weapon and the original crime cartridge cases in the laboratory of the National Bureau of Investigation in the presence of their people. Some of the crime cartridges positively matched test cartridges fired in this second suspect weapon.

Outcome: The four men originally tried and sentenced were freed. Eventually, two of the three real assassins were apprehended, tried, and convicted.

CASE NUMBER 2

Contributed By: Massachusetts State Police Firearms Investigation Laboratory

Nature of Case: The original Sacco-Vanzetti Case is in many ways the most interesting politically and psychologically that ever occurred in America. It is also of extreme importance ballistically when one realizes the relatively primitive nature of the science at that time compared to the progress that was made within the next decade. We are reviewing this case as regards firearms evidence in accordance with the original records of the Massachusetts State Police and in the light of later investigation of this evidence by Captain Charles Van Amburgh, Jr., son of Captain Charles Van Amburgh, for many years head of the Massachusetts State Police Laboratory and who testified as the principal prosecution expert more than thirty years ago.

Crime: Multiple murder and robbery

Principals, Time, and Locality: Five armed men held up Paymaster Parmenter and Payroll Guard Berardelli on the afternoon of 15 April 1920 in South Braintree, Massachusetts. These two had in their custody the payroll for the shoe factory of Slater and Morrill and were in the street in front of it. Both Parmenter and Berardelli were shot and killed. The five robbers escaped in an automobile. The $16,000 payroll was never recovered.

Circumstances and Evidence: In order to understand this crime in its larger sense, it is necessary to realize the political conditions at the time and some phenomena of trials which attain widespread public interest. Sacco and Vanzetti were charged with the crime of 5 May 1920 and were indicted on 14 September 1920. The original trial began on 31 May 1921 and lasted for a month and a half. Numerous appeals were made; finally a special committee upheld the original findings in 1927. Throughout this entire period, there was great feeling throughout the country both for and against Sacco and Vanzetti, who were admittedly not very patriotic. They had been political radicals in Italy and here. Strong efforts were made by various organizations, many of which were Communistic, to arouse sentiment for these two individuals. There was equally strong feeling on the other side. A few went practically as far as saying that Sacco and Vanzetti should be executed whether or not they were guilty of the

"FINGER PRINTS" IN THE HISTORIC SACCO-VANZETTI CASE

The two sets of pictures shown above and on the opposite page illustrate perfectly the method of identifying weapons from the markings left on the primer and the cartridge head by the irregular tool or file marks on the breech face. X, at the top of the left page, shows a test cartridge in the Sacco-Vanzetti case; Y, at the top of the right page shows the evidence cartridge in this same famous trial. The photographs furnished by Col. Goddard.

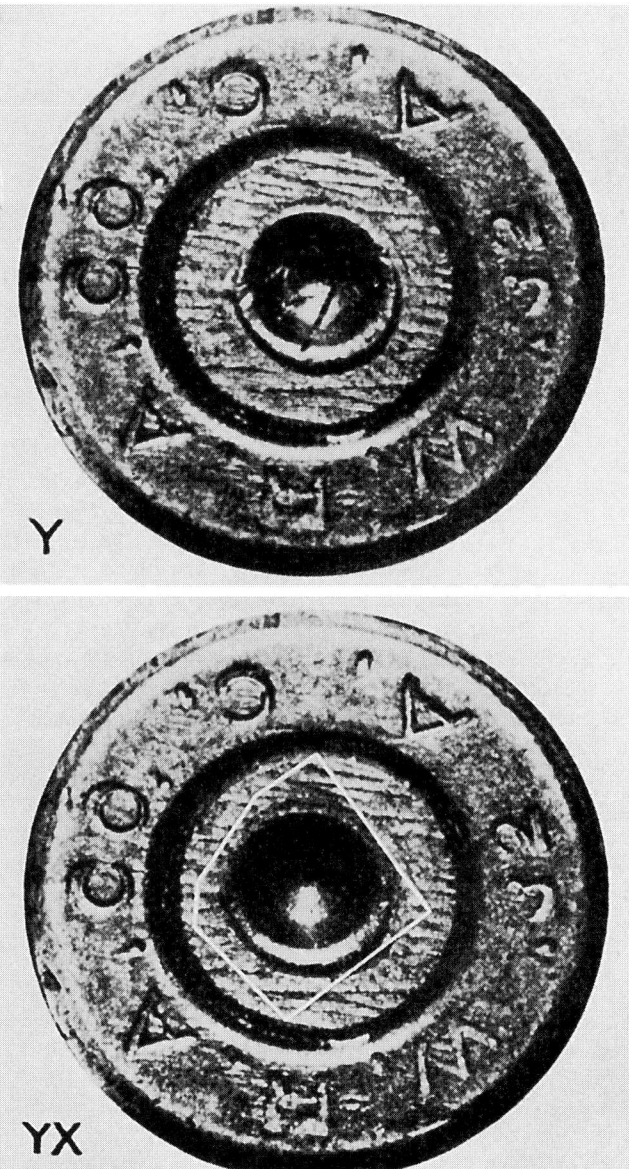

XY at bottom of left page is a duplicate photograph print of Test shell X with an irregular polygonal area from Evidence shall Y superimposed and pasted in place —note the exact matching of markings from the breech face; YX is a duplicate photographic print of evidence shell Y with an irregular polygonal area from Test shell X superimposed and pasted in place—note that markings of breech face match exactly.

actual shooting. This was not an atmosphere conducive to impartial justice and judicial review.

At the original trial, witnesses for both sides were guilty of perjury, both intentional and unintentional. A great deal of evidence was set aside that would have been conclusive if the time interval had been shorter. As it happened, people's minds became confused. Further, they discussed the case so much that they crystallized in their minds details which others discredited.

The firearms evidence was extensive, although some of it proved to be of little value. There were five cal. .32 ACP bullets taken from the bodies of the two victims for which a matching automatic pistol was never found. From the right-hand twist evidence bullets and cartridge case markings, it is thought that these were fired in a Savage automatic pistol never produced. Further, a cal. .38 Harrington & Richardson revolver, very similar to that carried by Berardelli, was taken from Vanzetti. It was never possible, however, to tie this revolver definitely to the victim. A total of 32 cal. .32 ACP cartridges similar to those used in the crime were taken from Sacco at the time of his arrest.

The clinching evidence, however, was a cal. .32 Colt Pocket Model automatic pistol taken from Sacco at the time of his arrest, a single bullet taken from the body of Berardelli, and one cartridge case found at the scene of the crime.

Captain Van Amburgh, as well as Lieutenant William H. Proctor, of the Massachusetts State Police, indicated in their original testimony that the pistol, the bullet, and the cartridge case all matched. The reading of the testimony of a firearms nature is bewildering. They went on hour after hour speaking of stuff that is trivial and unimportant to us today. It is the opinion of the present personnel of the Massachusetts State Police Firearms Investigation Laboratory, as well as the authors, that occasionally everyone in the court, including both the prosecution and defense attorneys and the Firearms Experts, became confused.

Everything imaginable was discussed in connection with groove width, pitch, manufacturing procedures, bullet cannelures, and the like. Almost every Expert in the East was called, including some with no very profound knowledge of what they were discussing. There was even an unsuccessful attempt by one of the Defense Experts to substitute a different barrel in the evidence Colt automatic.

However, later examination of the pertinent evidence, after the adaptation of the comparison microscope to bullet examination and modern procedures in connection with cartridge case examination, showed that there can be no doubt that Sacco's pistol fired one cartridge case and one of the fatal bullets.

Outcome: After extreme court turmoil and an unprecedented amount of publicity, public feelings, and even riots both here and abroad, both Sacco and Vanzetti were executed on 23 August 1927, more than seven years after the original crime. The other three men who took part in this robbery were never apprehended.

CASE NUMBER 3

Contributed By: New Jersey State Police and Union County, New Jersey Prosecutor's Office

Nature of Case: A crime in which the previous confession of a murderer was denied in court but made to stand up by the restoring to operating condition and firing for test purposes of a weapon buried in moist ground for eleven months.

Principals: John Edwards vs. Jesse Fitchett

Crime: Murder

Time: 11:30 p.m., 29 November 1951

Circumstances: In an alley in Elizabeth, New Jersey, several men were shooting dice. One of them was winning with astonishing regularity. Apparently a substitution of dice was suspected by one of the other players. A shot was fired and Jesse Fitchett was found dead a few minutes later, when the sound was investigated by the police.

The restored revolver in the Jesse Fitchett Case. This weapon lay in a swamp for eleven months and resembled more a large irregular piece of rust than anything else when brought into the laboratory. It was possible, however, after alternately scrubbing and soaking the revolver, to restore it to operating condition and fire test bullets through it which matched positively crime bullets.

Because of the nature of this crime, there was nothing that could be done immediately; however, competent detectives started to piece together information that they gathered from various sources. Eleven months later, John Edwards was taken into custody and accused of the crime. He knew that a confession means very little so he did not really give much trouble. He confessed freely to committing the crime, sure that he would be able subsequently to deny the confession in court. There was actually very little positive evidence against him. Apparently the shot could have been fired by any participant in the original dice game.

However, in the course of his confession, he admitted burying the crime weapon. He directed detectives to the place where he had done this burying. They recovered a weapon in a terribly bad state of preservation.

A POSITIVE COMPARISON

The evidence and the test bullets in the Edwards-Fitchett case which showed a positive comparison.

Evidence: The original crime bullet—a .32 Smith & Wesson Long recovered from the body—and the suspect weapon.

Laboratory Procedure: The gun was allowed to soak in solvent, made of Kotol and Magnusite, until most of the rust could be removed with brushes and scrapers. On 16 November 1952 the New Jersey State Police Laboratory was able to open the cylinder and unload it. The gun was a .32 Smith and Wesson Hand Ejector with 5-inch barrel, and had been nickel plated. There were three empty shells in the cylinder, and three loaded cartridges: a .32 Smith & Wesson Long Rem-UMC, and two .32 Smith & Wesson Long Western with Lubaloy coated bullets. One of the empty shells was a Western, one a Rem-UMC, and the other a W.R.A. An obstruction in the bore proved to be a lead bullet. It was pushed out of the barrel to the rear, retracing its path, so to speak.

Examination of the bore showed it to be in surprisingly good condition; there was almost no rust in the barrel. While the outside of the cylinder and frame was originally so thickly encrusted as to make it impossible to even open the cylinder, let alone operate the gun, once this was removed and the side plate taken off, the operating parts were relatively easy to clean. The gun was re-assembled and test shots fired, using fresh ammunition, since the cartridges found in the cylinder would probably not fire.

Examination of the evidence bullet, removed from the body by the Medical Examiner, and the bullet found lodged in the barrel showed positive results: they were both fired in the same barrel. The two test shots which were fired in the same gun also matched the evidence bullet and the one found in the gun.

Outcome: An unsupported confession may be useless; however, a confession leading to the recovery of a specific weapon in the New Jersey marshes, based solely and entirely on information supplied by a suspect, is another matter. The positive matching of the original crime bullet with a weapon recovered in this manner was conclusive. A second-degree murder conviction was obtained and the defendant sentenced to life imprisonment.

CASE NUMBER 4

Contributed By: Police Commissioner Thomas J. Gibbons and the Philadelphia Police Department
Nature of Case: Three (3) robberies were connected and solved by bullet and cartridge case identification and final positive matching with test bullets and cases from weapons picked up with suspects.
Crime: Armed Robbery
Time: Winter 1950-1951
Circumstances and Evidence: 29 December 1950, David Berkowitz, 353 Dickinson Street, was held up and robbed, inside his drugstore, at same address, by two unknown white men, one of whom in his escape fired a shot from an automatic pistol. No one was injured, and a fired cartridge case was found in the laboratory of the drugstore.

One (1) fired cartridge case, Western manufacture, cal. .32 ACP, with the outer rim of the case badly mutilated, apparently caused by the cartridge having been inserted into the magazine approximately twelve to fifteen times. The markings apparently caused by the two lips of the magazine.

23 January 1951 about 9:30 A.M., Samuel Schwart, 1412 N. Marshall Street, bartender, taproom, 100 W. Allegheny Avenue, was held up by three unknown white men, one of whom in his escape fired several shots from an automatic pistol. No one injured. A bullet was removed from the west wall of the taproom, about 4½ feet above floor, 20 feet from the south wall, imbedded to a depth of 1½ inches.

One bullet, cal. .32 ACP, full jacket, weighing 70.2 grains, slight loss in weight, fired from a domestic smokeless powder cartridge case. Bullet considerably mutilated, flattened on one side of nose, deep foreign parallel scratches on flat surface, some plaster adhered to nose, the entire surface of circumference covered with foreign scratches, apparently caused when bullet penetrated plaster wall. The general rifling or class characteristics on surface of bullet are similar to those produced by the German Ortgies Patent automatic pistol. The possibility of other types of pistols being used cannot be eliminated due to the similarity of the Ortgies pistol with other foreign or Baltic countries weapons. The specimen is in poor condition; all individual characteristics are destroyed.

17 February 1951 about 2:00 P.M., Charles Perchick, 40 years, white, residence 2133 N. 33rd Street, manager of American Store, 700 Pine, was held up by three unknown white men. In their getaway, one of the bandits stopped an automobile operated by Joseph Fisher, 4733 N. Marshall Street, at 6th and Lombard Streets, and fired one shot into the dashboard. The bullet was recovered and the fired cartridge case was found on the highway, Addison Street, west of 6th Street, by Patrolman Rice, 19th Police District.

One bullet, cal. .32 ACP, full metal jacket, weighing 70.0 grains, slight loss in weight, fired from a Western smokeless powder cartridge case. Bullet mutilated, one-half of bullet flattened and distorted. Lead core of bullet mutilated. From the general rifling or class characteristics on the surface of the bullet, would indicate bullet was fired from a foreign weapon of German manufacture. The possibility of other types of pistols being used cannot be eliminated due to close similarity of class characteristics with other pistols, including an American pistol.

One (1) fired cartridge case of Western manufacture, cal. .32 (auto). The outer rim of the case is badly mutilated; it appears cartridge has been inserted in magazine numerous times. The markings apparently caused by the two lips on the magazine.

25 February 1951, Thomas Parkowitz, 26 years, white, 230 Spruce Street, and Morris Turganski, 31 years, white, 310 S. Philip Street, were arrested on the highway, at 760 N. 45th Street, and charged with Violation of the Uniform Firearms Act and Conspiracy to Commit a Hold-Up. The following two weapons were confiscated from Thomas Parkowitz:

German (Ortgies) automatic pistol, cal. .32 (auto), Serial No. 4-999, containing six Western, .32 (auto) full metal jacket cartridges.

German (Ortgies) automatic pistol, cal. .32 (auto), Serial No. 117-542, containing six Western, .32 (auto) full metal jacket cartridges.

Laboratory Procedure: Microscopic comparative examination of fired cartridge case found in the drugstore where David Berkowitz was held up and robbed, and test case fired from German (Ortgies) automatic pistol, cal. .32 (auto), Serial No. 4-999, confiscated from Thomas Parkowitz, has shown conclusively that both were fired from the same weapon. The firing pin impression on the head of the case is similar to the markings on the test case, as well as extractor markings on the rim of the cases.

Microscopic comparative examination of crime bullet removed from wall of taproom, 100 W. Allegheny Avenue, where Samuel Schwartz was held up, and test bullet fired from German (Ortgies) automatic pistol, .32 (auto), Serial No. 117-542, confiscated from Thomas Parkowitz, has shown that while numerous similarities appear, no conclusion could be reached as to whether or not they were fired from the same weapon. This is due to the fact that crime bullet was found imbedded in plaster wall to a depth of $1\frac{1}{2}$ inches. There are not sufficient points of comparison for Court.

Microscopic comparative examination of crime bullet recovered after the hold-up of Charles Perchick, manager of American Store, 700 Pine Street, and the test bullet fired from German (Ortgies) automatic pistol .32 ACP Serial No. 117-542, confiscated from Thomas Parkowitz, has shown conclusively that they were both fired from the same weapon.

Examination of the cartridge case found on the scene and test cartridge fired from German (Ortgies) automatic pistol, .32 (auto) Serial No. 117-542, show numerous similarities.

Outcome: On 18 April 1951, Thomas Parkowitz and Morris Turganski were tried before Judge Milner and found guilty on all counts. Sentence deferred pending motion for a new trial. The motion was withdrawn by defense attorney Lemisch on 15 June 1951, and the two defendants were sentenced to serve 2½ to 5 years in the Eastern State Penitentiary.

CASE NUMBER 5

Contributed By: Texas Department of Public Safety.

Nature of Case: Murder of a wife by her husband, who attempted to make the murder appear as an accident.

Circumstances and Evidence: The suspect in this case stated that he had returned home after having several drinks at a local bar and had been reading the paper. He then decided to clean his rifle before putting it away for the winter, as his job was going to necessitate his being out of the city for several months. He went to a bedroom, got the rifle, a .30/40 Krag, and as he was entering the living room where his wife was sitting facing him, the gun went off, accidentally killing the woman.

His story was that the bolt slipped out of his hand, slid forward and in some way closed and fired the gun. The rifle was submitted to the laboratory to substantiate or deny this explanation. The rifle was examined and found to be in good condition and perfect working order. By pulling the bullets and dumping the powder charges from the cases and snapping the primers only, we were able to demonstrate to the district attorney, in his office, exactly the condition and position the bolt would have to be in to make this gun fire. Which was in direct opposition to the story of the defendant.

During the trial this evidence was used in rebuttal, showing that the bolt could not slide forward, close itself and fire. We were able to prove that the defendant's story was untrue. He was convicted and given a 20-year sentence.

Comment: This case serves to illustrate the necessity for proper experimentation and testing with the firearm involved and for the ability to give an actual courtroom demonstration of the conditions indicated—which demonstration is often better than thousands of words.

CASE NUMBER 6

Contributed By: Scientific Laboratory, New York State Troopers
Nature of Case: An atrocious assault of the proprietor of an isolated filling station was solved completely by careful restoration of the broken parts of the stock of a revolver and one of the most astonishing bits of Interstate Police cooperation known to the authors.
Principals: The proprietor of the filling station vs. three young and fairly inexperienced holdup men.
Crime: Atrocious assault and robbery
Time: 9 February 1942
Circumstances and Investigation: On the evening of 9 February 1942, at approximately 6:50 P. M., an automobile containing three men pulled into a gas station, located about two miles south of the village of Fort Ann, New York, on Route 4. The driver of the car got out and walked into the gas station; the proprietor was in the rear of the station. The customer struck the owner over the head, causing only a slight injury. The owner turned around and the man said: "Hand me your money you . . ." and at the same time pointed a nickel plated revolver at him. Not having any intention of complying with such request, the owner grabbed a club from beneath the counter and went after the man. The man fired a shot and then proceeded to use the butt of the gun as a club. Finally, the owner was struck in a vulnerable spot on the head and knocked unconscious. The assailant then picked him up and threw him through the glass window of the front door. He regained consciousness just in time to see the men drive away from the station, and observed only that the car was similar to a 1934 Ford Coach.

The State Police detail at the South Glens Falls station were notified and immediately proceeded to the scene of the crime. Meanwhile, medical aid had been summoned and although at first it appeared that the filling station owner had been shot and was in a very serious condition, subsequent X-ray examination at the hospital revealed that his condition was very good and that he would be incapacitated for only a few days.

A search of the scene of the crime revealed that a violent struggle had taken place and there were blood stains over most of the equipment and furniture in the station proper. Further examination at the scene of the crime by the State Police detail resulted in the finding of several small pieces of a revolver grip, four loaded cartridges, one exploded cartridge and a lead bullet.

The only identification that the owner was able to give the investigators was that the assailant was wearing a dark-colored cap, dark suit and dark-colored shirt; that he was of a dark complexion and had several days' growth of beard on his face. He also advised that approximately $30.00 was taken by the robber, that one of the dollar bills was torn just to the right of the center through the eye of Washington and was patched with cellophane tape.

In the course of the investigation by the B.C.I. men, photographs of all known burglars, thieves and robbers in that locality and nearby cities were submitted to the proprietor of the filling station. He failed to identify any of them. All of the suspects who were picked up and questioned were able to substantiate their alibis.

From an examination of the evidence found at the scene of the crime, namely, broken pieces of a revolver grip, four unexploded and one exploded cartridge as well as the lead bullet, the B.C.I. men assumed and rightly so, that the robber had used the revolver with such degree of force that the grips had broken and that the cartridges had fallen from the chamber of the gun. Careful reconstruction and examination of the small pieces of broken revolver grips by the B.C.I men resulted in the finding of scratched markings on the inner surface of the revolver grips which when pieced together in jig-saw puzzle fashion revealed a name and address.

"L. A. Martin
Brandon, Vermont
Jan. 31, 1941"

The B.C.I. men immediately contacted the Attorney General's office, State of Vermont, which office stated that Mr. L. A. Martin was well known to them as they had previously arrested him for burglary and at the present time he was on probation. They further advised that Martin was employed at Waterbury, Conn. In view of this, the B.C.I. men, in company with a detective from Vermont, proceeded to Waterbury, Conn., where the Connecticut Police Department was contacted. Mr. Martin was finally located at Southbury, Conn., and when questioned about this gun stated that he recalled having had the gun but had left it in the home of a friend at Rochester, Vermont. Martin gladly agreed to accompany the B.C.I. unit to Rochester since, he stated, he wished to clear himself of this holdup. Upon questioning the friend, he stated that the gun in question was sold by his son sometime during the spring of 1941. Upon questioning the purchaser, the B.C.I. men learned that the gun had been sold to another man sometime during the latter part of September, 1941. When this man was contacted, he advised that the gun in question had been traded to a man living in Windsor, Vermont for an automobile tire.

The B.C.I. unit then contacted the Sheriff of Windsor County, State of Vermont, who recalled that he knew the ultimate recipient and several other lads who associated with him. One of these had a criminal record,

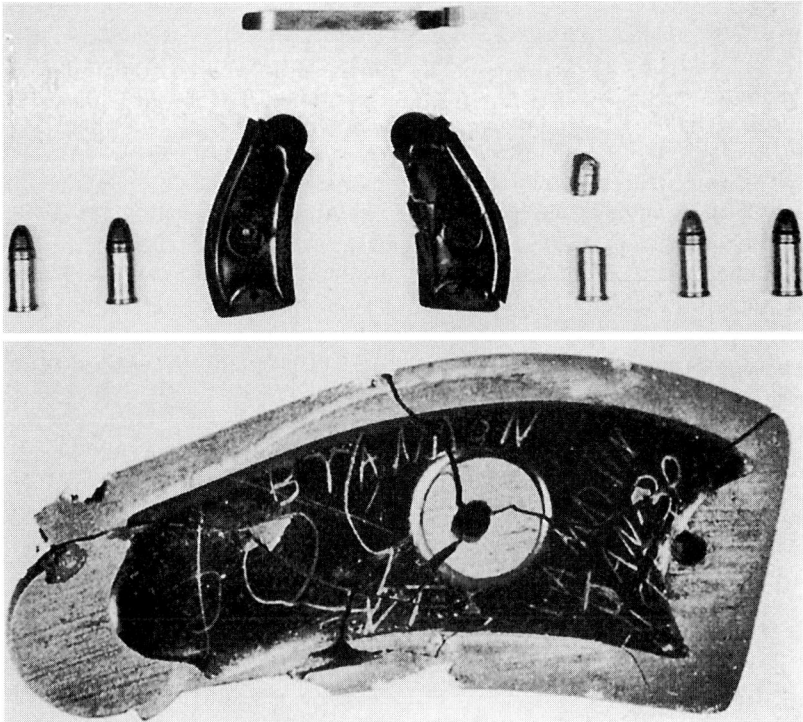

The evidence as found and reconstructed in this filling station assault case.

having served time in Windsor Prison for Grand Larceny. The detail proceeded to Windsor, Vermont, and picked up the last named lad, who after a short period of questioning, finally admitted his part in the holdup, implicating two other young men.

One of these, when picked up and questioned at Windsor, admitted his part in the holdup and also implicated the third. The detail was then obliged to proceed to Lebanon, New Hampshire in order to pick up this last man, who upon questioning readily confessed his part in the holdup.

The detail proceeded to Woodstock, Vermont, with the three prisoners where they were taken to court in Windsor County and there signed waivers of extradition, after which the prisoners were returned to the South Glens Falls station where they were fingerprinted and photographed. They were identified by the filling station owner, who picked out one of them as the man who had struck and shot at him.

In confessing their crime, the three men said—they drove into New York State on February 8, 1942, in a 1934 Ford Coach owned by one of their girl friends. They had in mind sticking-up a gas station near Whitehall and Fort Ann; however, they did not pull the job that day but returned on Monday, the 9th, around 4:30 or 5:00. As they were driving past the gas station, one said, "That was the place" only they were afraid because a car was parked there. They drove past the station but returned in a few minutes and, seeing the car gone, pulled into the station. One got out and entered the station. The other two heard a shot and breaking of glass. The man came running out and said, "Get going." They disposed of the gun in the Champlain Canal and drove into Lebanon, Vermont where they divided the loot, each receiving about $9.00.

The close observation of the investigating officers as was used in the reconstruction of the broken pieces of revolver grips resulting in the finding of the scratched name and address is to be commended, as well as information and close cooperation of the several law enforcing agencies of the other states. Without that information and close cooperation perhaps the case never would have been closed.

Outcome: On 16 June 1942, the three men plead guilty to robbery second degree and were sentenced.

CASE NUMBER 7

Contributed By: Department of Oregon State Police.

Nature of Case: Assault with intent to kill with a foreign automatic pistol, in which the accused claimed an accidental discharge, yet laboratory evidence revealed it required more than twelve pounds of trigger pressure to fire the firearm. The firearm proved to be in a safe mechanical condition.

Principals: Raymond Leo Helton, victim; Harold Sidney Bush, accused.

Crime: Assault with intent to kill.

Time: Approximately 7 p.m. July 22, 1954.

Circumstances: According to the victim and his wife, Harold Bush was a friend of theirs who had been preparing to stay with them for the night and was about to retire to his sleeping bag when an argument arose between the victim and the accused. Bush drew a .32 automatic pistol from his holster and shot Helton through the leg. Bush later contended that the gun had discharged accidentally as he removed it from the holster.

Laboratory Procedure: The .32 automatic pistol was submitted to the laboratory together with the bullet and cartridge removed from the floor of the cabin. Routine examination revealed they were fired in the pistol. The mechanical condition of the pistol was studied carefully and it was found to be in satisfactory condition. The trigger pull was measured and found to require more than twelve pounds to fire the gun. This information was relayed to the investigating authorities.

Outcome: The subject changed his plea to guilty and was sentenced to the Oregon State penitentiary.

CASE NUMBER 8

Contributed By: Princeton Township Police Department; New Jersey State Police Firearms Identification Laboratory

Nature of Case: An accidental shooting at extremely long range. Although extremely improbable, shootings do sometimes occur in which an innocent citizen is hit by parties unknown. The solution and conclusive proof of this crime were well handled by the Departments involved.

Principals: Weller vs. innocent victim of accidental shooting

Crime: Possible Manslaughter

Time: 7:00 P.M., 11 April 1947

Circumstances: At time of shooting, Jac Weller was explaining to a friend the operation of some ammunition that the friend had brought back with him from overseas. This ammunition was a standard type of tracer in cal. .30-06. A round was fired at high elevation towards an uninhabited mountainous section lying to the north of the Weller home, in order to view the entire trajectory of burning. The tracer composition, however, is extremely corrosive if allowed to remain in contact with a rifle bore, so a ball round was immediately fired, at the base of a stone wall, to clear out the tracer residue.

This wall had just been constructed. An extra piece of building stone, photographed herewith, was lying about three feet from the base of the wall. The bullet struck the surface of this stone fairly closely. Further, the bullet had considerably more velocity remaining than is normal in cases of this type, then it traveled a distance of more than 1400 yards, passed through a copper mesh screen and a light canvas shade of a window in a dwelling and hit the victim in the upper abdomen.

The wound was at first thought to be serious, but upon examination was found to be relatively minor. The bullet entered the body of the victim at an angle and traveled along for a distance of perhaps three inches between the surface of the skin and the abdominal cavity, almost parallel to these two surfaces and midway between them. It did not enter the abdominal cavity.

Evidence: The crime bullet, the rifle, and the piece of building stone. The line of flight could be determined by the positions of the holes in screen

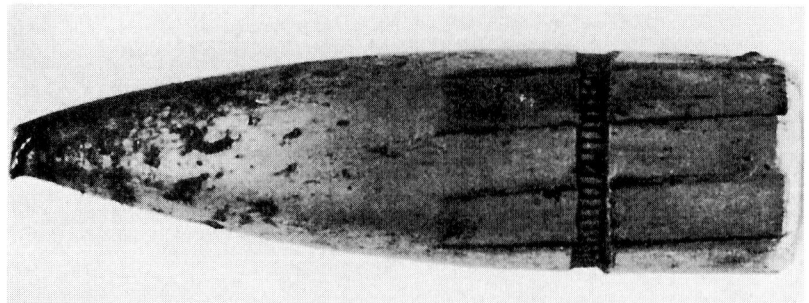

A view of the bullet from the accidental shooting by one of the authors, described in detail above. Plenty of evidence left hereon.

and shade and the chair occupied by the victim.

Preliminary Investigation: Chief Clausen, upon arriving upon the scene, determined the line of flight as indicated above. This led towards the only approved rifle range in the area. Upon investigation, Chief Clausen was told the entire circumstances insofar as they were known to Weller at the time.

Whatever doubt there was in the minds of both Clasen and Weller was dissipated upon examination of the crime bullet at the hospital, with a jeweler's loupe. The rifle used was a 1917 Enfield with a relatively loose bore and left-hand twist rifling. The bullet not only conformed to those specifications, but told the story of its unusual ricochet quite plainly.

The evidence mentioned above was sent along to the New Jersey State Police Firearms Identification Laboratory for examination.

Laboratory Procedure: It was first determined that the crime bullet came from the suspect rifle. A comparison of a test bullet and the crime bullet under the comparison microscope was positive, although the stretching and warping of the bullet jacket makes photographic proof difficult.

Finally there were found on the surface of the bullet three particles of sand similar in composition to the stone.

Since the penetration of the bullet in screen, shade, and human tissue was slight, the velocity at that time was low. Therefore, the bullet deformation of the bullet occurred at the beginning of its trajectory. Particles of bullet metal on the stone and stone in the bullet confirmed this and proved the ricochet. The shooting was obviously unintentional. Reasonable precautions had been taken to prevent such a happening; the extra building stone caused the accident. Its shape led to the slow change of direction so that the bullet was not smashed, nor its velocity greatly diminished. It carried about a mile in a reasonably straight line.

Outcome: The victim recovered completely in a relatively short time. Jac Weller assumed responsibility immediately. The New Jersey State Firearms Laboratory proved the accidental nature of the happening and the unusual circumstances causing the final bullet destination. Full compensation was paid the victim; no action, either criminal or civil, was taken.

CASE NUMBER 9

Contributed By: New Jersey State Police, Elizabeth, New Jersey Municipal Police; and Union County, New Jersey Prosecutor's Office

Nature of Case: A double murder of a most atrocious type in which conviction was obtained almost entirely on firearms evidence alone, which was of an unusual character. The crime weapon was destroyed. Three fragments only of it were recovered, none of which were of a type to influence either fired cartridge cases or bullets. The case hinged on the identification of the crime cartridge cases with cartridge cases proved to have been fired by the defendants in their home long before they destroyed the weapon.

Principals: Eugene and Michael Monahan v. Sebastian Wielandics, an elderly bartender, and a tavern customer by the name of Diskin.

Crime: Murder

Time: 11:20 a.m., 7 March 1953

Circumstances: Eugene Monahan went into the Shamrock Bar and Grill, located on South Broad Street in Elizabeth, about 10:30 Saturday morning. He ordered a beer and was waiting for his son to report to the Probation Authority diagonally across the street in the Court House. When the boy, aged 15, returned, the father sent him out to their automobile to get a 9mm Luger automatic pistol. The two Monahans then

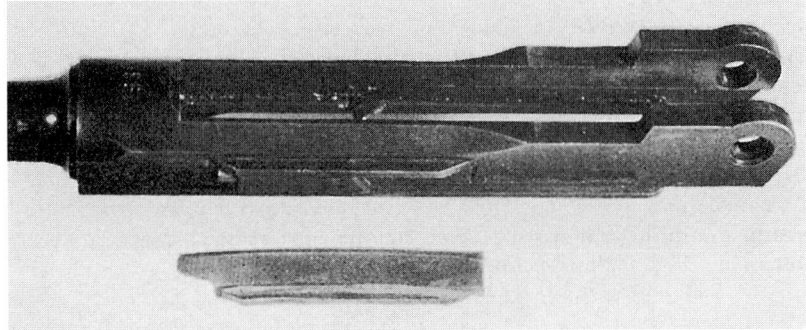

The single fragment of metal found in Monahan cellar turned out to be a portion of the Luger receiver.

BREECHFACE SIGNATURES IN MONAHAN TRIAL

The Wielandics-Diskin murder was an atrocious crime and the direct evidence connecting the two killers with the shooting was mainly circumstantial. Hence, the firearms experts went to every extreme to be certain of their testimony in the case.

(Continued at bottom of opposite page)

shot the bartender, took about $53.00 from the cash register, and before they left shot a would-be customer. They fired in all seven times. Seven fired cartridge cases were recovered and six bullets, most of which had passed through the bodies of the two victims.

This crime made a very considerable stir in the press and among co-operating police departments. A total of 362 comparisons were made with other bullets and cartridge cases used in similar crimes. All were negative.

However, the Elizabeth Police and the Union County Prosecutor's Office continued to investigate. The detectives from the two departments, working in close cooperation, came to the conclusion that the murders were committed during a hold-up and robbery because the perpetrators were known to either or both of the victims, or were afraid they might be identified by them later. There was also a clue of sorts in the fact that a man, a boy, and a dog, which were later identified as the Monahans and their police dog, had been in the neighborhood sometime around the time the crimes were committed.

On 7 April 1953, at about midnight, a Verona New Jersey Police cruiser was patrolling the back streets and spotted a boy on the roof of a house. They apprehended him, and immediately afterward picked up a man and dog trying to get away in a car. They were taken to Verona Police Headquarters. The circumstance of man—boy—dog struck a responsive chord in the mind of the patrolman and he questioned the boy, out of hearing of the man. The boy immediately confessed that he and his father had held up the Shamrock Bar and had killed the two men in it.

That set things rolling. The Essex County (in which Verona is located) Prosecutor; Union County Prosecutor; Union, New Jersey Police (the home town of the prisoners); and Elizabeth Police were notified, and the boy, Michael Monahan, 15, of 3 Byron Terrace, Union, New

As no evidence pistol was available in which to fire test cartridges, the case hinged upon the comparisons of the empty cases found at the scene of the crime with other fired cases found in the Monahan cellar and also with an empty case thrown away by the younger Monahan along with other rubbish.

This evidence resulted in a collection of splendidly matching breechface signatures, as the Luger pistol used in firing these cartridges was most violent in its action, positively marking the base of each case fired with distinctive markings clearly visible to the examiners. Under the comparison microscope other positive markings were found on both case heads and primers.

In this illustration, the top pair of case heads is: Left (or No. 1) a fired case from the scene of the killings; right (or No. 2) a fired case from the Monahan cellar. Middle pair: Left, the above No. 2 case exposed for comparison along with (at right) the fired cartridge case thrown away in the vacant lot by young Monahan. Bottom row: A pair of fired cases from the floor of the tavern in which the killings took place.

Notice that all five of these case heads are clearly and distinctively battered at the bottom edge of the rims (6 o'clock) and (at 12 o'clock) on the upper edges of the primer pockets. Other markings existed which were not brought out in these low-magnification photographs.

These fired cases were photographed in pairs, as shown here. The print of the bottom pair used here was a bit too dark for proper study and comparison. It is used here to impress upon the reader the importance of using the correct grade of paper and right exposure in making such prints for examination and study. Generally, it pays to make several prints in varying degrees of contrast and choose and study accordingly.

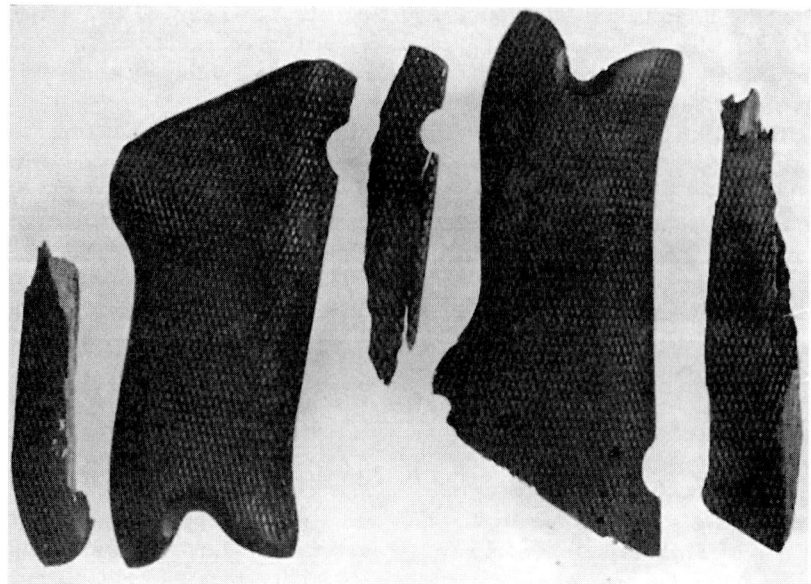

The Luger stock fragments found in the Monahan cellar, as positioned and compared with perfect stocks from another Luger pistol.

One of the evidence fragments as fitted into place on a different Luger pistol used for test purposes.

Jersey, was questioned and gave a statement. He did some bragging and talked about how easy it was to kill. He insisted that he and not his father, Eugene Monahan, had done the shooting. At one point, he said he had pointed the gun at Sebastian Wielandics, the elderly bartender, and told him to hand over the money. He said Wielandics laughed at him and said he was too nice a boy to shoot anyone. Then he said, "I fired three shots at him and said 'Now laugh, you old —.'"

The elder Monahan was then confronted with the statement and admitted the hold-up and murder, but said he, and not the boy, had done the shooting.

There remained, however, very little actual proof. A search was made of the Monahan's home. They had done target practice in the cellar. Two 9mm Luger cartridge cases were found in this cellar as well as a small fragment of metal and three fragments of wood.

Evidence: The seven crime cartridges, the six crime bullets, the two fired cartridge cases from the Monahan's cellar, all 9mm Luger, a fragment of a Luger receiver, and three fragments of a Luger stock.

Laboratory Procedure: A positive identification was made between all nine cartridge cases. The four fragments were positively identified as portions of a Luger.

Further Investigation: From confessions of the two Monahans and various other data, the Luger was traced from the time the Monahans stole it, from a small shop, until they threw it in pieces into streams nearby. Further, they recovered one additional 9mm Luger fired cartridge case, which matched positively the other nine, from a place where the younger Monahan claimed to have thrown at random cartridge cases cleaned out of the Monahan's cellar. In other words, the Monahans had been firing the murder weapon in their cellar. This tied them positively to the crime.

Outcome: Eugene Monahan was tried, convicted, sentenced to death and executed. The younger Monahan was sentenced as a juvenile.

Supplementary Ballistics Evidence: Although not used in the trial itself, a .32 rimfire rifle was stolen by the Monahans at the time they stole the Luger. This weapon was recovered from the Monahans. Fired .32 caliber rimfire cartridge cases were found in the Monahans' cellar, which matched positively test cases from this weapon.

CASE NUMBER 10

Contributed By: Texas Department of Public Safety.
Nature of Case: Game violation, involving the mutilation of gun parts to destroy evidence.
Circumstances and Evidence: Not all firearms investigation cases will include homicide, suicide or such serious crimes. Here is an investigation covering a less serious crime which had some very interesting aspects.

The suspect in this case left a fired .22 Hornet cartridge case at the scene of an illegal shooting of a deer. The bullet, being the type that it is, was completely destroyed when it came in contact with flesh. The investigating officer recovered the fired cartridge case at the scene and submitted it for examination.

Several weeks passed before the suspect could be located and a test case fired from his rifle. When these tests cases were received in the laboratory it was noted that the firing pin impressions in their primers showed a very definite marking which did not appear on the evidence cases.

Above left section shows base of evidence case submitted to the department. Right section shows base of a test case fired in the suspect's rifle after the firing pin profile had been altered and a bit of emery cloth used around the striker hole in the bolt face.

Three more months elapsed before the suspect rifle itself was recovered. When it was submitted to the laboratory, it was found that the suspect had completely worked over or had someone rework this rifle—a Savage .22 Hornet—even to the point of filing a groove in the nose of the firing pin so as to make an obvious dissimilarity. However, the bolt face markings had mainly been left intact.

When test cartridges were fired in this rifle, a perfectly matching set of primer embossings was found on both evidence and test cases, even though the firing pin imprints were dissimilar. We were also to show the defendant and all persons concerned the reason for, and to give an explanation of, the difference in the firing pin impressions, simply by removing the firing pin from the bolt assembly.

Comment: The unusual feature of this interesting case was the extent to which the suspect went in his attempts to destroy identification, yet still overlooked one of the most obvious means of comparison.

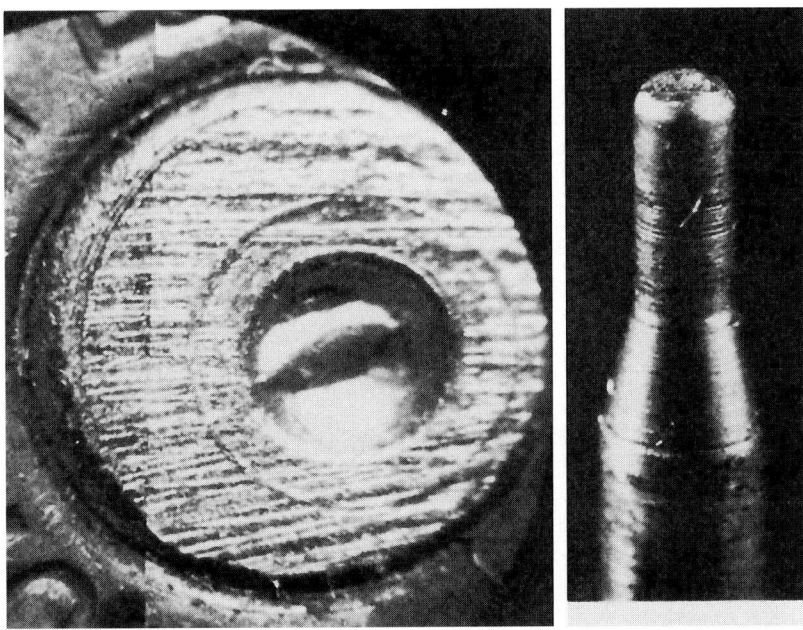

Left view shows a positive matching of primer embossings on evidence (left) and test (right) cases. Obviously, the emery treatment did not go deep enough nor did it cover sufficient area. The right section shows the firing pin after being altered, it appears to have had its tip filed off before being grooved.

CASE NUMBER 11

Contributed By: Department of Oregon State Police.

Nature of Crime: Negligent homicide committed during illegal hunting of deer, and an attempt to mislead the investigating officers as to what actually happened.

Principals: Joseph Warren Ferguson, accused; Charles Edward Eaton, victim.

Crime: Negligent homicide (manslaughter).

Time: December 26, 1954.

Circumstances: The accused and the victim, Charles Edward Eaton, were hunting deer out of season in a heavily wooded mountainous area the day following Christmas. The victim was mistaken for a deer by the accused who shot him with a .30 caliber rifle. After Joseph Ferguson found that he had killed Eaton, he became panic-stricken and left the area to return home alone. He hid his Winchester Model 1895 under the davenport.

When Eaton failed to return in the late afternoon, his family became alarmed and started a search of the hunting area. At 3 a.m. on December 27, his body was found in a brushy area by a group of local residents.

At the time Eaton's body was found, he had a .22 Marlin rifle with him. Examination of the rifle revealed a fired cartridge in the chamber. At this time it was uncertain if he had died from a self-inflicted wound or had met with foul play.

Photomicrograph illustrating the identification of rifling impressions on the bullet removed from the tree near the spot where the body was found. Note the near perfect condition of the evidence bullet on the left; the mushrooming of the tip protected the rear portion of the bullet.

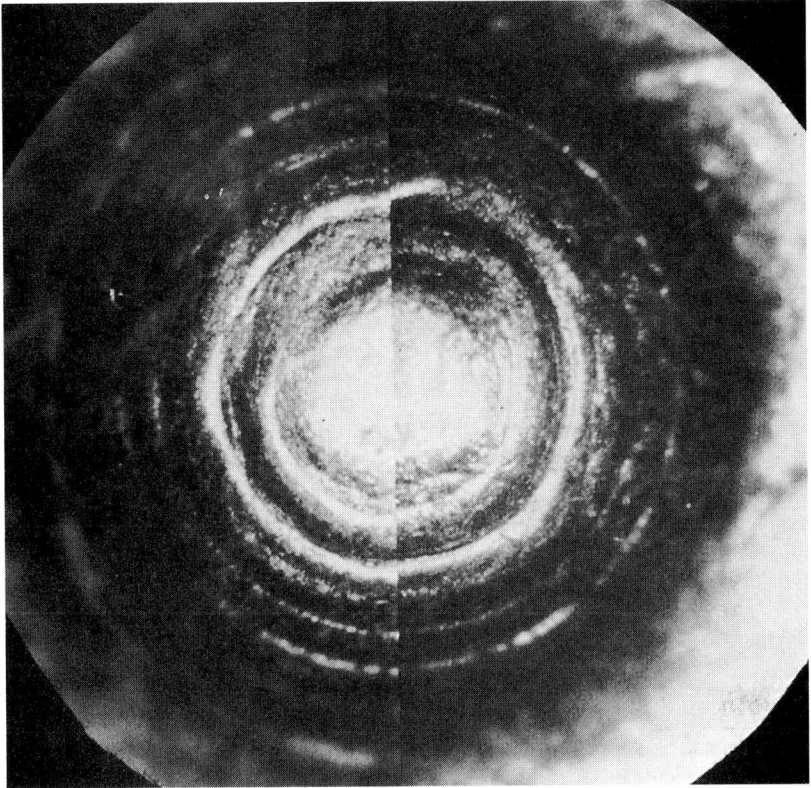

Photomicrograph illustrating the identification of the firing pin impression on one of the fired .30 USA cartridge cases found at the scene. The evidence cartridge is on the left.

Evidence I: At this point, the evidence consisted of the body, the .22 Marlin, the fired cartridge, and the .22 rifle that Ferguson claimed to have been carrying during the hunt.

Laboratory Procedure I: Examination of the fired .22 cartridge indicated it had been fired in the .22 caliber Marlin rifle. Nothing remarkable was noted about Feguson's .22 rifle.

Examination of the body revealed an entrance bullet hole just forward of the right ear with the exit behind and slightly above the left ear. Chemical examination of the wounds showed no evidence of gun powder residue.

Post-mortem examination revealed the bone in the area of the wound was badly fragmented and several pieces of lead and copper jacket were found in the brain. There were no pieces of the jacket complete enough to identify rifling markings. However, it did indicate the victim had come to his death by a high-powered rifle bullet rather than by .22 caliber.

Further Investigation: When the above information was relayed to the officers in the field, they searched the area again. Their search revealed

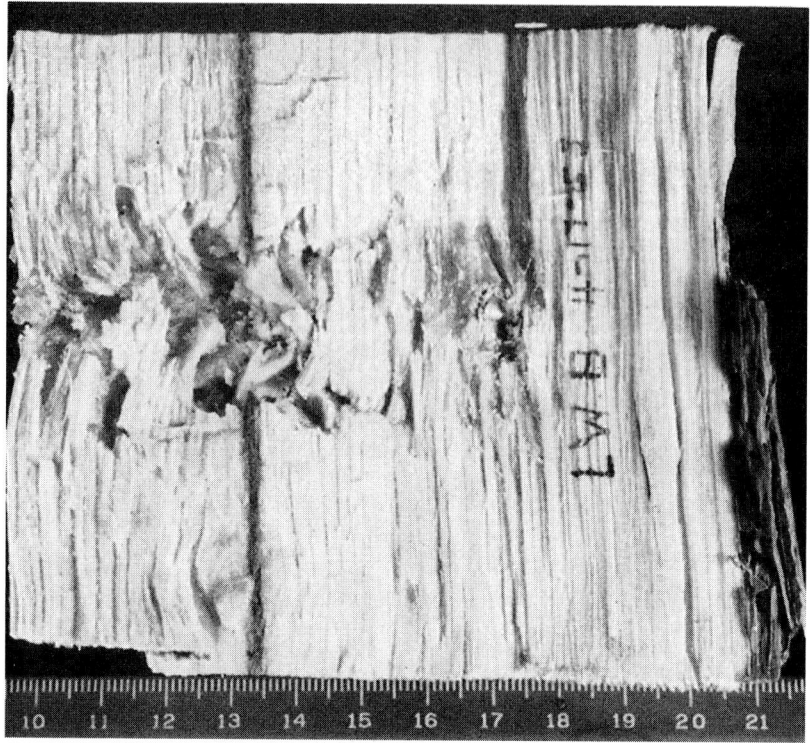

Photograph of the tree trunk showing the path and location of the bullet which was found at the scene.

two fired .30 U. S. A. rifle cartridges approximately 140 feet from where the body was found.

Ferguson was again questioned by the officers concerning his firearms and denied having a high-powered rifle. Further interrogation revealed that he once owned a .30 caliber rifle but claimed to have sold it two years previously to an unknown person. He finally admitted that he had a .30 U. S. A. Winchester Model 1895 rifle and produced it from its hiding place beneath a davenport in the living room. On producing the rifle, the subject admitted that he had shot and killed his brother-in-law, thinking he was shooting at a deer.

On April 17, 1955, an additional search of the scene revealed a tree bearing a hole which appeared to have been made by a high-powered rifle bullet. A section of the tree was removed and brought to the laboratory.

Laboratory Procedure II: The bullet was removed from this tree trunk and identified with test bullets fired in the .30 U. S. A. Winchester rifle. The fired cartridge cases had been matched with the rifle shortly after it was given to the officers.

Outcome: The accused was tried for manslaughter in Wasco County, Oregon, and found guilty.

CASE NUMBER 12

Contributed By: Scientific Laboratory, New York State Troopers
Nature of Case: Active, successful, and wide-awake Firearms Identification coupled with standard and efficient police procedure solved a most difficult homicide where there were virtually no clues save a bullet in the body of the victim.
Principals: An independent druggest vs. a solitary hold-up man.
Crime: Murder
Time: 29 December 1939
Circumstances and Evidence: A young man entered a small drugstore in the City of Buffalo on a cold and wintry night. The time was just 9:40 P. M., and the purpose of the young man's entrance was to purchase some ice cream. Entering the store, he noticed two small girls busily engaged in conversation while awaiting the appearance of the proprietor. This young man, being very familiar with the druggist and not seeing him present, walked into the prescription room looking for him. However, the prescription room as well as the living quarters in the rear of the store was quiet, and there was no indication of the presence of the owner nor any of his family. Not finding anyone in the rear of the store, the young man started back into the main part of the store. As he was passing the soda fountain, he noticed a foot sticking out past the end of the fountain. Investigating, he found the proprietor lying on the floor, quite still. The young man sent the two girls to run for help.

A doctor was summoned and immediately pronounced the druggest dead. The Medical Examiner's office of Erie County was notified, and the Medical Examiner arrived. Preliminary examination did not reveal the cause of death, but in the course of opening the deceased's shirt and underwear, a small hole was noticed just above the nipple of the left breast with a small blood-stain around the hole. The body was removed to the City Morgue, where an autopsy revealed that death was caused by a bullet which entered the left chest anteriorly about one inch to the left and above the nipple, between the fourth and fifth ribs, and took a diagonal course from the left side to the center back, passing through the left lung, left pericardial sac, rupturing the left ventricle, laying the heart wide open, and finally lodging in the tenth dorsal vertebra.

The bullet was removed and turned over to the Firearms Identification Office of the Buffalo Police Department, who placed the bullet in a small box and took it to the laboratory for examination. After washing the bullet, weight and caliber were determined; it was found to be a .22 long rifle bullet by virtue of its weight of 39.1 grains and diameter of .221″. Further examination revealed the possible presence of six lands and six grooves. Only three lands and grooves were visible, but from a calculation of the periphery occupied by these three lands and grooves, an accurate calculation that six lands and six grooves were originally present was made. The angle of rifling was left hand, which immediately indicated the firearm as being a Colt or some foreign make revolver, if the bullet was fired in a revolver.

It is most important to note that in the investigation of homicides involving .22 caliber firearms, both rifles and revolvers are to be included as possible weapons. Often only a study of the width of the lands and grooves and angle of twist of rifling will indicate which type of firearm was used in the commission of such crimes.

Police investigation was without success for many weeks. The wife of the deceased, who was absent the evening of the tragedy, was unable to furnish any possible motive for the crime other than robbery. However, a check of merchandise and cash of the store indicated that nothing had been taken.

On 20 February 1940, detectives, in the course of investigation of a liquor store burglary, questioned a suspect; a search of his room at 464 Clinton Street was made. In the course of this search, there was found a .22 Colt revolver.

Since routine police procedure of all police departments calls for a check of all firearms found or confiscated against open homicide cases; and more so, since the drug store homicide was still relatively fresh, an immediate examination of this revolver was made. The Firearms Identification Officer was much surprised when his microscopic comparison revealed that this was the drug store murder gun.

The suspect was carefully questioned and finally confessed that the gun was given to him by a friend, who was promptly picked up and questioned. The second suspect steadfastly denied any knowledge of the gun. However, a search of his residence revealed the presence of a pouch of .22 cartridges, a rifle, and many other articles which had been reported stolen from a hardware store. An exhaustive questioning finally produced results when the second suspect confessed:

"That on the night of 29 December 1939, after playing pool for several hours, he went to his home and took the .22 Colt revolver, which he said he bought from another boy for $5.00, and, loading it, proceeded to the drug store for the purpose of committing robbery. He had never been in the drug store and apparently did not know the proprietor. He loitered around outside waiting for a customer to leave. After the customer did leave, he waited until the druggist walked to the rear of the store. Then, he entered. Upon the druggist's question, 'Yes, sir, what can

I do for you,' he answered, 'I want a pack of cigarettes.' With that, the druggist proceeded to the cigar stand, which was located next to the soda fountain. He got the cigarettes and handed them to the suspect, who put them in a paper bag which he had brought with him. He then held the bag towards the proprietor and, pulling out the gun, said, 'Fill it up.' At this moment the gun went off, and the druggist fell to the floor. Suspect grabbed the cigarettes and ran out of the store. A few days prior to his arrest, the suspect gave the gun to a friend just in case he was picked up; then the cops would have nothing on him."

On 26 February 1940, all of the evidence was brought to the New York State Police Laboratory for corroboration of scientific examination. The results of examination by the Scientific Laboratory indicated that the fatal bullet was fired in the .22 Colt Officers Model Revolver, Serial No. 4474, which was taken from the first suspect's possession. Identification was made by comparison of an individual and peculiar "fanning" out of several striations along a land mark. This characteristic was the basis of comparison. The fatal bullet was identical to the ammunition found in second suspect's possession; namely, Remington .22 Long Rifle High-Speed Kleankote cartridges. It was also determined that the fatal bullet was fired from a distance of not less than three feet nor more than five feet. This was accomplished by applying Walker's Test or "C" acid test to the clothing of the deceased. In this test, photographic paper which has been desensitized and then immersed in a warm solution of naphthylamine—4, 8 disulphuric acid and dried; this treated paper is then placed over the bullet hole in the clothing and a towel moistened with 20% acetic acid is placed over it and then another towel, all of which is then pressed with an electric iron for about ten minutes. The nitrates and nitrites of the powder residue will cause red spots to appear on the treated paper. This gives an accurate pattern of the powder dispersion. Then, test shots at various distances are fired at blotting paper targets and the dispersion patterns are matched. Of course, it is essential that the identical firearms and ammunition be used.

Outcome: Second suspect was found guilty of first degree murder, the jury recommending leniency due to the fact that the defendant was an honor student in the Opportunity School in Buffalo. He was sentenced on 3 May 1940 to life imprisonment.

CASE NUMBER 13

Contributed By: Texas Department of Public Safety.

Nature of Case: Murder with a shotgun and a determination of the distance, made by means of shot trajectory and pattern.

Circumstances and Evidence: The scene of this crime was in a large, heavily-timbered game preserve area, in charge of a state wildlife biologist. On the day of the crime, some pipeline workers had talked with the biologist just prior to his entry into the reservation. Shortly thereafter, they heard the report of a gun shot and never did see the victim leave the area. A three-day search finally revealed the body of the victim, which had been removed from the spot where he had been shot, floated out into the swamp and hidden under a log. The suspect, after shooting the victim, had returned home, changed clothing, and had given away several freshly-killed ducks to neighbors.

Another of the evidence photographs taken for courtroom use. This shows twigs clipped or perforated by the passage of the shot charge.

One of the evidence photographs used to obtain a conviction during the trial of this case. It shows the place amongst trees and dense underbrush where the killing took place; the stretched cord shows the path of the charge of shot, easily determined by clipped twigs, pellets in bush and tree trunks, bloodstains, and a considerable number of pellets of duck shot which had missed the victim and lodged in the large tree trunk at the end of the cord.

The victim had caught the suspect hunting in the area where hunting was illegal, and had followed him until both had reached the opposite sides of a small pond. The suspect had then shot the victim, leaving him there until the next day when he had returned, taken the body off some distance and then hidden it under a log.

After the body had been found an autopsy was ordered. During this autopsy a measurement of the shot pattern was taken, pellets removed from the body for evidence; later a cartridge case found at the scene of the shooting and from the suspect gun was recovered. The size of the shot was determined, and the gun identified by the usual method of breech-block identification. The shotgun was a 12-gauge Winchester, Model 12.

The shot pattern area was found to be 12 by 14 inches. It became necessary to determine the distance from the victim that the gun had been fired. Test patterns were fired into large sheets of paper and the distance determined to have been approximately $37\frac{1}{2}$ feet; the report was given as at some point between 35 and 40 feet.

It then became necessary to return to the scene of the shooting and determine where the victim and suspect had been at the time of the shot, by locating leaves, twigs, branches and tree bark chipped or perforated by the shot, and checking the point where blood was found. The exact path of the shot was found and the position of the shooter located. The underbrush and trees were so dense that it was relatively easy to trace the path of the shot charge by means of freshly-clipped twigs and shot pellets in tree trunks.

During the trial it was shown to the jury, by means of stretched string in the courtroom, the height of the gun when fired, the ascending angle of the path of the shot; then a person of identical proportions to the victim walked beside this string until the charge of shot would have struck him in the face at the same point as the victim—placing the distance at the point the test pattern showed to be approximately 38 feet and matching the exact area and pellet density of the wound.

The defendant was given the death penalty and executed for his crime. **Comment:** This case was a most unusual and interesting one and represented a great deal of investigation. It illustrates the value of supporting evidence that may be obtained if the investigators are qualified and act promptly while such evidence is fresh and available.

As the principles and technique of firearms identification become generally known by the public, there is a tendency on the part of some persons guilty of a crime to alter various parts of their firearm in an effort to avoid detection. The commonest alteration is to change the shape of the firing pin.

The above photomicrograph shows the tip of a firing pin that has been altered very crudely—and in a manner that could be serious to the shooter. A sharp and burred profile such as this would distinctively imprint the fired case and it might also result in a pierced case or primer, permitting an escape of gas into the eye of the shooter. No firearm manufacturer would permit a gun with so dangerous a fault to leave his plant.

CASE NUMBER 14

Contributed By: Police Commissioner Thomas J. Gibbons and the Philadelphia Police Department
Nature of Case: This is an outstanding example of a routine street-recovered spent bullet identification, leading to the solution of a murder in which the bullet had not figured.
Principals: Isaac Teitelbaum vs. Grover Alexander Edwards
Crime: Murder
Time: 31 August 1952
Circumstances and Evidence: Isaac Teitelbaum, 64 years, white, 4612 Osage Street, found in his home about 3:55 A.M. (EST), 31 August 1952, shot several times about the body. The motive was robbery. He had notified the police by telephone, and, enroute to the hospital, he died.

Investigation disclosed that he was a retired baker and owned a chain of bakery shops throughout the city. Before his death, he disclosed that the burglar, who surprised him while he was in bed, demanded his money. After giving him the money he had in his wallet, the burglar shot him. The burglar was described as a white man. There was no other information.

Spent bullet specimen, Marked 1, removed from wall in bedroom. Bullet .32 ACP, full metal jacket, weighing 71.0 grains, no loss in weight, fired from a domestic smokeless cartridge case. Nose of bullet mushroomed diagonally, plaster and red brick adhered to flattened surface. The remaining surface circumference, approximately one-half of bullet, covered with foreign scratches.

The general rifling characteristics on the surface of the bullet are those produced by barrels of foreign automatics, i.e. Walther (old), German Dreyse, etc. The specimen bears rifling of right-hand twist, four lands and four grooves. Specimen in very poor condition for comparative purposes.

Spent bullet specimen, Marked 2, removed from wall in bedroom. Bullet .32 ACP same as above. Bullet considerably flattened diagonally from nose to base, rupturing metal jacket and exposing lead core. Bullet covered with foreign scratches and some foreign substance adhered to flattened portion of specimen. Specimen in very poor condition.

Spent bullet specimen, Marked 3, found in bedroom. Bullet .32 ACP same as above. Specimen in fair condition.

Three cartridge casings found on scene. Remington, .32 ACP, brass casings, nickeled primers.

Laboratory Procedure: Bullet, Marked 1, when compared with Bullet, Marked 2, showed similar class characteristics, indicating that the two specimens were fired from the same make of pistol. However, due to the considerable mutilation of Bullet, Marked 2, not possible to make a conclusive identification.

Microscopic comparative examination of bullet specimens (Exhibits 1 and 3) showed some similarities, indicating the two bullets were fired from one and the same pistol.

Casings—Depressions on the primer caps of the three casings are identical in characteristics. The three casings were identified as having been fired from the same automatic pistol.

A spent bullet was recovered by an unknown person from the surface of the street at 9th and Bainbridge Streets, and surrendered to a Police Officer at 11:30 P.M. (EST), 31 August 1952. When this arrived at the Firearms Identification Laboratory, it was found to be a .32 ACP, full metal jacket, weighing 68.6 grains, original weight 71.0 grains, considerably mutilated, fired from an automatic pistol having four lands and four grooves, right-hand rifling.

Three land impressions on the remaining surface of the spent bullet, which, when compared with the bullet specimen (Exhibit 3), in the case of Isaac Teitelbaum, were found to be identical in size to the land impressions on the Teitelbaum bullet, indicating that the two bullets were fired from a weapon or weapons having the same type rifling.

The individual microscopic characteristics of the two bullets were compared, and although numerous similarities were noted, no conclusion could be reached. Not a positive, but a very interesting comparison.

More General Investigation: Detectives investigating the second location found that one Grover Alexander Edwards, No. 149994, 34 years, white/ negro, of 702 S. Darien Street, was displaying pistol, in an endeavor to sell the weapon, in playground at 9th and Bainbridge Streets. 5 September 1952, Grover Alexander Edwards was arrested and informed of the evidence in possession of the Police Department. He confessed as follows:

1—He shot Isaac Teitelbaum with a German Walther (old) pistol.

2—He fired the same weapon in playground, 9th and Bainbridge Streets.

3—He had stolen a German Walther .32 auto pistol from the home of Paul Gruenfelder, 4516 Osage Avenue, near home of the slaying, during the week of 26 August, through unlocked apartment door.

4—He was formerly employed by Isaac Teitelbaum.

He stated that he had disposed of the murder weapon by throwing it into the Schuylkill River, giving location. The weapon was recovered with a large magnet and identified by Edwards.

Further Laboratory Work: Test bullets and cartridge cases from the recovered weapon were compared with crime bullets and cartridge cases, with positive results.

Outcome: Convicted of 1st degree murder and sentenced to death.

CASE NUMBER 15

Contributed By: Department of Oregon State Police.
Nature of Case: Fatal shooting with a shotgun.
Principal: Thomas J. Coyle.
Crime: Suspected suicide.
Time: Exact time unknown, believed to be sometime between 8:45 a.m. and 3 p.m. on March 11, 1955.
Circumstances: At approximately 3 p.m. on March 11, 1955, a man's body was found along a country road near the junction of Highway No. 97 and No. 216 in Sherman County, Oregon. Police officers were notified and started an investigation to determine the details.

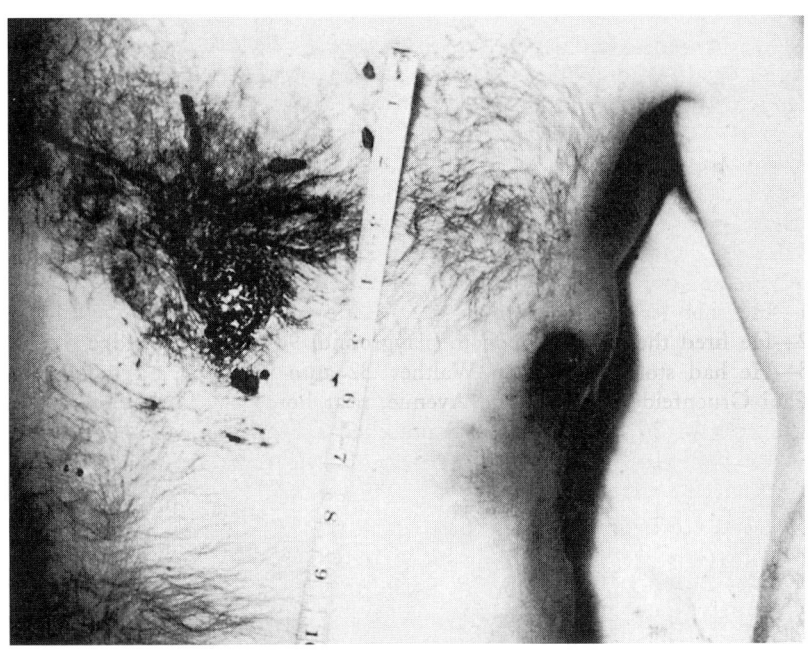

Photograph showing the relative position of the two shot gun wounds.

Upper: Ordinary black and white photograph of dark forest green jacket. Adjacent to the scale and the snap fastener are the entrance holes.

Lower: Infra-red photograph of the same areas indicate the true size and shape of the powder patterns.

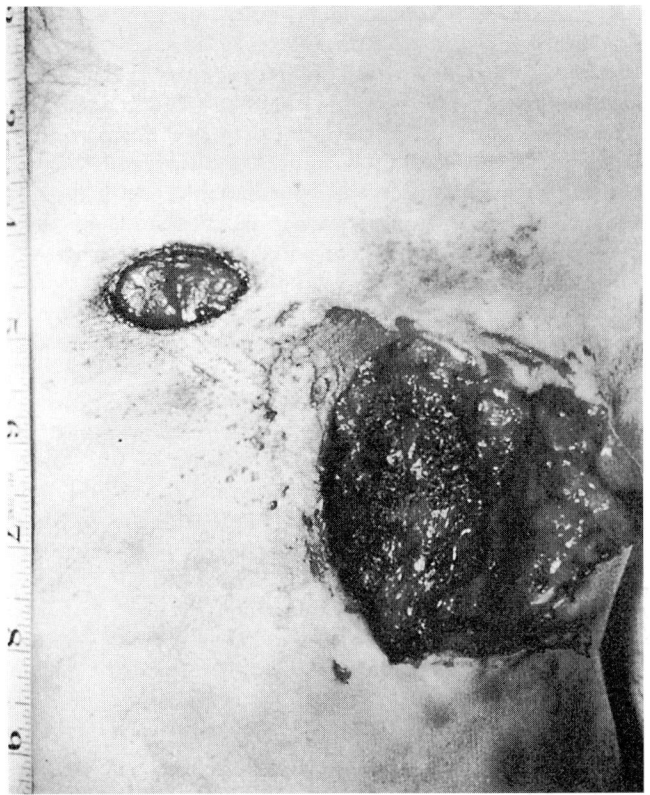

Close up of the superficial wound on the left side. Note the small, oval shaped entrance and large exit holes.

Evidence: The evidence consisted of the victim's body, clothing, Model 1897 Winchester shotgun, three fired shotgun shells, shot and wadding removed from the body, a partially filled box of 12 gauge shotgun shells found in the victim's pickup truck.

Laboratory Procedure: Examination of the body revealed two large, irregular wounds in the left front area of the chest. One was a superficial wound that penetrated the flesh over the ribs and made an exit in a downward direction of approximately 30° to the vertical axis of the body. Careful examination of this wound revealed it did not penetrate the chest cavity and was not necessarily of a fatal nature. The second wound was approximately in the midline of the chest and ranged slightly downward into the chest cavity. This shot charge caused extensive damage to the internal organs and was definitely a death-inflicting injury. The superficial wound along the rib cage was in all probability inflicted first, since it was not necessarily fatal. The second shot would logically have been the one which penetrated the chest.

The victim was wearing a dark forest-green jacket at the time of death, and it was subjected to close examination. There were two ragged holes in the front of the jacket showing some indication of a sooty deposit which was barely visible to the unaided eye. Infrared photographs were taken of these two areas and they indicated very clearly the size and shape of the powder patterns.

Test fire-patterns were made with the suspected weapon and some of the shells found in the victim's pickup. The most-similar patterns were made when the shotgun muzzle was held in contact with the test targets.

There were numerous fresh scratches on the stock of the shotgun which could have been made by the barbed wire fence where the victim was found. The distance and range of the shotgun blasts and the fresh scratches on the stock indicated that the victim could have placed the stock of the shotgun across the fence, held the barrel toward himself with one hand, and pushed the trigger with the other.

Three fired shells found on the scene were easily identified as having been fired from the victim's shotgun. The presence of the third fired shell was not explained by the evidence at hand.

Shot pellets removed from the body were examined and found to be size No. 5, the same as those from in the victim's pickup. Wadding found in the chest cavity was found to be similar in size, construction, and material to the 12 gauge shells from the pickup.

Further Investigation: Further investigation revealed that there were sufficient reasons to believe the victim had taken his own life. His mental condition had been poor since the death of his wife the previous year. The victim had left his shotgun with his son-in-law until March 8, 1955. On March 2, he had gone to his son-in-law's ranch and asked to obtain the shotgun. When this request was refused, he remarked that there were lots of ways he could take his life. Shortly after this he did somehow obtain the gun.

One of the local stores reported that on March 7 or 8 he purchased a box of Remington 12 gauge shotgun shells, No. 5 shot. On the morning of March 11, he informed his daughter that he was feeling too nervous to drive into town and "would just drive around for awhile." That was the last time he was seen alive.

Outcome: In view of the laboratory evidence and field investigation, it was determined that the victim had died of self-inflicted wounds.

CASE NUMBER 16

Contributed By: Scientific Laboratory, New York State Troopers
Nature of Case: This is an astonishing Firearms Investigation of hand-loaded ammunition and the solution and proof of a crime from extremely good work with modern apparatus.
Principals: The proprietor of a miscellaneous business establishment vs. unemployed tenant in debt to him
Crime: Murder
Time: 14 December 1948
Circumstances and Investigation: Two weeks before Christmas, 1948, the proprietor of a successful gas station located in up-state New York, which included four tourists' cabins and accommodations for trailers, was murdered. There were twelve trailers occupying this site at the time.

It was his custom to open his business at approximately eight o'clock in the morning and to close about seven o'clock at night. His wife usually met him at the gas station and then returned with him to their home after he had closed his place of business.

On the evening of the homicide, his wife came to the gas station and found the lights on but could locate no one around the station. She obtained the assistance of one of the trailer occupants and began a search of the cabins and premises. They discovered the body of her husband in one of the cabins, shot to death. The scene of the crime was a cabin with a small lavatory and a separate shower room attached. This particular cabin had been closed for the winter. Investigation developed that the proprietor did not rent this cabin under any circumstances, as he used it for personal needs. The wife of the victim also stated that he was of such pugnacious nature that he would have put forth extreme resistance to any person attempting to force him into the cabin. This indicated that the proprietor had entered the cabin willingly before his death. It was common knowledge that the subject carried large sums of money and as no money was found on the body, robbery was indicated.

X-ray pictures taken before the autopsy disclosed numerous spaghetti-like particles dispersed throughout the brain. These particles, when recovered, had the appearance of small pieces of wire. Several pieces of blood-matted paper were also taken from the brain and this evidence was delivered to the State Police Scientific Laboratory for analysis. The evidence all pointed to a shotgun having been used.

Each of the occupants of the trailers was questioned. One of these persons who occupied a trailer located directly behind the cabin in which the crime was committed was an unemployed carpenter. It was known that he had been very friendly with the deceased. He denied having been in the trailer or having seen the proprietor since about 3:00 P. M. The time of the murder was established as having been between 5:00 P. M. and 6:00 P. M. on that day. This suspect denied that he had ever possessed a gun.

A check of this subject disclosed that he was a heavy drinker and that he was behind in his trailer rent. One of the occupants of another trailer, when questioned, stated that she had observed the suspect standing near the station at approximately 5:45 P. M. and that she had seen him walking toward the cabin in which the body was found.

The Laboratory examination of the material removed from the brain of the deceased showed the spaghetti-like particles to be pieces of small-diameter rosin-core lead solder, such as is used in radio repair and electrical work. The blood-matted paper when cleaned and washed was identified as pieces of advertisements torn from newspaper. Both of these evidences indicated that the fatal shell had been home loaded. For pellets, the wire solder was employed and instead of wads, the matted newspaper was used. This information was forwarded at once to the Troopers making the investigation.

The trailer occupied by the suspect was immediately searched and all torn newspaper and magazines were collected and transported to the State Police Laboratory for comparison with the blood-matted paper taken from the brain of the victim. This led to an almost unbelievable discovery. Laboratory examination revealed definite proof that the pieces of paper taken from the brain of the deceased had been torn from a December 12th issue of a local newspaper found in the defendant's trailer. The edges and wordings matched perfectly.

The suspect was again questioned regarding his actions prior to the crime and, when confronted with the torn paper that was used for the wadding in the home-made shell, he admitted that on 11 December 1948 while in the home of a friend, he stole a 20 gauge Western Field bolt-action shotgun. He tried to sell this gun but could find no buyer. He took the gun to his trailer as well as some shotgun shells that he had stolen. On the night of the 12th, he took one of the shells apart, removing the pellets and wadding and reducing the powder charge. He tore up pieces of newspaper and made wadding and cut up pieces of fine solder for the pellet charge and thus reloaded the shell. On the 14th he went downtown, had a few drinks and then returned to the gas station. There he had a couple of drinks with the deceased. In the course of the conversation, he suggested to him that since it was a cold night outside they should check the toilets in the cabins so that they would not freeze and break. They both proceeded to the cabin. Meanwhile, he had picked up the gun and, holding it behind his back, let the proprietor go into the bathroom to check the toilet. Just as the owner reached the bathroom door, he shot him in the back of the head. He then left the cabin and hid the gun down an embankment behind his trailer. He took the money from the coveralls of the deceased; leaving the cabin he then picked up his gun and, about a half mile away, threw it into a swamp from which it was later recovered. He hid in a cherry tree most of the money he had stolen.

Outcome: Based on this evidence, a first degree murder indictment was returned by the county Grand Jury. The defense, the court, and the district attorney each caused a psychiatric examination of the defendant. The three examinations indicated a mental psychosis. The defendant was, therefore, committed by the Supreme Court to the Mattewan State Hospital for the criminally insane.

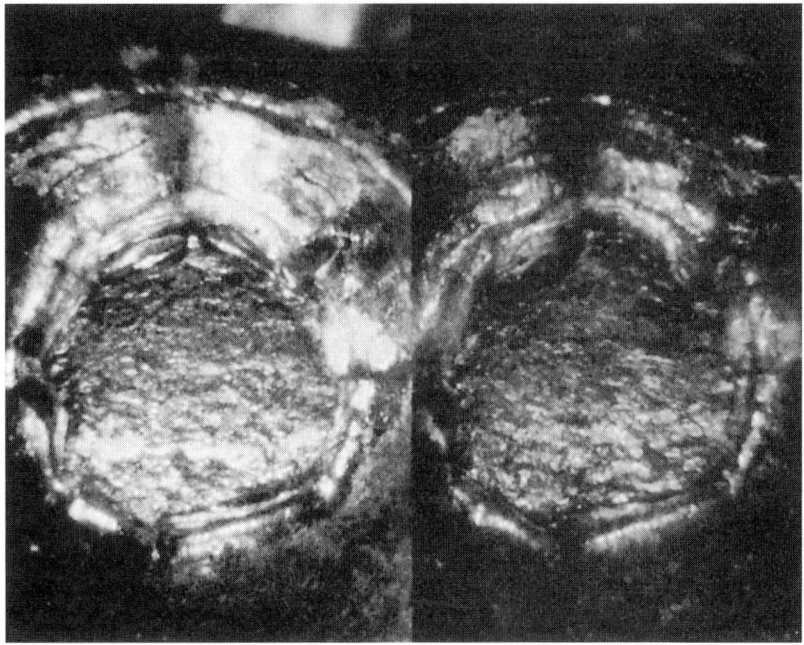

Photomicrograph of firing pin impressions made by a Winchester Model 63, semiautomatic rifle. (The evidence cartridge is on the left. This photomicrograph was taken by vertical illumination.)

Examination of the firing pin revealed that it was a few thousandths too short for good ignition. At some time, the user had removed the firing pin and had "peened" the striking end so as to lengthen it. The pounding formed a pronounced lip on the striking end which was not removed. This condition has not resulted in any pierced cases, but microscopic examination reveals that the metal about the impressions is very thin.

This is an example of a repair job that results in changing the impression of a firing pin in a very peculiar manner. An identification of the resulting impression is very easy.

CASE NUMBER 17

Contributed By: Department of Oregon State Police.

Nature of Case: This is a case where firearms examination discredited the story of the victim.

Principals: Verl Gray Jenkins, age, 18, West Fall, Oregon.

Crime: Alleged assault with a dangerous weapon.

Time: Between 11 a.m. and 11:30 a.m. June 16, 1954.

Circumstances: The police were called to investigate a case where an eighteen-year-old boy had been shot while irrigating on his father's farm. Examination of the scene revealed that a bullet had been fired through the windshield of the boy's pickup truck. The bullet holes lined up with the ridge 400 yards away from which the shots were alleged to have come. The boy stated that he had been down in the field attending the irrigation ditch and was facing back toward the north with the shovel over his left shoulder when the first shots were fired. He stated that the shovel was hit twice while on his shoulder and that he was wounded in the left forearm. He ran back to the pickup about 100 yards over rough terrain, put the shovel in the pickup and threw his own pistol on the floorboards. As he started to get into the vehicle, a bullet went through the windshield and out through the open right door. The boy then drove away from the field at high speed and went to his grandfather's house for treatment.

Examination of the shovel handle revealed two holes approximately an inch apart. This was explained as being a coincidence, because he was running when the second shot was fired. He could not explain how he was injured in the arm and not in the back, if he was carrying the shovel over his left shoulder.

Two teen-age youngsters, found in the area, were questioned because they were armed with a scope equipped .300 Savage rifle and a .22 Hi-Standard pistol. Investigation revealed that they had not been near the Jenkins ranch.

Laboratory Procedure: The only firearms evidence submitted in this case consisted of the shovel handle. Around the periphery of both holes was noted a dark discoloration and several partially burned grains of powder. Chemical testing of the region about the holes revealed the presence of

Photograph of shovel handle showing the gun powder pattern around one of the entrance holes. Each division on the scale is one-sixteenth of an inch.

nitrates. Chemical testing indicated the presence of lead in both holes. The laboratory examination indicated that both of these bullet holes were made at a close distance.

Further Investigation: This information was relayed to the officers in the field. Shortly afterward, it was reported that the alleged victim had confessed that the wound was self-inflicted and deliberate. The only motive that he would give was that he and his father had not been getting along. He admitted that he had shot a hole in the windshield of the pickup while at the upper part of the ranch and driven beyond the gate, stopping about a mile away where he fired two more rounds into the shovel handle, then laid the gun across his left forearm so that the bullet would tear the skin, not fire directly into his arm. However, the self-inflicted wound bled more than he intended.

Outcome: This case illustrated the value of powder patterns to refute testimony of a person alleging a crime committed against him.

CASE NUMBER 18

Contributed By: New Jersey State Police and Mercer County, New Jersey Prosecutor's Office

Nature of Case: A typical Saturday night gun whose identification as the weapon used in a previous crime solved the crime, but with inconclusive results.

Principals: Black vs. Brown. Since this case has not as yet come to trial, these are obviously fictitious names.

Crime: Possible attempted murder and carrying concealed weapons

Time: Shooting of Brown, 18 April 1953. Apprehending of Black with a concealed weapon, 6 June 1953.

Circumstances: Brown was shot in unusual and mysterious circumstances and brought into the Princeton Hospital to have a bullet removed from the rear of his hip after it had passed through his lower abdomen. A routine investigation of this matter was made by detectives in the Mercer County Prosecutor's Office. Brown accused a young woman of shooting him. However, the detectives, upon investigating the accusation, found no evidence whatever to support Brown's statements. In fact, they found the young woman to be of excellent character and to have publicly refused to have anything to do with Brown.

The State Police Patrol at Princeton, New Jersey apprehended Black and a friend in a car. Black had in his possession a .32 S&W revolver. They were brought to the station for questioning, denied any knowledge of any crime or intent to commit one. They knew it is against the law to carry a gun in a car without a permit. They had no explanation for the fact that they were carrying a loaded revolver. They were arraigned before the magistrate for the crime of carrying concealed deadly weapons, and sent to the County Jail in default of $1,000 bail. The routine Police Information teletype was sent out. The gun and the three cartridges found in it were forwarded to the Firearms Laboratory together with the five copies of the Lab Request Report form, properly accomplished. Examination requested was the routine: "Fire test shots and compare against pending cases. Check files for any record on gun."

Evidence: A cal. 32 S&W crime bullet removed from Brown; a cal. .32 S&W revolver carried by Black

Laboratory Procedure: In routine fashion, the gun was examined, as were the cartridges. Two of these turned out to be not the correct ones for the .32 S&W, but were a .32 Short Colt, a somewhat different .32 cartridge, smaller in diameter of shell. The third was a .32 S&W with no manufacturer's stamp on the head, probably a foreign make, but they were unable to identify it further. All three shells showed firing pin impressions but had misfired; one of the .32 Short Colts had been struck twice.

Two shots, using Rem-UMC .32 S&W ammunition, were then fired in the gun without misfiring. Two .32 Short Colt also Rem-UMC (since the two identifiable cartridges were of this make) were tried, both of which misfired on first try but fired on second try. The two .32 S&W shells were then compared in the comparison microscope and found to match perfectly. Then the two .32 Short Colt shells were compared and found to match; then one of the .32 S&W and one of the .32 Short Colt were compared and also found to match. The same procedure was followed with the bullets. All were found to match. Of course, they were sure they would since they had personally fired all of them in the same gun. These preliminaries out of the way, the best specimen bullet was put on the right stage of the microscope, and the pending file referred to.

Fifteen cases involving .32 S&W with five lands and five grooves, right-hand twist, and two involving .32 Short Colt bullets with five lands and five grooves, right-hand twist, were pulled from the file and arranged, with jackets unopened, in chronological order. Since there were two .32 Short Colt cartridges in the gun, and also since there were only three cases involving this type of bullet in the pending file, it was decided to process these first.

The first two were negative and were returned to the files, with that notation on a slip inserted in the jacket. (This is done to verify at some future date that the case was checked.)

The third case, that of Brown, of Trenton, New Jersey, who was shot in the abdomen and a lead bullet removed from his buttocks in the Princeton Hospital on 18 April 1953, was then put on the left side of the microscope and compared against the test bullet fired from the gun. It was found to match on every land and on every groove.

Further Investigation: Since Princeton and Trenton are quite close by, and to expedite matters, a phone call was made by the Firearms Expert to the trooper who had arrested Black and his companions, informing him of the fact that the gun he had taken from the two prisoners was the gun used in the Brown shooting. The trooper immediately phoned the Mercer County Prosecutor's Office and the Trenton Police Detective Bureau; they all went to the County Jail to question Black and his companion. They denied everything.

Outcome: Black was convicted of a concealed weapons charge. Brown would not testify or cooperate in any way, so the atrocious Assault & Battery or Attempted Murder charge was not even made.

CASE NUMBER 19

Contributed By: Texas Department of Public Safety.

Nature of Case: An accidental death while cleaning a pistol which had been taken apart.

Circumstances and Evidence: Early on the morning of November 28, 1954, the department was called upon to assist in the investigation of one of the most unusual cases involving firearms which our experts have ever had an opportunity to work upon. A man had been shot and killed while cleaning his pistol. This is by no means an unheard-of accident; however upon examining the room in which the tragedy occurred it was found that the pistol had been taken apart and was in four pieces when the accident happened.

Details of the affair revealed that the man had been sitting on a stool and to have received a fatal shot through his stomach, the bullet passing through the body and into the wall behind. After falling off the stool, the victim had walked into an adjoining room where he took the telephone from its hook but died before he could use it.

The disassembled parts comprised a 9mm Luger automatic pistol and subsequent identification proved it to be the fatal weapon. The side plate

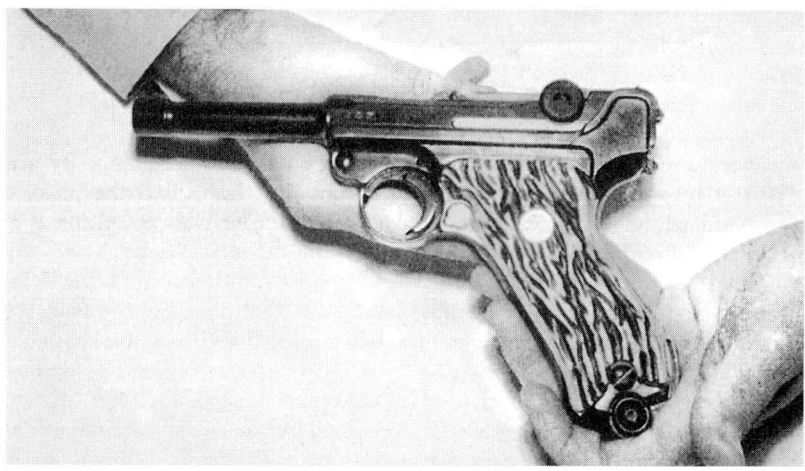

The Luger 9mm found at scene of accident.

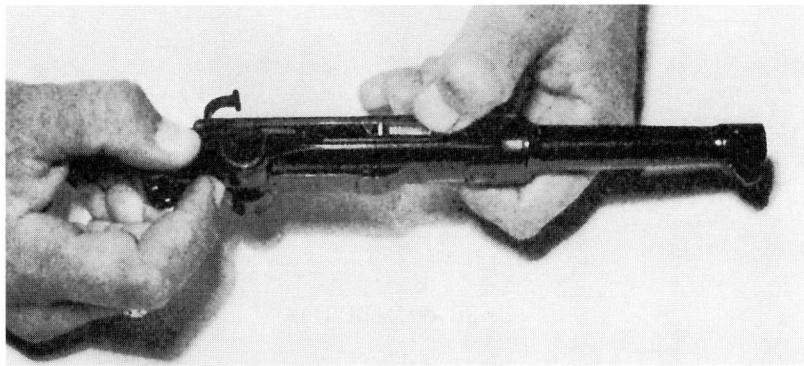

The Luger barrel, receiver and action assembly. By pressing the exposed end of the sear bar, underneath thumb, the firing pin can be released and the assembly fired.

and locking bolt had been removed and placed upon the bed; the frame was lying on the floor. The barrel with its attached receiver, breechblock and firing mechanism were found together—with a partially-extracted fired cartridge case in the chamber.

The ensuing investigation revealed what some owners of the Luger pistol may know—that the barrel and receiver assembly alone can fire a cartridge loaded into the chamber. This is due to the peculiar construction of the Luger mechanism. In this pistol the firing pin with its coil spring are housed in the breechblock (bolt) and a projection or hook-like portion of the firing pin is exposed on the left-hand side of the breechblock. This hook rides in a groove in the barrel extension, on the left side of which there is mounted a pivoted lever the rear end of which engages with the hook above mentioned, holding the firing pin cocked and the firing pin spring compressed when the breech is closed. When the front end of this lever is depressed the rear end raises and releases the firing pin. Normally this is accomplished by the trigger acting through a right-angled member contained in the side plate; but when the barrel is removed from the frame a simple pressure of the thumb on the lever will fire the pistol.

In this particular case the victim had obviously left a cartridge in the chamber and then started to 'take down' the pistol for cleaning. By accidentally pressing on the exposed sear mechanism he caused the pistol to fire—and what first appeared to be a possible murder was settled for as an accidental shooting.

Comment: This case is unique in that it serves to illustrate the point that a firearms technician should make a thorough study of any firearm submitted in a case, in order to determine any peculiarities of construction which might give rise to irregularities in his interpretation of the evidence, due to a lack of experience with the arm. It often takes quite a bit of experience, at that, before a person thoroughly knows even the one firearm he may be using—and a good firearms investigator is supposed to know them all.

CASE NUMBER 20

Contributed By: Police Commissioner Thomas J. Gibbons and the Philadelphia Police Department

Nature of Case: Cases and bullets from a Spanish .25 automatic linked several crimes and led to the solution of a murder.

Principals: Michael DiCriscio vs. Three Juveniles

Crimes: Murder, Attempted Murder and Robbery

Time: 10 October 1951; 12 October 1951

Circumstances and Evidence: 10 October 1951, Michael DiCriscio, 41 years, white, residence 1928 S. 10th Street, Yellow Cab Driver, was found, shot through the chest, while seated in his cab, with the door open, which was parked in the middle of 34th Street, 180 feet south of Girard Avenue. He was pronounced dead at the Philadelphia General Hospital by Dr. Clark at 1:55 A.M., 10 October 1951.

Found on the front seat of the Yellow Cab, one (1) fired cartridge case, marked Western, .25 ACP, brass case, nickel primer with some unburned powder present in interior of case. The depression on the primer cap caused by the firing pin or striker shows a distinctive form; the base of the depression shows a flat surface, irregular in area. There were several marks on the rim caused by the extractor and ejector.

13 October 1951, the Firearms Laboratory received from the Homicide Division, one bullet removed from the body of Michael DiCriscio by Dr. Wadsworth, Coroner's Physician, 12 October 1951. This was a .25 ACP, full metal jacket, and weighed 50.3 grains (no loss in weight). The shape of the bullet was uniform, cylindrical with a conical nose. There were no mutilated or flattened surfaces. From the class or rifling characteristics, would indicate bullet was fired from an automatic pistol having six (6) lands and six (6) grooves, right-hand twist. In addition to the class characteristics, the whole of the surface of the bullet was marked with fine and course scratches, caused by minute roughness and slight imperfections in the barrel. The general rifling characteristics on the bullet surface were those produced by foreign automatics, with regard to the number of lands and grooves, direction, pitch, and size of grooves. Due to the close resemblance of foreign pistols manufactured in Spain, Germany, and Austria, it was impossible to determine the make of pistol having fired the fatal bullet.

15 October 1951, the Firearms Laboratory received Spanish (Model Automatic Pistol), cal. .25 ACP automatic pistol, found 13 October 1951 at 10:00 A.M., in rear yard, 1010 Melon Street, by one Charles Davis, 17 years, Negro, of that address. Investigating officers stated that this may have been the weapon used in an attempt hold-up at 5th and Callowhill Streets, 12 October 1951.

Charles Davis, who was supposed to have found the weapon in the rear yard of his home, was questioned; subsequently three other boys, friends of Charles Davis, were seized and questioned by the Police relative to the grocery store hold-up at 4th and Buttonwood Streets. They were identified by the proprietor and his wife as the three boys who entered their store on 12 October 1951, demanding money with one of the young bandits firing two shots on the outside of the front door. They were also questioned in connection with the slaying of the cab driver. They admitted getting into the cab at Broad and Columbia Avenue and directing the driver to proceed to 33rd and Girard Avenue where the slaying occurred.

15 October 1951, the Firearms Laboratory received two (2) fired cartridge casings, found outside the front door of grocery store, 501 N. 4th Street, where, on 12 October 1951 at 8:20 P. M., three unknown colored boys attempted to hold up the owner, Mr. Charles Berger, who pushed the boys from the store, shutting the door. One of the boys fired two shots through the glass door, and made their escape. The bullets were never recovered.

Laboratory Procedure: Microscopic comparative examination of fired cartridge case, found on the front seat of the Yellow Cab, and test cases, fired from pistol, found in rear yard, 1010 Melon Street, showed that the fired cartridge case had been fired from the pistol under examination.

Microscope comparative examination of fatal bullet, removed from the body of Michael DiCriscio and test shot fired from pistol, found in rear yard of 1010 Melon Street, showed that fatal bullet had been fired from the pistol under examination.

Microscopic comparative examination of two (2) fired cartridge casings, found outside grocery store, 501 N. 4th Street, and test casings, fired from pistol found in rear yard, 1010 Melon Street, has shown that the two fired casings had been fired from the pistol under examination.

Outcome: The three boys were tried in Common Pleas Court; all pleaded guilty. They were sentenced to life imprisonment.

CASE NUMBER 21

Contributed By: Texas Department of Public Safety.

Nature of Case: Murder of a white male, shot and killed while attempting to prevent the theft of his automobile.

Circumstances and Evidence: In the very early hours of a foggy fall morning, the victim in this case was aroused from sleep when he heard someone trying to start his old-model Buick car, parked in front of his house. As he hurriedly approached the automobile to stop the thief, he was killed instantly by a shot fired from within the car. The now-murderer was seen by a neighbor to flee into the early morning fog, however this witness was not close enough to ascertain whether the killer was a white man or a negro. A comb found along his line of flight had the light brown hairs of a white man in its teeth; this small clue caused the officers to pursue a wrong course of investigation for a week or so.

On the seat of the car was found a fired 8mm Mauser rifle cartridge case which had been ejected from the rifle by the killer. Immediately following the murder, a complete radio report was circulated by the local radio stations, requesting that anyone seeing a stranger promptly report it. In answer to this broadcast, an individual contacted the officers to advise them that an 8mm Mauser rifle had been stolen from his home a night or so previously; he also stated that he reloaded his own ammunition for this rifle and had in his possession several fired cases, which might be suitable for comparison with the evidence cartridge case.

The writer contacted this individual and secured several of those fired cases. A prompt comparison of the extractor and breech block markings on the base of this evidence case with the cases submitted by the owner of the rifle definitely identified his stolen rifle as the one used to commit this crime.

At this point, the officers knew the caliber, make and serial number of the rifle used but could not locate it. However, even before it was found they knew what gun had been used.

A leather jacket was meanwhile found in a nearby park. This jacket was identified by a white man as the one which had been stolen from him some days prior to the crime. He also identified the comb found as being his; the hairs in the comb were found to be his also. This jacket had been dropped by the killer in his flight. A search for a negro was now begun.

A month passed, during which time a reward of $100.00 was offered by the family of the victim, to be given to anyone finding the wanted rifle. Shortly thereafter, a young negro 'found' the wanted rifle and was given the reward money. Subsequent investigation revealed that the 'finder' was actually the murderer.

The rifle was taken to the laboratory for further test purposes and comparison with the evidence cartridge case. Positive identification was made

The positive matching evidence by which this criminal was convicted and executed. Upper section shows the extractor striations caused on evidence case (lower) and a splendid matching by those from the test case (upper). At bottom are the bolt face comparisons (evidence case at the left). Observe how clearly seen are the imprints on these primers of the rusted and corroded patches on the face of this bolt; these were so pronounced they could be picked out and matched by eye alone.

through the extractor scrapings plus numerous bolt face markings found on the primer; these primer embossings were very pronounced as the bolt face was somewhat corroded and eroded away and were visible to the eye alone.

It was conclusively proven that the cartridge case found on the car seat was fired in the submitted rifle. The negro confessed, was tried, found guilty, assessed the death penalty, and has been executed for his crime. **Comment:** The ironical part of this case was that the family of the victim paid for the defense of the man who killed him. The case illustrates the necessity of looking about for evidence previously made that may be of use in obtaining a conviction, such as locating and using those previously-fired cartridge cases.

CASE NUMBER 22

Contributed By: Department of Oregon State Police.

Nature of Case: A premeditated murder in which the identification of gun powder residue on the victim's clothing and a thorough knowledge of firearms identification led to an early solution of the crime.

Principals: Mary Ellen Campbell, 18-year-old deaf-mute, and Elmer Harlan Belcher, 15-year-old youth.

Crime: First degree murder.

Time: Approximately 6 p.m. on April 4, 1952.

Circumstances: The victim was a mentally retarded deaf-mute. Although she had the mentality of an infant, she had the body of a woman and reportedly was very fond of the company of men. In the latter part of March, it became quite apparent that she was pregnant. The teen-age children in the neighborhood soon became aware of her condition and began teasing two of the neighbor boys who had been seen in her company at various times. One of these boys was Elmer Belcher, age 15. He became enraged over the accusations of his schoolmates and it was believed that he may have become apprehensive that he might be forced to marry the victim.

In any event, he began a plan to eliminate her. He observed that every evening the parents drove about a quarter of a mile away to milk the cow, taking the younger children with them, and leaving the victim alone in the house. He noticed they were gone for a period of fifteen to twenty minutes.

Each afternoon after school and on Saturdays, the accused was employed to help a man build a house next door to the Campbell residence, which was across the road from the Belcher farm. The presence of this man, whom he was assisting, interfered with carrying out of his plan. However, on Friday, April 4, this man became ill and went home early. The accused seized this opportunity to put his plan into effect. He went home, put on a pair of knee-length rubber boots, took his father's .22 Colt automatic pistol from the hiding place in his father's bedroom, and loaded it. He then returned to the new house and continued working.

At 5:45 p.m., he observed the victim's family leave in their car. He immediately went to the Campbell home where he found Mary Ellen alone. Having only about ten minutes to carry out his plan, he quickly persuaded Mary Ellen to accompany him. He took her up the hillside about 300 yards from the house where there was a small level spot. When she bent over to pick a wild flower, he took the .22 caliber automatic from his boot and fired two shots at her. One of these shots apparently missed, the other struck her in the back and she died immediately. Later he stated that when she fell forward on her face, he turned and ran down the hill to the new house, where he was found working when the Campbells returned. After talking awhile with the victim's family, he went home and returned the pistol to its hiding place in his father's bedroom; he then removed his rubber boots and took them to the swamp where he threw them into a deep hole, as he feared that his footprints might be detected at the scene of the crime. When the family became aware that Mary Ellen was missing, they organized a search. Elmer Belcher participated in the search but was careful not to discover the body.

When the body was initially found, it was unknown whether it was murder or a death caused by a stray bullet. However, careful examination of the scene revealed the presence of a fired .22 caliber cartridge case. The case and the girl's clothing were turned over to the Crime Detection Laboratory for examination.

Evidence: The evidence consisted of the victim's clothing, the fired .22 cartridge, and the bullet recovered from the body.

Laboratory Procedure: Laboratory examination of the clothing revealed the presence of gunpowder residue, thus eliminating any possibility of the girl's death having been caused by a stray bullet. From an examination of the firing pin impression and other marks on the cartridge case and rifling impressions on the bullet, the firearms examiner concluded they were from a .22 Colt Woodsman automatic pistol.

Further Investigation: Meanwhile, all of the .22 caliber firearms in the neighborhood—with the exception of an automatic pistol from the Belcher home and a revolver owned by the girl's grandfather—were picked up for examination. The grandfather denied possession of a .22 caliber pistol until it was revealed to him that the neighbors had stated he owned such a weapon; he then produced a .22 caliber H & R seven-shot revolver from his dresser drawer. The grandfather's revolver and the other .22 caliber weapons were forwarded to the laboratory for examination. All of them were definitely eliminated as the murder weapon. Other findings inferred that the suspected firearm was a Colt Woodsman automatic pistol.

One of the officers then recalled that the Belcher boy's father owned such a pistol. When the investigating officers were making a canvass of the neighborhood for all .22 caliber guns, the Belcher family had not been home. At approximately 2 a.m. on the morning of April 6, 1952, police officers arrived at the Belcher residence and requested to see the .22 caliber Colt Woodsman. This was turned over to the officers by the boy's father. The boy was again interrogated and asked to repeat his story. After numerous discrepancies were pointed out to him, he admitted he was the one that had shot the victim and related the circumstances. To corroborate his story, the rubber boots were located and extracted from the swamp where he had thrown them. The .22 pistol received from the father was immediately forwarded to the laboratory and conclusively linked to the bullet removed from the body and the cartridge case found at the scene.

Outcome: The accused was tried and convicted of murder. The crime had taken place at approximately 6 p.m. on April 4 and by 2:45 a.m. April 6 the accused had confessed to the murder.

CASE NUMBER 23

Contributed By: New Jersey State Police

Nature of Case: A triple murder committed with a foreign automatic pistol converted to fire different ammunition by substituting four parts from a U. S. Model 1911 A1 .45 automatic.

Principals: Garner Powers vs. his wife, sister-in-law, and mother-in-law

Crime: Murder and suicide

Time: 7:25 p.m., 23 January 1952

Circumstances: Garner Powers, 32, had a long record of petty indictments but had escaped ever serving any time in prison. He had received several suspended sentences and fines for breaking and entering, improper operation of motor vehicles and the like. He was employed as a civilian mechanic in the Government work shops at Fort Dix, New Jersey.

He had purchased a Spanish "Llama Extra" automatic pistol chambered for the 9-mm Luger cartridge. This weapon is, in appearance and operation, almost identical with the U. S. Model 1911 A1 service automatic. He had approached various individuals working at the Ordnance Shop who had discussed with him the possibility of converting this weapon by the substitution of a new barrel, barrel bushing, receiver, and link to fire the .45 ACP cartridge. Powers stole these components from the Ordnance Shop; this theft was not detected until after the crime. He then made the substitution and had an automatic pistol with which he could shoot both cartridges.

After considerable argument with the three women, he shot them all with the Llama Extra automatic pistol, killing them instantly. At the time of the shooting, however, this weapon had been converted to use, as pointed out above, .45 ACP ammunition.

Immediately following the crime, he disassembled the weapon and put back the original parts. He now loaded the weapon with 9-mm Luger cartridges. However, a patrol car had been in the vicinity of his home, where the murder had been committed. A neighbor had called the State Police, who in turn had immediately alerted the patrol car by radio and instructed the two troopers in it to proceed to the scene of the crime. They arrived within five minutes of the shooting. Powers did not have a chance to get rid of the 1911 A1 parts in a satisfactory manner and therefore killed himself.

Evidence: Seven .45 ACP expended cartridge cases; seven cal .45 ACP full metal-jacketed bullets, some of which at least had passed through the bodies of three women. One 9-mm Luger expended cartridge case and one 9-mm Luger full metal-jacketed bullet which had passed through the brain of Powers before embedding itself in the ceiling were also found. Finally, there was the 9-mm Llama Extra automatic pistol.

Laboratory Procedure: The seven ACP bullets were compared and found to have come from the same barrel. The seven .45 ACP cartridge cases

Here is the Llama automatic, chambered for the 9mm Luger cartridge, used in a triple murder and suicide. It is thought that the intention of the murderer was to commit the crime with .45 ACP ammunition by substituting a slide barrel and magazine from a Colt 1911-A1 and then throw the components of the Colt away and reassemble the gun as it was made in Spain. This photograph shows the weapon assembled to fire .45 ACP ammunition with the Llama components below. It will function satisfactorily in either condition.

similarly were found to have been fired in the same weapon. The one 9-mm Luger cartridge case and bullet matched positively test bullets and cartridge cases from the 9-mm Llama. However, from the evidence submitted, there was no identification possible in connection with the crime bullets and cartridge cases used to murder the three women.

Further Investigation: Upon receipt of a verbal call from the Firearms Investigation Laboratory, the scene of the crime was subject to an extremely careful search for another weapon; namely, an automatic pistol firing the .45 ACP cartridge. This weapon was not found, although four meaningless firearms components suddenly assumed far more importance than they had before. These were now sent to the laboratory where the facts began to fall into place.

Outcome: This particular Llama Extra automatic could be disassembled and reassembled with the 1911 A1 components. It would function so assembled. Actually, it fired extremely accurately with the .45 ACP cartridges; the only complaint was that occasionally the slide would stick slightly and have to be pushed forward by hand. Had Powers been able to dispose of the 1911 A1 barrel, slide, etc., used to fire the fatal shots, it would probably have been extremely difficult to convict him of the crime without explaining in some way this ballistic discrepancy.

CASE NUMBER 24

Contributed By: Texas Department of Public Safety.

Nature of Case: Murder of a white female by the use of a bullet of the wrong caliber and the matching of a sweater weave pattern.

Circumstances and Evidence: Two itinerant families had come to this state to seek employment. After having moved about considerably they finally settled in a small community, where the suspect began and continued to abuse the victim in this case, accusing her of various and sundry actions and becoming violently angry when they were denied.

During a particularly violent argument one day, one of the neighboring couples, who was downstairs in the back yard, called to the suspect to stop his noise and abuse. The suspect then became enraged and, having an Italian make revolver and some .32 Smith and Wesson ammunition which was too small for the bore of the revolver, came to a window and fired several shots at the couple below, then turned and killed the victim. He then left in his car, proceeded several miles to another town and hid the revolver in a tree, where it was later recovered.

The bullet was too small to fit closely the bore of the revolver and, instead of striking nose on, had 'keyholed' when it struck the victim. The bore fit was so poor that no striae identification was possible. However, it was noted during the investigation and examination that a checkerboard pattern was impressed on one side of the evidence bullet, naturally an explanation of this was desired. Test shots were fired into the sweater worn by the victim at the time of her death and a similar weave pattern, obtained on the test bullets, was found to match with the evidence bullet.

The suspect was tried, found guilty and given a penitentiary sentence. This case illustrated the necessity for not overlooking any extraneous markings found on evidence bullets, or for supporting evidence which may help break down a suspect and assist in the solving of a crime.

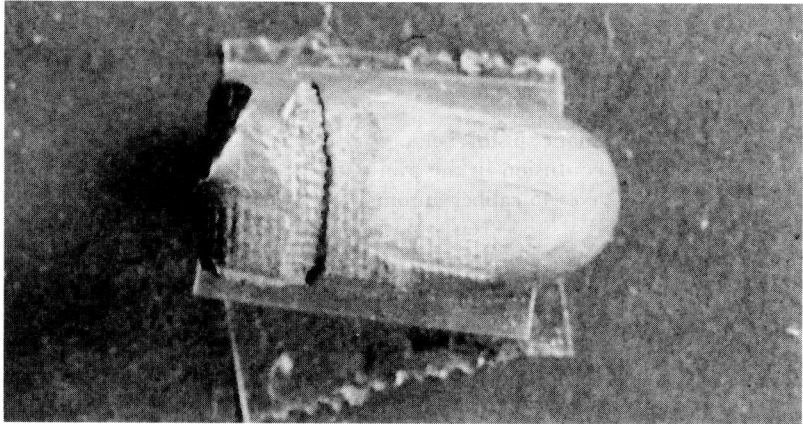

The .32 S. & W. bullet after having been fired in an oversized bore, striking sideways and impressing the weave of sweater into its side.

CASE NUMBER 25

Contributed By: Texas Department of Public Safety.

Nature of Case: Armed robbery of a white male and the subsequent murder of a white female, involving the elimination of one gun at the scene and the later identification of the fatal weapon.

Circumstances and Evidence: The suspect in this case had stolen an automobile in a neighboring state, then later burglarized a home in this state in which he secured a .45 caliber Smith & Wesson Model 1917 revolver. Leaving the scene of this robbery during the night, he came upon a car parked along the road with its occupant asleep. He moved cautiously upon the sleeper, but the man awoke, during the struggle following the suspect slugged and shot the man, taking his billfold. This wounded man staggered blindly out into the highway, hoping to find aid, and fell directly in front of an oncoming automobile, compelling it to stop. The suspect, by then in pursuit, began firing at the stopped car, which was occupied by a man and his wife. The husband had in his possession a .45 caliber Colt automatic pistol and he immediately began to return the fire of the suspect. During the exchange of shots the wife fell over dead, a bullet having penetrated her neck. The suspect then fled from the scene.

The bullet from the woman's neck, several from the car in which she had been riding, together with one bullet from the car of the man who was first wounded and robbed, were submitted for examination, together with the .45 automatic pistol belonging to the husband. At this point, there was a bit of confusion in the minds of the investigating officers, since all of the bullets were .45 caliber automatic type. However, it was promptly observed that all of these .45 automatic bullets had six grooves and lands with a right hand twist, which eliminated the husband's gun as having been involved in the firing of these bullets.

Several weeks later, the suspect was arrested while attempting to escape from officers who were questioning him regarding a different but local matter, and from his possession was taken a .45 caliber Smith & Wesson Model 1917 revolver. This revolver was originally designed for firing the rimless .45 automatic cartridge, but only in connection with the use of a 3-shot, half-moon clip, in order to effect simultaneous ejection.

Photomicrograph of .45 Auto evidence bullet fired from a Model 1917 S. & W. revolver, showing the skid marks (on upper section) plus identifying markings within it and blending into test bullet (lower), which was the 30th bullet fired before a skidding test bullet could be obtained.

An examination of the evidence bullets revealed a perfect skid mark in addition to clear striations. We were immediately concerned with the bullet taken from the neck of the victim, which also showed a particularly definite skid mark and clear striae. It was necessary for us to fire 30 test shots from the suspect gun before this skid mark was reproduced and a positive identification obtained.

The suspect was tried, found guilty and is now serving a life sentence.

CASE NUMBER 26

Contributed By: Texas Department of Public Safety.
Nature of Case: Murder of a policeman while attempting to apprehend an armed robber; elimination of many guns; and the obtaining of a conviction mainly on ballistic testimony.
Circumstances and Evidence: Early one winter evening the police dispatcher in one of our larger cities broadcast a message to all units in a certain area to be on the lookout for a man who had just committed two armed robberies and was thought to be in a certain area. In response to this call, many units cruising in that vicinity converged on this section. One particular unit, with two men, arrived just as the suspect had completed his third robbery and was making his 'get away' on foot. Seeing that he was about to be taken into custody, the suspect, flourishing a .38 Colt Super automatic pistol, fled to the rear of a building and opened fire on the patrolmen. As more units arrived, more men became involved and many handguns were fired. When the firing had ceased and the suspect taken into custody, with the .38 Super still in his possession, it was learned that one of the patrolmen had been shot in the head and was dead.

The bullet was removed from the dead patrolman's head and the laborious task began of determining from which of all the guns involved had been fired the fatal bullet. All but seven of the guns were quickly eliminated, since they were revolvers and the evidence bullet was definitely a .38 automatic type projectile. Tests were fired from the remaining .38 Colt Super automatics. The gun taken from the suspect was definitely identified as the one having fired the evidence bullet taken from the head of the patrolman.

There also entered into the case the question of the pistol having repeatedly jammed during the exchange of shots. The suspect had two .38 clips in his possession and it was known that the pistol had jammed several times during the shooting. The district attorney desired to know which, if either, of the two clips was defective, or both, causing the gun to jam. A series of test shots was fired, using both clips, and one was found to be improperly fitted, causing the gun to jam every time it was fired.

During the trial of this case, prospective jurors were qualified almost solely on whether or not they would accept as definite proof the testimony of a ballistic expert, as the outcome of the case hinged solely upon firearms identification. After several hours of most rigorous cross-examination of the firearms examiner by the defense counsel, the defendant was found guilty, assessed the death penalty, and has paid for this crime with his life.

Comment: The matter of this .38 Super having jammed consistently with one of the magazines brings up an interesting point that we speak of for the possible benefit of those peace officers armed with an automatic pistol who may carry extra, charged magazines. The fit and the adjustment of the lips of those extra magazines is of great importance and whenever extra magazines are obtained, the pistol they are to be used in should be sent to the factory, or taken to a qualified gunsmith and those spare magazines properly fitted to the pistol. Unless properly fitted, especially the lip adjustments, they are almost certain to jam and hang-up frequently.

CASE NUMBER 27

Contributed By: Texas Department of Public Safety.

Nature of Case: Murder of a negro female by negro boy friend.

Circumstances and Evidence: The circumstances of this case involved the deceased and her boy friend, who had gone for an outing in a secluded wooded area of the country early in the spring. During this outing they quarreled and the boy friend shot the deceased in the back of the head. Many months later, a small boy searching for stray cattle found the bones of a human being, later identified as a negro female. A search of the surrounding area and the debris near the skeleton revealed a fired Remington .32 caliber automatic cartridge case. Several articles such as glasses, rings and other personal items were found, by means of which definite identification was established.

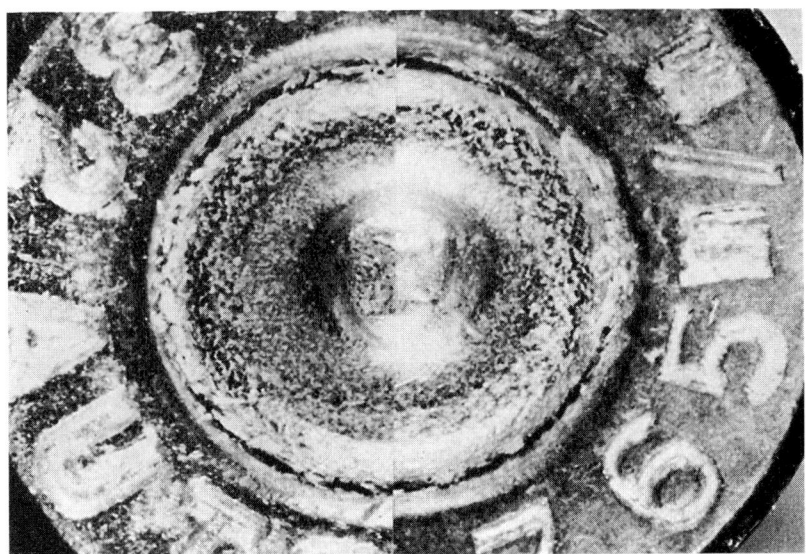

Photomicrograph showing the identifying markings in the firing pin imprint and some of the breechblock markings on this 7.65mm automatic pistol case which had been lying out in the elements for several months. Evidence case at left, test at right.

The cartridge case, which had been lying out in the elements and woods for many months, was submitted to the laboratory in an attempt, if possible, to determine the make of gun used to fire it. In the meantime, a search for a gun known to have belonged to the last person seen with the deceased, was begun. This gun was traced from person to person, until the fifth person contacted actually had the gun in his possession. However, the ownership of the gun was definitely established as belonging to the suspect.

This gun, a Belgian 7.65mm automatic pistol, was received by the laboratory and compared with the evidence cartridge case. Positive identification was established by means of numerous individual markings appearing in the firing pin imprint, breech block markings on the primer, and certain irregularities on the rim of the test and evidence cartridge cases.

The suspect admitted the crime when confronted with this evidence and later plead guilty at the trial.

Comment: The unusual features to this case were a cartridge case which had lain out in the elements for months, yet was still identifiable; the tracing of the gun through many hands; and the final positive identification made primarily on the firing pin imprint.

CASE NUMBER 28

Contributed By: Texas Department of Public Safety.

Nature of Case: Definite proof of a murder, by exhumation of a body in order to determine location and extent area of shot pellets.

Circumstances and Evidence: The victim in this case had been on a coon hunting expedition prior to returning home. He suffered from asthma and slept propped up in bed on a large pillow. His wife stated that he had seemed depressed upon returning home and requested her to make him a sandwich before going to bed. While in the kitchen preparing the sandwich she heard a shot fired and rushed to the bedroom to find her husband dead from a shotgun blast which had blown off the right side of his head. A shotgun was clasped in both hands, making it appear to have been a suicide. At that time, the case was assumed to be suicide, but a verdict was withheld and the grand jury passed it by, due to lack of evidence.

However, one of the local investigating officers was not satisfied with this verdict and later, when elected to public office, he reopened the case.

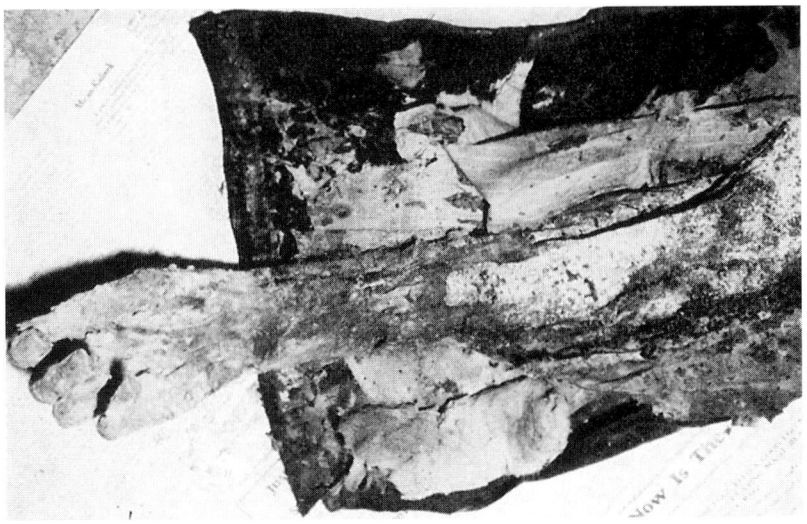

Embalmed right arm of victim, photograph taken some two years after burial. There were some 25 pellets of shot recovered from the wrist and forearm area, thus disproving the suicide story.

The undertaker who had prepared the body for burial had noticed some shot pellet wounds on the underside of the right arm. Two years after burial the body was exhumed, the right arm and skull removed, and taken to a hospital where photographs and X-rays made revealed the presence of approximately 25 pellets of shot in the wrist; these were later identified as No. 6 shot. The victim had been shot in the right eye, blowing out the eye and taking off the top of the head; however No. 6 shot were also found in the right forearm, starting at the heel of the hand and extending toward the elbow for some two or three inches. At the time of embalming, no powder burns had been detected. Test patterns with the shotgun used showed that no visible powder burns extended over three feet. Apparently the victim had been shot while propped up on the pillow and having his right hand resting under the back of his head.

It was proved to the satisfaction of the jury that the victim had not committed suicide but had been shot while asleep, by a person standing further than three feet away. The defendant was assessed a prison term. **Comment:** An interesting case, one proving the length of time during which conclusive evidence may exist; and the manner in which such evidence may be correctly interpretated and evaluated by skilled investigators, even years after the crime has been committed.

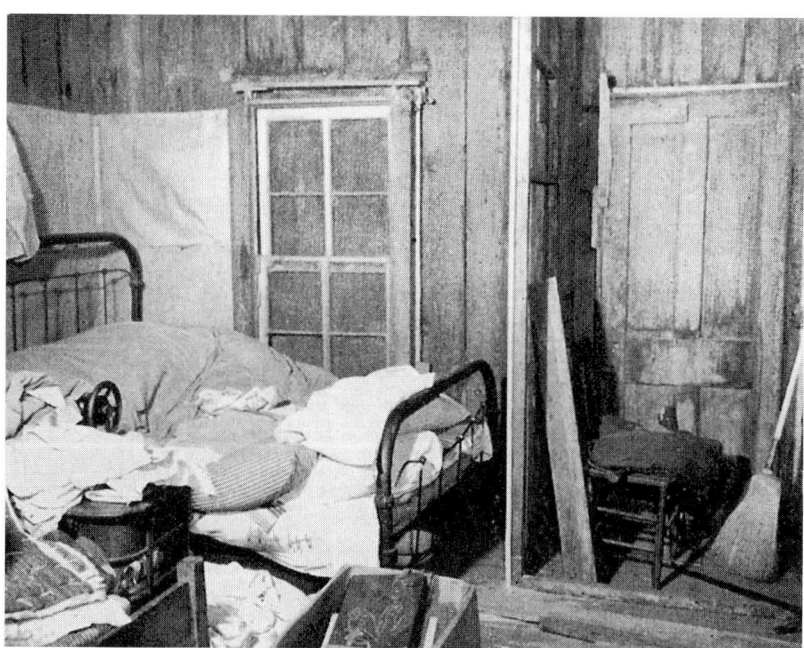

An evidence photograph showing the interior of the bedroom and the bed on which victim was lying when shot.

INDEX

527

Samworth Books on Firearms

This pair of books presents to the shooter an essential and basic education in firearms knowledge, understanding and adaptability.

SMALL ARMS DESIGN AND BALLISTICS
By Townsend Whelen

Volume I—Design, Construction and Operation; Price $6.00

This book covers the entire field of modern small arms and ammunition, with special stress on design, construction and operation. The author, Colonel Whelen, an outstanding rifleman and small arms authority and one of the U. S. Army's leading ordnance experts, has made the study and use of firearms his hobby and profession for more than half a century.

Although replete with technical notes and data, Volume I is devoid of higher mathematics. The vast subject is covered in a clear and masterful manner by the writer, easily understood by the average man, thereby affording an opportunity for the layman to grasp and apply solutions to his problems, which, unfortunately, all too often is not the case where technical or semi-technical works are concerned. This is a book the beginner will easily understand and is a splendid work to start off with.

Volume I not only thoroughly covers the construction, design and operation of small arms—including rifles, pistols, revolvers, shotguns and submachine guns—but gives minute technical details of cartridges and ammunition components. This concise, authentic information regarding design, manufacture, functioning and use of small arms and ammunition constitutes a most liberal education for any shooter and the entire subject is presented in a manner that he can absorb and utilize in a practical way in his everyday shooting problems.

Volume II—Interior and Exterior Ballistics; Price $6.00

In Volume II, Colonel Whelen thoroughly covers the important and hitherto vague subjects of Interior and Exterior Ballistics. This work clearly and graphically depicts just what happens from the time the trigger is let off until the bullet strikes the target.

Volume II differs from other textbooks on small arms ballistics in that the writer treats the subject in plain language that the average reader can understand, rather than in cold erudite terms and a maze of advanced mathematics which oft-times puzzle and bewilder even learned scholars and scientists.

This book, complete in itself, presents extensive explanations and studies relative to pressure, velocity, combustion, recoil, jump, vibration, flight of the bullet, air resistance, force of gravity, drift, yaw, trajectory, accuracy, killing power, testing methods, bench and machine rests, care, cleaning and storage of materials and the necessary methods of computing and keeping proper records and scorebooks.

The entire subject is approached from the practical, rather than the scientific angle, and the text will be found of particular benefit to the average shooter and hunter. The set comprises a basic foundation of ballistic knowledge for arms students, ordnance technicians, sportsmen and firearms lovers in general.

Price, $6.00 per volume, delivered

Gunsmithing

By Roy F. Dunlap

A 1950 work on gunsmithing—the most complete ever written, containing information on every phase of gunwork from selection of the stock blank on through to the metal engraving and blueing. It is *thorough!*—with not only the "what" and "how" but also the "why." This work was instigated by the publisher and written at his request by Roy Dunlap with a view of replacing Baker's *Modern Gunsmithing*, which was written back in 1927. Everyone who has read Dunlap's manuscript says it is "better'n Baker."

The average user of firearms will find this book to be of value in its technical information on barrels, action bedding, accuracy adjustments and trouble corrections. The shotgun information is exceptionally complete and practical. One can read exactly how the job he wants should be done and how to have the guns fitted properly to himself so that he can get the most out of them.

The targetshooter will be interested in the chapters on modern target rifles, their barreling and chamber work, their special furniture and fittings. Dunlap is a hard-boiled, competent rifleman and his information on rifles of precision is backed by trial, experience, quite a few medals and trophies, and many a head of big game.

The general gunsmith will gain information he has never been able to find except through trial and error. Individual shotguns, rifles, revolvers and pistols are covered in detail, their weak points mentioned, and instructions given as how to fix them without the use of a fully equipped shop and special, expensive machinery.

For the first time, complete and official cartridge and chamber specification drawings are published, with headspace data and barrel threadings, on modern cartridges from the .22 long rifle to the .375, including the more popular wildcats. All barrel shank and thread data is shown by drawings as well as dimensions. Barrel specifications and rifling information in all calibers is listed and analyzed.

Above all—although complete and thorough—this book is not written over the shooter's head. Written and published with the definite aim of turning out the most possible up-to-date information and instruction under one cover—*Gunsmithing* is the best one-book buy that can be obtained today. Its pages are crammed with instruction and formulae necessary in all phases of the gunsmithing art. Professional, amateur, or just plain shooter—this Dunlap work is by far the best shooting-buy offered at the start of the half-century. It has been four years in the making and will prove a milestone in gun literature equal to Baker's famous work of three decades back.

Gunsmithing is sold under the guarantee that it is better and more applicable today than any other published work on that craft. 800 pages—200 illustrations—36 chapters of the most modern, most complete, best all-around book on gunwork published. Price $7.50

The Book of The Springfield

By Edward C. Crossman
and Roy F. Dunlap

If you are a bolt-action rifleman—and most of us are—you will need this book in your library. Although written around the 1903 Springfield rifle and its 1906 cartridge, this is the most complete and attractively written work on modern rifles and their ballistics that is available today. Its text and teachings will stand for another generation or so.

Book of the Springfield was originally written by the late Captain Edward C. Crossman—the foremost writer-rifleman of his day and an author whose style of writing has never been surpassed. The first edition was published in 1931 and has been out of print for several years. Continued and insistent demand has necessitated reprinting in toto the original material compiled by Ned Crossman plus some 160 additional pages of text written by Roy F. Dunlap, a firearms authority and writer of ability equal to that of the original author. His new material brings all subject matter up to the year 1952.

This 1952 edition is a volume of 567 pages, well illustrated throughout. It is the most exhaustive and thorough work in existence on modern bolt-action rifles—military, target and sporting models. The 16 lengthy chapters differentiate at length between the rifle—the cartridge—sights—and on technical tolerances, adjustments, and care.

Its rifle text consists of some 153 pages covering the various military, target and sporting models.

The chapters on metallic sights comprise 65 pages—treating fully all types of military, target and hunting sights.

When it comes to the subject of the modern telescopic rifle sight there is actually a book in itself in the 152 pages in three chapters of up-to-the-minute material on military, target, hunting and varmint scopes—with special emphasis placed upon these last two types. Included also are comprehensive instructions for the proper fitting and adjustment of the many various makes and models. The practising gunsmith will find this material of particular value in his daily work. To the custom gunmaker it will prove a *must* work of reference.

The ammunition section is particularly complete; military, target and hunting ammunition is fully analyzed and discussed, with many illustrations of sectioned bullets included for classification and study.

Roy Dunlap took up where Crossman left off and their combined effort is a work no rifleman or ballistic student can afford to pass up. The hundreds of thousands of owners and users of Springfield rifles will find in this book everything that is today known in practical experience and ballistic science relative to .30 '06 rifle and cartridge matters. *Book of the Springfield* has long been one of the classics in firearms literature and this revised and greatly enlarged edition will assure its top position for years to come. **Price $6.00.**